BUILDING BROADBAND NETWORKS

RELATED TITLES

ABCs of IP Addressing
Gilbert Held
ISBN: 0-8493-1144-6

Application Servers for E-Business
Lisa M. Lindgren
ISBN: 0-8493-0827-5

Architectures for E-Business Systems
Sanjiv Purba, Editor
ISBN: 0-8493-1161-6

A Technical Guide to IPSec Virtual Private Networks
James S. Tiller
ISBN: 0-8493-0876-3

Building an Information Security Awareness Program
Mark B. Desman
ISBN: 0-8493-0116-5

Computer Telephony Integration
William Yarberry, Jr.
ISBN: 0-8493-9995-5

Cyber Crime Investigator's Field Guide
Bruce Middleton
ISBN: 0-8493-1192-6

**Cyber Forensics:
A Field Manual for Collecting, Examining, and Preserving Evidence of Computer Crimes**
Albert J. Marcella and Robert S. Greenfield, Editors
ISBN: 0-8493-0955-7

Information Security Architecture
Jan Killmeyer Tudor
ISBN: 0-8493-9988-2

Information Security Management Handbook, 4th Edition, Volume 1
Harold F. Tipton and Micki Krause, Editors
ISBN: 0-8493-9829-0

Information Security Management Handbook, 4th Edition, Volume 2
Harold F. Tipton and Micki Krause, Editors
ISBN: 0-8493-0800-3

Information Security Management Handbook, 4th Edition, Volume 3
Harold F. Tipton and Micki Krause, Editors
ISBN: 0-8493-1127-6

Information Security Policies, Procedures, and Standards: Guidelines for Effective Information Security Management
Thomas Peltier
ISBN: 0-8493-1137-3

Information Security Risk Analysis
Thomas Peltier
ISBN: 0-8493-0880-1

Information Technology Control and Audit
Frederick Gallegos, Sandra Allen-Senft, and Daniel P. Manson
ISBN: 0-8493-9994-7

New Directions in Internet Management
Sanjiv Purba, Editor
ISBN: 0-8493-1160-8

New Directions in Project Management
Paul C. Tinnirello, Editor
ISBN: 0-8493-1190-X

A Practical Guide to Security Engineering and Information Assurance
Debra Herrmann
ISBN: 0-8493-1163-2

**The Privacy Papers:
Managing Technology and Consumers, Employee, and Legislative Action**
Rebecca Herold
ISBN: 0-8493-1248-5

**Secure Internet Practices:
Best Practices for Securing Systems in the Internet and e-Business Age**
Patrick McBride, Joday Patilla, Craig Robinson, Peter Thermos, and Edward P. Moser
ISBN: 0-8493-1239-6

Securing and Controlling Cisco Routers
Peter T. Davis
ISBN: 0-8493-1290-6

Securing E-Business Applications and Communications
Jonathan S. Held and John R. Bowers
ISBN: 0-8493-0963-8

**Securing Windows NT/2000:
From Policies to Firewalls**
Michael A. Simonyi
ISBN: 0-8493-1261-2

TCP/IP Professional Reference Guide
Gilbert Held
ISBN: 0-8493-0824-0

The Complete Book of Middleware
Judith Myerson
ISBN: 0-8493-1272-8

CRC PRESS

www.crcpress.com
To Order Call: 1-800-272-7737 • Fax: 1-800-374-3401
E-mail: orders@crcpress.com

BUILDING BROADBAND NETWORKS

Marlyn Kemper Littman, Ph.D.

Graduate School of Computer and Information Sciences
Nova Southeastern University
Fort Lauderdale, Florida

CRC PRESS

Boca Raton   London   New York   Washington, D.C.

Library of Congress Cataloging-in-Publication Data

Littman, Marlyn Kemper.
    Building broadband networks/
    Marlyn Kemper Littman.
        p. cm.
    Includes bibliographical references and index.
    ISBN 0-8493-0889-5 (alk. paper)
        1. Broadband communication systems. 2. Distance education--Communication
    systems. I. Title.

    TK5103.4 .L58 2002
    004.67'8--dc21

                                                                    2002017495

Visit the CRC Press Web site at www.crcpress.com

© 2002 by CRC Press LLC

No claim to original U.S. Government works
International Standard Book Number 0-8493-0889-5
Library of Congress Card Number 2002017495
Printed in the United States of America 1 2 3 4 5 6 7 8 9 0
Printed on acid-free paper

Dedication

For my husband Rabbi Lewis Littman, D.D.,
for making every day special

Preface

Accelerating demand for extendible, dependable, and scalable high-speed, high-performance networks with vast transmission capacities and potentially unlimited bandwidth contributes to the present-day popularity of broadband communications technologies. *Building Broadband Networks* is a comprehensive examination of recent developments and innovations in this dynamic field of study.

The text provides a foundation for understanding distinctive attributes and functions of broadband technologies and the support of these technologies in enabling development of high-performance, high-capacity, and high-speed networking configurations. Technical features and functions, standards activities, and approaches for enabling effective broadband network deployment are described. Practical considerations for building broadband networks that are extendible, flexible, available, scalable, and reliable are presented. Representative broadband tele-education initiatives that enable students and lifelong learners to participate in virtual classes, telecourses, and teleprograms, regardless of geographic location, are highlighted. National and international research and education networks that provision a diverse and powerful mix of broadband services are highlighted. Wireless solutions that support fast Web connectivity at any time and from any place are indicated. Advanced broadband network applications such as digital libraries and information grids are described. Capabilities of undersea networks and powerline configurations are explored.

Building Broadband Networks is written from a contemporary perspective. Emphasis is placed on exploring the distinctive characteristics of broadband technologies, architectures, and services and current and next-generation networking implementations in present-day environments.

This book begins with an examination of ISDN (Integrated Services Digital Network) and ATM (Asynchronous Transfer Mode) technologies. Optical network solutions based on SONET/SDH (Synchronous Optical Network and Synchronous Digital Hierarchy), WDM (Wavelength Division Multiplexing), and DWDM (Dense WDM) technologies are then described.

The text continues with an exploration of Ethernet, Fast Ethernet, Gigabit Ethernet, and 10 Gigabit Ethernet operations and services. Features and functions of Frame Relay and Fibre Channel networks are introduced. Distinctive attributes of DSL (Digital Subscriber Line) solutions and wireline and wireless cable networks in the residential broadband access arena are then delineated.

Capabilities of second- and third-generation cellular communications technologies, such as GSM (Global System for Mobile Communications) and UMTS (Universal Mobile Telecommunications Systems) in provisioning access to communications resources at any time and from any place, are reviewed. Distinguishing characteristics of wireless networking technologies and configurations such as Bluetooth and IEEE 802.11b Ethernet WLANs (Wireless Local Area Networks) in enabling multimedia applications in research and actual environments are described. An examination of satellite technologies and a description of broadband satellite network implementations are presented. Next-generation high-speed, high-performance network configurations such as Internet2 (I2) and GÉANT, the next-generation pan-European network, are explored as well. The text concludes with an exploration of network security problems and solutions.

Demand for fast, reliable, and secure access to bandwidth-intensive Web resources contributes to the development and implementation of a remarkable array of broadband networks and media-rich network applications and services. Although Internet addresses are subject to change, the Web remains a good source for monitoring developments in network technologies that are examined in this book. As a consequence, pointers to selected Web sites are provisioned at the conclusion of each chapter. An online component to this text at http://www.scis.nova.edu/~marlyn provides links to relevant Web sites, as they become available. Broadband networks capable of transmitting voice, video, data, and still-image traffic across localities, cities, regions, and continents are being implemented at an unprecedented rate. There is an expanding range of innovative options in the broadband network arena. A flood of specialized acronyms accompanies the rapid emergence of these technologies and configurations.

The process of building and deploying broadband communications networks is technically complicated. There is no single solution. Decisions are dependent upon multiple factors such as the mission, goals, and objectives of the sponsoring entity; capabilities of the in-place infrastructure; and application, security, and performance requirements.

Building Broadband Networks is about the mortar and bricks out of which broadband networks are built. My goal in writing this book is to provide a practical yet detailed explanation of major technologies, standards, applications, and solutions in the broadband network arena. The subject itself is complex. Sufficient technical detail and technical clarity are provided to remove the confusion and mystery surrounding the topic. Important broadband initiatives are described to provide readers with an understanding of practical implementations that distinguish this rapidly expanding field.

There are numerous networking configurations in use, each with distinctive performance characteristics, advantages, and limitations. Complex technological advancements, the remarkable increase in network capacity, the multiplicity of networking applications, and pressures to improve the quality and reliability of network services underscore the importance of developing and implementing effective broadband networking solutions.

Basic communications technologies, architectures, and protocols are examined in introductory undergraduate and graduate textbooks in the field of telecommunications and computer networks. However, there remains a need for a text in the academic arena that examines the distinctive attributes of high-performance broadband communications technologies and focuses specifically on current and next-generation wireless and/or wireline network implementations in real-world and research environments. This book accomplishes these objectives.

Building Broadband Networks is designed for senior undergraduate students and graduate students in the fields of education, information systems, and information science. It can also be used by faculty, corporate, and academic administrators and managers, network planners and consultants, information systems specialists, and librarians who want to learn more about the capabilities of broadband communications technologies and current and next-generation networking initiatives. This book can be readily employed as a textbook for advanced undergraduate and graduate courses in telecommunications and computer networks.

Material in this book has been examined in doctoral courses in telecommunications and computer networks taught by this author at the Graduate School of Computer and Information Sciences at Nova Southeastern University over the past 15 years.

Communications services, products, equipment, and solutions available from vendors, NSPs (Network Service Providers), and communications carriers mentioned in this text illustrate the features, functions, and capabilities of the technologies that are described. This information should not be interpreted as any kind of endorsement.

The Author

Marlyn Kemper Littman, Ph.D., is a Professor at the Graduate School of Computer and Information Sciences at Nova Southeastern University. Dr. Littman teaches doctoral courses and mentors doctoral candidates in the field of telecommunications and computer networks. Dr. Littman is the author of numerous professional publications in the telecommunications and computer network arena, beginning with the publication of her book entitled *Networking: Choosing a LAN Path to Interconnection* in 1987.

Marlyn Kemper Littman holds a Ph.D. with a specialization in telecommunications and computer networks from the Graduate School of Computer and Information Sciences at Nova Southeastern University, an M.A. in Anthropology from Temple University, and an M.S. in Information Science from the University of South Florida. Dr. Littman is a member of the Institute of Electrical and Electronics Engineering (IEEE), the Association for Computing Machinery (ACM), and the Phi Kappa Phi National Honor Society.

Acknowledgments

I especially wish to thank Dr. Edward Lieblein, my Dean at the Graduate School of Computer and Information Sciences (GSCIS), Nova Southeastern University, for his valuable contributions; and Dr. David S. Metcalf, II, for his outstanding technical illustrations throughout the text. I also wish to express my appreciation to Dr. Gertrude Abramson, Dr. Maxine Cohen, and Dr. Laurie Dringus, my colleagues at GSCIS, for their helpful suggestions. I am grateful to Dr. Jane Anne Hannigan, Professor Emerita at Columbia University, and Dean Kay Vandergrift, Director of Distance Education and Professor at the School of Communications, Information and Library Studies, Rutgers University for their mentorship.

I am indebted to Rich O'Hanley, President of Auerbach Publications, and Gerald Papke, Editor at CRC Press, for their constructive comments and enthusiastic support throughout the publication process. I also wish to acknowledge Gerry Jaffe, Project Editor, and Helena Redshaw, Supervisor of Editorial Project Development at CRC Press, for their diligent work on the book's production.

This book is dedicated with all my love to my husband Lew for his extraordinary encouragement and remarkable patience during the seemingly endless days I spent on the Web immersed in broadband technologies, services, applications, and initiatives.

Table of Contents

1 Integrated Services Digital Network (ISDN)

1.1 INTRODUCTION

Spiraling demand for transparent, affordable, and dependable access to media-rich Web-based voice, data, and video services contributes to persistent usage of Integrated Services Digital Network (ISDN) applications in the present-day networking environment. Also known as Narrowband ISDN (N-ISDN), ISDN supports digital transmission over ordinary twisted copper wire pair traditionally used for telephone service.

Distinguished by its projected ability to facilitate worldwide connectivity via the in-place twisted copper wireline infrastructure, ISDN technology was widely promoted by the communications industry as a universal global transport solution during the 1970s and 1980s. At that time, however, affordable, dependable, compatible, and easily implemented ISDN services for small business and residential networks were not readily available from telecommunications carriers, vendors, and equipment manufacturers.

By the 1990s, competitive residential and small business solutions based on DSL (Digital Subscriber Line) and wireline and wireless cable network technologies outpaced ISDN implementations at SOHO (Small Office/Home Office) venues and small-sized business establishments. In addition, ISDN implementations were also overshadowed by multiservice, high-capacity, high-performance broadband solutions based on ATM (Asynchronous Transfer Mode) technology.

As a consequence, ISDN is not currently viewed as a worldwide platform for provisioning access to voice, video, and data services over the local loop. Nonetheless, commitment to ISDN utilization in residential venues and in sectors that include education, medicine, and business is reflected in continued ISDN deployments. The primary attraction of ISDN in the present-day marketplace is its ability to provision affordable video, audio, and data services and dependable throughput over the same twisted pair copper communications lines in place for the Public Switched Telephone Network (PSTN).

1.2 PURPOSE

This chapter presents an examination of ISDN technology. Key ISDN concepts and recent research in the ISDN arena are introduced. Challenges associated with the incorporation of ISDN technology into the networking infrastructure for enabling reliable information transport are explored. Standards organizations and standards activities in the ISDN domain are reviewed. Guidelines for planning an ISDN implementation are described. Representative ISDN initiatives and applications in fields such as telemedicine, tele-education, and telebusiness are highlighted. The role of ISDN technology in enabling implementation of Web-based virtual communities is noted as well.

1.3 FOUNDATIONS

Developed in the 1970s to provision digital voice and data services over copper wire phonelines, ISDN technology was expected to replace conventional PSTN (Public Switched Telephone Network) technology. The Consultative Committee for International Telephone and Telegraph or CCITT (now known as the International Telecommunications Union or the ITU) completed the initial I.210 Recommendation for ISDN implementation in 1984.

ISDN was initially distinguished by its capabilities in enabling subscribers at SOHO (Small Office/Home Office) venues to access the Internet at faster rates than speeds supported by conventional analog voiceband modems. Despite ISDN capabilities in economically facilitating digital video, voice, and data delivery over the POTS (Plain Old Telephone System) infrastructure to the customer premise, ISDN services were not widely deployed or universally accepted.

The complexity of the ISDN ordering and service initialization process and poor technical support provisioned by communications carriers to ISDN subscribers adversely affected ISDN implementation. ISDN products from competitive communications carriers were not always interoperable and ISDN services were not universally obtainable. Moreover, ISDN subscribers encountered variations in vendor packages and were often frustrated in their efforts to determine the availability of ISDN services and applications at any given location.

Additional roadblocks to the realization of global ISDN included lack of uniform services and technical complexity in integrating ISDN into the existing telecommunications infrastructure. The relatively limited use of this technology contributed to cynical interpretations of the ISDN acronym that continue to circulate today.

For disenchanted ISDN customers, the ISDN acronym translates to "It Sure Does Nothing," and "It Sure Doesn't Network." By contrast, ISDN advocates maintain that the ISDN acronym stands for "It Sure Does Network" and "Information Services Delivered Now." Despite the accelerating popularity of competitive residential access technologies, ISDN remains a viable solution for facilitating access to previously inaccessible teleservices, particularly for subscribers in isolated locations and remote communities.

1.4 ISDN FEATURES AND FUNCTIONS

Depending on user needs and requirements, ISDN configurations enable diverse tele-applications and tele-activities. For example, ISDN deployments in metropolitan areas and rural locations facilitate LAN (Local Area Network) interconnectivity, teleworking, telemarketing, remote publishing, electronic commerce (E-commerce), videoconferencing, and voice telephony service. ISDN also fosters tele-instruction, telemedicine consultations, voicemail, and remote monitoring and surveillance services.

In addition, ISDN enables connectivity to enterprisewide intranets and extranets, and serves as an effective backup solution for networks employing ATM, Frame Relay (FR), and T-1 (1.544 Mbps or Megabits per second) or E-1 (European-2.048 Mbps) leased-line connections.

A multiservice technology, ISDN enables transmission of delay-sensitive and bursty data traffic via virtual links that can be shared with other subscribers. For example, ISDN employs circuit-switched connections for establishing a virtual pathway between two ISDN subscribers that is virtually fixed for the duration of the phone call. In addition to basic telephony service, ISDN circuit-switched connections also support caller ID (Identification), call forwarding, call hold, and automatic callback. ISDN employs packet-switched connections to facilitate desktop publishing, compressed video transmission, and bulk file transfer. ISDN works in concert with ITU-T (International Telecommunications Union-Telecommunications Standards Sector) Group 4-compliant facsimile (fax) implementations to facilitate dependable transmission of high-resolution images such as blueprints and medical scans. In addition, ISDN provisions non-switched service for information transport via dedicated leased lines.

ISDN is designed as a global public telecommunications network service. However, in reality, multiple ISDN networks are implemented for achieving interconnectivity within and across local, metropolitan, regional, national, and international boundaries. Interoperable links to out-of-state locations not served by ISDN technology are established with the use of Switched 56 services. Switched 56 solutions support data-only connections at speeds up to 56 Kbps (Kilobits per second). However, Switched 56 solutions are not capable of supporting out-of-band D (Delta) Channel signaling provisioned by ISDN configurations. In contrast to ISDN, Switched 56 services also cannot provision concurrent voice, video, and data transmissions.

In comparison to the always-on and always-available capabilities of competitor technologies such as DSL (Digital Subscriber Line) and cable networks, ISDN readily enables connections to be established and discontinued. In an ISDN implementation, after the transmission ends and the communications link is idle, the ISDN connection is automatically terminated.

ISDN supports development of an end-to-end digital network by converting every standard analog POTS (Plain Old Telephone Service) line into a high-speed digital connection for enabling information transport. With ISDN, multiple channels support diverse applications simultaneously on the same twisted pair circuit enabling POTS delivery. Prior to ISDN implementation, separate phone lines were required for accessing telephone calls, fax transmissions, and computing services. (See Figure 1.1.)

1.5   ISDN TECHNICAL FUNDAMENTALS

1.5.1   BASIC ISDN INSTALLATION REQUIREMENTS

ISDN implementation involves determining the number of locations and types of devices that will be attached to the configuration and bandwidth or transmission rate

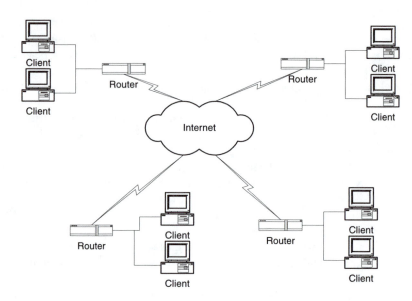

FIGURE 1.1 An ISDN configuration with multiple routers providing links to the Internet.

requirements. In addition, ISDN deployment requires reconfiguration of PC (Personal Computer) software to support ISDN links and rewiring or replacing a single phone jack with a dual port to enable ISDN connections. The installation of additional PSTN wiring at the subscriber premise may also be required. In addition, ISDN implementations also involve utilization of ISDN Terminal Equipment (TE), ISDN Network Termination (NT) devices, and ISDN Terminal Adapters (TAs).

1.5.2   ISDN TERMINAL ADAPTERS (TAs)

ISDN supports cost-effective digital information transport at considerably faster rates than conventional analog telephony service. As part of the ISDN implementation process, subscribers purchase ISDN-compatible equipment such as ISDN Terminal Adapters (TAs) instead of modems to facilitate conversion of analog voice signals into digital signal formats that are fully compatible with ISDN telephony service and ISDN TAs at the local telephone exchange. The local telephone exchange is also known as the telephone company central office (CO).

Available as stand-alone units or PC (Personal Computer) cards, Terminal Adapters (TAs) allow non-ISDN equipment to support operations via the in-place ISDN configuration. In addition to enabling digital data and analog voice devices to interwork via ISDN connections, TAs also distinguish between voice and non-voice signals so that voice calls and data frames can be directly routed to designated destination addresses.

1.5.3   ISDN TERMINAL EQUIPMENT (TE)

ISDN works in tandem with two types of terminal equipment (TE). Devices that employ ISDN directly and foster ISDN services are called Terminal Equipment Type

1 (TE1). By contrast, non-ISDN devices are called Terminal Equipment Type 2 (TE2). TE2 devices are not compatible with the ISDN specification. As a consequence, these devices require the use of Terminal Adapters (TAs).

1.5.4 ISDN NETWORK TERMINATION (NT) DEVICES

ISDN employs two types of network termination devices. Network Termination Type 1 (NT1) refers to a network terminal device situated at the customer premise for handling Physical Layer or Layer 1 and Data-Link Layer or Layer 2 connections. NT1 devices enable B (Bearer) Channel and D (Delta) Channel multiplexing activities. Moreover, NT1 devices also handle the physical link over the local loop extending from the subscriber site to the local telephone exchange and support network monitoring and performance assessment.

As with NT1 devices, Network Termination Type 2 (NT2) devices perform Physical Layer and Data-Link Layer functions. In addition, NT2 equipment enables voice and data switching and seamless aggregation or BONDING (Bandwidth On-Demand Interoperability Group) of multiple ISDN channels.

1.5.5 ISDN CODECS (CODERS AND DECODERS)

In ISDN implementations, a codec or chip performs digital-to-analog and analog-to-digital conversions. In addition, a codec supports compression to minimize redundancies in voice, data, and video transmissions for facilitating high-quality videoconferences. To enable clear and robust ISDN telephony services, a codec converts analog signals into digital formats at call setup for network transmission and then reconstructs the digital signals into analog formats at call reception.

1.5.6 ISDN DIGITAL PIPE

The access path from the local telephone exchange to the customer premise over the last mile or the local loop in an ISDN network is commonly called a digital pipe. The size of the digital pipe for ISDN transmission depends on variables such as customer application requirements and fees established by the communications carrier.

1.5.7 ISDN REFERENCE POINTS

For transparent transmission, ISDN defines reference points indicating protocols employed between different functional devices. R, S, T, and U are the commonly defined reference points for an ISDN configuration. The R reference point refers to communications between TE2 and TA devices. The S reference point refers to communications between TE1 or TA devices and Network Termination (NT) equipment or NT1 and NT2. The T reference point indicates links between customer premise switching equipment (NT2) and local loop termination (NTI) devices. Defined by the ISDN communications carrier, the U reference point refers to the link between the local telephone exchange and NT1 equipment. Every U interface frame consists of 240 bytes.

1.6   B (BEARER), D (DELTA), AND HYPER (H) CHANNELS

1.6.1   B (Bearer) Channel

The B (Bearer) Channel is the basic building block in an ISDN configuration. Capable of supporting circuit-switched and packet-switched connections, the B Channel carries digitized voice, video, and data at rates up to 64 Kbps and provisions asynchronous, synchronous, and isochronous services for dependable and reliable information transport. B Channel protocols include the Point-to-Point Protocol (PPP) for transporting diverse LAN traffic over telecommunications links and the multi-point PPP (ML-PPP) for extending PPP services.

BONDING (Bandwidth On-Demand Interoperability Group) enables the aggregation of six B channels into one H Channel or HyperChannel for achieving high throughput via the PSTN and provisioning bandwidth required by advanced ISDN voice, video, and data applications. Typically, ISDN employs BONDING (Bandwidth On-Demand Interoperability Group) or inverse multiplexing to combine separate B (Bearer) Channels into a single virtual wideband digital channel or digital pipe for supporting interactive videoconferences over the PSTN.

1.6.2   D (Delta) Channel

The D (Delta) Channel enables signaling and control capabilities such as call acknowledgment, call setup, and automatic number identification for each ISDN line installed. In terms of operations, the D Channel is a 16 Kbps or 64 Kbps circuit, depending on the specified network interface that supports communications between the ISDN device and the switch at the local telephone exchange. The D Channel also fosters asynchronous packet data transport at 9.6 Kbps (Kilobits per second) and works in concert with the X.25 protocol for facilitating access to PSTN services. The D Channel rarely employs all of its available bandwidth. As a consequence, the excess capacity typically supports data transport. Defined by National ISDN-1 (NI-Phase 1), ISDN switch protocols control the initiation and termination of telephone calls over D Channels.

1.6.3   H (Hyper) Channel

In an ISDN configuration, B Channels can be combined into H Channels or Hyper-Channels through BONDING or inverse multiplexing. H Channels typically enable high-performance applications such as Group 4 fax (facsimile) transmission, high-speed file transfer, and videoconferencing. As noted, six B Channels form a single H Channel or HyperChannel. A HyperChannel supports full-duplex rates to enable transmissions at H0 or 384 Kbps. By BONDING two B Channels and one D Channel, the H11 Channel supports transmissions at 1.544 Mbps (Megabits per second). This rate is equivalent to ANSI (American National Standards Institute) T-1 carrier line speed.

FIGURE 1.2 BRI and PRI connections.

1.7 BASIC RATE ISDN (BRI) AND PRIMARY RATE ISDN (PRI) SERVICES

ISDN service levels are called BRI (Basic Rate ISDN) and PRI (Primary Rate ISDN). B, D, and H Channels serve as the framework for establishing BRI and PRI solutions.

1.7.1 BASIC RATE ISDN (BRI) SERVICES

BRI (Basic Rate Interface Service) transforms a single twisted pair telephone line into the equivalent of two conventional telephone lines consisting of two independent 64 Kbps B Channels for user information and one 16 Kbps D (Delta) Channel for call signaling, call control, and slow data transfer. As a consequence, BRI service is also known as 2B+D. BRI employs echo cancellation to eliminate noise and 2B1Q (2 Binary 1 Quaternary) data encoding methods for enabling relatively high-speed transmission rates over the local loop via a single copper pair telephone line. The local loop refers to the distance between the customer premise and the local telephone exchange. The BRI Physical Layer specification is defined by the ITU-T I.430 Recommendation. For BRI service, the U interface or reference point supports two-wire and four-wire links.

With BRI service, the two B Channels and one D Channel can be consolidated for remote LAN connectivity and Web exploration at a rate of 144 Kbps. A single BRI connection supports as many as eight devices and 64 separate phone numbers.

Equipment typically employed for BRI service depends on application requirements. For example, an inverse multiplexer is required for BRI videoconferencing sessions. Initially designed for telephone operations, BRI service is also a popular solution for enabling fast Internet connectivity and desktop videoconferencing in SOHO (Small Office/Home Office) venues, small-sized enterprises, public and private K–12 schools, and post-secondary institutions. As with ADSL (Asynchronous Digital Subscriber Line) implementations, ISDN subscribers must be within 18,000 feet of the local telephone exchange in order to utilize BRI services effectively. Coverage beyond this distance requires installation of ISDN signal repeaters. (See Figure 1.2.)

1.7.2 PRIMARY RATE INTERFACE (PRI) SERVICES

1.7.2.1 North America and Japan

A sophisticated solution for bandwidth-intensive applications, Primary Rate ISDN (PRI) service is popularly known as 23B+D in North America and Japan. PRI service supports utilization of 23 independent B Channels that are capable of supporting 23 simultaneous digital telephone calls and one D (Delta) Channel. Each channel supports data transmission at 64 Kbps. PRI service in North America and Japan supports bi-directional transmission rates at 1.544 Mbps or T-1 speeds. (See Figure 1.3.)

1.7.2.1.1   T-1 Fundamentals

Established by ANSI, the North American T-1 digital hierarchy serves as the basis for defining ISDN PRI rates and services. The basic building block for T-1 transmission is a single 64 Kbps DS (Digital Signal) Channel or digital voice circuit. In terms of operations, the letter "T" refers to hardware that generates signals for transmission and the letters "DS" refer to transmission rate and signal structure. The terms "DS-1" and "T-1" are used interchangeably.

A T-1 transmission line consists of 24 DS Channels, with each channel operating at DS0 or 64 Kbps. T-1 and E-1 private carrier lines require modification of the in-place infrastructure including the installation of repeaters to regenerate signals every 3,000 to 5,000 feet. Two sets of twisted copper pair wires for transmission are required as well.

1.7.2.2   Australia and the European Union

Based on TDM (Time-Division Multiplexing), the E-1 (European-2.048 Mbps) digital hierarchy established by the Conference of European Postal and Telecommunications Administration (CEPT) serves as the basis for defining PRI rates and services in Australia and member states in the European Union. E-1 is the European counterpart to T-1. The basic building block for E-1 transmissions is 2.048 Mbps. An E-1 transmission line consists of 30 digital voice channels and enables bi-directional rates at 2.048 Mbps. As a consequence, in Australia and the European Union, PRI is also called 30B+D. PRI service supports applications in medium- and large-sized enterprises. The PRI Physical Layer specification is defined by the ITU-T I.431 Recommendation.

1.8   ISDN FRAMES

ISDN frame formats differ, depending on whether the frame is inbound from the network to the ISDN terminal at the subscriber premise in the downstream direction or outbound from the ISDN terminal at the subscriber premise to the network in the upstream direction. An ISDN frame typically consists of 48 bytes, with 36 bytes allocated for data and 12 bytes designated for overhead functions such as synchronization, device activation, adjustment of byte value, and contention resolution in the event that several nodes contend for channel access simultaneously.

ISDN employs specially designed equipment for transmission of ISDN formatted frames. ISDN frames conform to V.120 encapsulation specifications and carry payloads and sequencing information for ensuring error-free delivery. ISDN signaling specifications determine frame setup and the pathway for frames to move through the network.

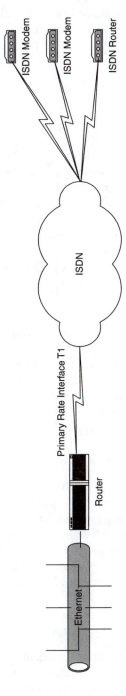

FIGURE 1.3 A campus networking solution based on an ISDN PRI platform.

Because certain ISDN devices function with only one particular switch, ISDN subscribers must identify the vendor and type of switch supporting their own ISDN service in order to ensure ISDN equipment interoperability with devices used by other ISDN subscribers. Equipment in use for ISDN service from one communications carrier does not necessarily interwork with ISDN devices in use by a competitive communications carrier.

1.9 ISDN PROTOCOLS

1.9.1 STATISTICAL TIME-DIVISION MULTIPLEXING (STDM)

ISDN provides a digital framework for voice, text, video, and still-image transmission by utilizing Statistical Time-Division Multiplexing (STDM). Also called intelligent TDM, STDM is a sophisticated form of TDM (Time-Division Multiplexing). Conventional TDM divides bandwidth into fixed timeslots so that information from each channel is transported in a predetermined rotation. Multiplexing refers to the process of combining multiple information channels that consist of numerous analog or digital signals into a single, high-capacity transmission link.

In an ISDN configuration, STDM divides available bandwidth on a single ISDN line into B (Bearer), D (Delta), and H (Hyper) Channels. These channels or circuits function as timeslots for transmission of data, video, and audio signals generated by devices linked to the ISDN configuration. STDM enables multiple ISDN devices to receive or transmit video, voice, and data concurrently by assigning a fixed amount of time for transmission to each ISDN node.

With STDM, numerous ISDN signals are combined into composite signals that transit the communications channel via fixed timeslots at specified intervals. Individual signals are subsequently separated from composite signals and routed to designated termination points. In comparison to TDM, STDM facilitates more effective utilization of available bandwidth capacity.

1.9.2 LAP-D (LINK ACCESS PROTOCOL-D CHANNEL)

Operating above the Physical Layer or Layer 1 of the OSI Reference Model, the D Channel employs the Link Access Protocol-D Channel (LAP-D) to enable acknowledged and unacknowledged information transfer services that support Layer 2 or Data-Link Layer operations. The LAP-D frame format features a 2-octet address field, a 2-octet CRC (Cyclic Redundancy Check) field for determining data errors, a 7-byte terminal endpoint identifier field, and a 6-byte SAPI (Service Access Point Identifier) field. ISDN Data-Link Layer capabilities are defined by the ITU-T Q-Series of Recommendations.

1.9.3 ISDN USER-TO-NETWORK SIGNALING PROTOCOL

The ISDN User-to-Network Interface (UNI) signaling protocol defines Layer 3 or Network Layer operations. This protocol enables the establishment, maintenance, and termination of network connections via circuit-switched or packet-switched B

FIGURE 1.4 The seven layers of the OSI (Open Systems Interconnection) Reference Model.

Channel connections. ISDN Layer 3 signaling specifications are defined in the ITU-T I.43 and the ITU-T I.431 Recommendations.

1.9.3.1   Open Systems Interconnection (OSI) Reference Model

ISDN features a layered protocol stack that conforms to the format developed by the Open Systems Interconnection (OSI) Reference Model established by the ISO (International Standards Organization). (See Figure 1.4.) ISDN operations take place at the Network Layer or Layer 3, the Data-Link Layer or Layer 2, and the Physical Layer or Layer 1 of the seven-layer OSI Reference Model. At the Physical Layer or Layer 1, ISDN supports Basic Rate Interface (BRI) and Primary Rate Interface (PRI) service levels.

Developed by the International Standards Organization in the 1980s, the OSI Reference Model describes the way in which voice, video, and data are transmitted between any two points in a telecommunications network. Communications technologies such as ISDN describe functions in terms of their relationship to the seven layer OSI Reference Model. The Application Layer or Layer 7, the Presentation Layer or Layer 6, the Session Layer or Layer 5, and the Transport Layer or Layer 4 delineate the process of transmitting a message to or from a network user. The Network Layer or Layer 3, the Data-Link Layer or Layer 2, and the Physical Layer or Layer 1 establish the process for enabling message transmission across a physical medium such as coaxial cable, optical fiber, hybrid optical fiber and coaxial cable (HFC), and twisted copper pair. Messages can be directly forwarded to another network or passed via the upper OSI Layers to the designated recipient.

1.9.3.1.1   OSI Reference Model and TCP/IP Protocol Suite

Developed by DARPA (United States Department of Defense Advanced Research Agency) in the 1960s, the TCP/IP Protocol Suite is an open system that serves as the framework for the present-day Internet. An affordable, flexible, and dependable interconnect solution, TCP/IP describes standardized protocols for enabling internetworking services between computers that vary in size, feature diverse operating systems, and enable functions in all types of government, research, educational, and corporate networks worldwide.

Like the OSI Reference Model, the TCP/IP Protocol Suite consists of a layered communications architecture with each layer responsible for a particular facet of the communications process. In contrast to the seven layers defined by the OSI Reference Model, the TCP/IP Protocol Suite describes four layers, specifically the Application Layer or Layer 4, the Transport Layer or Layer 3, the Network Layer or Layer 2, and the Physical or Media-Access Layer or Layer 1. The Application Layer defines services performed by protocols such as HTTP (HyperText Transfer Protocol), FTP (File Transfer Protocol), SMTP (Simple Mail Transfer Protocol), and SNMP (Simple Network Management Protocol). Two of the most widely used TCP/IP protocols, TCP provides Transport Layer services and IP enables Network Layer operations and functions as a network overlay in conjunction with technologies such as ISDN and ATM.

TCP/IP is a streamlined architectural model that supports layers that are functionally equivalent to the Application, Transport, Network, and Physical Layers of the OSI Reference Suite. As with OSI, upper TCP/IP Layers employ the functions provisioned by the lower layers for enabling reliable telecommunications applications and services.

1.10   AMERICAN STANDARDS ORGANIZATIONS AND ACTIVITIES

Standards organizations in the United States, including the American National Standards Institute (ANSI), the National ISDN Council, the North American ISDN Users Forum (NIUF), and the Access Technologies Forum (ATF), originally called the Vendors ISDN Association (VIA), establish specifications that accelerate ISDN deployment. These organizations also develop recommendations for streamlining the ISDN ordering and installation process and specifications for promoting the use of ISDN with spread spectrum, microwave, and satellite technologies in mixed-mode wireline and wireless networks.

The popularity of ISDN technology in the United States is reflected in the proliferation of state-based ISDN Users Forums. These coalitions address ISDN equipment incompatibility problems and network interoperability constraints by supporting the development of uniform specifications and implementation of standards-compliant ISDN equipment.

1.10.1   AMERICAN NATIONAL STANDARDS INSTITUTE (ANSI)

In the United States, the American National Standards Institute (ANSI) endorses ISDN specifications that describe management operations and principles for ISDN operations. In addition, ANSI specifications clarify ISDN functions that accommodate specific requirements of ISDN implementations in the United States. In contrast to ANSI, the European Telecommunications Standards Institute (ETSI) defines EuroISDN specifications to meet the specific needs of European installations.

1.10.2 NATIONAL ISDN COUNCIL

Established by Bellcore and the Corporation for Open Systems (CoS) in 1991, the National ISDN Council is a forum for telecommunications service providers and switch suppliers. The National ISDN Council establishes the framework for NI-1 (National ISO-Phase 1) and NI-2 (NI-Phase 2), and supports development of NI-3 (NI-Phase 3). In addition, this Council promotes utilization of standardized ISDN products, distributes documents containing descriptions of current and projected ISDN applications, and provisions recommendations for ISDN service enhancement.

1.10.2.1 SPID (Service Profile Identifier) and AutoSPID (Automatic SPID)

The National ISDN Council maintains a registry of customer equipment and ordering codes to facilitate ISDN deployment. ISDN ordering codes enable service providers to configure their switching equipment to match the ISDN capabilities of customer premise equipment (CPE). The National ISDN Council also publishes reports clarifying BRI and PRI functions, TA and TE applications, and Service Profile Identifier (SPID) operations.

Each SPID consists of a unique ten-digit telephone number for each ISDN line, as well as a prefix and suffix indicating service features. The communications carrier or service provider assigns a SPID to an initializing ISDN terminal when the user places an order for BRI service. The SPID enables the Stored Program Control Switching System (SPCS) to identify the initializing ISDN terminal at Layer 2 or the Data-Link Layer of the OSI Reference Model. This information is a prerequisite for provisioning ISDN service. In cases where multiple terminals are assigned to a single BRI, SPIDs enable identification of terminals experiencing problems. Developed by the National ISDN Council through its ISDN enhancement program, AutoSPID automates the terminal initialization process.

1.10.2.2 National ISDN-Phase 1 (NI-1)

The National ISDN Council sponsored an industrywide effort that culminated in the establishment of National ISDN-1 (NI-Phase 1). Adopted in 1996 and 1997, NI-1 supports terminal portability for enabling ISDN subscribers to use NI-1 equipment at any ISDN-compliant location.

1.10.2.2.1 SS7 (Signaling System 7)

NI-1 employs SS7 (Signaling System 7) for provisioning seamless circuit-switched and packet-switched services. Endorsed by the ITU-T and the American National Standards Institute, SS7 is an out-of-band signaling system that facilitates call setup, data routing, billing services, and information exchange via the PSTN. Moreover, SS7 defines message transfer protocols and signaling operations in support of switched-voice and non-voice ISDN services, and enables applications that range from e-mail and voicemail to remote meter reading and teleshopping.

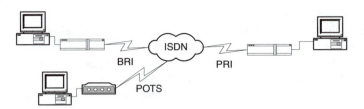

FIGURE 1.5 An ISDN small-sized business configuration supporting BRI and PRI services via a POTs leak.

1.10.2.3   National ISDN-2 (NI-Phase 2) and NI-3 (NI-Phase 3)

National ISDN-2 (NI-Phase 2) was adopted by the National ISDN Council in the late 1990s. NI-2 supports service uniformity in ISDN operations. NI-2 also standardizes Basic Rate ISDN (BRI) and Primary Rate ISDN (PRI) services, clarifies uniform billing methods, and enables more sophisticated data applications than N1-1. Specifications for National ISDN-3 (NI-Phase 3) are in development. (See Figure 1.5.)

1.10.3   ACCESS TECHNOLOGIES FORUM (ATF)

Originally called the Vendors ISDN Association (VIA), the Access Technologies Forum (ATF) is a nonprofit corporation that promotes rapid deployment of standards-compliant ISDN services and products and accelerates the availability of interoperable ISDN solutions. ATF members include 3Com Corporation, Adtran, Ascend Communications, Cisco Systems, Intel Corporation, and Virtual Access.

1.10.4   NORTH AMERICAN ISDN USERS FORUM (NIUF)

Sponsored by the National Institute of Standards and Technology (NIST), the North American ISDN Users Forum (NIUF) became operational in 1988. As with the National ISDN Council, NIUF promotes implementation of standardized ISDN installations. ISDN subscribers, vendors, service providers, and manufacturers are encouraged by the NIUF to take part in the ISDN design, development, and implementation process in order to ensure deployment of standards-compliant ISDN products, equipment, services, and applications.

Moreover, the NIUF supports utilization of generic forms for ordering ISDN services, standardized ISDN ordering packages, and flat fees for ISDN usage. The NIUF also encourages communications carriers and service providers to adopt national ISDN ordering codes and SPIDs (Service Profile Identifiers) for further streamlining the ISDN provisioning process.

In addition, the NIUF promotes development of encryption mechanisms, network management operations, and remote diagnostics specifically designed for identifying and correcting problems with ISDN implementations at the customer premise. The NIUF also supports specifications enabling ISDN to interwork with IP (Internet Protocol), Ethernet, Frame Relay, ATM, cable modem, and DSL technologies.

1.11   INTERNATIONAL STANDARDS ORGANIZATIONS AND ACTIVITIES

1.11.1   INTERNATIONAL TELECOMMUNICATIONS UNION-TELECOMMUNICATIONS STANDARDS SECTOR (ITU-T)

Recommendations developed by the ITU-T define standardized applications and services in the telecommunications and computer network domain. Participants in ITU-T Study Groups represent government agencies, communications carriers, software companies, corporations, vendors, educational institutions, and regulatory authorities. ITU-T Study Groups develop Recommendations for current and emergent technologies. These Study Groups work in concert with other standards development organizations (SDOs) such as the Internet Engineering Task Force (IETF) and the International Standards Organization (ISO) to eliminate replication of work effort and facilitate the standards development process.

1.11.1.1   ITU-T H.261 Recommendation

The ITU-T H.261 Recommendation for video codecs (coders/decoders) supports ISDN services and applications such as videoconferencing and videotelephony. Procedures for video codecs to transport as many as 30 frames-per-second over multiple 64 Kbps lines are clarified. Prior to the ITU-T H.261 Recommendation, videoconferencing vendors employed proprietary solutions. The ITU-T H.261 Recommendation also describes approaches for compression in order to eliminate latencies between frames, and delineates specifications for point-to-point and point-to-multipoint videoconferencing, multipoint bridging, internetwork security, and seamless multimedia transmission.

1.11.1.2   ITU-T H.320 Recommendation

An extension to the ITU-T H.261 Recommendation, the ITU-T H.320 Recommendation clarifies technical requirements facilitating audiovisual transport via narrowband transmission systems at rates ranging from 128 Kbps to 1.544 Mbps (T-1) and establishes the framework for ISDN videoconferencing and videotelephony services. Furthermore, the ITU-T G.711 Recommendation clarifies the basic audio encoding specification for narrowband digital telephony operations via ITU-T H.320 devices.

The ITU-T H.320 Recommendation complements the ITU-T H.324 Recommendation. The ITU-T H.324 Recommendation describes approaches for transmitting compressed video via the PSTN (Public Switched Telephone Network). (See Figure 1.6.)

1.11.1.3   ITU-T H.323 Recommendation

Adopted in 1996, the ITU-T H.323 Recommendation provisions a framework for implementation of video, audio, and data teleservices and tele-applications across multivendor IP (Internet Protocol) networks in point-to-point, point-to-multipoint, and multipoint-to-multipoint configurations. This Recommendation establishes approaches

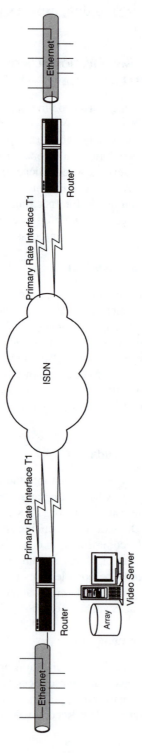

FIGURE 1.6 The addition of a video server and array to an in-place ISDN network to support additional applications.

for enabling multimedia services via LANs (Local Area Networks) based on technologies such as ISDN that do not provision QoS (Quality of Service) guarantees. ITU-T H.323-compliant terminals can support any combination of data, video, and voice traffic ranging from a single person-to-person voice-only call to multiparty multimedia interactive videoconferences.

Moreover, the ITU-T H.323 Recommendation describes packet telephony architecture and protocols that facilitate call signaling, call management, and call control. Compression and decompression algorithms and functions supported by the ITU-T H.323 Recommendation are also clarified in the ITU-T G.711 and ITU-T G.729 Recommendations.

ITU-T H.323 services support RTP/RTCP (Real-Time Protocol/Real-Time Control Protocol) for video and audio signal management, synchronization of video and audio streams, and transmission of real-time content. H.323-compatible services work in concert with RSVP (Resource Reservation Protocol) for provisioning dedicated bandwidth to specified applications.

The ITU-T H.323 Recommendation establishes a foundation for multicast services and bandwidth management to enable seamless traffic flows. In addition, the ITU-T H.323 Recommendation establishes a framework for enabling point-to-point, point-to-multipoint, and multipoint-to-multipoint videoconferences that function independently of in-place network operating systems and hardware configurations. ITU-T H.323-compliant solutions support tele-education, room-based and desktop videoconferencing, workgroup computing, electronic whiteboarding applications, IP telephony or voice-over-IP (VoIP), and interactive teleshopping services.

1.11.1.4   ITU-T T.120 Recommendation

The ITU-T T.120 Recommendation provisions a scalable and extendible framework for multipoint data communications services that operate in interactive multimedia videoconferencing environments. This Recommendation also complements the ITU-T H.323 Recommendation and enables streaming voice and video transmission and multimedia applications such as videoconferences, teleseminars, teleworkshops, teletraining sessions, and simulations. In addition, the ITU-T T.120 Recommendation describes procedures for distributing IP multicasts; defines methods for secure transmissions that operate in concert with ISDN, PSTN, ATM, and Frame Relay (FR) solutions; and specifies procedures for data and file exchange and transmission of still images.

1.11.1.5   ITU-T I-Series of Recommendations

The ITU-T I.100-Series of Recommendations clarify fundamental ISDN principles and concepts. The ITU-T I.200-Series of Recommendations reference telecommunications services supported by ISDN. The ITU-T I.300-Series of Recommendations describes ISDN-based network functions, performance objectives, network protocols, and numbering and addressing principles. The ITU-T I.400-Series of Recommendations indicates attributes of ISDN User-to-Network Interfaces (UNIs).

The ITU-T I-Series of Recommendations also refers to Recommendations in the ITU-T V-Series, the ITU-T G-Series, and the ITU-T Q-Series that affect ISDN deployment. The ITU-T V-Series of Recommendations address issues associated with the provision of data communications and message handling services via the PSTN. The ITU-T G-Series of Recommendations clarifies processes relating to telephone connections and transmission systems. The ITU-T Q-Series of Recommendations specifies switching and signaling functions.

1.11.2 EUROPEAN ISDN USERS FORUM (EIUF)

Regarded as the counterpart to the North American ISDN Users Forum, the European ISDN Users Forum (EIUF) encourages utilization of standardized EuroISDN implementations by every member state in the European Union. EuroISDN offers BRI (Basic Rate Interface) and PRI (Primary Rate Interface) services, and supports supplementary applications that include caller ID, direct dialing, and terminal portability. EuroISDN also facilitates utilization of standardized network services and applications. Prior to EuroISDN implementation, proprietary European ISDN installations were incapable of supporting interoperable applications and transborder services.

The European ISDN Users Forum (EIUF) encourages EuroISDN deployment throughout member states in the European Union. Created by the European Commission (EC) in 1990, the EIUF establishes testbeds for assessment of EuroISDN applications and services prior to full-scale implementation. EIUF also sponsors the BIRD (Better Infrastructure for Rural Development) initiative. This project demonstrates the economic and practical value of providing EuroISDN services in remote communities. EIUF works in concert with the European Telecommunications Standards Institute (ETSI) in fostering implementation of EuroISDN-compliant products and services.

1.12 ISDN SERVICE ENHANCEMENTS

ISDN is no longer regarded as a universal network solution. However, ISDN technology continues to overcome physical and electrical local loop impairments, thereby enabling additional bandwidth capacity for accessing multimedia services at the customer premise. To encourage widespread ISDN utilization, vendors, standards organizations, and user groups promote routine deployment of AO/DI (Always On/Dynamic ISDN), CPE (Customer Premise Equipment) Diagnostics, and ISI (Initialization Simplification Initiative) service enhancements.

1.12.1 ALWAYS ON/DYNAMIC ISDN (AO/DI)

1.12.1.1 AO/DI Functions

Always On/Dynamic ISDN (AO/DI) refers to network applications that utilize the ISDN D (Delta) Channel for X.25 packet data service and maintain always-on connectivity between the communications carrier and the ISDN subscriber. With an

AO/DI constant virtual connection, ISDN enables low-bandwidth transmissions at speeds up to 9.6 Kbps.

When additional bandwidth is required to support information access and delivery, AO/DI automatically provisions use of the B (Bearer) Channel to support network operations at 64 Kbps. With two B Channels, ISDN enables rates reaching 128 Kbps without compression and rates up to 512 Kbps with compression. When extra bandwidth is no longer required, one or both of the B Channels are dropped from the connection. The D Channel remains in place for packet-switched services.

1.12.1.2   AO/DI Supporters

In the United States, AO/DI implementations are supported by ATF (Access Technologies Forum) members that include Ascend, Cisco Systems, 3Com, and Microsoft. National ISDN Council participants such as Ameritech, Verizon, BellSouth, and SBC (Southwestern Bell and Pacific Bell) endorse AO/DI functions as well.

Participants in the European AO/DI Interest Group include Deutsche Telekom, France Telecom, Swiss Telecom, TeleDanmark, BT (British Telecom), Telia, and Telenor. Organized in 1999, the European AO/DI Interest Group participates in the Global ISDN Industry Forum (GIIF). The GIIF promotes EuroISDN implementation by each member state in the European Union.

1.12.2   CUSTOMER PREMISE EQUIPMENT (CPE) DIAGNOSTICS

CPE (Customer Premise Equipment) Diagnostics is a program for monitoring ISDN operations in subscriber equipment. CPE loopback, fault management, and state management tests aid in the identification and correction of equipment problems at the subscriber premise that adversely impact ISDN performance.

1.12.3   INITIALIZATION SIMPLIFICATION INITIATIVE (ISI)

To facilitate ISDN implementation, the Initialization Simplification Initiative (ISI) provides mechanisms for automating ISDN installation procedures. A component of the ISI, autoSPID (Automatic Service Profile Identifier) enables automatic SPID detection by the communications carrier. AutoSPID eliminates the tedious and time-consuming process of providing detailed ISDN information by the subscriber to the communications carrier. Generic SPIDs are also in development to further streamline the initialization process.

1.13   ISDN MARKETPLACE

Regional ISDN communications providers and interexchange carriers in the United States include AT&T, Ameritech, Verizon, BellSouth, GTE, SBC, Sprint, Southwestern Bell, and U.S. West. ISDN is available worldwide in countries that include the United Kingdom, Germany, Italy, Canada, France, Switzerland, the Netherlands, Israel, New Zealand, Australia, Indonesia, and Japan. (See Figure 1.7.)

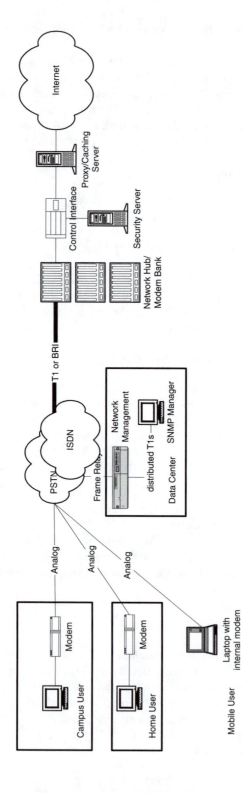

FIGURE 1.7 A NSP (Network Service Provider) ISDN solution provisioning data center, proxy-caching, and security services for campus and SOHO subscribers. Supports remote access via a network hub/modem bank.

1.13.1 SIEMENS

Available from Siemens, the Optiset series of ISDN desktop terminals is equipped with snap-in Terminal Adapter (TA) modules for interlinking multiple analog devices. Optiset ISDN desktop terminals also support a TA for MPD (Multi-Purpose Data) to facilitate access to Web resources at rates up to 115.2 Kbps. In addition to provisioning cost-effective ISDN services, the Optiset series of desktop terminals streamlines the ISDN implementation process at the customer premise by eliminating the need for every ISDN device to be equipped with an interface and an application-specific TA.

1.13.2 TELCORDIA TECHNOLOGIES

As a service to ISDN equipment manufacturers, Telcordia conducts a series of tests to verify the capabilities of standards-compliant ISDN customer premise devices such as modems, phones, multiport LAN bridges or routers, and TAs and software products.

1.14 ISDN COMPETITOR TECHNOLOGIES

1.14.1 BROADBAND RESIDENTIAL ACCESS SOLUTIONS

Demand for fast access to the Web contributes to the popularity of wireline and wireless broadband residential access solutions based on cable modem, LMDS (Local Multipoint Distribution System), MMDS (Multichannel Multipoint Distribution System), and DSL technologies. As with ISDN implementations, these competitor solutions provide dependable connections to voice, video, and data services via the local loop or the first mile between the subscriber premise and the local telephone exchange.

1.14.2 N-ISDN (NARROWBAND-ISDN) AND B-ISDN (BROADBAND-ISDN) INSTALLATIONS

ISDN is the core N-ISDN (Narrowband-ISDN) technology. ATM is the core B-ISDN technology. In comparison to ATM implementations, ISDN deployments are limited in enabling access to high-performance Web-based multimedia applications. Moreover, ISDN is viewed as an interim step in the evolution to multi-service, high-speed, and high-capacity boradband networks. ISDN is also regarded as an outmoded network technology without much promise for the future.

Nonetheless, as demonstrated by the initiatives that follow, ISDN solutions continue to support current and emergent applications in sectors that include education, medicine, and business. ISDN videoconferences interlink geographically dispersed individuals in virtual classes for enabling interactive tele-instruction, tele-training sessions, and telecollaborative workgroup activities in public and private K–12 (Kindergarten through Grade 12) schools and post-secondary institutions. Moreover, ISDN technology also supports teleconsultations, telesurgery, and virtual grand rounds in the field of telemedicine.

1.15   UNITED STATES (U.S.) ISDN TELE-EDUCATION INITIATIVES

The Federal Communications Commission (FCC) in 1997 encouraged renewed interest in ISDN implementation in the educational sector with the endorsement of the E-Rate (Education-Rate) program. The E-Rate program enables K–12 public and private schools and libraries following federal guidelines to receive significant discounts on telecommunications equipment, services, and deployments.

Despite the soaring popularity and availability of broadband networking solutions, ISDN continues to accommodate a diverse range of tele-education requirements. ISDN facilitates implementation of innovative networking configurations such as virtual schools, virtual libraries, virtual universities, and virtual communities. In school and university venues, ISDN telecommuting initiatives enable institutional administrators and staff to effectively balance work and home responsibilities. Because ISDN deployments also support flexible class scheduling, telecollaboration among peers, telementoring after class hours and on weekends, and Web exploration, students and faculty benefit from ISDN solutions as well. In addition, ISDN distance education implementations foster collaborative teleprojects between schools, school districts, and postsecondary institutions such as community colleges, four-year colleges, and universities.

The following national and international initiatives demonstrate ISDN capabilities in enabling communications between students, subject experts, tutors, teachers, and administrators. These deployments also illustrate ISDN functionality in fostering development of global classrooms. Capabilities of ISDN videoconferencing in promoting student achievement, professional development, and curricular enrichment are demonstrated in these initiatives as well.

As noted, the appeal of ISDN is its ability to optimize capabilities of in-place twisted copper pair wiring that also supports PSTN services. By enabling video, still image, audio, and data transmission via twisted copper pair connections, ISDN dependably fosters access to tele-education applications such as desktop videoconferencing and Web browsing via the local loop.

1.15.1   UNITED STATES GOVERNMENT

1.15.1.1   U.S. Department of Agriculture (USDA)
Rural Utilities Services (RUS)

Tele-education applications and programs sponsored by the U.S. Department of Agriculture (USDA) Rural Utilities Services (RUS) enable rural residents to access Web resources in libraries, museums, training centers, and vocational schools throughout the country via technologies that include ISDN. ISDN distance learning projects supported by the USDA RUS provision access to K–12 activities and enrichment teleclasses, vocational teletraining sessions, and college degree and adult education telecourses and teleworkshops.

1.15.2   ALABAMA

1.15.2.1   Huntsville School System

ISDN facilitates a variety of administrative functions in the telelearning environment, including physical security and video surveillance. At the Huntsville School System, an ISDN configuration supports transmission of video images from remote cameras at more than 40 public schools to a centralized facility for review by system personnel monitoring the safety of the campus community and the security of on-site facilities.

1.15.3 CALIFORNIA

1.15.3.1 Statewide Tele-education Initiative

In 1994, Pacific Bell, now known as SBC Communications, initiated a $100 million program called Education First to promote the installation of four ISDN lines in every public school, library, and community college in the State of California. ISDN services were designed to enable voice, video, and data distribution, and thereby support curricular enhancement and enrichment. After providing free installation of ISDN lines and service for one year, Pacific Bell supported a flat-rate pricing plan for ISDN utilization by Education First sites. Pacific Bell also provisioned PRI services in larger library systems and school districts.

Discontinued in the late 1990s, the Education First Program helped teachers, librarians, principals, and administrators effectively use the in-place ISDN telecommunications infrastructure in fostering access to tele-education applications. To facilitate effective videoconferencing deployment, the Education First Program supported Web access to ISDN videoconferencing resources, tutorials, and teaching guides. In addition, the Education First Program provided Web links to instructional strategies for helping librarians and educators plan, organize, conduct, and evaluate ISDN videoconferences. Schools, libraries, community colleges, and community organizations throughout California that used ISDN videoconferences for tele-education projects were also identified. Although the Education First Program was terminated, schools and school districts throughout California continue to use ISDN solutions.

As an example, the Colton, Chico, Fresno, San Francisco, and Santa Ana Unified School Districts implement ISDN solutions for enabling access to teleclassroom activities, Web resources, and interactive tele-instruction programs, and supporting administrative and technical services such as inventory control. In addition, ISDN also facilitates e-mail exchange and desktop publishing and provisions links to attendance records, student transcripts, and administrative policies and reports.

1.15.3.2 Butte College

Butte College uses ISDN technology for facilitating access to library resources and interactive videoconferencing sessions between students at distant locations and faculty in classrooms on the main campus. Students working full-time participate in ISDN videoconferencing sessions at home or in the workplace.

1.15.3.3 Los Angeles Harbor and West Valley Community Colleges

At the Los Angeles Harbor and the West Valley Community Colleges, ISDN supports teletraining and teleconferencing sessions and enables place-bound students to access diverse tele-education courses and programs.

1.15.3.4   Mendocino Unified School District

The Mendocino Unified School District employs an ISDN platform to inform students about career opportunities in technology. This platform supports access to applications that enable students to develop skills in communications, problem solving, critical thinking, and information literacy. In addition, ISDN-based video-conferencing enables students to study real-world issues with their peers at remote locations and experts in the field. During these interactive videoconferences, Mendocino students learn to justify positions, ask critical questions, and draw conclusions based on the information presented.

ISDN-based videoconferences also support virtual explorations beyond the confines of the school setting. As an example, Mendocino Middle School students met virtually with NASA (National Aeronautics and Space Administration) scientists at a remote location and participated in a virtual tour of a space shuttle mockup. The Mendocino Unified School District also owns and operates the ISDN-supported Mendocino Community Network.

1.15.3.5   Pasadena and Sacramento Public Libraries

Library patrons and local residents utilize ISDN teleservices at the Pasadena and Sacramento Public Libraries for Web browsing, searching reference indexes, and accessing electronic library (E-library) digital resources.

1.15.4   FLORIDA

1.15.4.1   Broward County School District

The Broward County School District employs an ISDN network that interlinks district high schools and supports interactive videoconferencing sessions, team-teaching, in-service training, and expanded course offerings. The Broward County School District ISDN configuration also enables high school students to participate in special programs and tele-enrichment activities and enroll in virtual classes sponsored by local community colleges. Moreover, high school teachers participate in real-time ISDN-supported administrative telemeetings, teleworkshops, and telesymposia via the Broward County School District ISDN implementation.

1.15.4.2   Florida International University (FIU)

The Florida International University (FIU) Career Planning and Placement Center uses ISDN to support tele-interviews between upcoming graduates and prospective employers throughout the United States.

1.15.4.3   Nova Southeastern University (NSU)

Regardless of location, working professionals enrolled in advanced degree programs in fields that include education, public health, and pharmacy at Nova Southeastern University (NSU) participate in ISDN-supported interactive graduate class telesessions, teleworkshops, and videoconferences with peers, subject experts, and NSU professors.

1.15.5 ILLINOIS

1.15.5.1 North Suburban Higher Education Consortium (NSHEC)

The North Suburban Higher Education Consortium (NSHEC) sponsors implementation of an ISDN interactive videoconferencing network for high schools, hospitals, research laboratories, community colleges, and universities in the Northern Chicago suburbs.

1.15.5.2 Northwestern University

Northwestern University uses ISDN technology to link videoconferencing facilities at its Evanston and Chicago campuses for supporting interactive telecourse sessions, videoconferences, and telecollaborative research projects. The ISDN facilities also interoperate with NSHEC ISDN videoconferencing installations.

1.15.6 INDIANA

1.15.6.1 Indiana Higher Education Telecommunications System (IHETS)

The Multicampus Technology Project sponsored by the Indiana Higher Education Telecommunications System (IHETS) supports ISDN videoconferences between Indiana University campus sites. This ISDN configuration enables interactive virtual classes and telecollaborative research projects.

1.15.6.2 Notre Dame University

The University of Notre Dame at South Bend employs a mixture of T-1 and ISDN technologies for interactive videoconferencing and distance education course delivery in graduate programs in business and management to off-campus students. Transmission rates at 384 Kbps are supported.

1.15.7 KANSAS

1.15.7.1 Fort Hays State University (FHSU)

Fort Hays State University (FHSU) employs ISDN point-to-multipoint videoconferences for real-time tele-instruction. The FHSU ISDN configuration also enables connections to the Kansas University Medical Center, Wichita State University, the State Board of Education, and the High Southwest Plains Network for supporting teleseminars and curricular enhancement programs. Additionally, FHSU participates in the A+ network. This network employs ATM technology for enabling high-speed, high-capacity voice, video, and data transmission via optical fiber links.

1.15.8 MICHIGAN

1.15.8.1 Dexter Community Schools and MichNet (Michigan Network)

Dexter Community Schools use ISDN technology for interlinking community school sites to MichNet (Michigan Network). A statewide educational and research network for academic institutions, MichNet provisions access to Web applications, digital library resources maintained by state-supported research universities, and electronic library records documenting holdings at local public libraries.

1.15.9 Mississippi

1.15.9.1 Mid-Mississippi Delta Consortium

The Mid-Mississippi Delta Consortium uses ISDN videoconferences to compensate for teacher shortages and enrich learning opportunities for homebound, at-risk, and underserved students and residents in rural Washington, Humphreys, Leflore, and Golivar Counties. In addition, this Consortium utilizes the Mississippi Educational Television (ETV) Network for provisioning access to courses in fine arts, language arts, mathematics, social studies, science, and foreign languages.

1.15.9.2 Mississippi–Alabama School District Consortium

Sponsored by Jefferson and Green Counties in Mississippi and Sumpter County in Alabama, the Mississippi–Alabama School District Consortium utilizes an ISDN configuration to facilitate interactive videoconferencing, tele-instruction, and curricular enrichment. This configuration also provisions access to lifelong telelearning programs. Designed for rural schools and isolated communities, this ISDN initiative overcomes barriers to learning associated with the geographic isolation.

1.15.10 New York

1.15.10.1 Long Island Educational Enterprise Zone (LIEEZ)

The Long Island Educational Enterprise Zone (LIEEZ) is a consortium of public and private schools in Suffolk and Nassau Counties. LIEEZ employs an ISDN infrastructure to enable the participation of school library media specialists in tele-workshops. This infrastructure also provisions access to multimedia resources and NASA tele-education programs, and enables interactive videoconferencing between directors at major art museums and students and their teachers in school classrooms.

1.15.10.2 New York Institute of Technology (NYIT)

ISDN videoconferences are powerful complements to on-site instruction. As an example, the New York Institute of Technology (NYIT) employs ISDN as a platform for provisioning interactive videoconferencing sessions between classes at the NYIT campus and subject experts at sites in Korea, Taiwan, and Chile.

1.15.10.3 New York University

At the New York University Virtual College, students working at their own pace use ISDN connections to submit coursework online to instructors at a variety of locations and virtually access multimedia course-related materials and video-on-demand (VOD) programs at their convenience.

1.15.11   NORTH CAROLINA

1.15.11.1   Appalachian State University

Appalachian State University delivers interactive ISDN videoconferences, teleworkshops, and tele-education courses to high school students at public schools in the western part of the state. Appalachian State University also provides E-learning (electronic learning) opportunities at ISDN sites statewide for graduate and undergraduate students in the field of teacher education.

1.15.12   RHODE ISLAND

1.15.12.1   Brown University

The Brown University Computer and Information Services Department provides technical support for ISDN videoconferences between the campus studio in Providence and subject experts at national and remote locations. As an example, ISDN facilitates telecollaborative research, formal presentations, and interactive question-and-answer sessions between faculty and students at Brown University and NASA scientists and engineers. Rates at 384 Kbps are supported. Moreover, the in-place ISDN infrastructure provides low-cost connections for enabling telecollaborative research applications between faculty and students at the Brown University campus and scientists at the Institute for Space Research, a research center affiliated with the Russian Academy of Sciences in Moscow.

1.15.13   TENNESSEE

1.15.13.1   DIANE (Diversified Information and Assistance Network)

Tennessee State University sponsors an innovative distance learning configuration called the Diversified Information and Assistance Network (DIANE). Project DIANE employs a mix of ISDN and ATM technologies for interlinking schools, universities, museums, libraries, and community centers in Tennessee, Alabama, and Maryland. Project DIANE participants include K–12 teachers and their students, librarians, guidance and business counselors, disadvantaged youth, inner-city community residents, senior citizens, and the hearing impaired.

In addition to provisioning access to tele-education programs and Web resources, Project DIANE supports telelectures and teleworkshops on senior healthcare, interactive telefield trips, online library reference services, faculty teleresearch, and student teletutorials. Moreover, Project DIANE facilitates collaborative initiatives between business development agencies, educational institutions, and community service organizations.

1.15.14   TEXAS

1.15.14.1   City of Houston School District

The City of Houston School District employs an ISDN configuration for enabling district teachers to access digitized student records and monitor student progress.

This network also enables students in schools with class sizes that do not justify salaries for on-site instructors to participate in virtual Advanced Placement (AP) classes.

1.15.15 WASHINGTON

1.15.15.1 K–20 Educational Telecommunications Network

The K–20 Educational Telecommunications Network in the State of Washington employs ISDN and satellite technologies for supporting curricular enrichment at public primary and secondary schools and post-secondary institutions. Additionally, this network interconnects community and technical colleges, regional education service districts, and libraries to the State of Washington Intergovernmental Network.

1.15.15.2 University of Washington

The University of Washington promotes ISDN deployment in schools in the greater Seattle area for enabling K–12 students and their teachers to participate in the Live from Earth and Mars initiative sponsored by NASA. This initiative supports real-time videoconferencing and dissemination of curricular modules and resources associated with space exploration, meteorology, the space sciences, and mathematics.

1.15.16 WISCONSIN

1.15.16.1 Nicolet Distance Education Network (NDEN)

The Nicolet Distance Education Network (NDEN) supports interactive ISDN applications and services for K–8 and K–12 rural school districts, the Nicolet Area Technical College, and the Stone Lake Native American Reservation. In addition, NDEN enables rural residents to access job teletraining sessions and personal enrichment courses. Schools in Vilas, Iron, Oneida, Forest, Rusk, Florence, and Lincoln Counties in Northeastern Wisconsin offer NDEN tele-education programs.

NDEN also facilitates interactive teleconferences between students and their teachers and subject experts, and delivery of high school physics, calculus, and analytical geometry telecourses to local schools. Students are required to complete these courses in order to graduate from high school. However, these subjects are not taught in traditional classrooms as a consequence of low enrollments and fiscal constraints.

1.15.16.2 University of Wisconsin–Extension Educational Telecommunications Network

Developed by the University of Wisconsin–Extension Educational Telecommunications Network, ISDN configurations provision access to teletutorials for athletic teams on the road and delivery of tele-education programs to private colleges and businesses. The University of Wisconsin–Extension Educational and Telecommunications Network also supports teleconsultations between university hospital personnel and incarcerated offenders in an initiative that provides delivery of low-cost health-care services to local prisoners.

The University of Wisconsin–Extension Instructional Communications Systems schedules videoconferences at university venues, government offices, and state facilities to enable staff and administrators to participate in lifelong learning programs and college credit courses. Video quality reflects the speed at which the video is transported via the ISDN configuration. In addition to ISDN, Switched 56 services are also available.

1.15.17 WYOMING

1.15.17.1 Natrona County School District

Located in Casper, the Natrona County School District employs a mix of technologies including dedicated T-1 lines and ISDN technology for enabling high school students and their teachers to participate in collaborative educational programs such as the SAXophone (Students All Over the World Exchanging Over Phone) initiative. Sponsored by Nova Southeastern University (NSU), the SAXophone initiative also enables high school students to participate collaboratively in interactive videoconferences and teleprojects and supports virtual field trips, virtual musical concerts, and virtual class sessions.

1.16 INTERNATIONAL ISDN TELE-EDUCATION INITIATIVES

1.16.1 AUSTRALIA

1.16.1.1 South Australia Tele-Learning Consortium

The South Australia Tele-Learning Consortium employs ISDN videoconferencing for supporting interactive class sessions in accounting, foreign languages, law, tourism, business, and automotive studies. Between 15 and 25 students participate in class sessions that are held in rural community centers.

1.16.2 FINLAND

1.16.2.1 Helsinki University of Technology

The Helsinki University of Technology utilizes a EuroISDN platform for provisioning interactive videoconferences, telelectures, distance learning activities, teletraining services in programs leading to a Master's of Science Degree, and telecourses in environmental management and foreign languages.

1.16.3 JAPAN

1.16.3.1 Children's Network

Developed by Nippon Telephone and Telegraph (NTT) and sponsored by the Japanese Ministry of Education, the Children's Network employs an ISDN multipoint-to-multipoint videoconferencing configuration for interlinking Japanese teachers and their students in virtual classroom environments.

1.16.4 UNITED KINGDOM

1.16.4.1 National Council for Educational Technology (NCET)

Sponsored by the National Council for Educational Technology (NCET), the Whitby Project evaluates capabilities of EuroISDN videoconferences in supporting interactive instructional telesessions and team teaching initiatives in schools in North Yorkshire. In another NCET initiative, the Whitby Primary Videoconferencing Project employs EuroISDN videoconferences for enabling students in primary schools in North Yorkshire to become proficient in conversational French by participating in real-time interactive teleclasses with their peers attending primary schools in France. Also an NCET initiative, the Libraries of the Future Project supports EuroISDN deployment in academic libraries for enabling library patrons to access digital information resources.

1.16.4.2 Project Connect

Project Connect employs EuroISDN technology for interconnecting primary and secondary schools to Internet resources via the Joint Academic Network (JANET) and its successor the Super Joint Academic Network (SuperJANET). Now in Phase 4, SuperJANET is a high-speed, high-performance research and education network that enables research centers, K–12 schools, and post-secondary institutions throughout the United Kingdom to access multimedia applications and services.

1.16.4.3 University of Exeter

A participant in the JANET and the SuperJANET initiatives, the University of Exeter employs the EuroISDN platform to deliver tele-education courses in foreign languages, teletutorials, interactive videoconferences, and teletraining sessions to students in the workplace.

1.16.4.4 University of Ulster

Developed by the University of Ulster, the ACTOR (Applications for ISDN Communications Technologies to Extend OutReach) initiative facilitates course delivery to an expanded student population in Northern Ireland. ACTOR participants include women at home in rural and remote communities and employees working at computer software firms. EuroISDN telecourses supported by the ACTOR initiative are designed for students with job and family responsibilities, childcare obligations, and/or transportation problems that prevent their enrollment in traditional classes. Mixed-mode EuroISDN and satellite configurations also enable investigators at the University of Ulster to participate in telecollaborative research and tele-instructional projects with their peers at Letterkenny Regional Technical College in the Republic of Ireland and Hokaido University in Japan.

## 1.17	EUROPEAN COMMISSION (EC) TRANS-EUROPEAN TELECOMMUNICATIONS NETWORK (TEN-Telecom) PROGRAM

Sponsored by the European Commission (EC), the Trans-European Telecommunications Network (TEN-Telecom) Program provisioned an array of telecommunications initiatives in fields that included transportation, the environment, and healthcare. Operational between June 1997 and December 2001, TEN-Telecom telematics projects in tele-education enabled lifelong learners, business executives, agricultural workers, and K–12 and post-secondary students, faculty, staff, and administrators to access teletraining and tele-education courses at home, in school venues, and at the workplace. Communications carriers, service providers, public authorities, and educational institutions supported TEN-Telecom initiatives. Representative TEN-Telecom implementations that employed EuroISDN for teletraining and telelearning are highlighted in this section.

### 1.17.1	ADVANCED SOFTWARE FOR TEACHING AND EVALUATION OF PROCESSES (ASTEP)

ASTEP supported development of self-paced course modules featuring video, audio, text, and animation in job teletraining sessions that were distributed via the EuroISDN infrastructure. In addition, ASTEP used the EuroISDN platform for provisioning access to tele-education courses, teletutorials, videoconferencing, and telesymposia. Participants included employees in the manufacturing sector. In addition, West Lothian College and Buskerud College took part in this initiative.

### 1.17.2	DYNAMIC AWARENESS RAISING PROCESS REGARDING TELEMATICS IN THE FRAMEWORK OF NATURA (DART)

The DART initiative used an EuroISDN platform for provisioning access to tele-education programs, telecourses, and teletraining applications. Designed for students enrolled in agriculture and veterinary medicine programs, the DART project also promoted development of a Virtual University of Agriculture and Veterinary Medicine.

### 1.17.3	EXTRANET EDUCATION (EXE)

Based on an EuroISDN infrastructure, the EXE initiative fostered utilization of an extranet for training teachers to effectively employ new technologies in support of classroom activities. Educators in Bologna, Barcelona, and London accessed EXE to participate in interactive videoconferencing and telecollaborative sessions.

### 1.17.4	HARMONIZED ACCESS AND RETRIEVAL FOR MUSIC ORIENTED NETWORK INFORMATION CONCERTED ACTION (HARMONICA)

The HARMONICA initiative supported access to digitized music resources ranging from symphonic, operatic, and concert recordings to collections of sheet music via the EuroISDN infrastructure. Participants included the National Library of France and the Music Information Center in Denmark.

1.17.5 Students Across Europe Language Network (SAELN)

The SAELN project employed EuroISDN for enabling videoconferences in trans-European virtual classroom environments. Students developed language skills in German, French, Spanish, and English through interactive conversations with their peers in member states throughout the European Union.

1.18 VIRTUAL COMMUNITY NETWORKS

Virtual community networks varying in size, scope, and sophistication appear with increasing frequency on the Web. These virtual networks correspond to actual physical communities and support online data interchange, diverse telecollaborative applications, and timely and economical access to vast reservoirs of electronic resources. Their proliferation is made possible by a convergence of ongoing technical advances in telecommunications, multimedia, and computer hardware and software. As indicated in the initiatives that follow, ISDN supports collaborative networking among participants in Web-based virtual community networks and provisions access to a wide range of tele-activities, tele-applications, and teleservices.

1.18.1 Illinois

1.18.1.1 Champaign County Network (CCNet)

Champaign County Network (CCNet) employs a mix of communications technologies, including POTS (Plain Old Telephone Service) and ISDN, to facilitate videoconferencing and dependable exchange of images and large data files among local businesses, libraries, and government agencies. CCNet also provisions access to Internet services for students and faculty at local schools and post-secondary educational institutions. Community residents participating in CCNet Task Forces on Education, Healthcare, Agribusiness, Small Businesses, and Libraries guide CCNet development.

1.18.2 Virginia

1.18.2.1 Blacksburg Electronic Village (BEV)

One of the first virtual community networks in the United States, Blacksburg Electronic Village (BEV) employs an array of network technologies, including ISDN, Ethernet, and ATM. Entities such as businesses, public schools, and local public libraries in Blacksburg and Montgomery County take part in the BEV initiative. The BEV Web site features online surveys and opinion polls for school renovations and road construction and contains online referendums for electing local civic leaders. BEV also supports access to e-mail accounts, listservs, and chat rooms, and features collections of online resources in disciplines such as engineering, computer science, philosophy and ethics, psychology, and political science.

Designed to foster local economic development and civic participation, Blacksburg Electronic Village maintains a storefront office for user registration. Connectivity to BEV is available at the regional public library. School libraries in Montgomery County enable access to Web resources via BEV connections as well.

BEV provides training, documentation, and technical assistance to local residents through videoconferencing, teletutorials, and online help sessions, and builds community support and participation by demonstrating the practical benefits of Web services and applications. Moreover, BEV sponsors teleseminars for teachers, medical professionals, and corporate employees. Researchers associated with the BEV initiative develop approaches for leveraging collaborative networking services to support instructional enhancement and promote development of virtual schools that effectively function within electronic villages.

Established as a partnership between the Town of Blacksburg, Verizon, and Virginia Polytechnic Institute and State University (Virginia Tech), BEV also solicits the participation of businesses, schools, public libraries, museums, community nonprofit organizations, and government agencies as information providers. BEV services were initiated in 1993. In 1994, ISDN and Ethernet network connections became available. Currently, BEV belongs to the New Century Communications Network, Inc., a consortium that supports implementation of next-generation educational networking solutions in schools, colleges, universities, and local businesses.

1.19　U.S. TELEMEDICINE INITIATIVES

An increasingly popular solution for telehealthcare delivery around the globe, telemedicine provisions access to telehealthcare resources and medical services and supports voice, video, and data transmission from one location to another location via a telecommunications infrastructure that features technologies such as ISDN and ATM. Moreover, telemedicine initiatives foster access to continuing tele-education and teletraining courses for physicians, nurses, and clinical staff.

An ISDN infrastructure supports point-to-point, point-to-multipoint, and multipoint-to-multipoint interactive telemedicine videoconferences and transmissions at rates ranging from 64 Kbps to 1.544 Mbps (T-1) and higher, depending on the numbers of ISDN lines and types of applications that are employed. In addition to ISDN and ATM, cable modem, DSL, Ethernet, Fast Ethernet, and Gigabit Ethernet also enable access to telemedicine services and initiatives.

ISDN-based telemedicine applications and projects accommodate the needs of a diverse population. As an example, these initiatives enable medical students working at remote healthcare centers to participate virtually in interactive clinical videoconferences with medical specialists at major teaching hospitals.

1.19.1　U.S. Federal Government

1.19.1.1　U.S. Department of Agriculture (USDA) Rural Utilities Services (RUS)

Funding for rural telehealth initiatives employing an ISDN infrastructure is provisioned by the U.S. Department of Agriculture (USDA) Rural Utilities Services

(RUS). USDA RUS sponsors ISDN telemedicine projects that enable residents in rural communities to access the same quality of education and healthcare services that are available to individuals in major cities across the United States. For example, ISDN-based telemedicine links facilitate teleconsultations between patients and their primary care physicians at rural hospitals and isolated medical clinics and medical specialists at major metropolitan medical centers.

USDA RUS also supports ISDN tele-education initiatives for enabling remote clinicians to participate in medical staff training sessions, delivering continuing education programs to patients and their caregivers, and provisioning advanced certification courses for healthcare personnel and public safety officials. USDA RUS encourages rural institutions such as libraries, schools, hospitals, and healthcare clinics to form partnerships with their metropolitan counterparts to support implementation of sophisticated telemedicine applications and services. As noted, USDA RUS also provisions distance education services in order to expand educational and occupational opportunities available to residents at far-flung locations.

1.19.2   U.S. Armed Forces

1.19.2.1   Fort Gordon Teledentistry Initiative

At Fort Gordon, dental personnel take part in teledentistry trials that involve utilization of intra-oral video cameras and videoconferencing equipment for increasing the quality of dental treatment provisioned to soldiers in the field. This infrastructure consists of a hybrid network that employs INMARSAT (International Maritime Satellite) services and ISDN technology. In addition, this mixed-mode satellite and ISDN platform enables transmission of continuing education courses to professional dentists and dental assistants at rates reaching 128 Kbps.

1.19.3   Arkansas

1.19.3.1   Arkansas Rural Medlink (ARM) Consortium

Consisting of medical professionals and staff at five hospitals, the ARM (Arkansas Rural Medlink) Consortium provisions healthcare services to rural residents in the Arkansas Delta region. The ARM Consortium also sponsors interactive videoconferences between medical specialists at the University of Arkansas and general practitioners at rural medical clinics via ISDN connections. In addition, the ARM ISDN platform enables community outreach telehealthcare programs and delivery of continuing education telecourses to nurses and paraprofessional medical personnel.

1.19.4   California

1.19.4.1   Charles R. Drew University of Medicine and Science

Situated in Los Angeles County, the Charles R. Drew University of Medicine and Science sponsors local tele-ophthalmology clinics for individuals in underserved communities where on-site physicians are not available. Each clinic uses ISDN links for enabling transmission of text, still images, video, and voice signals at 128 Kbps and higher rates.

1.19.5 ILLINOIS

1.19.5.1 Children's Memorial Hospital

A hybrid satellite and ISDN platform that enables transmission rates at 384 Kbps links cardiologists at Children's Memorial Hospital in Chicago and cardiologists at Aghia Sophia Hospital in Athens, Greece. This ISDN platform enables interactive videoconferencing, teleconsultations, and telediagnoses of complex heart problems in children. In addition, ISDN connections support links between Children's Memorial Hospital and rural community hospitals for enabling teleradiology teleconsultations between hospital radiologists and remote primary care physicians.

1.19.6 IOWA

1.19.6.1 Midwest Rural Telemedicine Consortium (MRTC)

The Midwest Rural Telemedicine Consortium (MRTC) employs ISDN for enabling interactive videoconferences and supporting implementation of teleradiology and telepathology applications. This ISDN system facilitates teleconsultations between medical specialists in dermatology, mental health, neurology, and orthopedics on staff at major urban hospitals and patients at primary healthcare clinics and long-term nursing facilities in remote locations. In addition, the ISDN infrastructure enables hospital staff and medical professionals in rural communities to participate in distance learning programs and professional enrichment teleseminars. The Mercy Hospital Medical Center, Mercy Health Services, and the North Iowa Mercy Health Center participate in MRTC activities.

1.19.6.2 University of Iowa

In the telehealth and tele-education domain, the University of Iowa supports ISDN delivery of continuing education courses in telemedicine, nursing, biomedical engineering, health informatics, and library and information sciences to primary care physicians at remote locations. The University of Iowa also uses an ISDN platform for enabling students at outlying sites to enroll in a distance education program provisioning Physician Assistant (PA) certification.

1.19.6.3 University of Iowa Hospital and Clinics (UIHC) and the Iowa Communications Network (ICN)

The University of Iowa Hospital and Clinics (UIHC) provisions delivery of telehealthcare services to patients and their caregivers in rural communities statewide via the Iowa Communications Network (ICN) ISDN platform. The ICN configuration also employs ISDN for distribution of educational programming to K–12 schools and public and private colleges and universities, telecourse delivery, and telecollaborative research initiatives.

In addition, the ICN ISDN platform facilitates real-time videoconferencing among physicians at rural hospitals and clinics in Iowa with staff physicians at the UIHC. The ICN ISDN platform at the UIHC also supports telepsychiatry, teleradiology, and telecardiology services and teleconsultations in orthopedics, dermatology, and internal medicine for treating Iowa residents at underserved rural locations. Moreover, the ICN ISDN configuration facilitates teleconsultations in pediatric echocardiology and psychology, provisions interactive genetic counseling services, and enables individuals with diabetes to access online healthcare resources.

Medical practitioners on staff at the UIHC conduct ISDN teleconsultations via the ICN platform with prisoners at the Iowa State Penitentiary and correctional facilities. This program eliminates costs associated with transporting prisoners to medical clinics and potential risks to patients and their families at public healthcare facilities where inmates are treated.

The UIHC initiated an ISDN ResourceLink networking project for provisioning home healthcare assistance to patients in Little Rock, Sheldon, Davenport, Cedar Rapids, Independence, Olin, Urbandale, Little Sioux, and Council Bluffs. Through the ISDN-supported Healthnet and the Virtual Hospital Program, the UIHC enables patients, caregivers, and medical practitioners to access current and authoritative digital library resources.

As a participant in the National Library of Medicine Rural Medicine Program, the UIHC employs the ICN ISDN infrastructure for interactive videoconferencing. This platform supports dependable data, video, and voice delivery to at least one Point of Presence (PoP) in each county in the state.

In addition to ongoing support of ISDN, the ICN currently employs a SONET (Synchronous Optical Network) fiber-optic dual-ring configuration and Ethernet, T-1, and Frame Relay (FR) technologies. The UIHC also broadcasts telemedicine programs on public television networks and cable network systems via satellite uplinks. Mixed-mode satellite and ISDN connections enable medical students and medical specialists on staff at the UIHC, West Virginia University, and the Kameda Medical Center in Japan to participate in interactive videoconferences. Approaches for enabling migration from the ICN ISDN platform to an ATM infrastructure are under consideration.

1.19.7 Mississippi

1.19.7.1 Delta Rural Health Network (DRHN)

The Delta Rural Health Network (DRHN) consists of small independent rural hospitals situated in Humphreys, Montgomery, Yazoo, Sunflower, and Tallahatchie Counties. The DRHN supports ISDN teleconsultations between hospital-based medical specialists and primary care physicians for enabling treatment of patients in rural intensive care and pediatric acute care facilities. The DRHN also uses ISDN technology to provision interactive geriatric psychological counseling service and support an ISDN teleradiology network.

1.19.8   MONTANA

1.19.8.1   Regional Health Information Network (RHIN)

The Regional Health Information Network (RHIN) operates an ISDN telemedicine network that enables teleradiology and teleconsultations. Participants include medical specialists on staff at regional hospitals and general practitioners provisioning on-site healthcare treatment to underserved and aging patients at St. Joseph Hospital and the Arlee Family Medical Clinic on the Flathead Indian Reservation in Western Montana. Additionally, the RHIN platform enables general practitioners at rural locations to access telemedicine courses and programs via the ISDN infrastructure.

1.19.9   PENNSYLVANIA

1.19.9.1   Children's Telemedicine Network

The Children's Telemedicine Network supports ISDN videoconferences between primary care physicians at regional hospitals in Western Pennsylvania, Eastern Ohio, and Northern West Virginia and specialists at the Children's Hospital of Pittsburgh. This network also enables teleconsultations between emergency room physicians and medical specialists and delivery of tele-education courses to healthcare professionals in rural Western Pennsylvania.

1.19.9.2   Geisinger Healthcare System

The Geisinger Healthcare System sponsors a telemedicine project between the Department of Emergency Medicine at the Bucktail Medical Center in Renovo and the Department of Emergency Medicine at the Geisinger Medical Center in Danville. This initiative demonstrates the feasibility of utilizing ISDN connections for enabling interactive videoconferencing between medical experts and primary care physicians provisioning treatment and trauma care to patients in hospital emergency rooms (ERs). Transmission rates at 384 Kbps are supported.

1.19.9.3   Pennsylvania (PA) HealthNet

Sponsored by the Pennsylvania Department of Health, the PA HealthNet project employs ISDN, satellite, T-1, and Frame Relay technologies for interlinking rural physicians and their patients at remote locations with medical specialists at urban hospitals and university medical schools. ISDN videoconferences support teleconsultations in cardiology, dermatology, pediatric neurology, and nephrology. In addition, satellite systems broadcast Grand Rounds originating from the University of Pittsburgh Medical Center and a nurse midwifery program originating from the University of Pennsylvania Medical Center to medical practitioners in real-time at distant locations.

1.19.9.4   University of Pennsylvania Medical Center

The Department of Radiology at the University of Pennsylvania Medical Center employs a mixed-mode network configuration that includes satellite and ISDN technologies to support teleradiology projects. This configuration also enables teleconsultations between medical specialists associated with the TeleQuest Teleradiology Consortium in Horsham and primary care physicians at the Allentown Osteopathic Medical Center.

1.19.9.5   University of Pittsburgh Medical Center

The University of Pittsburgh Medical Center utilizes an ISDN configuration to facilitate teleconsultations between hospital-based radiologists and primary care physicians at rural locations. In addition, specialists at the University of Pittsburgh Medical Center assist primary care physicians providing emergency medical treatment to patients at community hospitals in Eastern Ohio and Western Pennsylvania via ISDN videoconferencing configurations.

1.19.10   TEXAS

1.19.10.1   Sam Houston University (SHU)

Sam Houston University (SHU) operates an ISDN network called RESNET (Residential Network) for provisioning teleservices to residents on the Alabama-Coushatta American Indian Tribal Reservation outside Houston. RESNET also supports interactive teleconsultations between SHU medical specialists and primary care physicians and their patients on the tribal reservation. In addition to facilitating access to healthcare services, RESNET delivers distance education classes to students in three rural school districts.

1.19.10.2   University of Texas Medical Branch (UTMB)

The University of Texas Medical Branch (UTMB) employs an ISDN infrastructure for enabling real-time videoconferences and teleconsultations between specialists at urban hospital Emergency Rooms and primary healthcare providers treating patients at rural trauma centers. This infrastructure also supports delivery of healthcare treatment to individuals working at offshore gas and oil facilities.

1.19.11   VERMONT

1.19.11.1   Fletcher Allen Health Care (FAHC) and the University of Vermont College of Medicine (UVM)

Fletcher Allen Health Care (FAHC), the largest tertiary healthcare center and teaching hospital in Vermont, and the University of Vermont College of Medicine (UVM) sponsor joint ISDN telemedicine programs that support interactive videoconferences, teleradiology applications, and teleconsultations in emergency medicine, vascular surgery, cardiology, gastroenterology, and nephrology. The ISDN infrastructure also facilitates delivery of distance education telecourses and teleprograms originating at UVM to FAHC pharmacists, nurses, and laboratory technicians on the FAHC Burlington campus at rates reaching 384 Kbps. The FAHC Burlington campus features ISDN installations in the main auditorium, conference rooms, the emergency room, the radiology and pathology clinics, and the endoscopy and angiography suites. FAHC and UVM also participate in a regional ISDN telemedicine network called VTMEDNET (Vermont Medical Network). VTMEDNET provisions telemedicine services to patients in Vermont and upstate New York.

In 1999, FAHC and UVM conducted the VTMEDNETPLUS initiative. VTMEDNETPLUS participants included Hanoi Medical College, Hanoi area hospitals, UVM, FAHC, the Canton-Potsdam Hospital in New York, and the George Washington University Medical Center in Washington, D.C.

VTMEDNETPLUS supported implementation of a Virtual Wide Area Network (VWAN) that employed global ISDN services for enabling primary care physicians, medical specialists, medical students, and faculty at participating institutions to take part in real-time videoconferences, teleconsultations, medical teletraining sessions, and tele-education symposia. Uplift International, a humanitarian organization promoting the advancement of healthcare and medical education in Asia, provided technical assistance for VTMEDNETPLUS participants at Vietnamese medical facilities.

1.19.12 VIRGINIA

1.19.12.1 University of Virginia (UVA)

The University of Virginia (UVA) Office of Telemedicine provisions videoconferencing services at rates reaching 384 Kbps via an ISDN platform that supports telecollaborative activities at campus locations including the Camp Heart Auditorium, the Telemedicine Command Center, and the Urology Conference Room. The UVA Office of Telemedicine also supports utilization of a sophisticated diagnostic stethoscope that transmits heart sounds via ISDN connections for enabling UVA pediatric cardiologists to listen to a full range of audio frequencies and accurately diagnose heart murmurs in children at remote locations. In addition, the UVA ISDN configuration facilitates teleconsultations between UVA physicians and inmates at local jails.

1.20 INTERNATIONAL TELEMEDICINE INITIATIVES

1.20.1 CANADA

1.20.1.1 Canadian Rural Medicine Network (CARMEN)

The Canadian Rural Medicine Network (CARMEN) ISDN infrastructure supports interactive videoconferences and teleradiology teleconsultations between radiologists at metropolitan hospitals and primary care physicians at rural clinics. The CARMEN ISDN platform also enables transmission of patient x-rays and CAT

(Computerized Axial Tomography) scans from rural clinics to major urban hospitals for enabling telediagnoses and determining treatment plans.

1.20.1.2 Quebec Interregional Telemedicine Network

The Quebec Interregional Telemedicine Network evaluates the medical and economic impacts of telemedicine on continuing medical education and healthcare delivery services. This ISDN network supports point-to-point, point-to-multipoint, and multipoint-to-multipoint videoconferencing to enable teleconsultations, telediagnoses, and international staff meetings between medical healthcare professionals at university teaching hospitals in Montreal and their peers at the Cochin Hospital in Paris. In addition, the ISDN infrastructure enables real-time digitized ultrasound imaging transmission at 512 Kbps. Research studies documenting the financial benefits of telemedicine and its capabilities in supporting on-site clinical treatment are also conducted.

1.20.1.3 Telemedicine Centre of Newfoundland

The Telemedicine Centre of Newfoundland provides affordable healthcare services to remote communities via communications networks based on technologies that include ISDN. Established in 1976, the Telemedicine Centre of Newfoundland is the oldest continuously operational telemedicine center in the world.

1.20.2 NORWAY

1.20.2.1 Telemedicine Department in Northern Norway

The Norwegian Telemedicine Department in Northern Norway enables delivery of clinical telemedicine services to isolated populations. Specialists routinely conduct teleconsultations with primary care physicians via EuroISDN videoconferencing systems. Rates at 384 Kbps are supported.

1.20.3 UNITED KINGDOM

1.20.3.1 University College London (UCL)

The University College London (UCL) sponsors a EuroISDN network implementation to interlink medical schools throughout the European Union. This WAN (Wide Area Network) supports the telecollaborative participation of students, faculty, and medical specialists in virtual classes and virtual clinics. ATM utilization as a replacement for EuroISDN is under consideration.

1.21 EUROPEAN COMMISSION TELEMATICS APPLICATIONS PROGRAM (EC-TAP) TELEMEDICINE PROJECTS

Projects in the healthcare domain sponsored by the European Commission Telematics Applications Program (EC-TAP) are highlighted in the material that follows.

1.21.1 DiabCare Q-Net

The DiabCare Q-Net project enabled patients with diabetic retinopathy to participate in interactive videoconferences with medical specialists at urban hospitals, receive quality healthcare treatment, and access healthcare resources via the EuroISDN network platform. Participants included the World Health Organization (WHO), the Technical University of Ilmenau, Tromso University, and Stavinger Technical College.

1.21.2 European Pathology Assisted by Telematics for Health (EUROPATH)

The EUROPATH initiative employed a combined Ethernet and EuroISDN infrastructure for enabling teleconsultations between pathologists at university medical centers and medical practitioners at private laboratories and district hospitals. Participants included the Universities of Edinburgh and Oxford and the Nijmegen Faculty of Medical Sciences.

1.21.3 Home Rehabilitation Treatment-Dialysis (HOMER-D)

The HOMER-D initiative supported utilization of the EuroISDN platform for remotely monitoring isolated patients in end-stage renal failure who were in need of uninterrupted and continuous home dialysis. This platform also enabled interactive videoconferences between medical specialists at metropolitan hospitals and caregivers provisioning home healthcare treatment. Project sponsors included the University of Athens and York District Hospital.

1.21.4 Travel Health Information Network (THIN)

Designed for travel agents and their clients, the THIN initiative employed a hybrid EuroISDN and ATM infrastructure for provisioning access to current data on health risks and epidemics. Participants included the Liverpool School of Tropical Medicine and the Swiss Tropical Institute.

1.22 EUROPEAN COMMISSION (EC) ELECTRONIC COMMERCE (E-COMMERCE) INITIATIVES

A major enabler of telemedicine, tele-education, and job training, EuroISDN also supports teleworking in SOHO venues through its dependable delivery of digitized voice, video, and data services. Sponsored by the European Commission, teleworking initiatives demonstrated EuroISDN capabilities in provisioning networking services for small- and medium-sized enterprises. EuroISDN initiatives supported by the European Commission also enabled an assessment of E-commerce capabilities in facilitating an array of online activities between virtual buyers and virtual sellers in the global marketplace. Representative EuroISDN projects in the E-commerce sector sponsored by the European Commission are highlighted in this section.

1.22.1   A NETWORK OF SMALL–MEDIUM ENTERPRISE NETWORKS USING TELEMATICS (AGORA)

AGORA was a multilingual initiative that supported interactive videoconferencing, remote file access, groupware applications, and information transport via the EuroISDN platform. The AGORA Web site included online resources on market opportunities, teletraining programs, and descriptions for using information technology (IT) in business transactions. Participants in this business network included artists, farmers, and small business owners. Local Chambers of Commerce, municipal agencies, and research organizations took part in AGORA teleservices and tele-applications as well.

1.22.2   ADVANCING RURAL INFORMATION NETWORKS (ARIN)

The ARIN initiative delivered information services and E-business programs via the EuroISDN platform to homebound students in rural Northwestern England and Central Finland. Local schools, colleges, and businesses took part in ARIN-supported activities. In addition to EuroISDN, ARIN also provided DSL service.

1.22.3   AQUATIC RESEARCH INSTITUTIONS FOR THE DEVELOPMENT OF USER-FRIENDLY APPLICATIONS IN TELEMATICS (AQUARIUS)

The AQUARIUS project promoted the utilization of diverse telecommunications technologies to increase market competitiveness in the aquaculture sector. EuroISDN and cable modem networks were among the technologies employed by AQUARIUS for providing access to tele-instruction courses, teletraining sessions, teleseminars, and telesymposia. Participants included aquaculture students and professors, and fish farmers working in small- and medium-sized enterprises.

1.22.4   ONLINE TRANSACTION DATABANK OF AGRICULTURAL, WOOD, AND BREEDING PRODUCTS (AGRELMA)

The AGRELMA initiative employed the EuroISDN platform for enabling agricultural producers and livestock breeders to interact virtually with traders and representatives of processing industries and explore firsthand E-commerce capabilities, benefits, and constraints.

1.22.5   TELELOPOLIS

Based on the EuroISDN infrastructure, the TELELOPOLIS initiative supported development of a commercial framework for implementation of Web-based virtual businesses. The TELELOPOLIS project also facilitated electronic marketing of virtual goods and services and online trading. Guidelines for central billing, site security, and revenue sharing were developed. Tactics for ensuring compliance with national tax and commerce regulations were explored. Security mechanisms for safeguarding electronic payments were also examined.

1.23 EUROPEAN COMMISSION ADVANCED COMMUNICATIONS TECHNOLOGIES AND SERVICES (EC-ACTS) PROGRAM

Operational between 1994 and 1998, the European Commission Advanced Communications Technologies and Services (EC-ACTS) Program supported EuroISDN-based research projects. Representative EC-ACTS initiatives in the EuroISDN arena are highlighted in this section.

1.23.1 RECONSTRUCTION USING LASER AND VIDEO (RESOLV)

The RESOLV project employed EuroISDN and ATM technologies to promote the use of VRML (Virtual Reality Modeling Language) for enabling the telecollaborative development of realistic three-dimensional (3-D) models of buildings and housing interiors slated for reconstruction.

1.23.2 SCALABLE ARCHITECTURES WITH HARDWARE EXTENSIONS FOR LOW-BIT RATE VARIABLE BANDWIDTH REAL-TIME VIDEOCOMMUNICATIONS (SCALAR)

The SCALAR project utilized scalable architectures that worked in concert with EuroISDN and the PSTN to support interactive videoconferences. The SCALAR platform enabled delivery of tele-instruction to individuals in remote communities in Northern Sweden and provided forest surveillance and fire-detection services in Portugal.

1.23.3 TEAM-BASED EUROPEAN AUTOMOTIVE MANUFACTURE (TEAM) USER TRIALS

The TEAM initiative promoted virtual teamworking and telecollaboration among workers in the European Union automotive manufacturing sector via a hybrid EuroISDN and ATM infrastructure. Trials in the United Kingdom and Italy evaluated capabilities of network applications and the effectiveness of EuroISDN and ATM performance. The Berlin University of Technology and the University of Warwick participated in this initiative.

1.24 ISDN IMPLEMENTATION CONSIDERATIONS

Strengthened demand for ISDN services in the tele-education domain reflects the increasing availability of ISDN fixed-rate fees for monthly service and the proliferation of affordable products and equipment. Current and emergent video, audio, and data applications require more bandwidth than the analog telephone network can provide. Modems that support transmissions over the PSTN typically support transmission at rates of 14.4, 28.8, or optimally 56 Kbps. By contrast, a single ISDN B Channel supports transmission at 64 Kbps (DS-0).

With the advent of ATM and Gigabit Ethernet, ISDN is no longer viewed as a panacea for provisioning universal services at the highest possible speed. Rather, the attraction of ISDN is based on its ability to facilitate dependable connectivity via the PSTN infrastructure for enabling interactive videoconferences and provisioning access to Web-based resources.

In the distance education domain, interactive ISDN videoconferences provide increased educational opportunities for students in remote or isolated communities and save commuting time and expenses for faculty. Moreover, an ISDN infrastructure enables educational institutions to offer a broad array of tele-education courses and degree programs to students regardless of location and deliver telecourses to diverse populations in support of lifelong learning. In addition to the technical features and functions of ISDN videoconferencing systems, factors that impact the effectiveness of ISDN videoconferences in the tele-education environment include the size and composition of the learner group and telecollaborative activities employed to enrich the learning experience.

Despite competition from broadband residential wireline and wireless networking solutions, ISDN remains a viable solution for supporting a broad spectrum of applications in schools and universities. In assessing the workability of ISDN implementations, specific goals and expectations for ISDN deployment should be clarified, user participation in evaluating ISDN benefits and drawbacks should be encouraged, and the advantages and limitations of ISDN usage in terms of institutional mission should be determined.

A feasibility study can establish with greater accuracy the capabilities of ISDN in accommodating institutional requirements. On the basis of the outcomes, a determination can be made to implement ISDN or investigate another communications solution. If the decision is made to go forward with ISDN, a RFP (Request for Proposal) can be distributed to communications carriers and Network Service Providers (NSPs). Another option is outsourcing or establishing a contract with an outside vendor for planning, designing, and implementing an ISDN solution.

Successful transition to ISDN technology in K–12 schools and post-secondary institutions requires development of ISDN telelearning networks for maximizing student motivation, problem-solving skills, and accomplishments. Effective ISDN deployment depends on its capabilities in facilitating exploratory learning and knowledge-building competencies, focused research and quality education, and instructional creativity and innovation.

1.25 SUMMARY

In this chapter, ISDN technical features, functions, capabilities, and enhancements are reviewed. ISDN initiatives in sectors that include tele-education, telebusiness, and telemedicine are described. ISDN support of virtual community networks is explored. As noted, ISDN enables faster connection rates than POTS (Plain Old Telephone Service) over twisted-pair copper lines. ISDN technology promotes enterprisewide network interconnectivity; LAN, MAN, and WAN deployments; and dependable access to intranets, extranets, and the Internet. Because ISDN systems work over regular phonelines, special wiring is not necessary. However, installation of additional equipment at the customer premise is required. ISDN standards are subject to diverse interpretations. Generally, UTP (Unshielded Twisted Pair) supports residential links over the local loop to the local telephone exchange, and STP (Shielded Twisted Pair) enables business connections to the local telephone exchange.

It is important to note that ISDN standards are subject to diverse interpretations. As a consequence, ISDN equipment from multiple vendors is not always interoperable.

ISDN provides a migration path to high-speed, high-performance ATM networks. However, as a consequence of its slow start and piecemeal acceptance, ISDN is sometimes viewed as a transitory solution. In addition to ATM, ISDN competes with residential broadband wireline and wireless solutions such as Ethernet, DSL, cable modem, LMDS (Local Multipoint Distribution System), MMDS (Multichannel Multipoint Distribution System), and VSAT (Virtual Satellite Aperture Terminal) implementations.

Despite the lack of universal ISDN services and ongoing problems with determining the availability of ISDN services and applications, ISDN technology enables basic and sophisticated applications and promotes creation of dynamic and interactive virtual communities and tele-education, telebusiness, and telemedicine environments. ISDN is an enabler of telementoring, real-time interaction with subject experts at remote locations, bulk file transfer, and multimedia services and applications. Successful implementation of ISDN applications requires an understanding of networking trends and developments, an analysis of specifications and requirements, clearly articulated policies and procedures, management commitment, training sessions, and dependable and reliable technical support and assistance.

1.26 SELECTED WEB SITES

Alliance for Telecommunications Industry Solutions (ATIS). ATIS-Sponsored Committees and Forums: Overview.
Available: http://www.atis.org/
AT&T. Global ISDN.
Available: http://www.att.com/isdn/prod1.html
Becker, Ralph. ISDN Tutorial. Last modified on July 27, 2001.
Available: http://www.ralphb.net/ISDN/
Committee T1.
Available: http://www.t1.org/html/geninfo.htm
Eicon Technology. The ISDN Zone.
Available: http://www.isdnzone.com/
European Commission. TEN-Telecom: A Community Support Program for Trans-European Telecommunications Networks.
Available: http://www.ten-telecom.org/en/context.html
European Commission. The ACTS Information Window.
Available: http://www.de.infowin.org/
Fletcher Allen Health Care in Alliance with the University of Vermont. The Telemedicine Program at Fletcher Alan Health Care.
Available: http://www.vtmednet.org/telemedicine/index.htm

International Telecommunications Union (ITU). Welcome to the International
    Telecommunications Union.
    Available: http://www.itu.int/home/index.html
National ISDN Council. Home Page. Last modified on March 20, 2001.
    Available: http://www.nationalisdncouncil.com/
North American ISDN Users Forum (NIUF). What is ISDN? What is the
    NIUF?
    Available: http://www.niuf.nist.gov/
Project DIANE. Executive Summary. Last modified in February 2001.
    Available: http://www.diane.tnstate.edu/
Qwest. ISDN. Last modified on July 12, 2001.
    Available: http://www.qwest.com/products/data/isdn/
U.S. Department of Agriculture (USDA) Rural
    Utilities Services (RUS). Tele-education and Telemedicine Program.
    Available: http://www.usda.gov/rus/telecom/dlt/dlt.htm
University of Portsmouth and The British Library Telemedicine Information
    Service.
    Available: http://www.tis.bl.ui/

2 Asynchronous Transfer Mode (ATM) Networks

2.1 INTRODUCTION

Demand for fast and dependable access to Web-based applications and real-time delivery of multimedia transmissions via an integrated network infrastructure drives implementation of ATM (Asynchronous Transfer Mode) broadband solutions. ATM is a high-speed, high-performance multiplexing and switching technology that provides bandwidth on-demand for seamless transport of full-motion video, audio, data, animations, and still images in local and wider area environments.

A flexible and extendible telecommunications solution, ATM interlinks distributed networks and heterogeneous technologies into integrated configurations, thereby eliminating the need for multiple network overlays. ATM interworks with diverse narrowband and broadband wireline architectures, protocols, and technical solutions such as SONET/SDH (Synchronous Optical Network and Synchronous Digital Hierarchy), WDM (Wavelength Division Multiplexing), and DWDM (Dense WDM). ATM also interoperates with FDDI (Fiber Data Distributed Interface), Ethernet, Fast Ethernet, Frame Relay, ISDN, and IP (Internet Protocol) and wireline and wireless residential broadband access networks employing cable modem, DSL (Digital Subscriber Line), and VSAT (Very Small Aperture Terminal) solutions. In addition, ATM works in conjunction with second-generation GSM (Global System for Mobile Communications) and third-generation UMTS (Universal Mobile Telecommunications Systems) cellular communications technologies.

The ATM platform enables fast access to basic and sophisticated tele-education, telemedicine, electronic commerce (E-commerce), and electronic government (E-government) services. ATM networks are reliable, dependable, and scalable, and flexibly accommodate an array of topologies, applications, and services.

2.2 PURPOSE

ATM is a complex cell multiplexing and switching technology. This chapter provides a high-level introduction to ATM technical attributes, features, and functions. Representative ATM implementations that support a diverse and powerful mix of applications are examined. Wireless ATM (WATM) configurations are described, and the capabilities of next-generation ATM networks are explored. Research initiatives in the ATM arena are highlighted.

2.3 ATM FOUNDATIONS

2.3.1 ATM Development

The ATM platform enables multimedia transmission via fixed-sized 53-byte packets called cells in network environments ranging from desk area networks (DANs) to global implementations. The term "Asynchronous" refers to ATM support of intermittent bit rates and traffic patterns in accordance with actual demand. The phrase "Transfer Mode" denotes ATM multiplexing capabilities in transmitting and switching multiple types of network traffic.

Bell Labs initiated work on ATM research projects in the 1960s and subsequently developed cell relay technology and cell switching architecture for handling bursty transmissions. Originally, ATM was called Asynchronous Time-Division Multiplexing (ATDM) and regarded as a successor to TDM (Time-Division Multiplexing). As with TDM, ATDM supports transmission of delay-sensitive and delay-insensitive traffic. TDM and ATDM assign each fixed-sized cell or information packet to a fixed timeslot. By contrast, ATM supports dynamic allocation of timeslots to cells on-demand. In comparison to ATM, TDM and ATDM protocols are limited in optimizing utilization of available bandwidth for effectively handling volume-intensive multimedia applications.

2.3.2 International Telecommunications Union (ITU)

During the 1980s, the CCITT (International Telegraph and Telephone Consultative Committee), now known as the ITU (International Telecommunications Union), developed a series of recommendations for Narrowband ISDN (N-Integrated Services Digital Network). In 1990, the CCITT approved a set of recommendations for Broadband ISDN (B-ISDN) and designated ATM as the core multiplexing and switching technology for B-ISDN services. Subsequently, the ITU-T (ITU-Telecommunication Standards Sector) developed Recommendations for enabling standardized ATM operations, network architectures, protocol layers, and User-to-Network Interfaces (UNIs) and clarified procedures for effective ATM transmission.

2.3.3 Joint ATM Experiment on European Services (JAMES)

The foundation for practical ATM implementations was established in the European Union with the JAMES (Joint ATM Experiment on European Services) Project. Operational from 1996 to 1998, the JAMES Project operated a trans-European network that validated ATM capabilities in supporting broadband applications such as videoconferencing and verified ATM interoperability with IP, SMDS (Switched Multimegabit Data Service), and Frame Relay technologies. Telecommunications operators in Finland, Germany, Greece, France, Switzerland, Denmark, Spain, Ireland, Portugal, and Norway participated in the JAMES initiative. The United Kingdom Education and Research Networking Association (UKERNA), SURFnet (National Research and Education Network or the NREN of the Netherlands), and RedIRIS (NREN of Spain) also supported JAMES experiments.

2.3.4 Delivery of Advanced Network Technology to Europe, Ltd. (DANTÉ)

DANTÉ (Delivery of Advanced Network Technology to Europe, Ltd.) provided technical service and managed network operations for the JAMES initiative and TEN-34 (Trans-European Network-34.368 Mbps) implementations. Established by National Research and Education Networks (NRENs) of member states in the European Union, DANTÉ provisioned technical support and operations management for the TEN-155 (Trans-European Network-155.52 Mbps) Program and allocated MBS (Managed Bandwidth Service) to specified user groups within the TEN-155 community. In addition, DANTÉ established virtual paths for VPN (Virtual Private Network) implementations with guaranteed bandwidth to support telecollaborative NREN activities. GÉANT, the next-generation Trans-European Network, succeeded TEN-155 on November 1, 2001. Operational between December 1998 and October 30, 2001, TEN-155 supported multimedia transmission at rates of 155.52 Mbps (OC-3) and 622.08 Mbps (OC-12). DANTÉ also manages the GÉANT Program.

2.3.5 Trans-European Network-34.368 Mbps (TEN-34) Initiative

As with the JAMES Project, the TEN-34 (Trans-European Network-34.368 Mbps) Program also established a foundation for ATM deployment throughout the European Union. Sponsored by the European Commission as a component in the TAP (Telecommunications Applications Program) initiative, the TEN-34 Program supported pilot tests and actual implementations over the JAMES infrastructure to benchmark capabilities of IP-over-ATM operations at rates ranging from 10 Mbps to 34.368 Mbps (E-3 or European-3). Operational from February 1997 to December 1998, TEN-34 initiatives validated the performance, availability, extendibility, scalability, and reliability of IP-over-ATM configurations and contributed to the development of advanced IP services. In December 1998, shortly after JAMES operations and the interim QUANTUM (Quality Network Technology for User-Oriented Multimedia) Project were discontinued, the TEN-34 Program was succeeded by the TEN-155 Program.

2.3.6 JAMES and TEN-34 Operations and Technical Contributions

The TEN-34 Program demonstrated the capabilities of an IP-over-ATM network in interlinking European research and academic networks or NRENs (National Research and Education Networks) and confirmed ATM switching capabilities. In addition, the TEN-34 Program authenticated SNMP (Simple Network Management Protocol) functions in supporting ATM network operations. Capabilities of the Classical IP with Address Resolution Protocol (CIP with ARP), the Next Hop Resolution Protocol (NHRP), and the Resource Reservation Protocol (RSVP) were substantiated in TEN-34 implementations supported by the JAMES infrastructure as well. Moreover, TEN-34 ATM experiments verified the effectiveness of Private Network-to-Network or Network-to-Node Interfaces (PNNIs) and ATM routing capacities, and confirmed the capabilities of ATM in enabling point-to-point, point-to-multipoint, and multipoint-to-multipoint connections and MultiProtocol Label Switching (MPLS) operations.

In addition to IP, technologies that operated over the JAMES ATM platform included SMDS (Switched Multimegabit Data Service), CBDS (Connectionless Broadband Data Service), and Frame Relay. SMDS is a public packet-switched networking service that fosters exchange of variable length data units at fast speeds, thereby eliminating call setup procedures. CBDS (Connectionless Broadband Data Service) is the ETSI (European Telecommunications Standards Institute) equivalent of SMDS. Frame Relay solutions support Permanent Virtual Circuits (PVCs) that are primarily used for high-speed, high-performance ATM LAN (Local Area Network) interconnections.

The absence of clearly defined ATM applications contributed to difficulties at the outset in substantiating the performance of ATM operations in TEN-34 networks. Problems with verifying QoS (Quality of Service) guarantees, benchmarking capabilities of SVCs (Switched Virtual Circuits) and PVCs (Permanent Virtual Circuits), and authenticating CBR (Constant Bit Rate), VBR (Variable Bit Rate), and UBR (Unspecified Bit Rate) functions were encountered as well. As a consequence of ATM complexity and the steep learning curve associated with ATM deployments, TEN-34 experiments and trials over the JAMES infrastructure took more time than initially scheduled.

TEN-34 findings contributed to ETSI (European Telecommunications Standards Institute) ATM specifications and ITU-T ATM Recommendations. TEN-34 outcomes also fostered development of the TEN-155 (Trans-European Network, 155.52 Mbps) Program.

The TEN-155 Program officially replaced the interim QUANTUM Program and the TEN-34 Program in 1999. The migration of NRENs such as GRnet (NREN of Greece), RedIRIS (NREN of Spain), RCCN (NREN of Portugal), POL-34, (NREN of Poland), and CESNET (NREN of the Czech Republic) from TEN-34 to TEN-155 marked this transition.

2.3.7   Quality Network Technology for User-Oriented Multimedia (QUANTUM) Program

As with the TEN-34 Program, the QUANTUM Program supported networking testbed experiments between February 1997 and December 1998. An intermediate pan-European network initiative culminating in the establishment of the TEN-155 infrastructure in 1999, the QUANTUM Program provisioned stable IP services and fostered transport rates extending to 155.52 Mbps (OC-3) for bandwidth-intensive teleservices and tele-applications.

Moreover, the QUANTUM Program supported NREN applications, MBS (Managed Bandwidth Services) for specified user groups, and enhancements to network protocols such as RSVP (Resource Reservation Protocol) and DiffServ (Differentiated Services). The QUANTUM Program also sponsored research trials that validated ATM capabilities in provisioning QoS (Quality of Service) guarantees.

Participants in the QUANTUM Program included ACONET (NREN of Austria), HEAnet (NREN of Ireland), and ARNES (NREN of Slovenia). In addition, BELnet (NREN of Belgium), DFN or Deutsche Forschungsnetz (NREN of Germany), HUN-GARnet (NREN of Hungary), NORDUnet2 (the Nordic Countries Network, Phase 2), and SuperJANET (Super Joint Academic Network) participated in the QUAN-TUM Program as well.

2.3.8 QUANTUM TEST PROGRAM (QTP)

The QUANTUM Program established the Quantum Test Program (QTP) for benchmarking the capabilities of technologies, products, and services such as IP multicasts and IPv6 (Internet Protocol version 6) and determining their suitability for enabling TEN-155 operations. NRENs such as RedIRIS and SURFnet took part in the QTP initiative. TF-TANT (Task Force-TERENA or the Trans-European Research and Education Network and DANTÉ) managed QTP operations.

2.3.9 TF-TANT (TASK FORCE-TERENA AND DANTÉ)

In addition to managing QTP operations, the TF-TANT participated in the QUAN-TUM Program and supported development of a high-quality, high-speed network infrastructure for enabling next-generation tele-education and teleresearch services. This Task Force supported the migration of NRENs (National Research and Education Networks) from TEN-34 to TEN-155; contributed to advances in IP-over-ATM, IPv6 (Internet Protocol version 6), and IP multicasting services; and developed enhancements to protocols such as MultiProtocol Label Switching (MPLS), DiffServ (Differentiated Services), and RSVP (Resource Reservation Protocol).

2.3.10 QUALITY OF NETWORK TECHNOLOGY FOR USER-ORIENTED MULTIMEDIA IN THE EASTERN MEDITERRANEAN REGION (Q-MED) PROJECT

A component in the European Commission Telematics Applications Program (EC-TAP), the Q-MED (Quality Network Technology for User-Oriented Multimedia in the Eastern Mediterranean Region) Project supported high-quality Internet services for NRENs in Cyprus, Greece, and Israel. Q-MED was the Mediterranean component of the QUANTUM Program.

As with the QTP Project, the Q-MED Project enabled telecollaboration among research and academic communities and implementation of tele-initiatives contributing to the development of next-generation technologies, protocols, architectures, and services. Q-MED also established NREN connections to TEN-155 for Q-MED Consortium participants, including the IUCC (Israeli InterUniversity Computation Center), the Universities of Cyprus and Athens, GRnet (NREN of Greece), and CYNET (NREN of Cyprus). GARR (NREN of Italy) and DANTÉ provisioned Q-MED network management services. The Q-MED Consortium facilitated migration from the Q-MED Project to the Mediterranean Network (MEDNET) as well.

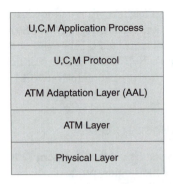

| U,C,M Application Process |
| U,C,M Protocol |
| ATM Adaptation Layer (AAL) |
| ATM Layer |
| Physical Layer |

FIGURE 2.1 The ATM protocol stack (U = user, C = control, and M = management).

2.3.11 MEDITERRANEAN NETWORK (MEDNET)

The Mediterranean Network (MEDNET) employs a high-speed ATM networking infrastructure that supports real-time videoconferencing, predictable QoS guarantees, and bandwidth on-demand. Lebanon, Syria, Jordan, Israel, Turkey, Egypt, Malta, Cyprus, Tunisia, and Morocco take part in this regional network initiative. The National Technical University of Athens and the University of Cyprus provide network services and technical support.

2.4 ATM STANDARDS ORGANIZATIONS AND ACTIVITIES

2.4.1 ATM FORUM

Established in 1991, the ATM Forum supports the development of ATM specifications for network protocols, network architectures, and network management services and promotes ATM implementations in a broad range of networking environments. The ATM Forum also facilitates the development of interoperable ATM applications and deployment of standardized ATM hardware components, software products, and peripheral devices. (See Figure 2.1.)

In addition, the ATM Forum sets standards for User-to-Network Interfaces (UNIs) and Network-to-Network or Network-to-Node Interfaces (NNIs) for virtually routing ATM signals between ATM network switches or between networks and PNNIs (Private NNIs) in private ATM networking configurations. Moreover, the ATM Forum defines voice-over-IP (VoIP) or Internet telephony functions, clarifies ATM LANE (LAN or Local Area Network Emulation) operations, and establishes parameters for QoS guarantees. In addition, the ATM Forum supports VoDSL (voice-over-Digital Subscriber Line) solutions, ATM-over-ADSL (Asynchronous DSL) implementations, Frame-based ATM Transport-over-Ethernet (FATE) installations, Frame-based ATM-over-SONET/SDH Transport (FAST) specifications, and utilization of elliptic curve cryptography for enabling secure network operations.

ATM Forum participants include communications service providers, equipment manufacturers, government agencies, research centers, and educational institutions.

Principal ATM members include Nokia, Nortel Networks, Korea Telecom, Cisco Systems, Lucent Technologies, Global One, Verizon, Sprint, Chunghwa Telecom Labs, and WorldCom.

ATM Forum affiliates include the Georgia Institute of Technology, the Hanoi University of Technology, the Jawaharlal Nehru Technological University, and Arizona State, George Washington, Johns Hopkins, Cairo, Florida State, and Kansai Universities. In addition, the Universities of Calcutta, Cape Town, Pretoria, Texas at San Antonio, Toledo, and the Philippines participate in the ATM Forum. The ATM Forum also works in concert with ATM alliances such as the New Zealand and the Austrian Interest Groups and the Belgium ATM Platform in promoting deployment of ATM research and educational projects, commercial ATM implementations, and global ATM standards.

2.4.1.1 Anchorage Accord

Developed by the ATM Forum, the Anchorage Accord contains a collection of approximately 60 specifications for building an ATM infrastructure. This Accord defines approaches for migrating to ATM multiservice networks and methods for ensuring interoperable ATM installations. The Anchorage Accord also establishes procedures for enabling ATM LANE operations and defines attributes of each layer in the ATM protocol stack.

2.4.2 EUROPEAN TELECOMMUNICATIONS STANDARDS INSTITUTE (ETSI)

The European Telecommunications Standards Institute (ETSI) establishes specifications clarifying PNNI, UNI, and NMI (Network Management Interface) operations for supporting uniform ATM services and applications. ETSI also supports development of signaling and switching protocols and specifications contributing to deployment of interoperable ATM networks throughout the European Union.

ETSI specifications are based on requirements established by the ATM MoU (Memorandum of Understanding) Working Group. Members in the ATM MoU Working Group include Siemens, Nokia, Ericsson, France Telecom, Alcatel, Nokia, and BT (British Telecommunications).

2.4.3 INTERNET RESEARCH TASK FORCE (IRTF)

Sponsored by the IETF (Internet Engineering Task Force) and the Internet Society (ISOC), the Internet Research Task Force (IRTF) promotes development of the next-generation Internet by encouraging innovations in network architectures, technologies, protocols, and applications. The IRTF supports implementation of standardized approaches for enabling IP multicast services via an ATM infrastructure and development of dynamic multicast protocols that work in conjunction with an ATM platform.

2.4.4 INTERNATIONAL TELECOMMUNICATIONS UNION-TELECOMMUNICATIONS STANDARDS SECTOR (ITU-T)

The ITU-T (International Telecommunications Union-Telecommunications Standards Sector) develops Recommendations for enabling deployment of global ATM solutions and ensuring the interoperability of ATM equipment in multivendor environments. ATM features, functions, and operations are described in the ITU-T I.356, the ITU-T I.371, and the ITU-T Q.2931 Recommendations. In addition, ITU-T Recommendations support implementation of a standardized ATM infrastructure that facilitates reliable CoS (Class of Service) transmission with QoS (Quality of Service) guarantees.

2.5 ATM TECHNICAL FUNDAMENTALS

An ATM broadband network enables dependable delivery of multimedia traffic in wireline and/or wireless environments. Prior to ATM implementation, individual networks carried data, voice, and video traffic separately on individual channels or circuits.

2.5.1 ATM CELL

ATM networks employ a standard, fixed-size 53-byte cell comprised of a 5-byte header and a 48-byte payload or information field as the basic unit of transmission. The 5-byte header includes an error detection field and a Virtual Channel Identifier (VCI) or Virtual Path Indicator (VPI) for transporting a cell payload to a destination address.

Through utilization of a common cell format, ATM enables real-time services, public and private network interconnectivity, and global interoperability. As noted in Chapter 1, ISDN employs STDM (Statistical Time-Division Multiplexing) for enabling transmission of frames via designated timeslots at specified intervals. In contrast to ISDN installations, the ATM protocol supports dynamic allocation of timeslots to cells on-demand for optimizing traffic throughput in high-performance network configurations.

ATM technology employs a priority switching technique for enabling ATM cells carrying delay-sensitive signals to access the first available timeslot. Because the ATM cell size is fixed and the buffer memory size is constant for each cell, switch queuing delays are predictable and jitter or the variation in signal delay is minimized. By contrast, signal delays degrade performance of real-time applications such as videoconferencing and interactive video-on-demand (IVOD) in networks such as Frame Relay (FR) and Ethernet that transport variable length packets.

The ATM Forum defines procedures for monitoring the effectiveness of network transmission based on cellular throughput. Cell Loss Ratio (CLR) describes the percentage of cells that are not transported to their destination addresses as a consequence of buffer overloads and network congestion. Cell Transfer Delay (CTD) refers to propagation and queuing delays experienced by cells transiting the network. Cell Delay Variation (CDV) measures variations in transmission delay between adjacent cells. Minimum Cell Rate (MCR) refers to the lowest cell rate supported by ABR (Available Bit Rate) service. In addition, metrics for Cell Delay Variation Tolerance (CDVT) and parameters for Maximum Cell Transfer Delay (MCTD) are also defined. The effectiveness of QoS delivery in ATM networks depends on such variables as Cell Transfer Delay (CDT) and Cell Delay Variation (CDV).

2.5.2   ATM APPLICATIONS

ATM is a connection-oriented virtual network transmission and switching technology that combines the low-delay of circuit-switched networks with the bandwidth flexibility and high-speed of packet-switched networks. ATM is an enabler of basic and advanced applications such as remote sensing, 3-D (three-dimensional) interactive simulations, tele-instruction, biological teleresearch, and medical teleconsultations. Edge devices at the boundary of an ATM network convert non-ATM traffic streams into standard ATM cells.

ATM technology is implemented in backbone, enterprise, and edge switches as well as hubs, routers, bridges, multiplexers, servers, server farms, and NICs (Network Interface Cards) in high-end Internet appliances. The ATM Data Exchange Interface (DXI) enables fast access to public network services. A flexible and extendible networking solution, ATM technology supports network configurations that include DANs (Desk Area Networks), LANs, MANs (Metropolitan Area Networks), WANs (Wide Area Networks), and GANs (Global Area Networks).

2.5.3   ATM TRANSMISSION RATES

ATM technology enables wireline transmissions via optical fiber, twisted copper pair, and hybrid optical fiber and coaxial cable media. To support bandwidth-intensive operations, the ATM platform multiplexes and relays a diverse mix of network traffic via optical fiber at rates that include 155.52 Mbps (OC-3 or Optical Carrier-Level 3) and 622.08 Mbps (OC-12). ATM also sustains speeds at 2.488 Gbps (OC-48), 10 Gbps (OC-192), and 13.21 Gbps (OC-255). Optical Carrier (OC) levels are sets of signal rates that describe digital transmission speeds over a fiber optic plant. In the United States, these levels are based on multiples of the base rate of 51.84 Mbps (OC-1).

In addition to optical fiber, ATM supports landline information transport via hybrid optical fiber and coaxial cable (HFC) connections, and ordinary twisted copper pair found in public networks such as the PSTN (Public Switched Telephone Network) at lower speeds. For example, ATM enables transmission at 1.544 Mbps or T-1 and DS-1 (Digital Signal-1) in terms of the North American digital hierarchy. In the European Union, ATM supports information transport at 2.048 Mbps or E-1 (European-1) and DS-1 (Digital Signal-1) in accordance with European Union digital specifications. ATM also supports rates at 44.746 Mbps or T-3 and DS-3 in North America, and 34.368 Mbps or E-3 and DS-3 in the European Union. By enabling transport of concurrent voice, still images, video, and data traffic in local, municipal, and wider area configurations, ATM technology promotes development of an integrated and scalable multi-service network infrastructure that optimizes resource sharing and user productivity.

2.6   ATM PROTOCOL STACK

ATM services are based on protocol layer operations. (See Figure 2.1.) The Physical Layer in the ATM protocol stack consists of the Physical Medium Sublayer and the Transmission Convergence Sublayer. These sublayers enable the use of diverse physical media, interfaces, and transmission speeds. In addition, these sublayers transform signals into electronic or optical formats, map and encapsulate IP packets into cells, and provision multiplexing services for transmitting cells over the same physical link. In addition, the Physical Layer defines the process for routing and switching cells in accordance with Virtual Path Identifiers (VPIs) or Virtual Channel Identifiers (VCIs). The ATM Physical Layer corresponds to Layer 1 or the Physical Layer of the Open Systems Interconnection (OSI) Reference Model.

Situated directly above the Physical Layer, the ATM Layer employs the 53-byte cell as the basic transmission unit. The ATM Layer operates independently of the ATM Physical Layer. At the ATM Layer, ATM switches route cellular streams received from the ATM Adaptation Layer (AAL) to destination addresses in accordance with the Virtual Channel Identifier (VCI) or Virtual Path Indicator (VPI) contained in each cell header.

Situated directly above the ATM Layer, the ATM Adaptation Layer (AAL) facilitates cell segmentation and reassembly by dividing media streams received from upper protocol layers into 53-byte ATM cells for transmission over the physical link. For reassembly, the AAL reconstitutes original media streams from cells that were received via the Physical Layer. The reconstructed media streams are then ready for transport by upper protocol layers to destination addresses.

The ATM Adaptation Layer (AAL) consists of five sublayers, ranging from AAL1 to AAL5. Each sublayer has specific and overlapping responsibilities for enabling dependable and robust transmissions. AAL1 supports cell segmentation and cell reassembly and the transmission of data, video, and high-quality audio signals for enabling real-time applications. AAL2 enables transport of connection-oriented VBR (Variable Bit Rate) packetized video-over-ATM and voice-over-ATM. Defined in the ITU-T I.363 Recommendation, AAL Sublayers 3/4 (3 and 4) handle connectionless and connection-oriented VBR transmissions. As with AAL1, AAL5 supports cell segmentation and cell reassembly operations.

The ATM Forum describes the features and functions of the ATM protocol stack and clarifies procedures for interworking ATM with technologies that include DSL (Digital Subscriber Line), SONET/SDH (Synchronous Optical Network and Synchronous Digital Hierarchy), WDM (Wavelength Division Multiplexing), and DWDM (Dense WDM). In addition, approaches for enabling ATM to interwork with SMDS, Frame Relay, cable modem, MMDS (Multichannel Multipoint Distribution System), and LMDS (Local Multipoint Distribution System) implementations are also clarified.

The ATM Forum works with other standards organizations such as the Digital Video Broadcasting/Digital AudioVisual Council (DVB and DAVIC) and the Full Service Access Network (FSAN) Consortium in developing specifications for ATM television broadcasts and APONs (ATM Passive Optical Networks). APONs interwork with FTTC (Fiber-to-the-Curb) and FTTH (Fiber-to-the-Home) broadband residential access networks.

2.7   ATM OPERATIONS

2.7.1   SVCs (Switched Virtual Circuits) and
PVCs (Permanent Virtual Circuits)

ATM networks sustain point-to-point links for direct connectivity, point-to-multipoint connections for broadcast and multicast services, and multipoint-to-multipoint

connections for applications such as interactive videoconferencing and telecollaborative teleresearch. Path specifications for moving traffic across ATM networks are
termed Switched Virtual Circuits (SVCs) and Permanent Virtual Circuits (PVCs).

SVCs are created virtually on a semi-permanent basis for enabling multimedia
transmission. SVCs establish connections on a call-by-call basis for accommodating
bursty transmissions and bandwidth on-demand. UBR (Unspecified Bit Rate) service
for SVC (Switched Virtual Circuit) connections supports information delivery on a
best-effort basis. SVC connections do not guarantee the availability of bandwidth
for enabling QoS (Quality of Service) transmissions.

In comparison to SVCs, PVCs are static virtual connections between network
endpoints that support always-available and assured bandwidth allocations for current and emergent network applications and services. As a consequence, PVCs enable
stable, dependable, and reliable transmission of voice, video, and data traffic with
QoS guarantees.

2.7.2   ATM SWITCHES

ATM networks consist of routers, servers, switches, and endpoint devices such as
network nodes and stations. The ATM switch family includes workgroup, campus,
enterprisewide, and next-generation switches that provide services in a variety of
LAN, MAN, and WAN environments. For example, ATM switches enable LATMs
(Local Area ATM Networks) to provision services to legacy workstations and support
sophisticated network backbone operations for advanced academic and research
networks.

ATM multiservice switches provide the underlying physical infrastructure for the
network configuration and control network processing speed. These devices uniformly
facilitate cell relay operations, sustain throughput and end-to-end network performance,
interlink nodes on ATM networks, and route multiple cells concurrently to destination
addresses. It is important to note that ATM switches also support diverse applications,
services, and operations, and vary in structure, capacity, value-added capabilities,
interoperability support, and traffic management functions in order to accommodate a
wide range of E-government (electronic government), E-business (electronic businesses), telemedicine, teleresearch, and/or tele-education requirements.

2.7.2.1   User-to-Network Interfaces (UNIs), Network-to-Node and
Network-to-Network Interfaces (NNIs), and Private
Network-to-Node or Network-to-Network Interfaces (PNNIs)

ATM installations consist of a set of ATM switches or internetworking devices that
are interconnected by point-to-point ATM interfaces. ATM interfaces or virtual
connections include User-to-Network Interfaces (UNIs) and NNIs (Network-to-
Node Interfaces or Network-to-Network Interfaces). UNIs are ATM protocols that
define standard interfaces between customer premise equipment (CPE) and the
network switch. For example, FUNI (Frame UNI) clarifies parameters for integrating
legacy devices with ATM switching equipment in mixed-mode Frame Relay and
ATM network configurations.

PNNIs (Private Network-to-Node or Private Network-to-Network Interfaces) are NNI protocols that define ATM interfaces within and between private networks. PNNIs determine approaches for routing ATM connection-oriented requests across an ATM network or between ATM networks.

Moreover, PNNIs employ signaling technologies to support SVCs and PVCs in multivendor environments, provision QoS guarantees, and foster distribution of reserved bandwidth. PNNIs also establish the format for the Broadband-Intercarrier Interface (B-ICI) between public networks for enabling seamless multicarrier multivendor multiservice ATM implementations.

2.7.3   ATM Class of Service (CoS) and Quality of Service (QoS)

ATM networks employ Classes of Service (CoS) for optimizing network performance and supporting applications with specified bandwidth or throughput requirements. ATM service classes resolve congestion problems and traffic management issues in order to ensure seamless transmission in multivendor environments.

A Class of Service (CoS) refers to a category of ATM connections that features identical traffic patterns and resource requirements. Each class provisions a distinct level of service and associated QoS guarantees. Depending upon the format of the QoS service requested, the ATM network defines a series of CoS categories.

The Variable Bit Rate (VBR) Class of Service consists of applications with specific requirements for delays and throughputs such as packetized voice and data applications. The real-time Variable Bit Rate (VBR-rt) Class of Service requires real-time support for provisioning applications such as video-on-demand (VOD) and voice-over-IP (VoIP). VBR-rt bandwidth requirements vary over time. However, delay and delay variance limits are clearly established.

The non-real-time variable bit rate (VBR-nrt) Class of Service eliminates the need for guaranteed delivery of applications such as multimedia e-mail, bulk file transmissions, and business and educational database transactions with minimal service requirements. Bandwidth for VBR-nrt applications varies within a specified range. However, delay and delay variance requirements are not fully defined.

The Available Bit Rate (ABR) Class of Service requires the use of flow control mechanisms for ensuring allocation of bandwidth on-demand for non-real-time, mission-critical applications. With ABR applications, guaranteed minimum transmission rates are specified for the duration of the connection. In addition, ABR also establishes peak transmission rates for data bursts when bandwidth is available. As a consequence, the ABR service class tolerates delay variations. Applications grouped into this category allow priority traffic to consume bandwidth first. ABR applications include LAN emulation (LANE), file and data distribution, and LAN interconnections.

The Unspecified Bit Rate (UBR) Class of Service is equivalent to best-effort delivery in IP networks. Delay-tolerant UBR applications include Web browsing and IP transmissions. Because UBR applications require minimal network support, QoS guarantees and pre-established throughput levels are not defined.

The Constant Bit Rate (CBR) Class of Service (CoS) requires utilization of a virtual channel with constant bandwidth for seamlessly transporting applications in accordance with pre-defined response time requirements. CBR applications include videoconferencing, telephony services, and television broadcasts.

In conjunction with establishing a CoS, ATM networks define cell rates and burst size to facilitate seamless network performance. For example, Peak Cell Rate (PCR) indicates the maximum rate at which cells transit the network for brief time periods. Sustainable Cell Rate (SCR) refers to the cell rate that is sustained for a specified period of time. Maximum Burst Size (MBS) defines the maximum number of back-to-back cells that transit the network.

2.7.4   ATM AND MPEG-2 (MOVING PICTURE EXPERTS GROUP-2)

ATM networks interwork with the MPEG-2 (Moving Picture Experts Group-2) specification to enable interactive high-quality videoconferences, optimize bandwidth utilization for delay-intolerant applications such as telesurgery, and facilitate real-time multimedia delivery with QoS guarantees. ATM transmission of MPEG-2-compliant voice, video, and data traffic follows the format specified in the ITU-T J.82 Recommendation. The ITU-T and the International Standards Organization (ISO) approved MPEG-1, the original MPEG standard for voice, video, and data compression, in 1992.

2.8   IP-OVER-ATM

The popularity of IP (Internet Protocol) applications contributes to implementation of IP overlays on top of multiservice ATM networks. IP-over-ATM solutions employ protocols such as MPOA (MultiProtocol-over-ATM) and MPLS (MultiProtocol Label Switching) for leveraging IP enhancements. Moreover, IP packets can be mapped to ATM service classes to transport, for instance, IP-based voice and video traffic via CRC and VBR-rt. Additionally, IP-over-ATM implementations support VPNs (Virtual Private Networks) and ATM emulated LAN (ELAN) protocols that work in concert with the Network Layer or Layer 3 and the Transport Layer or Layer 4 of the OSI (Open Systems Interconnection) Reference Model. Sponsored by the ATM Forum, the ATM-IP Collaborative Working Group (AIC) develops specifications for coordinating the provision of IP services with ATM technology. Approaches for mapping ATM QoS to the IP DiffServ (Differentiated Services) protocol are also in development.

In order to interoperate with IP packet-switched services, ATM defines a framing structure for carrying IP packets as sets of ATM cells. ATM PVCs (Permanent Virtual Circuits) support virtual connections within an IP network. In provisioning IP integrated services over an ATM infrastructure, a portion of the available bandwidth is reserved for specified CoS transmissions. By employing fixed capacity virtual connections for designated CoS transmissions, the ATM infrastructure guarantees the availability of reserved bandwidth on-demand.

2.8.1   IP-over-ATM Standards Organizations and Activities

2.8.1.1   ATM Forum IP-over-ATM Working Group

Sponsored by the ATM Forum, the IP-over-ATM Working Group supports development of IP applications and services that optimize ATM cell transmission capabilities and provision Quality of Service (QoS) assurances. Moreover, the ATM Forum also describes methods in its UNI specification for enabling point-to-point or unicast transmissions between routers and hosts within logical IP subnetworks and procedures for provisioning IP multicasts. It is interesting to note that approaches for supporting an IP network overlay that operates directly on top of SONET/SDH connections are also in development in order to eliminate transmission delays caused by ATM cellular overhead.

2.8.1.2   Forum for IP Networking in Europe (FINE)

The Forum for IP Networking in Europe (FINE) defines IP-over-ATM operations in enabling broadband services and clarifies procedures for IPv6 implementations.

2.8.1.3   IEEE 802.16 Specification

The IEEE 802.16 specification supports ATM cell relay service and provisions mechanisms for translating ATM addresses to IP formats. Designed primarily for IP applications and services, the IEEE 802.16 specification clarifies procedures for transport of variable-length IP datagrams, approaches for achieving virtual point-to-point connections, and strategies for IPv6 implementation.

2.8.2   CIP-over-ATM (Classical IP-over-ATM)

Protocols for preserving in-place infrastructure investments in ATM environments include Classical IP-over-ATM (CIP-over-ATM). A CIP-over-ATM solution employs Permanent Virtual Circuits (PVCs) or dynamic Switched Virtual Circuits (SVCs) for transporting IP packets to ATM addresses. Moreover, CIP-over-ATM deployments enable access to ATM services and connectivity to legacy IP applications. CIP-over-ATM implementations require modification of the IP Address Resolution Protocol (ARP) in order to establish ATM connections that correspond to IP addresses.

2.8.2.1   IETF (Internet Engineering Task Force) Network Working Group

The IETF (Internet Engineering Task Force) Network Working Group delineates approaches for enabling IPng (IP next-generation) tunnel and network architecture to work in concert with CIP-over-ATM installations. In addition, this Working Group clarifies procedures for supporting operations at AAL5 (ATM Adaptation Layer 5) of the ATM protocol stack.

2.8.3   MultiProtocol-over-ATM (MPOA) PROTOCOL

Endorsed by the ATM Forum, the MPOA (MultiProtocol over ATM) protocol defines Network Layer or Layer 3 services for enabling ATM implementations. IMPOA employs Next Hop Resolution Protocol (NHRP) for mapping IP packets to ATM cells at AAL5 of the ATM protocol stack. In addition, MPOA routes ATM traffic directly between ELANs (Emulated LANs) and employs SVCs (Switched Virtual Circuits) to ensure reliable and dependable voice, video, and/or data delivery to destination addresses. Robust transmissions are achieved by reducing the number of nodes participating in the internetwork transmission process.

With MPOA, network stations or nodes on different subnetworks establish Permanent Virtual Connections (PVCs) or shortcuts, thereby eliminating the need for intermediate cell segmentation and cell reassembly. In contrast to MPOA, LANE (LAN Emulation) and CIP protocols use intermediate routers for enabling intercommunications between subnetwork nodes. This process limits ATM transmission rates and the amount of voice, video, and data throughput transported via the network by requiring intermediate cell segmentation and reassembly.

2.8.4 MULTIPROTOCOL LABEL SWITCHING (MPLS)

Developed by the IETF (Internet Engineering Task Force), the MPLS (MultiProtocol Label Switching) protocol enables the provision of merged IP and ATM services within the same networking environment. To accomplish this objective, the MPLS protocol interlinks the IP Layer and the ATM Layer and interconnects IP routers and ATM switches, thereby enabling IP transmissions to take advantage of ATM traffic management capabilities in provisioning CoS assurances. IP also benefits from ATM broadband transmission rates for enabling high-speed and dependable multimedia delivery.

MPLS technology enables operations at the Data-Link Layer or Layer 2 of the OSI Reference Model, supports connection-oriented switching based on IP routing and control protocols, and employs fixed-length labels for rapidly routing transmissions to destination addresses. The MPLS protocol works in concert with its own LDP (Label Distribution Protocol) in establishing links and shortcuts in accordance with IP addresses, ATM CoS requirements, and ATM QoS guarantees.

MPLS implementation requires development of a Label-Switching Path (LSP) for handling volume-intensive traffic that takes a specific destination route over the network and supporting identification of a communications channel with high capacity and minimal congestion to accommodate application bandwidth requirements. MPLS solutions optimize network performance, control network operating costs, minimize congestion, decrease the number of information packets dropped as a consequence of network instability, and provision preferential service for delivery of priority transmissions. The MPLS protocol works in concert with IPv4 (Internet Protocol version 4) and supports migration to IPv6 (Internet Protocol version 6) operations. In addition to ATM, the MPLS protocol optimizes performance of network configurations based on POS (Packet over SONET/SDH), Frame Relay, Ethernet, Fast Ethernet, and Gigabit Ethernet technologies.

2.8.4.1 Internet Engineering Task Force (IETF) MPLS Working Group

The IETF MPLS Working Group defines requirements for effective MPLS utilization and develops standards for label switching technology.

2.9   INTERNET PROTOCOL VERSION 6 (IPv6)

2.9.1   IPv6 TECHNICAL FEATURES AND FUNCTIONS

2.9.1.1   IPv6 Capabilities

Created by the IETF (Internet Engineering Task Force), IPv6 (Internet Protocol version 6) is a next-generation networking protocol designed as a replacement for IPv4 (Internet Protocol version 4). Currently the predominant version of the Internet Protocol, IPv4 supports best-effort packet delivery service. In comparison to IPv4, IPv6 enables improved security and features a simplified header format, expanded addressing functions, and advanced autoconfiguration and routing procedures.

IPv4 employs a 32-bit addressing scheme that is limited to supporting several billion uniquely identified Web addresses. In contrast to IPv4, IPv6 includes a streamlined 64-bit header, followed by two 128-bit addresses indicating source and destination in order to make vast numbers of uniquely identified Web addresses available.

2.9.1.2   WIDE (Widely Integrated Distributed Environment) Project

In June 2001, the WIDE (Widely Integrated Distributed Environment) Project initiated the implementation of an Internet Exchange Point (IXP) called NSPIXP-2 in metropolitan Tokyo. This broadband IXP (Internet Exchange Point) supports IPv6 operations over a Gigabit Ethernet backbone network and transmission rates ranging from 4 Gbps to 8 Gbps. NSPIXP-2 is the first production Internet Exchange Point (IXP) to enable IPv6 traffic exchange among peer-level corporate entities. Initial NSPIXP-2 participants include WorldCom, NTT, the Tokyo Telecommunications Network Company, and the Mitsubishi Electric Information Network Corporation.

In addition, the WIDE Project supports IPv6 research efforts, development of practical applications for MBone (Multicast Backbone) technology, E-commerce services, IP-over-ATM implementations, and next-generation IP security mechanisms based on the IPSec (Internet Protocol Security) specification. Partners in the WIDE project include Cisco, Hitachi, Fujitsu, GlobalOne, IBM, NTT, AboveNet, the Yamaha Corporation, and Nortel Networks.

2.9.1.3   IPv6 Forum

A global consortium of Internet vendors, the IPv6 Forum supports implementation of IPv6 applications and services. Hewlett-Packard, Hitachi, Compaq Computer Corporation, Apple Computer, IBM, and Microsoft sponsor IPv6 Forum initiatives.

2.9.2   DANTÉ AND IPv6

To promote IPv6 development, DANTÉ (Delivery of Advanced Network Technology to Europe, Ltd.) supported implementation of multicast technologies via the TEN-155 (Trans-European Network-155.52 Mbps) configuration. In addition, DANTÉ validated the performance of BGMP (Border Gateway Multicast Protocol). Developed by the Internet Engineering Task Force (IETF) DANTÉ validated the performance of BGMP (Border Gateway Multicast Protocol), a scalable multicast routing protocol, BGMP enables widescale distribution of multicast services in an interdomain networking environment. BGMP supports an interdomain multicast routing solution for distributing messages within multicast groups. By employing TCP (Transmission Control Protocol) to streamline message exchange among every multicast group within a specified domain, BGMP eliminates the need for such time-consuming processes as message fragmentation, acknowledgement, and retransmission. DANTÉ also established procedures for enabling TEN-155 NRENs (National Research and Education Networks) to migrate from IPv4 to IPv6, managed trials enabling IPv6 assessment, and encouraged development of interoperable IPv6 products and services.

2.9.3 EUROPEAN COMMISSION INFORMATION SOCIETY TECHNOLOGIES (EC-IST) PROGRAM IPV6 INITIATIVE (6INIT)

The European Commission Information Society Technologies (EC-IST) Program continues the work of the European Commission Advanced Communications Technology and Services (EC-ACTS) Program. Like the EC-ACTS Program, the EC-IST Program supports next-generation research projects that are carried out by academic institutions and research centers and contribute to standards development.

Sponsored by the EC-IST Program, the IPv6 Initiative (6INIT) clarifies operational procedures for enabling migration from IPv4 to IPv6 implementations. Approaches for defining uniform operational procedures, establishing a trans-European IPv6 packet delivery service, and implementing a platform consisting of four IPv5 (Internet Protocol version 5) and IPv6 clouds that support connectivity to native IPv6 access points are in development. Moreover, procedures that enable high-quality, high-performance broadband IP-over-ATM applications such as IP telephony with QoS guarantees, videotelephony, and video-on-demand (VOD) are also evaluated.

2.9.4 IPV6 RESEARCH AND EDUCATION NETWORK (6REN)

The IPv6 Research and Education Network (6REN) supports assessment of IPv6 capabilities and worldwide deployment of IPv6 services. 6REN participants include RENATER (NREN of France) and vBNS+ (very high-performance Backbone Network Service Plus). CAIRN (Collaborative Advanced Internet Research Network), CERNET (NREN of China), and DANTÉ take part in 6REN as well. Additionally, SingAREN (Singapore Research and Education Network), CA*net II (Canadian Network for the Advancement of Research, Industry, and Education, Phase 2), and SURFnet5 (NREN of the Netherlands, Phase 5) support IPv6 research trials and initiatives. Communications carriers such as NTT, Chunghwa Telecom, and Sprint participate in 6REN projects.

2.9.5 IPV6 TRANSIT ACCESS POINT (6TAP)

To foster interconnectivity among 6REN participants, the Energy Sciences Network (ESnet) and the Canadian Network for the Advancement of Research, Industry, and Education (CANARIE) sponsor an IPv6 exchange called IPv6 Transit Access Point (6TAP). 6TAP provisions IPv6 routing services at the STAR TAP (Science, Technology, and Research Transit Access Point) NAP (Network Access Point) in Chicago. In addition, 6TAP employs an IPv6 router co-located at the STAR TAP NAP for testing the effectiveness of IPv6 operations.

6TAP provisions IPv6 routing services at the STAR TAP (Science, Technology, and Research Transit Access Point) NAP (Network Access Point) in Chicago. The STAR TAP NAP, also know as the Chicago NAP, is formally called the SBC/AADS (Ameritech Advanced Data Services) NAP in Chicago.

2.9.6 IPv6 BACKBONE NETWORK (6BONE)

Sponsored by the IETF and developed by the IETF NGtrans (Next-Generation Transition) Working Group, the IPv6 Backbone Network (6Bone) is a virtual IPv6 backbone network overlay that supports transmission of IPv6 packets via virtual point-to-point links or tunnels. Tunnel endpoints are situated at sites enabling 6Bone implementations. Special 6Bone routers transport IPv6 datagrams or packets to destination addresses because this function is not yet incorporated into each production router on the commodity or public Internet. In 2000, NTT became the first major Internet Service Provider (ISP) to implement IPv6 services on the commodity or public Internet.

Participants in the 6Bone initiative include Duke, North Carolina State, Carnegie Mellon, and Texas A&M Universities. The Universities of Massachusetts, Illinois at Chicago, Pennsylvania, and North Carolina and the National University of Singapore also participate in the 6Bone initiative. Networks that support 6Bone services include the Trans-Pacific Network (TRANS-PAC), STAR TAP, ESnet, the National Aeronautics and Space Administration Research and Education Network (NASA-NREN), the North Carolina Network Initiative (NCNI), the Abilene Network, vBNS+, and CA*net II.

2.9.7 6POP.CA (IPv6 POINT OF PRESENCE, CANADA)

CANARIE (Canadian Network for the Advancement of Research, Industry, and Education) supports IPv6 implementations by CA*net II Regional Area Networks (RANs) and sponsors RAN participation in the 6POP.CA project. A 6Bone tunnel that facilitates 6POP.CA functions operates in conjunction with the CA*net II backbone network and enables interconnections to international IPv6 initiatives.

2.9.8 IP MULTICASTS

2.9.8.1 MBone (Multicast Backbone) Operations

IP multicasts enable point-to-point, point-to-multipoint, and multipoint-to-multipoint multimedia distribution via the Multicast Backbone (MBone). The MBone is an enabler of diverse multicast applications, including real-time broadcasts from NASA space missions, theatrical productions, operas, and rock concerts.

Multicasting refers to the ability of a sender to simultaneously transmit a single packet to recipients at numerous destination addresses. To conserve bandwidth, recipients do not acknowledge reception of IP multicasts, and IP multicasts are sent only once and replicated in the network as needed. If only unicast or point-to-point services are available, an IP multicast is sent once to every destination address. Because multicast packets can be lost as a consequence of Web congestion, IP multicast transmissions over the commodity or public Internet are inherently unreliable.

The MBone is a virtual network overlay that operates on top of the physical Internet. It consists of multicast routers that forward multicast packets via tunnels or virtual point-to-point links to multicast group addresses. In ATM networks, SVCs (Switched Virtual Circuits) support point-to-multipoint virtual links for transporting MBone traffic.

Network overlays such as the MBone require time-consuming, labor-intensive management services. As a consequence, the University of Southern California Information Science Institute (ISI) developed an XBone tool for automatically managing IP overlay services. Capabilities of this tool are evaluated in pilot tests conducted over the CAIRN (Collaborative Advanced Internet Research Network) ATM testbed. In addition, an overlay registry supporting resource management is also in development.

2.9.8.2   Multicast Protocols

The Real-time Transport Protocol (RTP) is a multicast routing protocol that supports sequenced packet delivery for MBone applications. The Distance Vector Multicast Routing Protocol (DVMRP) and the Multicast Extensions to the Open Shortest Path First (MOSPF) Protocol enable distribution and delivery of IP multicast packets to destination endpoints.

The Reliable Multicast Transport Protocol II (RMTP, Phase II) supports the dependable transport of IP multicasts by enabling a small group of senders to transmit packets to large groups of recipients concurrently via multipoint-to-multipoint connections, thereby overcoming the inherent unreliability of traditional IP multicast transmissions. Developed by the University of Southern California (USC), the University of Michigan, and the Massachusetts Institute of Technology (MIT), the Protocol Independent Multicast (PIM) algorithm also enables fast and dependable multicast delivery over best-effort IP networks.

2.9.8.3   IP Multicast Initiative (IPMI) Forum

The IP Multicast Initiative (IPMI) Forum is a multivendor alliance that supports implementation of IP multicast services and development of procedures for enabling seamless multicast operations in IP-over-ATM configurations. Participants include Alcatel, AT&T, Cable & Wireless, Hewlett-Packard, Hughes Network Systems, and Microsoft Corporation.

2.9.8.4   IETF Inter-Domain Multicast Routing (IDMR) Working Group

The IETF Inter-Domain Multicast Routing (IDMR) Working Group fosters development of the IMDR (Inter-Domain Multicast Routing) protocol. This next-generation protocol supports IP multicast operations over advanced networks that include ESnet and vBNS+.

2.9.8.5   IETF DiffServ (Differentiated Services) Working Group

The IETF DiffServ (Differentiated Services) Working Group facilitates development of network architectures in support of differentiated services to enable utilization of IP CoS (Class of Service) with Quality of Service (QoS) assurances. The IETF DiffServ protocol categorizes IP traffic flows in terms of rates and burst size. Methods for mapping ATM QoS (Quality of Service) guarantees to DiffServ (Differentiated Services) IP (Internet Protocol) CoS (Class of Service) assurances are also in development.

DiffServ utilization enables available bandwidth to be divided between a number of pipes. Each pipe features a different Quality of Service (QoS) assurance, depending on whether applications are categorized as high-priority or low-priority. The IETF DiffServ protocol optimizes network performance and fosters development of priority queuing algorithms for seamless transmission of streaming media in IP-over-ATM environments.

DiffServ operations are based on rule statements that indicate how a packet is transported over a network. RedIRIS (NREN of Spain), GRnet (NREN of Greece), and CERN (European Organization for Nuclear Research) maintain DiffServ testbeds. The Universities of Bologna, Stuttgart, Twente, and Utrecht participate in DiffServ testbed initiatives.

2.9.8.6   IP Multicast Security

Because a host can anonymously join a multicast group and receive information distributed to authenticated group members without their knowledge, standards organizations such as the IETF (Internet Engineering Task Force) develop approaches for resolving IP multicast security problems. In addition, the IETF IP Multicast-over-ATM specification clarifies techniques for provisioning secure IP multicast services between multiple sites in local and wider area ATM environments.

2.10   ATM INTERWORKING CAPABILITIES

2.10.1   ATM Interoperability with Optical Network Technologies

ATM supports transport rates ranging from multimegabits to multigigabits. The ATM Forum defines 622.08 Mbps (OC-12) and 2.488 Gbps (OC-48) Physical Layer specifications for enabling ATM cell transmission directly over physical media. These specifications establish the foundation for ATM-over-SONET/SDH, ATM-over-WDM (Wavelength Division Multiplexing), and ATM-over-DWDM (Dense WDM) implementations.

2.10.2   European Services ATM Interoperability (EASI) Initiative

Sponsored by the European Telecommunications Standards Institute (ETSI), the EASI (European Services ATM Interoperability) initiative establishes specifications for interoperable multivendor multiprotocol ATM network implementations. EASI specifications describe procedures for permanent and semi-permanent VC/VP (Virtual Channel/Virtual Path) links and SVC (Switched Virtual Circuit) connections.

2.10.3 INVERSE MULTIPLEXING FOR ATM (IMA)

Endorsed by the ATM Forum, the IMA Protocol supports implementation of ATM networks that operate at 1.544 Mbps (T-1) and E-1 (2.048 Mbps) rates and enables ATM technology to interwork with residential broadband access deployments including ADSL (Asynchronous Digital Subscriber Line) and cable network solutions. Moreover, the Inverse Multiplexing for ATM (IMA) Protocol defines a new Physical Layer protocol for a User-to-Network Interface (UNI) that enables ATM cells to be spread across several T-1 and E-1 lines during transmission. These cells are subsequently reassembled at the destination point so that voice, video, and data transmissions arrive in the correct order.

2.10.4 ATM AND ASYNCHRONOUS DIGITAL SUBSCRIBER LINE (ADSL) OPERATIONS

Mixed-mode ATM-over-ADSL networks support high-speed digital services on existing twisted pair copper networks without interruption of traditional analog telephone service. Moreover, ATM-over-ADSL configurations enable access to broadband applications without large-scale infrastructure investments and facilitate delivery of broadband service to SOHO (Small Office/Home Office) venues over local loops.

2.10.5 ATM AND FRAME RELAY

Protocols such as ATM-DXI (ATM-Data Exchange Interface) and FUNI (Frame User-to-Network Interface) facilitate Frame Relay-over-ATM implementations for enabling real-time voice, video, and data transmission in corporate networking environments.

2.11 ATM LAN EMULATION (LANE)

2.11.1 ATM LANE FUNDAMENTALS

ATM LAN emulation (LANE) enables virtual LAN (VLAN) implementations across ATM backbone networks that reflect the logical associations of workgroups regardless of the physical location of workgroup participants. (See Figure 2.2.) Modifications in virtual ATM LANE topologies are accomplished by redefining workgroups in the network management system and reconfiguring software in ATM switches. MPOA (MultiProtocol-over-ATM) enables direct transmission of virtual ATM LANE traffic over the ATM Physical Layer or Layer 1 of the OSI Reference Model. The Cells-in-Frames (CIF) Alliance supports implementation of ATM desk area networks (DANs) that operate in concert with the virtual ATM LANE infrastructure. ATM LANES are also called ATM ELANs (Emulated LANs). (See Figure 2.3.)

2.11.2 ATM EMULATED LANs (LANEs) IN ACTION

ATM LANEs are scalable and flexible, feature sophisticated network management and control capabilities, and perform functions equivalent to those supported by

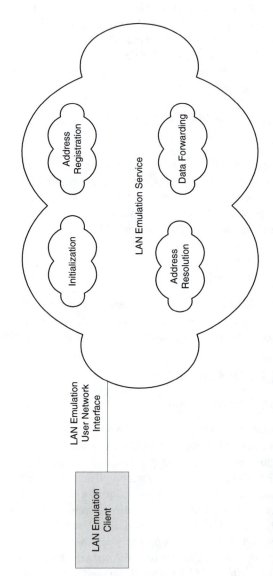

FIGURE 2.2 Local Area Network Emulation (LANE) services.

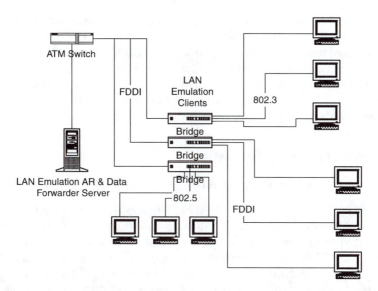

FIGURE 2.3 An ATM LANE client/server configuration allows interoperability with IEEE 802.3 and 802.5 LANs and FDDI technology.

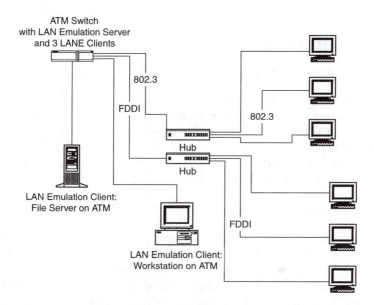

FIGURE 2.4 An ATM LANE client/server configuration.

conventional Ethernet and Token Ring VLANs (Virtual LANs). ATM LANES enable each participant in a logical workgroup to take part in collaborative networking activities. Internetworking devices such as bridges and routers support voice, video, and data exchange between participants in enterprisewide ATM LANEs.

In an ATM LANE, local networking applications access an ATM network configuration via IP protocols. IP packets are transported in ATM cells. ATM LANEs use LAN Emulation User-to-Network Interfaces (LUNIs) and LAN Emulation Network-to-Node or Network-to-Network Interfaces (LNNIs) to provision QoS guarantees, Internet telephony, and connectionless unicast and multicast delivery. ATM LANEs support MAC (Medium Access Control) operations at the Data-Link Layer or Layer 2 of the OSI Reference Model. (See Figure 2.4.)

2.12   WIRELESS ATM (WATM)

2.12.1   ATM FORUM WIRELESS ATM (WATM) WORKING GROUP

Sponsored by the ATM Forum, the WATM (Wireless ATM) Working Group encourages utilization of the AMES (ATM Mobility Extension Service) solution for interlinking ATM technologies with broadband fixed wireless access networks such as LMDS (Local Multipoint Distribution System) and MMDS (Multichannel Multipoint Distribution System). In addition, the WATM Working Group develops wireless ATM specifications for transporting voice, video, and data via second-generation and third-generation cellular communications configurations that employ GSM (Global System for Mobile Communications), 3GSM (Third-Generation GSM), PCS (Personal Communications Services), and UMTS (Universal Mobile Telecommunications Service).

2.12.2   WATM UNIVERSITY RESEARCH PROJECTS

2.12.2.1   Columbia University

The Control, Management, and Telemedia (COMET) Group at the Center for Telecommunications Research at Columbia University supports development and implementation of WATM networks that enable broadband applications featuring a wide array of QoS requirements. Moreover, the COMET Group also examines capabilities of the Real-Time Transport Protocol (RTTP) in provisioning seamless Internet telephony services, video-on-demand (VOD), and videoconferencing.

2.12.2.2   University of Kansas

At the University of Kansas, research initiatives verify the performance of WATM networks in enabling access to MRI (Magnetic Resonance Imaging) and CAT (Computerized Axial Tomography) scans. Moreover, capabilities of the WATM platform in supporting telecollaborative videoconferencing at rates and service levels that are equivalent to those provisioned by wireline ATM networks are evaluated.

The University of Kansas supports implementation of Rapidly Deployable Radio Networks (RDRNs) to support access to WATM applications and services. The performance of ATM wireless adaptive voice and data networks in facilitating location-independent access to information resources and delivery of multimedia traffic to remote users is also examined in pilot testbed initiatives. The University of Kansas is a founding member of the MAGIC-I (Multidimensional Applications and Gigabit

Internetwork Consortium-Phase I) and MAGIC-II (MAGIC-Phase II) initiatives. In addition to WATM initiatives, the University of Kansas sponsors the design of ATM-based networked virtual reality (VR) simulations and development of a Standard Active Switching Node (SASN) that interworks with ATM and Fast Ethernet technologies.

2.13   ATM NETWORK MANAGEMENT

2.13.1   Simple Network Management Protocol (SNMP)

The ATM Forum endorses the use of the Simple Network Management Protocol (SNMP) for performing network management functions in local and wider area ATM environments. SNMP services become operational with the exchange of messages between managed network devices or network nodes and the network management system. Each device or node is equipped with a network software module that provisions management information detailing device or node status to the network management system for accomplishing this process.

Network administrators monitor distributed SNMP operations and resolve network problems remotely from central consoles. SNMP provisions administrative services at the Application Layer or Layer 7 of the OSI Reference Model. Developed by the ATM Forum, ATM Management Information Bases (MIBs) foster implementation of standardized SMNP management functions defined by ITU-T Recommendations and IETF RFCs (Request for Comments).

2.13.2   Protocol Implementation Conformance Statement (PICS)

Developed by the ATM Forum, the Protocol Implementation Conformance Statement (PICS) enables ATM clients to verify compliance of multivendor products with ATM specifications.

2.14   ATM TESTBED IMPLEMENTATIONS

In the 1990s, ATM capabilities in enabling internetworking, distribution of multiple computing resources, reliable network management, and dependable transport of advanced applications were examined in testbed demonstration projects.

2.14.1   Initial ATM Testbeds

ATM functions in enabling high speeds, reliable operations, dependable delivery of multimedia applications, and effective resource allocations were initially evaluated at the Aurora, Blanca, Nectar, and Vistanet gigabit testbeds. The Corporation for National Research (CNR) supported ATM testbed initiatives.

2.14.1.1   *Aurora*

The Lawrence Livermore National Laboratory, the Massachusetts Institute of Technology (MIT), and the Universities of Pennsylvania and Arizona conducted trials of gigabit network applications and examined the performance of distributed networks for scientific applications over the ATM-based Aurora testbed.

2.14.1.2 *Blanca*

The National Center for Supercomputing Applications (NCSA) and the Universities of California at Berkeley, Illinois at Urbana-Champaign, and Wisconsin at Madison enabled development of simulations, multimedia digital libraries, and medical imaging applications in Blanca ATM testbed initiatives.

2.14.1.3 *Nectar*

Carnegie Mellon University (CMU) and the Pittsburgh Supercomputing Center conducted ATM trials at the Nectar testbed that contributed to the implementation of high-speed broadband computational network services and applications. Demand for high-performance networks with multimedia capabilities resulted in migration from the Nectar testbed to Gigabit Nectar, a next-generation testbed installation.

2.14.1.4 Vistanet

The University of North Carolina at Chapel Hill (UNC-CH), MCNC (formerly known as the Microelectronics Center of North Carolina), and North Carolina State University (NCSU) conducted telecollaborative ATM projects over the Vistanet testbed to demonstrate achievements that are possible when geographically dispersed research teams work together at gigabit speeds.

2.14.2 NASA ACTS (ADVANCED COMMUNICATIONS TECHNOLOGY SATELLITE) ATM INTERNETWORK (AAI)

The NASA ACTS (Advanced Communications Technology Satellite) ATM Internetwork (AAI), a DARPA (U.S. Department of Defense Advanced Research Projects Agency) implementation, functioned as a nationwide ATM testbed for benchmarking ATM throughput and performance. In addition, the AAI (ATM Internetwork) supported assessment of IP multicasts and negotiated QoS guarantees. The AAI also established IP gateways to non-ATM LANs, interoperated with the U.S. Department of Defense Research and Education Network (DREN), and provided operational support for research conducted at the Aurora, Blanca, Vistanet, and Nectar testbeds.

2.14.3 MAGIC-I (MULTIDIMENSIONAL APPLICATIONS AND GIGABIT INTERNETWORK CONSORTIUM-PHASE I) AND MAGIC-II (MAGIC-PHASE II)

2.14.3.1 MAGIC-I

The Mayo Clinic in Rochester, Minnesota, utilized the AAI (ATM Internetwork) platform to access the MAGIC-I testbed and conduct telemedicine trials and multicast experiments via MAGIC-I testbed facilities. Capabilities of NASA ACTS satellite implementations in fostering transmission of bandwidth-intensive medical imaging scans from the Mayo Clinic to ATDNet (Advanced Technology Demonstration Network) sites were also examined.

2.14.3.2 MAGIC-II

The MAGIC-II testbed supported utilization of an advanced information system for enabling high-performance computing storage services, ultra-fast access to multimedia applications, and quick and dependable information retrieval at rates reaching 2.488 Gbps (OC-48). MAGIC-II solutions also enabled real-time processing of massive datasets and utilization of a large-scale and extendible IP-over-ATM-over-SONET internetwork for providing access to simulations, satellite imagery, digitized maps, and aerial photographs.

MAGIC-II research supported military and near-shore naval operations, natural disaster response solutions, and intelligence imagery analysis. Approaches for implementation of distributed parallel storage systems (DPSS) were also investigated. Operational from 1996 to 1999, the MAGIC-II infrastructure interoperated with the ATDNet and the National Transparent All-Optical Network (NTON). The U.S. Geological Survey EROS (Earth Resources Observation System) Data Center and the Lawrence Livermore National Laboratory participated in MAGIC-II testbed initiatives.

2.15 FEDERAL ATM INITIATIVES

2.15.1 Collaborative Advanced Internet Research Network (CAIRN)

Research and experimentation in testbed environments remain necessary to extend and refine ATM applications for optimizing information access and facilitating cross-institutional collaborations. As an example, the Collaborative Advanced Internet Research Network (CAIRN) is a high-speed, high-performance ATM network testbed that interoperates with T-1, Ethernet, and Fast Ethernet installations. Sponsored by DARPA, CAIRN fosters assessment of advanced applications and network protocols such as IPv6, DiffServ, and SecureDNS (Secure Domain Name System). CAIRN also supports utilization of tunneling architecture for provisioning access to network resources across non-CAIRN routing platforms.

CAIRN sponsors development of MPLS (MultiProtocol Label Switching) as an alternative solution to IPv4 tunnels, implementation of large-scale multicast video-conferences, and deployment of interdomain multicast routing solutions. Moreover, CAIRN promotes utilization of Java tools to enable SNMP (Simple Network Management Protocol) services and design of advanced network security mechanisms. The CAIRN infrastructure facilitates telecollaborative research to substantiate performance of QoS guarantees, RSVP (Resource Reservation Protocol) functions, and IP Security (IPSec) capabilities. CAIRN interoperates with SingAREN (Singapore Advanced Research and Education Network), vBNS+ (very high performance Backbone Network Service Plus), and ESnet (Energy Sciences Network). In addition, CAIRN provisions network services for SuperNet and GloMo (Global Mobile Information Systems) Program initiatives and supports 6REN (IPv6 Research and Education Network) and Internet2 (I2) applications and services.

2.15.2 ENERGY SCIENCES NETWORK (ESNET)

2.15.2.1 ESnet Operations and Services

Sponsored by the U.S. Department of Energy, ESnet promotes development of an advanced network infrastructure and sophisticated distributed computing applications supported by supercomputing resources at the Lawrence Berkeley Laboratory. The ESnet infrastructure enables geographically separated scientists to participate in an international telecollaborative network environment that features multicast services and desktop videoconferencing. Moreover, ESnet fosters research leading to the development of new energy sources and enables scientists to monitor global climate change and conduct advanced research in biology, physics, and chemistry. ESnet also supports a seamless transition from IPv4 to IPv6 and participates in IPv6 trials and testbed initiatives.

ESnet and the Metropolitan Research and Education Network (MREN) conduct research via the ESnet/MREN Regional Grid Experimental Network (EMERGE). EMERGE is a next-generation Internet testbed that enables assessment of leading-edge networking services and telecollaborative applications.

2.15.2.2 ESnet Infrastructure

ESnet employs a high-performance ATM backbone network that interlinks key national laboratories and collaborator sites. ESnet transmission rates range from 1.544 Mbps (T-1) to 622.08 Mbps (OC-12). ESnet also maintains peering relationships with Abilene, vBNS+, and CalREN-2 (California Research and Education Network-Phase 2), and supports interconnectivity to public networks such as the commodity Internet. ESnet enables high-resolution simulations and promotes telecollaborative use of computational resources among researchers at distributed locations. In 2000, Qwest Communications began upgrading the ESnet infrastructure to enable transmission rates at 1 Terabit per second (Tbps) by 2005.

Currently, the ESnet backbone links major research sites, including the National Oceanic and Atmospheric Administration (NOAA), Sandia National Laboratories, the Stanford Linear Accelerator Center, and the Savannah River Plant. In addition, the Universities of California at Los Angeles and Texas at Austin, the California Institute of Technology (Cal Tech), the Massachusetts Institute of Technology (MIT), and Brandeis, New York, Columbia, and Yale Universities access the ESnet backbone to conduct scientific research.

ESnet maintains connections to regional networks and wider area configurations such as the Chicago NAP (Network Access Point) and the Oakland, Albuquerque, and San Diego PoPs (Points of Presence). ESnet also provisions virtual connections to FIX (Federal Interagency Exchange)-WEST at the NASA Ames Research Center and FIX-EAST at the University of Maryland at College Park. In addition, ESnet interworks with international implementations such as the National Institute for Fusion Science Network in Japan, the NREN (National Research and Education Network) of Germany (DFN or Deutsche Forschungsnetz), and the European Organization for Nuclear Research (CERN).

2.15.2.3 ESnet Applications and Services

A participant in National Information Infrastructure (NII) projects, ESnet sponsors initiatives in basic energy sciences, nuclear physics, high-energy physics, and fusion energy, as well as in mathematical, information, and computational sciences. In addition, ESnet supports projects in healthcare, environmental research, and international nuclear safety; implementation of next-generation IP networks; and utilization of PKIs (Public Key Infrastructures) for secure E-commerce transactions. Capabilities of ATM VLAN (Virtual LAN) implementations in actual environments and the effectiveness of PGP (Pretty Good Privacy) and PEM (Privacy Enhanced Mail) for safeguarding e-mail integrity are also examined.

2.15.3 U.S. ARMY RESEARCH LABORATORY (ARL)

The U.S. Army Research Laboratory (ARL) supports collaborative development of broadband communications services that facilitate secure wireline and wireless voice, video, and data delivery with academic institutions such as Johns Hopkins, Princeton, and Morgan State Universities. Approaches for effective implementation of high-performance internetworking technologies and infrastructure solutions in noisy and hostile environments are examined in testbed experiments as well. In pilot initiatives, ARL interlinked Fort Gordon to the Secure Survivable Communications Network (SSCN) for assessment of ATM performance in a tactical setting. ARL also evaluates the capabilities of ATM in provisioning multimedia services in near real-time in digitized battlefield conditions.

2.16 INTERNET2 (I2)

2.16.1 UNIVERSITY CORPORATION FOR ADVANCED INTERNET DEVELOPMENT (UCAID)

The University Corporation for Advanced Internet Development (UCAID) is a nonprofit consortium that consists of American research centers and universities. This consortium works in partnership with corporations and affiliate members in developing next-generation network architectures, protocols, and technologies. UCAID also sponsors implementation of advanced network applications and initiatives including Internet2 (I2) and coordinates scientific, research, and academic networking investigations.

2.16.2 I2 TECHNICAL FUNDAMENTALS

An advanced high-volume, high-speed network initiative, Internet2 (I2) employs an IP-over-ATM-over-SONET platform for enabling real-time teleresearch and interactive distance learning applications with QoS (Quality of Service) guarantees. The I2 project also supports telemedicine initiatives and fosters telecollaborative research among members of computational, medical, research, scientific, and academic communities. In addition to ensuring always-on networking connectivity, I2 enables participating institutions to reserve high quantities of bandwidth at specified times for scientific research and/or educational investigations.

Charter Internet2 members such as the Universities of Wisconsin and Virginia, Carnegie Mellon University, the California Institute of Technology (Cal Tech), and the University of California at Berkeley verify ATM-over-SONET capabilities in sustaining multimedia integration and on-demand real-time interactive telecollaboration. I2 innovations contribute to the implementation of advanced network architectures, technologies, and protocols on the commodity or public Internet for everyday use.

2.16.3 INTERNET2 (I2) NETWORK AGGREGATION POINTS OF PRESENCE (PoPs)

Internet2 (I2) deployment is based on the formation of GigaPoPs (Gigabit Points of Presence). I2 GigaPoPs are high-capacity, multiservice, multifunctional, interconnection regional transfer and aggregation PoPs (Points of Presence) that move vast volumes of voice, video, and data between I2 sites. Designed for regional groups of I2 participants, Type 1 GigaPoPs route Internet2 traffic through one or two connections. Type 2 GigaPoPs provision access to next-generation federal networks and international configurations such as the Asia-Pacific Network (APAN) and the Nordic Countries Network, Phase 2 (NORDUnet2). Commercial GigaPoPs that route I2 traffic to destination endpoints and optimize bandwidth availability are also in development.

2.16.3.1 Michigan GigaPoP

I2 GigaPoPs enable groups of Internet2 participants in specified geographical regions to access interactive multimedia applications, evaluate current and emergent protocols and specifications, and implement sophisticated educational technologies. For example, Michigan State University, Michigan Technological University, the University of Michigan at Ann Arbor, Wayne State University, and the UCAID office in Ann Arbor use the Michigan GigaPoP for conducting teleresearch projects to facilitate middleware and application development and next-generation routing operations.

The Michigan GigaPoP also transports traffic to and from the Abilene and vBNS+ Internet2 backbone networks and to the ATM-based Chicago NAP (Network Access Point). In addition to the Chicago NAP, major ATM traffic exchange points for peer-level entities include the FloridaMIX (Florida Multimedia Internet Exchange) in South Florida and the MAE-WEST Exchange Point in California.

2.16.3.2 Mid-Atlantic GigaPoP (MAGPI)

The Mid-Atlantic GigaPoP (MAGPI) provisions networking services for I2 institutions such as the Universities of Pennsylvania and Delaware and Princeton and Rutgers Universities situated in the mid-Atlantic states along the Eastern seaboard.

2.16.3.3 Mid-Atlantic (Middle-Atlantic) Crossroads (MAX)

Net.Work.Virginia, the Southeastern Universities Research Association (SURA), the Washington, D.C. Research and Education Network (WREN), and the Maryland

GigaPoP are among the entities that contribute to the formation of the Mid-Atlantic Crossroads (MAX). Designed for communications carriers, universities, research centers, and Network Service Providers (NSPs) in the Mid-Atlantic States, MAX serves as the Washington, D.C. area aggregation point for bandwidth-intensive network traffic. In addition, MAX provides access to advanced networking initiatives such as ESnet and connections to the Abilene and the vBNS+ Internet2 backbone networks.

2.16.3.4   Mid-Atlantic MetaPoP

MAX also plays a pivotal role in the deployment of the Mid-Atlantic MetaPoP. The Mid-Atlantic MetaPoP is a major network switching and aggregation point that provides high-performance multimedia services and high-speed network connections to regional initiatives such as the East Coast GigaPoP in the Northeastern region of the United States and SoX (Southern Crossroads) in the Southeastern region of the United States.

2.16.3.5   Southern Crossroads (SoX)

Sponsored by SURA (Southeastern Universities Research Association), the Southern Crossroads (SoX) initiative facilitates access to current and emergent networking services. The SoX ATM infrastructure provides video, data, and voice services in an integrated multivendor environment and connects Internet2 and non-Internet2 participants to each other and to the Abilene Network, vBNS+, and the Next Generation Internet (NGI).

In addition to providing expanded opportunities for telecollaboration among university scientists, researchers, and educators at SoX-affiliated institutions, SoX also supports access to regional networks such as SEPSCoR (SouthEast Partnership to Share Computational Resources) and the Atlanta MetaPoP, the major network aggregation point for the Southeastern United States. Participants in the SoX initiative include the Universities of Alabama, Delaware, North Carolina, Richmond, and Texas; Emory, Florida Atlantic, Tulane, Mississippi State, and West Virginia Universities; and the Alabama and North Carolina Supercomputer Centers.

2.16.4   PEERING RELATIONSHIPS

Peering or reciprocal relationships enable I2 participants, NRENs (National Research and Education Networks), and NSPs (Network Service Providers) to exchange Internet traffic with destination addresses on each other's backbone network at regional exchange points and NAPs (Network Access Points). In addition, every participant in a peering relationship is required to transfer information to and from affiliated networks. Affiliated networks include networks established by scientific libraries, research centers, academic institutions, and local and regional consortia that are interconnected to a peer-level network backbone. Peering exchanges require utilization of current router information between the peering entities via the Border Gateway Protocol (BGP).

In a peering environment, NRENs (National Regional and Education Networks) and regional network configurations use large-scale caches that offload traffic from the commodity Internet to reduce Web congestion. This traffic is distributed via intermediate or local caches or servers to network nodes or endpoints. Internet2 maintains peering or reciprocal networking relationships with NRENs worldwide, including the NRENs in Ireland (HEAnet), Taiwan (TAnet2), Switzerland (SWITCH), Canada (CA*net II and CA*net3), Singapore (SingAREN), and the Czech Republic (CESNET). As with I2, NRENs support advanced telecollaborative research projects and implementation of high-performance, high-speed broadband network services.

An I2 network backbone and service provider, the Abilene Network sustains reciprocal networking relationships with vBNS+ and high-performance NRENs such as NRENs in Germany (DFN), Taiwan (TAnet2), and Italy (GARR). In addition, the Abilene Network supports peer-level information exchange with federal networks such as the U.S. Department of Defense Research and Education Network (DREN) and the NASA Integrated Services Network (NISN).

2.16.5   METROPOLITAN INTERNET EXCHANGES (IXs) AND EXCHANGE POINTS (XPs)

Internet Exchanges (IXs) and Exchange Points (XPs) are reciprocal traffic exchange points for networks in peer-level relationships. For example, the Boston Metropolitan Exchange Point (Boston MXP) enables NSPs (Network Service Providers) such as HarvardNet, communications carriers such as Sprint and AT&T MediaOne, and higher educational institutions including the Massachusetts Institute of Technology and Boston University to exchange voice, video, and data with one another. Additional North American metropolitan IXs and XPs include the Anchorage Metropolitan Access Point (AMAP), the Seattle Internet Exchange (SIX), the Dallas-Fort Worth Metropolitan Access Point (DFMAP), and the Denver Internet Exchange (DIX).

2.16.6   EUROPEAN BACKBONE (EBONE) NETWORK

Developed by GTS (Global TeleSystems), the European Backbone (EBone) is the largest backbone network in the European Union. EBone employs an IP-over-SDH infrastructure for enabling reciprocal traffic exchange among peer-level NSPs at specified interconnection points, or PoPs (Points of Presence), in Amsterdam, Brussels, Barcelona, Bratislava, Copenhagen, Dusseldorf, Geneva, Frankfurt, London, Munich, Madrid, Milan, Prague, Stockholm, Vienna, Zurich, and Paris. EBone also supports interconnections to PoPs in New York City and Pennsaukin, New Jersey.

2.17   vBNS+ (VERY HIGH-PERFORMANCE BACKBONE NETWORK SERVICE PLUS)

2.17.1   vBNS+ FOUNDATIONS

As with the Abilene network, vBNS+ is a high-speed, high-performance network infrastructure that serves as a network backbone for Internet2. The vBNS+ acronym also stands for "very high-speed Broadband Network Service Plus." In 1995, the National Science Foundation (NSF) initiated work on the vBNS+ implementation. At that time, vBNS+ was known as vBNS.

Originally, vBNS+ was an experimental testbed for resolving performance issues associated with the delivery of high-capacity Internet services. It was the first backbone network to support IP-over-ATM-over-SONET operations at rates reaching OC-3 (155.52 Mbps). In addition, vBNS+ was the first production network to offer native IPv6 multicasting services and MPLS (MultiProtocol Label Switching) support.

2.17.2 vBNS+ Operations

vBNS+ achieves high-speed transmission by carrying IP traffic in an IP network overlay that operates on top of an ATM-over-SONET infrastructure managed by WorldCom. vBNS+ transports voice, video, and data via PVPs (Permanent Virtual Paths). PVPs consist of PVCs (Permanent Virtual Circuits) with every network node connected to every other network node in a mesh topology. vBNS+ also supports SVCs (Switched Virtual Circuits) and RSVP (Resource Reservation Protocol) for providing reserved bandwidth service. Designed to facilitate scientific research, vBNS+ initially interoperated with NSF (National Science Foundation) supercomputing sites managed by the Cornell Theory Center. Links were also established with the National Center for Atmospheric Research (NCAR), the National Center for Supercomputing Applications (NCSA), and the Pittsburgh and the San Diego Supercomputer Centers. (See Figure 2.5.)

2.17.3 vBNS+ in Action

In parallel with the Abilene Network, vBNS+ maintains high-speed interconnections with NRENs. As with the Abilene initiative, vBNS+ also functions as a non-commercial research platform for facilitating development of high-speed applications and innovations in network technologies, topologies, architectures, and protocols. IPv6-over-vBNS+ service became available in 1998.

vBNS+ trials and experiments evaluate capabilities of network technologies such as ATM-over-SONET in enabling real-time collaboration, interactivity, multimedia integration, and QoS (Quality of Service) guarantees. Currently, vBNS+ supports high-speed peering relationships and interconnectivity with NSF federal research and education networks such as the Metropolitan Research and Education Network (MREN), ESnet, and DREN. Approximately 40 GigaPoPs across the United States interoperate with vBNS+. Authorized I2 entities support transmissions to I2 GigaPoPs that in turn direct traffic to and from vBNS+ at rates of 155.52 Mbps (OC-3) and 622.08 Mbps (OC-12).

Northwestern University uses the vBNS+ platform for videoconferencing and development of complex computational grids; the University of Chicago employs the vBNS+ infrastructure for investigating the climate of the Earth and other planets; and the University of Illinois at Chicago (UIC) sponsors development of advanced data mining applications in high-energy physics via the vBNS+ platform. Carnegie Mellon University (CMU) develops simulations for predicting earthquake occurrence

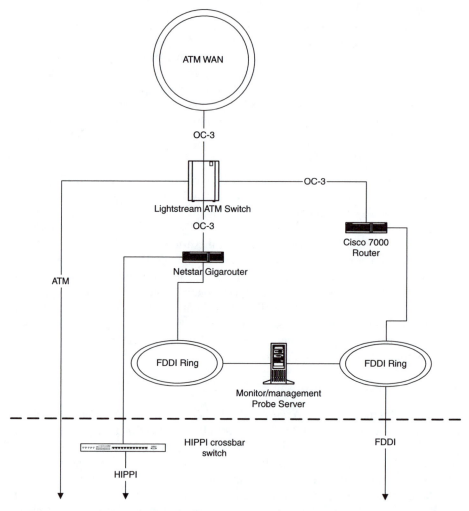

FIGURE 2.5 vBNS+ network segment featuring an ATM WAN, FDDI (Fiber Data Distributed Interface) dual-ring topology, HIPPI (High-Performance Parallel Interface connections), and ATM switching and routing equipment.

via the vBNS+ infrastructure. A participant in the Earth Systems Science Center (ESSC), Pennsylvania State University (Penn State) utilizes vBNS+ capabilities for determining water resource usage patterns. The University of Washington benchmarks vBNS+ performance in provisioning metropolitan area ATM-over-SONET transport services at 10 Gbps (OC-192).

2.17.4 vBNS+ IP Multicasting Service

vBNS+ supports a native IP multicasting service that enables direct and dependable delivery of MBone traffic, thereby eliminating MBone routing instabilities and the need for dedicated multicast routers to perform tunneling functions. vBNS+ multicast services enable traffic exchange between Web caches and MBone sessions. In 1999, vBNS+ supported approximately 60 multicast links to academic and research networks.

2.17.5   vBNS+ Features and Functions

An enhanced version of vBNS, vBNS+ is a nationwide network that provisions access to high-performance broadband applications. vBNS+ employs a dual backbone topology supporting ATM and packet-over-SONET (POS) technologies. In addition, vBNS+ supports innovations in IPv6 high-bandwidth multicast services, development of security filtering solutions, user-configurable routing policies, and implementation of SIP (Session Initiation Protocol) for voice-over-IP (VoIP) services. vBNS+ also facilitates access to sophisticated IPv6 applications, enables VPN (Virtual Private Network) implementation, and supports MPLS (MultiProtocol Label Switching) operations. Entities participating in vBNS+ track network usage by monitoring SNMP (Simple Network Management Protocol) Statistics and Measurement Services. I2 research centers and universities are selected through a peer review process to participate in vBNS+ initiatives.

2.18   NATIONAL ATM TELE-EDUCATION INITIATIVES

Accelerating global demand for distance education, dependable multimedia transport, and rapid access to sophisticated Internet resources and services contributes to the growing popularity and acceptance of ATM technology in school and university environments. Effective ATM deployment by educational and research institutions requires careful planning and a strategic commitment from administrators, faculty, and staff to utilize ATM applications to enhance the learning process. Representative ATM-based tele-education initiatives are explored in this section.

2.18.1   California

2.18.1.1   California Research and Education Network-Phase 2 (CalREN-2)

Developed by the Consortium for Education Network Initiatives in California (CENIC), CalREN-2 supports the establishment of a high-capacity, high-performance, next-generation network that interconnects higher education institutions statewide to each other and to major national broadband networking initiatives such as vBNS+, Abilene, and ESnet. Moreover, CalREN-2 employs an ATM-over-SONET infrastructure for enabling access to bandwidth-intensive telecollaborative services and teleresearch, telemedicine, and tele-education applications.

Each CalREN-2 campus employs IP technology and ATM switches to transmit voice, video, and data with CalREN-2 destination addresses to a virtual GigaPoP. CalREN-2 campus transmissions are then sent from the virtual GigaPoP to the CalREN-2 backbone network, and ultimately to recipient locations. Virtual GigaPoPs are located in key geographical areas throughout the state. CalREN-2 supports transmissions between member campuses and virtual GigaPoPs at rates ranging from 622.08 Mbps (OC-12) to 2.488 Gbps (OC-48). The CalREN-2 infrastructure enables connections between virtual GigaPoPs and vBNS+ at 155.52 Mbps (OC-3) and 622.08 Mbps (OC-12). In addition to ATM, SONET, and IP, CalREN-2 sites also support Fast Ethernet, Gigabit Ethernet, Frame Relay, and FDDI (Fiber Data Distributed Interface) services.

The CalREN-2 infrastructure provides a framework for implementation of the California Virtual University (CVU), features QoS assurances, and enables data collection for monitoring network performance. Moreover, CalREN-2 facilitates development of middleware, innovations in fields that include telemedicine and distance education, and implementation of advanced security solutions, IP multicasts, streaming media, and 3-D (three-dimensional) interactive simulations.

2.18.1.2   California State University at Monterey Bay (CSU Monterey Bay)

A CalREN-2 participant, California State University at Monterey Bay (CSU Monterey Bay) enables deployment of multimedia Geographic Information Systems (GISs) featuring high-resolution video, high-fidelity audio, and 3-D (three dimensional) imagery for creating Antarctic seafloor environments. Also a 3-D initiative, Salinas Valley 2020 simulates the impact of land-use practices and water resource policies on the local environment over time.

CSU Monterey Bay, the University of California at Santa Cruz (UCSC), and the Navy Post-Graduate School support implementation of a regional collaborative broadband tele-education network. This network enables distance education delivery from the main campus at UCSC to post-secondary institutions in the area surrounding Monterey Bay and facilitates collaborative development of K–12 (Kindergarten through Grade 12) tele-education enrichment projects for deployment in public schools situated in Santa Cruz and Monterey Counties.

2.18.2   FLORIDA

2.18.2.1   Florida International University (FIU)

A multi-campus institution in South Florida, Florida International University (FIU) employs an ATM infrastructure for provisioning high-speed access to data, audio, and video resources; museum holdings; and specialized department collections in architecture, music, and art history. In addition, this ATM configuration fosters interactive videoconferencing and delivery of real-time classroom lectures to various campus locations. Course grades are posted online and can be accessed by students via the FIU ATM platform as well. Voice, video, and data traffic is transported at 155.52 Mbps (OC-3).

2.18.3   GEORGIA

2.18.3.1   PeachNet and PeachNet2 (PeachNet Phase 2)

Sponsored by the State of Georgia, PeachNet employs a high-speed broadband ATM network infrastructure for distance learning programs and teleresearch projects. Within the State of Georgia, public colleges and universities including the University System of Georgia, vocational and technical schools, and public schools and school districts use the PeachNet infrastructure for enabling e-mail exchange, participation in interactive tele-education programs and videoconferences, Web browsing, and Internet research.

An enhanced version of the original PeachNet, PeachNet2 (PeachNet Phase 2) provisions access to digital library initiatives, enables telecollaborative research, and supports interactive videoconferencing. In addition, PeachNet2 enables IP telephony, video-on-demand (VOD), and advanced distance education initiatives.

PeachNet and PeachNet2 facilitate bandwidth-intensive transmissions via a distributed GigaPoP that works in conjunction with the in-place GigaPoP established by Georgia State University (GSU) and the Georgia Institute of Technology (Georgia Tech). The distributed GigaPoP supports links to research and education networks, including Abilene and vBNS+ at rates up to 155.52 Mbps (OC-3).

2.18.3.2   Georgia State University (GSU)

A PeachNet and PeachNet2 participant, Georgia State University (GSU) employs an ATM backbone network operating at 622.08 Mbps (OC-12) that works in concert with Fast Ethernet technology for enabling student, faculty, administrative, and staff applications. MultiProtocol-over-ATM (MPOA) services support connections between the GSU ATM backbone network and campus 100BASE-T Fast Ethernet segments. In addition, the GSU network platform supports I2 research, IP multicast services, Internet telephony, and seamless multimedia transmission.

2.18.3.3   University of Georgia

A participant in PeachNet and PeachNet2, the University of Georgia employs an extendible and scalable ATM-over-SONET backbone network. This platform fosters real-time telecollaboration between researchers at the University of Georgia Learning Performance and Support Laboratory, NASA, George Mason University (GMU), and the University of Houston. Moreover, this ATM-over-SONET infrastructure facilitates teleconsultations between veterinarians and students attending veterinary schools at the University of Georgia and Texas A&M University. The University of Georgia Virtual Electronic Network for University Services (VENUS) initiative enables LAN and WAN integration, provides direct links to bandwidth-intensive campus resources, supports virtual LAN (VLAN) implementations, and fosters high-speed voice, video, and data transmission.

2.18.4   MASSACHUSETTS

2.18.4.1   Boston University (BU)

Boston University (BU) employs a campus ATM configuration operating at 155.52 Mbps (OC-3) for enabling advanced scientific research and academic initiatives such as the MARINER (Mid-level Alliance Resource In the North East Region) project. This project fosters telecollaborative development of tele-instruction and teletraining programs for deployment in K–12 public schools.

The ATM network at Boston University (BU) enables multimedia applications and initiatives sponsored by the Departments of Physics and Chemistry, the College of Engineering, the Computer Graphics Laboratory, and the Center for Remote Sensing. In addition, BU initiated the establishment of a high-bandwidth ATM network infrastructure for interlinking local institutions in the Boston metropolitan area. With the aid of an NSF (National Science Foundation) grant in the DARPA (U.S. Department of Defense Advanced Research Projects Agency) Connections to the Internet Program, BU also established links between vBNS+ and the Boston MAN to support transmissions at 155.52 Mbps (OC-3).

2.18.5 MICHIGAN

2.18.5.1 Michigan Teacher Network (MichNet)

The Michigan Teacher Network (MichNet) fosters utilization of Internet resources in K–12 public schools and enables students in grades 4 through 9 to access Web resources and participate in tele-education programs. MichNet employs an IP-over-ATM backbone network that provisions multiple connections to the Internet via the Chicago Network Access Point (NAP) at rates ranging from 1.544 Mbps (T-1) to 622.08 Mbps (OC-12). MichNet maintains peering relationships with ESnet and the Ohio Academic Research Network (OARnet). Michigan State University (MSU), Wayne State University, and the University of Michigan (UM) participate in the MichNet initiative.

2.18.5.2 University of Michigan

The Center for Information Technology Integration (CITI) at the University of Michigan supports development and implementation of the Secure Distributed Video Conferencing (SDVC) initiative. This project employs cryptographic protocols and algorithms for smart-card key exchange to safeguard the integrity of video, audio, and data transmission to reception points on Internet2. The SDVC initiative operates over an experimental I2 ATM backbone network at the University of Michigan and supports connections to vBNS+.

2.18.6 MISSOURI

2.18.6.1 MOREnet3 (Missouri Research and Education Network, Phase 3)

A statewide initiative, MOREnet3 (Missouri Research and Education Network, Phase 3) employs an ATM backbone network to support multimedia applications and tele-education initiatives in K–12 public schools and post-secondary institutions. This ATM infrastructure works in concert with IP, Ethernet, Fast Ethernet, and Frame Relay technologies; enables IPv6 multicasts; and provisions MPOA and MPLS services.

2.18.7 NEBRASKA

2.18.7.1 Great Plains Network (GPN)

The Great Plains Network (GPN) employs an ATM backbone network for enabling broadband applications and telecollaborative scientific research in the field of earth systems science. GPN also facilitates connections to the Abilene Network at rates reaching 622.08 Mbps (OC-12). The initial GPN segment interconnects educational institutions and research centers in Kansas, Arkansas, Nebraska, North Dakota, South Dakota, and Oklahoma via a DS-3 (44.736 Mbps) link. The University of Nebraska at Lincoln provisions technical support and network management services for the GPN configuration.

2.18.8 NEVADA

2.18.8.1 NevadaNet

Sponsored by the University and Community College System of Nevada, NevadaNet provisions high-performance, high-speed ATM services statewide. Moreover, NevadaNet enables multimedia transmission, tele-education projects, and telecollaborative research. NevadaNet also supports high-speed Internet connections to K–12 public schools and public libraries and interconnects the University of Nevada at Reno and the University of Nevada at Las Vegas to the vBNS+ Network.

2.18.9 NEW JERSEY

2.18.9.1 Washington Township Public School System

Located in the Delaware Valley, the Washington Township Public School System utilizes an ATM backbone network to support videoconferences, tele-education services, and curricular delivery to multiple K–12 classrooms concurrently. In addition to ATM, the Washington Township Public School System employs Ethernet and Fast Ethernet segments for enabling access to Web applications, online coursework, and library resources. This configuration also supports television broadcasts and foreign language instruction. In addition, the township uses the public school system ATM platform for municipal operations; budgeting, purchasing, and payroll applications; and providing online access to titles of local library holdings.

2.18.10 NEW YORK

2.18.10.1 New York State Education and Research Network, Year 2000 (NYSERNet 2000)

The New York State Education and Research Network, Year 2000 (NYSERNet 2000) initiative has enabled implementation of an advanced IP-over-ATM network for provisioning next-generation networking services throughout the State of New York via a high-speed, optical fiber link extending from New York City to Buffalo. In addition, the NYSERNet 2000 platform supports connections to the Next-Generation Internet (NGI), vBNS+, Abilene, and Gemini 2000. Developed by IXC Communications, Gemini 2000, an advanced IP optical backbone network, transports commercial traffic and transmissions generated by NYSERNet 2000 research and educational institutions that are ineligible to use vBNS+ and Abilene facilities.

NYSERNet 2000 enables rates ranging from 622.08 Mbps (OC-12) to 2.488 Gbps (OC-48). The NYSERNet 2000 infrastructure employs a distributed GigaPoP with Points of Presence (PoPs) in Manhattan, Albany, Syracuse, Rochester, and Buffalo. Rensselaer Polytechnic Institute, the Universities of Rochester and Buffalo, and Columbia and New York Universities connect to the NYSERNet 2000 backbone via metropolitan SONET ring configurations. Moreover, NYSERNet 2000 utilizes sophisticated network management protocols and technologies such as MPLS (MultiProtocol Label Switching) and Carrier Scale Internetworking (CSI), a next-generation IP internetworking architecture developed by Siemens and Newbridge Networks for provisioning seamless broadband service.

2.18.11 Ohio

2.18.11.1 OARnet (Ohio Academic Research Network)

The Ohio Academic Research Network (OARnet) enables Ohio libraries, K–12 public schools, technical and vocational institutions, colleges, universities, research organizations, and state and local government agencies to access the commodity or public Internet. In addition, OARnet serves as a regional GigaPoP and enables the Ohio Supercomputer Center (OSC); the Universities of Cincinnati and Akron; and Kent State, Ohio, and Ohio State Universities to connect to I2 via the Abilene Network. OARnet maintains Points of Presence (PoPs) in Cincinnati, Cleveland, Akron, Columbus, Toledo, Dayton, and Detroit.

Currently, OARnet enables networking services for the Ohio Board of Regents' ATM testbed project called OCARnet (Ohio Communications and Computing ATM Research Network). Participants in this research testbed include Cleveland State, Kent State, Wright State, and Ohio State Universities and the Universities of Dayton, Cincinnati, and Toledo. OARnet also supports collaborative development of virtual environments (VEs) and simulations to facilitate scientific investigations. In addition, OARnet participates in the development of the NGI initiative.

2.18.12 Oklahoma

2.18.12.1 Advanced Research Network

Sponsored by institutions that include the University of Oklahoma and Oklahoma State University, the Advanced Research Network (ARN) received a National Science Foundation (NSF) grant in the Connections to the Internet Program for providing links to the Abilene Network at rates of 155.52 Mbps (OC-3). The ARN ATM network infrastructure supports telemedicine research, weather forecasting, simulations of lipid membranes, and distance education applications.

2.18.13 Oregon

2.18.13.1 Network for Engineering and Research in Oregon (NERO)

The Network for Engineering and Research in Oregon (NERO) is an ATM-over-SONET regional configuration that facilitates collaborative teleteaching and teleresearch. Oregon State and Portland State Universities use the NERO infrastructure to enable interactive multimedia applications and desktop videoconferencing. NERO also provides network services for the Oregon University System, the Oregon State Departments of Administrative Services and Education, local industry, the Hatfield Marine Science Center, and community colleges. Supported by NERO, the Oregon Public Education Network (OPEN) provisions links to curricular resources and educational services on the Web and enables connections to continuing tele-education courses and lifelong telelearning programs.

2.18.14 PENNSYLVANIA

2.18.14.1 Line Mountain School District

Situated in rural central Pennsylvania, the Line Mountain School District employs an ATM-over-SONET infrastructure for Web browsing and delivery of tele-education courses in advanced sciences. This network platform also provisions access to a broad array of telecourses and teleprograms to compensate for the lack of funding of on-site classes and shortages of qualified teachers. Approaches for linking the Line Mountain School District network to a regional ATM school network config- uration are under consideration.

2.18.14.2 Pittsburgh GigaPoP

Based at Carnegie Mellon University (CMU), the Pittsburgh GigaPoP is a regional NAP (Network Access Point) for PoPs (Points of Presence) in Central and Western Pennsylvania and in West Virginia. Operated by the Pittsburgh Supercomputing Center, the Pittsburgh GigaPoP enables the university community to access intranets and extranets, the Internet, the vBNS+ Network, and the Abilene initiative. The Pittsburgh GigaPoP features an IP-over-ATM infrastructure for enabling transmis- sions at 155.52 Mbps (OC-3) and provides networking services for K–12 schools and local government agencies.

2.18.14.3 University of Pennsylvania

An I2 participant, the University of Pennsylvania supports PennNet (University of Pennsylvania Network) operations. A multiservice, multifunctional university net- work, PennNet consists of network technologies that include Ethernet, Fast Ethernet, FDDI, and ATM. PennNet enables interactive videoconferencing, radio broadcasts, and real-time IP telephony; high-speed access to vBNS+; and telecollaborative research in biostatistics and clinical epidemiology. Plans for supporting Gigabit Ethernet implementation are under consideration.

2.18.15 VIRGINIA

2.18.15.1 Net.Work.Virginia (NWV)

Net.Work.Virginia (NWV) is a high-performance communications network that delivers ATM service statewide. Participants in NWV include the Virginia Community

College System (VCCS), Old Dominion University, and K–12 private and public schools. The Virginia State Library, the Institute of Marine Science, and state and municipal government agencies also participate in NWV. NWV dynamically allocates bandwidth on-demand. Rates of transmission from T-1 (1.544 Mbps) to OC-3 (155.52 Mbps) are supported. Net.Work.Virginia serves as a prototype for the next-generation Internet and a regional GigaPoP for enabling authorized institutions to access the Abilene Network, vBNS+, and ESnet.

Implemented in 2001, NWVng (Net.Work.Virginia next-generation), an enhanced version of NWV, supports high-capacity, data-intensive I2 applications featuring high-definition video, and provisions access to high-performance research applications. A consortium consisting of communications providers, Vision Alliance provides local access and switching services to enable seamless NWV and NWVng operations. Verizon-Virginia provisions technical support services.

2.18.15.2   George Mason University (GMU)

George Mason University (GMU) employs an ATM network that interconnects the main campus to satellite campuses in Arlington, Fairfax, and Prince William. In addition, this network facilitates connectivity to Net.Work.Virginia and provides MBone services. The GMU ATM platform supports initiatives in tele-education, space science, and telemedicine; enables testbed trials benchmarking the performance of IP multicasts; and provisions GigaPoP services for the I2 initiative.

2.18.15.3   Virginia Community College System (VCCS)

The Virginia Community College System (VCCS) employs an ATM solution for accommodating administrative and academic requirements of community college students and faculty at campuses throughout the Commonwealth of Virginia. The VCCS platform provisions access to synchronous and asynchronous tele-education programs and enables participants such as Lord Fairfax Community College to deliver distance education courses to local high schools.

2.19   INTERNATIONAL TELE-EDUCATION INITIATIVES

2.19.1   CANADA

2.19.1.1   Canadian Network for the Advancement of Research, Industry, and Education (CANARIE)

The Canadian Network for the Advancement of Research, Industry, and Education (CANARIE) promotes design and deployment of high-speed networking technologies and applications throughout Canada. In addition, CANARIE sponsors national network initiatives, including the National Test Network (NTN) — also called the original CA*net (Canadian Network for the Advancement of Research, Industry, and Education), and its successors CA*net II (CA*net, Phase 2) and CA*net3 (CA*net, Phase 3).

2.19.1.2   National Test Network (NTN)

Established in 1990 with the support of the National Research Council, the National Test Network (NTN) was distinguished by its early use of ATM technology. The NTN interconnected regional ATM test networks across Canada from St. John's, Newfoundland, to Vancouver, British Columbia. The NTN also interoperated with peer-level networks in the United States and the European Union. ATM field tests and pilot implementations were conducted by an alliance of NTN research and academic institutions.

Results contributed to the development of tele-education courses, telecollaborative videoconferences, digitized music applications, telemedicine services, and multipoint delivery of VRML (Virtual Reality Modeling Language) applications. The NTN also established a foundation for CA*net II (Canadian Network for the Advancement of Research, Industry, and Education, Phase 2), NTN backbone operations were terminated in 1997. At that time, NTN ATM connections were ported directly to CA*net II.

2.19.1.3   CA*net II (Canadian Network for the Advancement of Research, Industry, and Education, Phase 2)

The Canadian Network for the Advancement of Research, Industry, and Education, Phase 2 (CA*net II) is a high-speed network that initially functioned as a separate network apart from the public Canadian Internet. Developed by CANARIE and Bell Advanced Communications (BAC), the CA*net II infrastructure features an ATM-over-SONET backbone and an IP-over-ATM overlay network to facilitate concurrent voice, video, and data transmissions with differentiated QoS guarantees. In addition, the CA*net II infrastructure supports IPv6 multicasts, VPN implementations, utilization of 3-D workspaces, and delivery of real-time audio and video broadcasts.

As with Internet2, CA*net II promotes development of next-generation applications and multimedia services enabling tele-education programs, virtual classrooms, online learning environments, and virtual learning communities. In parallel with I2, CA*net II also fosters advanced telecollaborative research projects in disciplines that include science, biology, mathematics, zoology, astronomy, and high-energy physics. In contrast to NTN, CA*net II is an advanced academic research network that does not support links to the public or commodity Internet.

Participants in CA*net II include Canadian universities, scientific organizations, research centers, and provincial agencies. Community colleges, regional network consortia, small- and medium-sized enterprises that represent the IT (Information Technology) sector, corporations, and manufacturers also take part in the CA*net II initiative. Abilene, Internet2, and vBNS+ and international NRENs connect to CA*net II via transit points such as STAR TAP NAP in Chicago.

2.19.1.4   CA*net II RANs (Regional Advanced Networks) and GigaPoPs

To promote infrastructure development and implementation of high-performance applications, CA*net II sponsors Regional Advanced Networks (RANs) in every province in Canada. RANs route high-speed multimedia traffic at the provincial level, enable interconnectivity to other peer-level networks, and support high-speed connections to the CA*net II infrastructure. BCnet (British Columbia Network), ONet (Ontario Network), and RISQ (Quebec Scientific Internet Network) are examples of RAN implementations.

Canadian GigaPoPs are regional aggregation points or regional hubs that interlink educational institutions and research centers to CA*net II. RANs employ Coarse Wavelength Division Multiplexing (CWDM) technology. In comparison to WDM (Wavelength Division Multiplexing) and DWDM (Dense WDM), CWDM is a more affordable optical solution. However, CWDM is limited in supporting sophisticated optical functions and services.

2.19.1.5   London and Region Global Network (LARG*net)

Affiliated with ONet (Ontario Network), LARG*net (London and Region Global Network) is a regional area ATM network that supports distance learning, telemedicine, and videoconferencing applications. LARG*net participants include Fanshawe College, the University of Western Ontario, the Thames Valley District School Board, and the Thames Valley District Health Council.

2.19.1.6   Ottawa-Carleton Research Institute Network (OCRInet)

The Ottawa-Carleton Research Institute Network (OCRInet) is a regional area ATM network that interlinks Carleton University, Algonquin College, government agencies, local industries, and research libraries. Operational since 1994, this RAN facilitates delivery of interactive entertainment to residences on-demand and transmission of distance education courses to students in remote communities.

2.19.1.7   WURCnet (Western University Research Consortium Network)

The Western University Research Consortium Network (WURCnet) sponsors a high-performance ATM RAN called Wnet II that enables connectivity to regional and local computing centers and major university networks. Wnet II promotes delivery of IP multicasts via the WURCnet MBone implementation and provides access to advanced applications in telemedicine, tele-education, and the arts.

2.19.2   GERMANY

2.19.2.1   Research Institute for Open Communications Systems

The Department for Broadband Networks at the Research Institute for Open Communications Systems in Berlin tests the capabilities of the MobilAT (Mobile ATM) platform for providing dependable and reliable access to voice, video, and data applications and activities in mobile computing environments. MobilAT employs ATM switching to support the seamless integration of in-room, in-building, campus, metropolitan area, and regional area networks.

2.19.3 KOREA

2.19.3.1 Chonbuk National University

Chonbuk National University utilizes an ATM backbone network that supports rates at 622.08 Mbps (OC-12) for providing VOD (video-on-demand) services, Internet connectivity, and access to electronic library resources and E-learning applications. The Chonbuk National University ATM network also supports delivery of telecourses in law, the fine arts, and veterinary medicine, and provisions networking services for businesses and the national police agency.

2.19.4 UNITED KINGDOM

2.19.4.1 SuperJANET4 (Super JOINT ACADEMIC NETWORK, PHASE 4)

The SuperJANET4 ATM platform enables rates of transmission ranging from 155.52 Mbps (OC-3) to 2.488 Gbps (OC-48) and provisions QoS (Quality of Service) guarantees for bandwidth-intensive voice, video, and data services. This ATM platform interoperates with SMDS, IP, and SDH (Synchronous Digital Hierarchy) technologies. Moreover, SuperJANET4 supports pilot projects to substantiate the performance of ATM-over-WDM (Wavelength Division Multiplexing) and ATM-over-DWDM (Dense WDM) networks.

The SuperJANET4 platform enables IPv6 utilization, managed bandwidth services (MBS), VPN deployments, bandwidth allocations for bulk files transfers, streaming media, IP multicasts, IP telephony, scientific simulations, and tele-immersive applications. In addition, the SuperJANET4 platform provisions connectivity to digital libraries and digital film archives, and supports access to real-time video instruction, interactive vocational and lifelong distance education programs, asynchronous independent study telecourses, and IP videoconferencing. SuperJANET4 interoperates with NRENs throughout the European Union and maintains connections with next-generation initiatives such as Internet2, vBNS+, the Abilene Network, the National Grid for Learning (NGFL), CA*net II, and CA*net3.

2.19.4.1.1 SuperJANET4 Foundations

SuperJANET refers to the broadband or high-speed part of JANET (Joint Academic Network). The acronym JANET came into use in 1989. At that time, JANET supported EuroISDN videoconferences and data delivery services over an optical fiber infrastructure. SuperJANET1 (SuperJANET, Phase1) transformed JANET into a high-speed, high-performance broadband communications multiservice network capable of concurrently transmitting voice, video, and data. In addition, SuperJANET1 initiated the migration of academic and research networks from an SMDS platform to an ATM infrastructure.

In 1999, SuperJANET3 (SuperJANET, Phase 3), the successor to SuperJANET2 (SuperJANET, Phase 2), provisioned ATM multimedia services at rates reaching 155.52 Mbps (OC-3). A feasibility study conducted by SuperJANET3 participants established requirements for SuperJANET4.

2.19.4.1.2   SuperJANET4 Architecture

SuperJANET4 consists of Core Points of Presence (C-PoPs) or switching centers that perform routing functions. Located in Leeds, Bristol, Manchester, and London, C-PoPs employ fiber-optic cabling for high-speed broadband wireline transmissions and Backbone Edge Nodes (BENs) to extend SuperJANET4 services in England, Northern Ireland, Wales, and Scotland. C-PoPs support information delivery to and from BENs at 34.368 Mbps (E-3) and 155.52 Mbps (OC-3) rates.

BENs enable operations between SuperJANET4 and regional networks or MANs. BENs are typically situated at JCPs (JANET Connection Points) or network nodes where MANs or regional networks are linked to the SuperJANET4 backbone. Each regional network or MAN participating in SuperJANET4 enables information delivery to and from educational and research institutions in its domain. For example, the London MAN facilitates ATM multimedia transmissions between the University of London Computer Center and Imperial College London. The NorMAN (North East England MAN) employs a mixed-mode ATM wireless and wireline network platform to support communications services between Newcastle and Northumbria Universities and the Universities of Teesdale and Durham.

By 2003, SuperJANET4 will routinely enable transmissions at rates ranging from 155.52 Mbps (OC-3) to 2.488 Gbps (OC-48) between C-PoPs and BENs. Voice, video, and data transport at rates ranging from 2.488 Gbps to 80 Gbps between C-PoPs and BENs will be available via an optical network platform based on WDM and DWDM technologies by 2005.

2.19.4.1.3   SuperJANET4 in Action

The SuperJANET4 infrastructure provisions access to a diverse array of tele-education, teleradiology, telesurgery, and teleresearch projects and multipoint videoconferences. SuperJANET4 participants include the Universities of Glasgow, Manchester, Leeds, Newcastle, and Edinburgh. Imperial College, Trinity College Dublin, the Universities of Westminster and Wales, and the School of Slavonic and East European Studies participate in SuperJANET4 as well.

A SuperJANET4 participant, University College London (UCL) utilizes the SuperJANET4 ATM platform in combination with legacy configurations to support in-service teletraining programs for teachers and teleclasses for students who are disadvantaged in terms of location, distance, or disability. Also a SuperJANET4 participant, the School of Education at the University of Exeter uses the ATM infrastructure for enabling access to telecourses in English, mathematics, foreign languages, science, and art. In addition, Manchester University takes part in SuperJANET4 and employs the SuperJANET4 infrastructure for providing access to real-time audio and video presentations and broadcasts of theatrical performances.

The United Kingdom Education and Research Networking Association (UKERNA) manages the SuperJANET4 initiative. The SuperJANET4 Computer Emergency Response Team (CERT) distributes bulletins on security risks and implements solutions for safeguarding SuperJANET4 operations.

2.20 U.S. TELEMEDICINE INITIATIVES

Telemedicine networks support a broad range of configurations for interlinking such sites as hospitals, healthcare clinics, medical offices, and nursing homes. These networks also enable healthcare services in a patient's home in the event of budget cuts and hospital closures and provision access to diverse treatment options for patients at distant locations. Representative telemedicine initiatives in the ATM domain are highlighted in this section.

2.20.1 ALABAMA

2.20.1.1 University of Alabama (UAB)

An Internet2 participant, the University of Alabama (UAB) sponsors collaborative research, tele-instruction, and telemedicine services via an ATM infrastructure. The UAB ATM platform enables videoconferencing between UAB medical professionals and their colleagues at Stanford, Harvard, and Cornell Universities. Moreover, this infrastructure supports genetic telecounseling sessions between patients with hereditary cancers, their primary care physicians, and UAB medical specialists. The ATM platform enables students in the UAB nursing program to access medical images and multimedia medical resources. Approaches for developing and delivering distance education courses in music, history, macromolecular modeling, and anthropology, as well as tele-education programs for optometrists and public health professionals, are under consideration.

2.20.2 CALIFORNIA

2.20.2.1 Lawrence Berkeley Laboratory at the University of California at Berkeley and Kaiser Permanente Division of Research

A Health Maintenance Organization (HMO), the Kaiser Permanente Division of Research implements an IP-over-ATM NII (National Information Infrastructure) initiative in conjunction with the Lawrence Berkeley Laboratory (LBL) at the University of California at Berkeley (UC Berkeley). This broadband network initiative enables real-time multimedia transport and utilization of online tools for remote visualization. In addition, the IP-over-ATM platform supports transmission of x-rays, CAT (Computerized Axial Tomography), and MRI (Magnetic Resonance Imaging) scans, and video sequences of coronary angiograms from primary care physicians at remote healthcare centers to medical specialists at urban hospitals enabling teleconsultations and telediagnoses. Transmission rates at 155.52 Mbps (OC-3) are supported.

2.20.2.2 University of Southern California (USC)

The University of Southern California (USC) fosters implementation of a multiservice ATM testbed to support the provision of telehealthcare services by USC medical specialists to patients at remote locations. The USC Advanced BioTelecommunications and BioInformatics Center employs the ATM infrastructure for remote radiological teleconsultations, retinal image transmissions for diabetes screening, and staff videoconferences. In addition, this infrastructure provisions access to digital patient records and fosters connectivity to supercomputers that generate treatment plans and pharmacological guidelines. Transmission rates at 155.52 Mbps (OC-3) are supported.

2.20.3 OHIO

2.20.3.1 Ohio Supercomputer Center (OSC)

The Ohio Supercomputer Center (OSC) supports biomedical applications and VR (Virtual Reality) initiatives that integrate visual, speech, and haptic interfaces for surgical preplanning sessions and physician training via a high-speed, high-performance ATM infrastructure. This ATM platform also enables virtual simulations of temporal bone dissections and cranial tumors.

2.20.4 VIRGINIA

2.20.4.1 Southwest Virginia Alliance for Telemedicine

The Southwest Virginia Alliance for Telemedicine utilizes an ATM configuration for interlinking the University of Virginia (UVA) Office of Telemedicine, the Lee County and Norton Community Hospitals, the Thompson Family Health Center, and the Stone Mountain Health Services Clinic. This Alliance provisions telehealthcare services to patients at rural locations who are unable to travel to metropolitan medical facilities for treatment by medical specialists. Net.Work.Virginia provides technical support and manages network operations.

2.21 INTERNATIONAL TELEMEDICINE INITIATIVES

2.21.1 CANADA

2.21.1.1 Manitoba Telemedicine Research and Development Pilot Project

The Manitoba Telemedicine Research and Development Pilot Project supports utilization of an ATM-over-SONET infrastructure that works in conjunction with satellite technology for provisioning healthcare services. Sponsored by the University of Manitoba, this broadband initiative provides access to the Internet and videoconferencing services and supports delivery of continuing education courses to medical professionals in the Winnipeg communities of Norway House, Thompson, and Churchill. In addition, this platform enables nursing students to participate in a distance tele-education undergraduate program in nursing. The satellite network component supports information transport at 2.048 Mbps (E-1) rates via Ku-band frequencies. Earth stations are deployed at Norway House and in Ottawa where satellite operations are monitored.

2.21.1.2   Rnet (Research Network) of British Columbia

The Research Networking Association of British Columbia supports development of high-performance telecommunications and networking environments in fields that include distance learning, multimedia authoring systems, telemedicine research, and clinical practice. In addition, this association implements an advanced ATM testbed called Rnet (Research Network).

Rnet supports biomedical imaging, teleradiology, and teleconsultations between medical specialists and patients and their primary care physicians. In addition, Rnet enables videoconferencing and collaborative computing at sites in Montreal, Vancouver, Calgary, Toronto, and St. Johns. Rnet also facilitates access to Healthnet, a provincial information service that features an online organ donor registry and electronic pharmaceutical data. Rnet participants include the Universities of British Columbia and Victoria.

2.21.2   CHINA

2.21.2.1   Zhongshan University of Medical Science

An ATM installation at the Zhongshan University of Medical Science provisions delivery of tele-education courses to healthcare professionals and enables medical specialists to participate in videoconferences with patients and their primary healthcare providers at hospitals, medical centers, and medical schools throughout the region.

2.21.3   UNITED KINGDOM

2.21.3.1   University College London (UCL)

The University College London (UCL) sponsors a telecollaborative teaching project called INSURRECT (Interactive Surgical Teaching Between Remote Centers). This initiative supports undergraduate telesurgery instruction at six medical schools via the ATM network component of SuperJANET4. Because SuperJANET4 is a multi-point-to-multipoint network, each medical school is accorded the same status in terms of delivering and receiving lectures. In addition, the UCL sponsors implementation of the ESTVIN (European Surgical Teaching Using Video Interactive Networks) project via the ATM platform for transborder telesurgery tele-instruction.

2.22   E-GOVERNMENT INITIATIVES

2.22.1   COLORADO

2.22.1.1   City of Denver

The City of Denver deploys an ATM MAN that enables connections between government agencies, the municipal convention center, and the municipal airport. This network provisions links to an E-government intranet and supports videoconferencing between prisoners in jail and their attorneys at the courthouse. In addition, the City of Denver ATM MAN enables municipal residents to access real estate data, file permits, and pay fines for parking tickets online. Fiber-optic cabling throughout the state initially used for a statewide traffic signaling project serves as the foundation for the City of Denver ATM MAN configuration.

2.22.2 KENTUCKY

2.22.2.1 Fort Knox Municipal ATM Network

The Fort Knox Municipal ATM Network supports E-government services, telemedicine applications, remote surveillance, and teletraining sessions. Firefighters, police officers, and paramedics access this network to determine locations of fires, vehicular accidents, and healthcare emergencies. Rates at 622.08 Mbps (OC-12) are supported.

2.23   EUROPEAN COMMISSION TELEMATICS APPLICATIONS PROGRAM (EC-TAP) TELE-EDUCATION INITIATIVES

Sponsored by the European Commission (EC), the Telematics Applications Program (TAP) fostered implementation of state-of-the-art communications technologies. EC TAP initiatives also supported Web applications, development of multimedia toolkits, and telecollaborative activities. Representative EC-TAP projects are highlighted in this section.

2.23.1   ATM AND TELECONFERENCING FOR RESEARCH AND EDUCATION (ATRE)

The ATRE Program developed an IP-over-ATM platform for enabling telemeetings, teleconferencing, real-time broadcasts, and teleseminars. In addition, teleconsultations, teleresearch, teleteaching, and teleworking were supported. Designed for professionals in the Earth observation and nuclear physics communities, the ATRE platform provisioned IP multicast services and enabled point-to-multipoint and multipoint-to-multipoint videoconferences. Developed as part of the ATRE Program, an MBone (Multicast Backbone) toolset facilitated high-speed broadband applications and worked in concert with IPv6 services.

2.23.2   COLLABORATIVE BROWSING IN INFORMATION RESOURCES (CoBROW) AND COLLABORATIVE BROWSING IN THE WORLDWIDE WEB/DEPLOYMENT OF THE SERVICE (CoBROW/D)

The CoBROW initiative supported design of a real-time multimedia communications toolset called JVTOS (Joint Viewing and Tele-operation Services) for Web scientific research. CoBROW enabled utilization of IPv6 applications, intranets, and Web resources in conjunction with an IP-over-ATM platform. CoBROW/D validated capabilities of the CoBROW JVTOS toolset in implementations supported by an IP-over-ATM platform. As with CoBROW, the CoBROW/D platform enabled IPv6 services, IP multicasts, and real-time videoconferencing.

2.24 EUROPEAN COMMISSION (EC) TEN (TRANS-EUROPEAN NETWORK) TELEMEDICINE INITIATIVES

2.24.1 HAND ASSESSMENT AND TREATMENT SYSTEM (HATS)

Supported by an ATM infrastructure, the HATS project fostered development of advanced data acquisition assessment tools for use by hand therapists. By standardizing hand assessment and treatment protocols, the HATS initiative significantly improved the care and treatment of patients with hand injuries.

2.24.2 PATIENT WORKFLOW MANAGEMENT SYSTEMS (PATMAN)

The PATMAN initiative facilitated utilization of a standardized workflow system for enabling reliable and dependable access to healthcare resources via an ATM infrastructure. The PATMAN project also fostered telecollaboration among healthcare providers and teleconsultations between physicians and patients to enhance clinical treatment.

2.25 EUROPEAN COMMISSION ADVANCED COMMUNICATIONS TECHNOLOGY AND SERVICES (EC-ACTS) PROGRAM

Demand for a reliable high-speed ATM infrastructure to provide abundant support for bandwidth-intensive multimedia services contributed to the development of the European Commission Advanced Communications Technology and Services (EC-ACTS) initiatives in the ATM domain. As noted, from 1994 to 1998, the EC-ACTS Program supported development and implementation of advanced scalable, reliable, and dependable broadband ATM and WATM (Wireless ATM) networking services and applications. EC-ACTS projects confirmed the capabilities of ATM technology in facilitating high-speed, high-performance applications and services in sectors that included tele-education, telemedicine, E-government, and E-business.

2.25.1 A PLATFORM FOR ENGINEERING RESEARCH AND TRIALS (EXPERT)

The EXPERT project used an ATM backbone network for voice, video, and data transmission. This initiative demonstrated the capabilities of an ATM infrastructure in interworking with cable modem, DSL, Frame Relay, ISDN, and SDH technologies in SOHO environments and supporting QoS guarantees, video retrieval, multimedia videoconferencing, video-on-demand (VOD), and high-quality audio. Moreover, the EXPERT project also validated the performance of ATM technology in economically delivering broadband services in FTTH (Fiber-To-The-Home) and FTTC (Fiber-To-The-Curb) configurations.

2.25.2 INTERNET AND ATM: EXPERIMENTS AND ENHANCEMENTS FOR CONVERGENCE AND INTEGRATION (ITHACI)

The ITHACI project demonstrated capabilities of an IP-over-ATM configuration in supporting voice-over-IP (VoIP) or IP telephony, IP multicasts, and video distribution in trials conducted in Belgium, Greece, and Germany. Project outcomes contributed to development of a trans-European ATM WAN.

2.25.3 VIRTUAL MUSEUM INTERNATIONAL (VISEUM)

The VISEUM initiative demonstrated capabilities of an IP-over-ATM network in supporting a virtual museum, cross-cultural art exchanges, and real-time delivery of high-resolution images of famous North American and European artworks from distributed network servers. In addition, the VISEUM project also provided a foundation for implementation of transborder electronic commerce (E-commerce) services. A merged network consisting of CA*net II, SuperJANET4, and CanTat 3 (an undersea fiber-optic network that spanned the Atlantic Ocean) enabled ATM backbone connections between London (England) and Vancouver (Canada).

2.26 ATM IMPLEMENTATION CONSIDERATIONS

In the academic arena, ATM technology facilitates fast, reliable, and dependable access to an expanding array of Web initiatives and institutional resources. ATM enables tele-education, telementoring, and real-time interactions with subject experts in remote locations; multimedia applications; and curricular enhancement and enrichment. ATM also promotes deployment of virtual schools, virtual universities, virtual museums, and virtual communities.

ATM pilot trials and initiatives support the design and implementation of extendible, reliable, and scalable ATM configurations to accommodate current and anticipated network requirements. In addition, the ATM platform delivers high-capacity, high-speed multimedia services and applications. However, it is also important to note that major regulatory, technical, logistical, and economic issues associated with ATM deployment remain unresolved. As a consequence, the ATM acronym also stands for "All That Money."

ATM is an evolving technology. As a consequence, standards and testing methods are still in development. Congestion on ATM networks can lead to cell loss before traditional network tools detect problems. Problems associated with providing effective traffic management, seamless network performance, and network-level security for information integrity and high-speed interactive data, video, and voice delivery must be resolved through further research. ATM functions are also constrained by the lack of cross-vendor support.

Migration to an ATM solution typically requires acquisition of ATM products and services from a single vendor. The majority of ATM switches in use by early adopters of ATM technology are expected to be incompatible with next-generation ATM switches. As a result, replacement of expensive in-place ATM switches with costly next-generation ATM switches appears to be necessary for enabling ATM services.

Successful ATM deployment requires the use of carefully executed measures to manage traffic flows and accommodate application requirements. Inasmuch as ATM support of multiple QoS parameters contributes to difficulties in managing ATM configurations, development and implementation of network management policies are indispensable in facilitating realization of the full potential of ATM technology.

An understanding of ATM technical capabilities is essential in order to effectively address pedagogical challenges associated with ATM implementation. Although ATM supports multifaceted options for information delivery to the desktop, SOHO venues, and local and wider area environments, deployment of ATM technology does not automatically guarantee its effective utilization in the educational domain.

In implementing ATM applications and services in school and university environments, the capabilities of the proposed infrastructure must be determined. Requirements for a high-performance ATM infrastructure that is modular, reliable, secure, expandable, and available to accommodate bandwidth demands over time must be clarified. Effective ATM implementation in the tele-education milieu also involves developing ATM tele-learning paradigms for supporting problem-solving skills and accomplishment of learning goals and objectives. Effective ATM deployment in the telelearning environment ultimately depends on its ability to foster knowledge-building competencies and exploratory learning, quality education, and focused research and facilitate instructional innovation and creativity. Future research involving ATM deployment in school and university settings must also focus on the practical design and deployment of pedagogical strategies and collaborative instructional activities for optimizing student skills in broadband tele-education environments.

In the broadband networking arena, ATM's major competitor is Gigabit Ethernet technology. Gigabit Ethernet technology is compatible with the installed base of Ethernet and Fast Ethernet solutions in local area and wider area network environments. In comparison to ATM, Gigabit Ethernet does not provision information transport with QoS guarantees. However, Gigabit Ethernet leverages capabilities of newer technologies and protocols such as the Resource Reservation Protocol (RSVP) and the MultiProtocol Link Aggregation (MPLA) protocol to support scalable bandwidth, fault tolerance, network resiliency, and streamlined packet transmission for provisioning higher-level networking services. In addition, Gigabit Ethernet implementations are more affordable and easier to implement than complex ATM solutions.

2.27 SUMMARY

There is a growing consensus that ATM reliably and dependably accommodates requirements for high-speed, high-performance networking operations while also enabling a seamless migration path to the network of the future. Increasing numbers of ATM field trials and full-scale implementations demonstrate ATM capabilities in providing access to worldwide learning resources and supporting innovative tele-learning activities and applications.

This chapter describes ATM technical fundamentals and capabilities. Distinctive attributes of major national and international ATM initiatives and research efforts that contribute to establishing a global ATM infrastructure are examined. ATM systems featuring a mix of wireline and wireless technologies for enabling transborder interdisciplinary research and global connectivity to innovative instructional programs are explored.

ATM technology is uniquely suited for supporting error-free multimedia transport in high-speed network configurations. Moreover, ATM is an enabler of network traffic consolidation, thereby streamlining network management operations and optimizing utilization of high-speed network connections. In addition, ATM provisions networking services via twisted copper pair, optical fiber, and hybrid optical fiber and coaxial cable (HFC) wireline media and wireless technical solutions. National and international standards organizations such as the ITU-T, the Institute of Electrical and Electronic Engineers (IEEE), the American National Standards Institute (ANSI), and the European Telecommunications Standards Institute (ETSI) endorse ATM specifications.

ATM solutions are designed to function in multiservice, multivendor environments. However, debate persists about the suitability of ATM technology in accommodating mission, goals, and requirements economically and effectively in the academic arena. Potential barriers to ATM deployment include high costs, lack of universally accepted standards, restricted geographical availability, equipment incompatibilities, and insufficient research data on the capabilities of ATM in provisioning Quality of Service (QoS) guarantees.

Despite these constraints, ATM is regarded as a key enabler for tele-education, telebusiness, E-government, and telemedicine applications. ATM provisions dependable Internet, intranet, and extranet connectivity; facilitates implementation of Virtual Reality (VR) applications; and supports reliable access to broadband multimedia services.

ATM networks resolve problems associated with internetwork congestion and enable seamless voice, video, and data transmission over wireless, wireline, and hybrid wireline and wireless network configurations. In the distance education domain, ATM enables access to new student populations in remote locations, promotes transborder research and telecollaboration, and facilitates curricular enrichment.

Globally, ATM technology supports development and deployment of major research and education networks such as Abilene, vBNS+, Internet2, ESnet, CA*net II, and SuperJANET4. Moreover, ATM promotes incorporation of emergent network architectures, protocols, and transmission technologies into an integrated infrastructure. Continued research on the design and implementation of pedagogical approaches and methods for supporting student learning and achievement in ATM instructional settings is essential for achieving effective ATM implementation in school and university environments.

2.28 SELECTED WEB SITES

ATM Forum. Home Page.
    Available: http://www.atmforum.com/
Canadian Network for the Advancement of Research, Industry, and Education
    (CANARIE). About CANARIE. Last modified on June 26, 2001.
    Available: http://www.canarie.ca/about/about.html
Corporation for Education Network Initiatives in California (CENIC).
    CalREN-2.
    Available: http://www.cenic.org/CR2.html

Delivery of Advanced Network Technology to Europe, Ltd. (DANTÉ). TEN-34: The Information Superhighway for European R&D (Research and Development). Last modified on October 31, 2000.
Available: http://www.dante.net/ten-34.html

European Commission. The ACTS Information Window.
Available: http://www.de.infowin.org/

European Organization for Nuclear Research (CERN). Home Page. Last modified on May 22, 2001.
Available: http://welcome.cern.ch/welcome/gateway.html

IPv6 Forum. IPv6 Reports, Articles, and Papers.
Available: http://www.ipv6forum.com/

IPv6 Information Page. Networking for the 21st Century. Last modified on August 10, 2001.
Available: http://www.ipv6.org/

Lancaster University Computing Department. The U.K. IPv6 Resource Center.
Available: http://www.cs-ipv6.lancs.ac.uk/ipv6/

Mid-Atlantic Crossroads (MAX). Home Page. Last modified on January 22, 1999.
Available: http://www.networkvirginia.net/MAX/

National Aeronautics and Space Administration (NASA). Home Page.
Available: http://spaceresearch.nasa.gov/

Net.Work.Virginia. Home Page.
Available: http://www.networkvirginia.net/

New York State Education and Research Network (NYSERNet). Home Page.
Available: http://www.nysernet.org/network.html

Ohio Academic Research Network (OARnet). Home Page.
Available: http://www.oar.net/

Science, Technology, and Research Transit Access Point (STAR TAP). Home Page.
Available: http://www.startap.net/

Super Joint Academic Network, Phase 4 (SuperJANET4). Home Page.
Available: http://www.superjanet4.net/

University Corporation for Advanced Internet Development (UCAID). About Internet2.
Available: http://www.internet2.edu/html/about.html

University of Kansas. MAGIC Network. Home Page.
Available: http://www.ukans.magic.net/KU

University of Pennsylvania. Penn Computing: University of Pennsylvania Internet2 Project. Last modified on August 27, 2001.
Available: http://www.upenn.edu/computing/i2/

very high performance Backbone Network Service Plus (vBNS+). Home Page.
Available: http://www.vbns.net/

3 Optical Network Solutions

3.1 INTRODUCTION

Extraordinary demand for high-speed, high-performance networks with vast transmission capacities and potentially unlimited bandwidth contributes to the popularity of SONET/SDH (Synchronous Optical Network and Synchronous Digital Hierarchy) solutions. SONET/SDH are network transport technologies that use synchronous operations for facilitating real-time voice, video, and data transmission via fiber optic cabling at rates ranging from 51.84 Mbps (OC-1 or Optical Carrier-Level 1) to 13.21 Gbps (OC-255). The need for potentially unlimited bandwidth also fosters development and deployment of next-generation WDM (Wavelength Division Multiplexing) and DWDM (Dense WDM) optical network implementations in order to support bandwidth-intensive voice, video, and data transmissions at multigigabit and multiterabit rates.

In the 1980s, standardized rates and formats for optical transmissions in SONET/SDH networks were established. SONET/SDH installations replaced proprietary optical network implementations. Deployed by Regional Bell Operating Companies (RBOCs) and Interexchange Carriers (IXCs), proprietary optical network solutions were incapable of supporting internetworking services.

Present-day SONET/SDH deployments enable high-performance networking operations and interwork with diverse broadband technologies. ATM (Asynchronous Transfer Mode) and SONET/SDH are complementary technologies. In typical broadband installations, ATM functions in a switching and multiplexing capacity and SONET/SDH serve as the underlying Physical Layer transport technology.

ATM-over-SONET/SDH implementations support a rich array of tele-education and teleresearch initiatives. Despite their capabilities, SONET/SDH solutions exploit only a small fraction of the available capacity of the existing fiber optic plant. This operational constraint contributes to the development of next-generation WDM and DWDM optical networks.

WDM and DWDM systems employ multiple wavelengths for enabling transparent networking services that are scalable, extendible, modular, and available. As with SONET/SDH, WDM and DWDM implementations are flexible, dependable, and reliable; accommodate increased bandwidth requirements; and work directly as the support infrastructure for broadband communications technologies. In comparison to SONET/SDH solutions, WDM and DWDM technologies enable considerably faster speeds and vastly increased bandwidth and network capacity via the in-place fiber optic plant for transporting video, voice, and data traffic. (See Figure 3.1.)

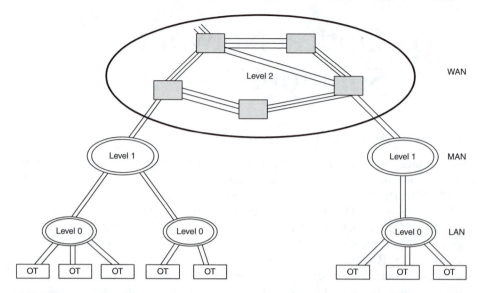

FIGURE 3.1 A WDM (Wavelength Division Multiplexing) implementation. Optical transports (OT) connect to three levels of WDM service, specifically, Level 0 (LAN), Level 1 (MAN), and Level 2 (WAN).

3.2   PURPOSE

In this chapter, optical networking capabilities, merits, constraints, and innovations are investigated. SONET/SDH attributes, configurations, and deployments are examined. Trends and advances in the utilization of ATM-over-SONET networks that support delivery of high-speed, high-performance multimedia applications are introduced. Recent innovations in Packet-over-SONET/SDH (POS) implementations are explored. Distinctive features of WDM and DWDM architectures, protocols, and configurations are described. WDM and DWDM capabilities in supporting sophisticated optical network configurations and a diverse range of applications and teleservices are reviewed. Innovations in undersea WDM and DWDM implementations are noted.

3.3   SONET/SDH FOUNDATIONS

In contrast to conventional electronic networks, SONET/SDH networks enable fast transmission rates for transport of multimedia teleservices by using a combination of optical-electronic solutions. The use of fiber optics for data transmission became popular in the 1970s when Illinois Bell, the predecessor to Ameritech, employed this medium for implementing network applications in downtown Chicago. In 1985, Bellcore proposed the SONET framework to promote interconnectivity between proprietary fiber optic systems.

The American National Standards Institute (ANSI) adopted the first of numerous SONET specifications in 1988. The ITU (International Telecommunications Union) endorsed the initial SDH specification in 1989.

Optical fiber installations provision support for current and emergent applications and services, require very little maintenance, and enable multimedia transport over extended distances. Optical transmitters such as lasers and LEDs (Light Emitting Diodes) transport photons or light pulses to optical receivers via multimode and single-mode optical fiber cabling consisting of ultra-thin strands of glass or plastic. Multimode fiber optic cabling solutions enable short-haul transmissions and carry multiple colors of light or wavelengths via a 62.5-micron fiber optic core. Thinner in structure than multimode optical fiber, single-mode optical fiber consists of an 8.3-micron fiber optic core and supports long-haul transmissions over a single light path.

Regardless of the size of the fiber optic core and distances over which transmissions travel, optical fiber configurations are robust and scalable. Distinguished by their flexibility and versatility, optical networks support the increasing mix of traffic generated by the commodity or public Internet and enable extremely fast transmission rates for accommodating the requirements of next-generation networking applications and services.

3.4   SONET/SDH TECHNICAL FUNDAMENTALS

SONET/SDH technologies accommodate bandwidth requirements for local and long-distance networks transporting bandwidth-intensive voice, video, and data applications. Differences between the SONET and SDH standards are very minor and occur only at the SDH STM-1 (Synchronous Transport Module-Level 1) sublayer. As a consequence, in the United States, SONET is also used as an umbrella term for both technologies. A distinction between SONET/SDH is typically made in referring to specific SONET and SDH communications rates and initiatives.

3.4.1   SONET STS (SYNCHRONOUS TRANSPORT SIGNALS) AND OC (OPTICAL CARRIER) LEVELS

SONET solutions support dependable and fast transmission of optical signals and enable internetworking operations among networks established by diverse communications carriers. SONET establishes interface specifications at the Physical Layer or Layer 1 of the OSI (Open Systems Interconnection) Reference Model.

SONET fosters information transport by defining optical carrier (OC) levels and electronically equivalent signal levels that are called Synchronous Transport Signal (STS) levels. STS-1 (STS-Level 1) features a base rate of 51.84 Mbps. As with STS-1, OC-1 (Optical Carrier-Level 1) is equivalent to 51.84 Mbps. OC-1 and STS-1 serve as the foundation from which higher signals in standard increments of 51.84 Mbps are derived. For example, levels at OC-3 and STS-3 (OC-3/STS-3) enable rates at 155.52 Mbps and OC-12/STS-12 at 622.08 Mbps. In addition, OC-48/STS-48 foster transmission at 2.488 Gbps and OC-192/STS-192 at 10 Gbps. At present, SONET/SDH support an optimum rate reaching OC-255/STS-255 or 13.21 Gbps. SONET/SDH operations are remarkably fast. Because STS and OC levels are equivalent, SONET/SDH rates are typically described in terms of Optical Carrier (OC) levels.

3.4.2 SYNCHRONOUS TRANSPORT MODULES (STM) AND OPTICAL CARRIER (OC) LEVELS

In contrast to SONET, SDH defines transmission rates in terms of Synchronous Transport Module (STM) levels. SDH transmissions at 155.52 Mbps are equivalent to STM-1 (STM-Level 1); SDH transmissions at 622.08 Mbps are equivalent to STM-4; and SDH transmissions at 2.488 Gbps are equivalent to STM-16. Furthermore, SDH transmissions at 10 Gbps are equivalent to STM-48, and SDH transmissions at 13.21 Gbps are equivalent to STM-64. SDH delineates STM building blocks in increments of 155.52 Mbps. SONET defines OC levels that describe transmission rates in increments of 51.84 Mbps.

In terms of comparing STM and OC levels, STM-1 and OC-3, STM-4 and OC-12, and STM-16 and OC-48 are optically equivalent and virtually identical. As an example, STM-1 and OC-3 indicate transmission rates at 155.52 Mbps. Optical multiplexers or digital cross-connects with multiplexing capabilities transform SDH STM frames into SONET OC formats for enabling STM voice, video, and data signals to transit a SONET network.

3.4.3 SONET/SDH FRAMES

SONET/SDH signals are transmitted in frames. Each SONET/SDH frame consists of 810 bytes and features a payload and an overhead. The frame payload envelope or service component contains user data and the SPE (Synchronous Payload Envelope) for ensuring dependable information transport.

The overhead component enables multiplexing functions, detects faults and errors resulting from jitter, and discovers time delays before these problems seriously degrade network performance. This component also provisions integrated support for network operations. SONET/SDH pointers located in the overhead component of each frame ensure seamless integration of SONET/SDH frames with Synchronous Payload Envelopes and minimize network jitter and time delays that adversely impact network throughput.

3.4.4 SONET/SDH OPERATIONS

In SONET/SDH configurations, information transmission via a fiber optic infrastructure is achieved using laser-generated light streams for carrying voice, video, and data traffic. SONET/SDH multiplexers transform electrical signals into optical signals or photons that then transit the network. In addition, optical transmitters, signal regenerators, and broadband digital cross-connects optimize bandwidth availability and transmission capacities of the fiber optic plant and route SONET/SDH frames to destination addresses.

By working in concert with the underlying fiber optic plant, SONET/SDH solutions feature low-bit error rates and immunity to noise, crosstalk, and other types of electromagnetic interference.

Traffic grooming on optical networks involves switching and consolidating traffic streams into faster streams. SONET/SDH configurations accomplish traffic grooming functions at 155.52 Mbps (OC-3) and higher rates for enabling dynamic

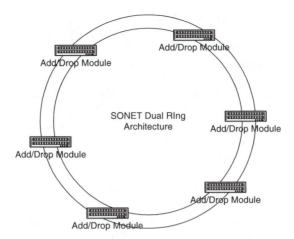

FIGURE 3.2 SONET (Synchronous Optical Network) dual-ring network architecture.

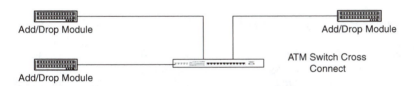

FIGURE 3.3 An example of a point-to-point SONET configuration.

network reconfiguration and remarkably efficient network operations. SONET/SDH networks also facilitate guaranteed service delivery; provide protection against network faults, timing problems, and equipment failures; and support point-to-point and point-to-multipoint connections.

3.4.5 SONET/SDH Architecture

Network nodes in SONET/SDH configurations are typically arranged in dual-ring topologies that support information transport unidirectionally and bi-directionally. SONET/SDH dual-rings are termed "self-healing" in recognition of their transmission reliability. Generally, two separate paths are used for transmitting digital signals in dual-ring installations. (See Figure 3.2.) In case of a node failure or a fiber cut on the primary ring path, network traffic is rerouted almost instantaneously to the alternate ring path until service on the primary ring path is restored.

SONET/SDH services provision interfaces necessary for working in concert with network protocols, architectures, and technologies such as Frame Relay (FR), Fast Ethernet, Gigabit Ethernet, and SMDS (Switched Multimegabit Data Service). Moreover, SONET/SDH implementations also work in conjunction with ATM, Fiber Data Distributed Interface (FDDI), FDDI-II (FDDI-Phase II), Fibre Channel (FC), WDM (Wavelength Division Multiplexing), and DWDM (Dense WDM) solutions. (See Figure 3.3.)

3.4.6 SONET/SDH PROTOCOL STACK

SONET/SDH solutions employ a four-layer protocol stack consisting of the Photonic Layer, the Section Layer, the Line Layer, and the Path Layer. The SONET/SDH four-layer protocol stack operates at the Physical Layer or Layer 1 of the Open Systems Interconnection (OSI) Reference Model. Situated at the lowest level, the Photonic Layer or Layer 1 converts STS electrical bits into optical bitstreams for transmission over the physical medium. The Section Layer or Layer 2 fosters signal regeneration at consistent time intervals and ensures dependable transport of SONET/SDH frames via the Photonic Layer. The Line Layer or Layer 3 enables transmission of SONET/SDH payloads and incorporation of these payloads into a series of STS frames. The Line Layer also performs multiplexing and frame synchronization functions. The Path Layer or Layer 4 at the top layer of the SONET/SDH protocol stack enables consistent and reliable network transmission services. The Path Layer also functions as the interface between the Physical Layer or Layer 1 and upper layer technologies and protocols such as ATM (Asynchronous Transfer Mode), IP (Internet Protocol), and Gigabit Ethernet.

3.5 SONET/SDH MULTIPLEXING

Multiplexing enables the combination of multiple data streams for transmission via the optical fiber link. In SONET/SDH network configurations, the multiplexing process optimizes utilization of available bandwidth for enhancing network performance.

3.5.1 SYNCHRONOUS TRANSMISSION MULTIPLEXING (STM)

SONET/SDH configurations employ STM (Synchronous Transmission Multiplexing), also called byte-interleaved multiplexing, for synchronizing transmission of voice, video, and data signals to a common external clock. STM enables signals in a SONET/SDH network to traverse the fiber-optic communications link in a steady and continuous stream. STM is also an enabler of signal rerouting after network failures, remote network management, and bi-directional or full-duplex information transmission. The Synchronous Transmission Multiplexing process is initiated in SONET networks with the generation of the OC-1 (Optical Carrier-Level 1) or the STS-1 (Synchronous Transmission Signal-Level 1) base signal at 51.84 Mbps, and in SDH networks with the generation of the STM-1 (Synchronous Transport Module-Level 1) base signal at 155.52 Mbps. SONET/SDH employ STM to enable dependable and reliable transmission in short-haul and long-haul network configurations and support synchronous signal transmissions to facilitate multigigabit network transmission rates.

3.5.2 TIME-DIVISION MULTIPLEXING (TDM)

SONET/SDH networks also employ TDM (Time-Division Multiplexing) for aggregating bitstreams into composite signals and assigning these signals for transmission to fixed timeslots in a predetermined rotation to optimize information transport over the fiber optic infrastructure. Developed by Bell Labs, TDM divides the entire bandwidth or frequency range of the transmission medium into a sequence of timeslots. Each timeslot uses the entire bandwidth but only for its assigned time interval. TDM flexibly adjusts time intervals to promote optimal use of available bandwidth and enables variation in the numbers of signals sent along the optical link.

TDM multiplexers convert signals from electronic-to-optical formats for enabling transmission over the physical medium. At the reception site, TDM demultiplexers convert signals from optical-to-electronic formats. It is important to note that processes associated with TDM electronic-to-optical and optical-to-electronic conversions impede the speed of SONET/SDH networks. Optical networks that employ electronic switching and conversion are generally limited to speeds of 13.21 Gbps (OC-255). To achieve higher rates, each signal must maintain the integrity of its photonic structure while transiting the network.

3.5.3 OPTICAL TIME-DIVISION MULTIPLEXING (OTDM)

SONET/SDH configurations also employ OTDM (Optical Time-Division Multiplexing) for enabling fast transmission rates. An emergent TDM option for optical network deployments, OTDM eliminates the cumbersome process of signal conversions by employing tunable lasers to generate very short optical pulses. With OTDM transmission, all-optical signals are transported via an all-optical fiber infrastructure over extended distances at rates reaching 160 Gbps. However, in comparison with WDM (Wavelength Division Multiplexing) and DWDM (Dense WDM), OTDM supports only a few wavelengths or channels of light on each fiber optic strand.

3.5.4 FREQUENCY-DIVISION MULTIPLEXING (FDM)

In contrast to TDM, FDM (Frequency-Division Multiplexing) assigns a discrete carrier frequency to each bit stream. Numerous bit streams are then combined for enabling effective transmission over the fiber optic medium. WDM and DWDM are regarded as optical equivalents of FDM.

3.6 SONET AND SDH STANDARDS ORGANIZATIONS AND ACTIVITIES

3.6.1 ALLIANCE FOR TELECOMMUNICATION INDUSTRY SOLUTIONS (ATIS) AND NETWORK AND SERVICES INTEGRATION FORUM (NSIF)

Affiliated with the Alliance for Telecommunications Industry Solutions (ATIS), the Network and Services Integration Forum (NSIF) develops solutions for enabling trouble-free transmissions in IP-over-SONET and ATM-over-SONET networks. The NSIF defines procedures for enabling bandwidth management and secure transmissions. In addition, the NSIF encourages utilization of interoperable SONET components in multivendor environments. The NSIF is a successor to the SONET Interoperability Forum (SIF).

3.6.2 AMERICAN NATIONAL STANDARDS INSTITUTE (ANSI)

The American National Standards Institute (ANSI) adopted the initial SONET standard for enabling optical transmission in 1988. This standard is based on the specification developed by the Exchange Carriers Standards Association (ECSA). In comparison to older telecommunications systems, SONET networks are flexible, reliable, scalable, and extendible.

Formally called ANSI T1.106, the ANSI SONET specification establishes guidelines for SONET operations in supporting short-haul, mid-range, and long-haul applications over fiber optic links. ANSI also defines a constellation of SONET specifications that clarify OC (Optical Carrier) levels and electronically equivalent Synchronous Transmission Signal (STS) levels. In addition, ANSI delineates a hierarchical set of transmission formats and transmission rates for use with SONET implementations and establishes the framework for deployment of SONET dual self-healing optical fiber ring topologies.

ANSI specifies procedures for evaluating the performance of dedicated digital services operating over SONET configurations at 51.84 Mbps (OC-1), 155.52 Mbps (OC-3), 622.08 Mbps (OC-12), and higher rates. ANSI also clarifies approaches for supporting SONET applications and network management operations, and establishes timing and synchronization specifications for SONET interfaces.

SONET standards compliance ensures equipment compatibility and interoperability between products from diverse manufacturers at both ends of a SONET link or pathway through the optical fiber infrastructure. Equipment that is compliant with ANSI SONET specifications is available in the United States from vendors such as Fujitsu, Alcatel, Lucent Technologies, 3Com, Nortel, and NEC. Communications carriers offering SONET services in the United States include Verizon, BellSouth, AT&T, SBC, and Ameritech.

3.6.2.1 ANSI T1X1 Subcommittee

Affiliated with the American National Standards Institute (ANSI) T1 Committee, the ANSI T1XI Subcommittee develops optical interface specifications.

3.6.3 INTERNATIONAL TELECOMMUNICATIONS UNION-TELECOMMUNICATIONS STANDARDS SECTOR (ITU-T)

The initial SDH (Synchronous Digital Hierarchy) Recommendation was endorsed in 1989 by the International Telecommunications Union-Telecommunications Standards Sector (ITU-T). Subsequently, the ITU-T described SDH support of STM transmission levels beginning with STM-1 or 155.52 Mbps. SDH support of transmission rates at STM-4 at 622.08 Mbps, STM-16 at 2.488 Gbps, STM-48 at 10 Gbps, and STM-64 at 13.21 Gbps were also indicated. In the G.707 to the G.709 Recommendations and the G.781 to the G.784 Recommendations, the ITU-T describes SDH Network-to-Node and Network-to-Network Interfaces (NNIs) for achieving effective synchronous multiplexing operations.

3.7   SONET/SDH IMPLEMENTATION CONSIDERATIONS

SONET/SDH technologies create a reliable high-speed transport infrastructure for multimedia traffic and offer abundant support for broadband services. SONET/SDH transmission capacities, management services, bandwidth optimization proficiencies, and flexibility in delivering basic and advanced applications affordably and ubiquitously contribute to the popularity of SONET/SDH solutions. Moreover, SONET/SDH infrastructures also accommodate current and emergent applications by employing a range of transmission rates for transporting digital bitstreams, work in concert with diverse technologies, and support transparent operations in multivendor environments.

Nonetheless, SONET/SDH implementations can require extensive infrastructure modifications, including installation or expansion of the in-ground fiber optic plant. SONET/SDH solutions are also limited in providing network support for accommodating accelerating bandwidth demands of high-performance networks such as CA*net3 (Canadian Network for the Advancement of Research, Industry, and Education, Phase 3) and vBNS+ (very high-performance Backbone Network Service Plus). SONET/SDH implementations are limited to provisioning bandwidth required for Abilene and SuperJANET4 (Super Joint Academic Network, Phase 4) operations. In addition, SONET/SDH implementations are restricted in delivering advanced multiservice support to meet sophisticated multifunctional requirements of diverse communities of users conducting advanced research, pilot studies, and experiments in the Next-Generation Internet (NGI) environments.

As noted, SONET/SDH deployments involve installation of dual or two separate fiber optic ring topologies. However, these deployments may only operate at 50% of the total network capacity, with one fiber optic ring supporting networking operations and the other fiber optic ring held in reserve in case of network node processing delays, optical fiber cuts, and equipment failures.

As a consequence, WDM (Wavelength Division Multiplexing) and DWDM (Dense WDM) mesh topologies are increasingly deployed as replacements for SONET/SDH dual-ring configurations. WDM and DWDM mesh architectures enable the integration of new networking installations into the existing infrastructure and support resilient operations. In WDM and DWDM mesh optical fiber configurations, damaged or compromised physical links do not necessarily bring all network operations to a halt. In addition, WDM and DWDM mesh networks free reserved bandwidth maintained by dual-ring topologies and use this excess capacity for supporting additional bandwidth-intensive multimedia teleservices and tele-applications via the in-place fiber optic plant.

3.8   SONET/SDH MARKETPLACE

The spiraling need for gigabit networks that support error-free transmission of broadband interactive services contributes to development and implementation of SONET/SDH initiatives. SONET/SDH technologies facilitate Local Area Network (LAN), Metropolitan Area Network (MAN), and Wide Area Network (WAN) integration; high-speed information transmission; and remote disaster protection and recovery services. Research and experimentation in testbed environments extend and refine SONET/SDH capabilities in supporting broadband applications and cross-institutional collaborations.

3.8.1 CISCO SYSTEMS

In 2000, Cisco Systems implemented an SDH optical transport extender for fostering delivery of broadband services via an optical network directly to SOHO (Small Office/Home Office) sites. This SDH extender provisions an integrated platform for creating FTTB (Fiber-to-the-Building) connections that deliver affordable teleservices to small- and medium-sized businesses in multi-tenant buildings. In addition to eliminating the need for overlay networks, this Cisco solution enables scalable applications and information delivery at E-1 (2.048 Mbps) rates via the in-place SDH infrastructure.

3.8.1.1 Cisco Systems, Lucent Technologies, and Alcatel

Cisco Systems, Lucent Technologies, and Alcatel support TDM SONET/SDH solutions that use electronic components for enabling transmission rates up to 13.21 Gbps (OC-255). TDM products that facilitate rates at 40 Gbps are also in development.

3.8.2 MANNESMANN MOBILFUNK AND ECI TELECOM

Mannesmann Mobilfunk, Germany's largest cellular operator, and ECI Telecom are building an optical network solution based on SONET/SDH and DWDM technologies that will enable approximately 17 million subscribers to access broadband applications and 3G (third-generation) cellular services. Rates ranging from T-1 (1.544 Mbps) and E-1 (2.048 Mbps) to STM-64 (13.21 Gbps) will be supported.

3.8.3 SPRINT INTERNATIONAL SONET RING

In the 1990s, Sprint initiated deployment of an international SONET ring network. This SONET configuration interlinks Canada and the United States through points in Buffalo, New York; Springfield, Massachusetts; and Montreal, Canada, and provisions communications coverage in an area that extends to 1,714 miles.

3.9 ABILENE NETWORK

Sponsored by the University Corporation for Advanced Internet Development (UCAID) in conjunction with Indiana University, Nortel Networks, Qwest Communications, and Cisco Systems, the Abilene Network enables telecollaborative research trials and next-generation computing initiatives. The term "Abilene" reflects the frontier spirit of residents in Abilene, Kansas, that was inspired by the establishment of the railroad and inauguration of transportation services in the 1860s.

3.9.1 ABILENE NETWORK INFRASTRUCTURE

As with vBNS+, the Abilene Network is an Internet2 (I2) backbone network that interconnects Internet2 GigaPoPs (Gigabit Points of Presence) and optimizes network performance using high-speed technologies such as ATM, SONET, and Packet-over-SONET/SDH (POS) technical solutions. Every UCAID I2 member is eligible to utilize the Abilene Network.

The fiber-optic infrastructure for Abilene is provided by Qwest Communications. Indiana University manages the Abilene Network Operation Center (NOC) and provisions technical support. In addition, Abilene uses high-speed SONET equipment from Nortel and Packet-over-SONET/SDH (POS) routers from Cisco Systems.

Implemented in 1998, the initial component of the Abilene Network supported high-speed services and high-performance applications for entities that included Indiana and Purdue Universities, the University of Washington, the Great Plains Network segment in Kansas City, and the City of San Francisco.

3.9.2 ABILENE NETWORK OPERATIONS

The Abilene Network features minimal queuing and propagation delays, low latencies, advanced capabilities, and dependable traffic throughput in support of broadband networking operations, applications, and initiatives. The Abilene Network also fosters implementation of innovative networking services such as IP multicasts and QoS (Quality of Service) guarantees. Technical devices, toolsets, and transmission techniques developed for the Abilene Network are evaluated in testbed trials prior to their incorporation into the Abilene infrastructure.

The Abilene Network supports transmission rates ranging from 622.08 Mbps (OC-12) to 2.488 Gbps (OC-48). Additional backbone links for provisioning transport rates at 13.21 Gbps (OC-255) via the Abilene Network are in development.

3.9.3 ABILENE NETWORK APPLICATIONS AND SERVICES

As with vBNS+, the Abilene Network supports interconnectivity with the National Aeronautics and Space Administration Research and Engineering Network (NASA-NREN), the U.S. Department of Defense Research and Engineering Network (DREN), interim DREN (iDREN), and the Energy Sciences Network (ESnet) sponsored by the U.S. Department of Energy. Moreover, the Abilene Network conducts tests for evaluating QoS delivery and IPv6 (Internet Protocol Version 6) functions and promotes development of advanced security solutions, authentication services, and cryptographic protocols.

The Abilene Network is also an enabler of next-generation applications and services such as telesurgery, broadcast quality video, and distributed terabit data mining. The National Laboratory for Applied Network Research (NLANR) monitors network operations for High-Performance Network Service Providers (HPNSPs) such as the Abilene Network, vBNS+, Internet2, STAR TAP (Science, Technology, and Research Transit Access Point), and the Next-Generation Internet (NGI).

3.10 ADVANCED TECHNOLOGY DEMONSTRATION NETWORK (ATDNet)

3.10.1 ATDNET TECHNICAL FUNDAMENTALS

ATDNet (Advanced Technology Demonstration Network) is an ATM-over-SONET and ATM-over-DWDM testbed that utilizes an extendible and scalable optical ring topology to facilitate information transport among government agencies in the Washington, D.C., area. A high-speed, high-performance metropolitan network, ATDNet was established by the U.S. Department of Defense Advanced Research Projects Agency (DARPA) to support telecollaborative projects and operations among federal agencies.

ATDNet participants, including the U.S. Department of Defense Information Systems Agency (DISA), NASA, and the National Security Agency (NSA), evaluate ATM-over-SONET and ATM-over-DWDM functions in ATDNet trials and experiments. The National Institute for Science and Technology (NIST) employs ATDNet for conducting research on the performance of advanced network technologies and architectures. Findings contribute to the development of NGI (Next-Generation Internet) projects and initiatives.

The ATDNet platform supports distributed computing; assessment of IPSec (Internet Protocol Security) and IPv6 (Internet Protocol version 6); and evaluation of software configurable architecture that supports diverse applications. In addition, this infrastructure enables testbed implementations to verify the capabilities of next-generation all-optical networks (AONs) and ATM-over-DWDM performance. Moreover, the performance of hardware and software products that can be used in future gigabit Internets are evaluated in ATDNet experiments as well.

The ATDNet infrastructure includes ATM switches and four bi-directional fiber optic rings. Each ring supports network transmissions at rates reaching 2.488 Gbps (OC-48). These rings also maintain terrestrial links to and from ground stations that interoperate with advanced satellite systems.

3.11 METROPOLITAN RESEARCH AND EDUCATION NETWORK (MREN)

3.11.1 MREN TECHNICAL FUNDAMENTALS

The Metropolitan Research and Education Network (MREN) employs an IP-over-SONET and ATM-over-SONET infrastructure that works in concert with legacy network; Ethernet, Fast Ethernet, Frame Relay, and FDDI technologies; and T-1 and E-1 leased line solutions. MREN supports video, voice, and data transmission at rates of 622.08 Mbps (OC-12). Upgrades to the MREN infrastructure that will support multimedia applications at rates reaching 2.488 Gbps (OC-48) via a SONET dual-ring topology are in development.

MREN maintains broadband links to advanced digital library initiatives and enables telecollaborative research at remote instrumentation sites such as the Chicago Air Shower Array, Sandia and Brookhaven National Laboratories, and the Goddard Space Center. MREN researchers employ satellite connections for enabling telecollaborative videoconferencing with their peers at remote locations such as the University of California NSF (National Science Foundation) Center for Astrophysics Research in Antarctica.

In addition, MREN defines procedures for the implementation of next-generation, high-performance networks. MREN also supports innovative, high-capacity applications and develops specialized facilities such as Multimode GigaPoPs (M-GigaPoPs). As with GigaPoPs, M-GigaPoPs support peer-to-peer information exchange and seamless transmission of multimedia traffic in a shared networking environment.

3.11.2 MREN Applications and Services

The MREN high-performance network infrastructure is an enabler of initiatives in high-performance computing, advanced medical imaging, computational biology, telemedicine, chemistry, astronomy, climatology, and astrophysics. The MREN platform also supports the assessment of advanced network architectures; the capabilities of IP, ATM, and SONET services; the performance of network protocols such as RSVP (Resource Reservation Protocol); and the evaluation of ATM multimedia transmissions with QoS (Quality of Service) guarantees.

In addition, the MREN infrastructure enables interactive CAVE (Cave Automated Virtual Environment) virtual reality (VR) projects and provisions sufficient bandwidth for advanced scientific supercomputing applications. The MREN platform also supports research enabling the interconnectivity of terabit storage facilities and high-performance networks; implementation of the CATS (Chicago-Argonne Terabyte System) initiative; and scientific visualization projects. Moreover, the MREN implementation facilitates seamless routing and exchange of voice, video, and data traffic at the SBC/AADS (Ameritech Advanced Data Services) NAP (Network Access Point) in Chicago.

3.11.2.1 MREN Digital Video Initiatives

The MREN infrastructure facilitates implementation of high-performance multimedia and digital video initiatives that feature IP multicasts, store-and-forward video delivery, advanced video streaming technology, and VR movies. Ohio State University and the University of Minnesota use the MREN platform to evaluate the performance of video-over-IP services that are compliant with the ITU-T H.323 Recommendation.

Research conducted over the MREN infrastructure in the digital video arena contributes to development of the Internet2 Digital Video Network (I2-DVN) initiative. This initiative supports creation of a nationwide digital video network that delivers live or stored streaming video and supports distribution of interactive, high-quality, digital video tele-education programs.

3.11.2.2 Multi-Modal Organizational Research and Production Heterogeneous Network (MORPHnet)

The MREN Multi-Modal Organizational Research and Production Heterogeneous Network (MORPHnet) project supports development of a very high-speed scalable network that provisions access to current and next-generation multimedia initiatives. The MORPHnet infrastructure supports QoS (Quality of Service) assessment of ATM service classes, enables SONET-over-WDM trial implementations, and fosters development of a next-generation Network Access Point (NAP) to optimize performance of local and regional institutional networks.

3.11.3 MREN Consortium

The MREN Consortium manages and administers MREN operations. MREN Consortium participants include Northwestern University, the University of Illinois at Chicago, the Argonne National Laboratory, the Fermi National Accelerator Laboratory, and the University of Chicago. The Canadian Network for the Advancement of Research, Industry, and Education (CANARIE), the Ohio Academic Research Network (OARnet), the U.S. Department of Defense Research and Engineering Network (DREN), and the National Center for Supercomputing Applications (NCSA) take part in MREN initiatives as well.

3.12 U.S. SONET INITIATIVES

3.12.1 California

3.12.1.1 California State University (CSU) Campuses and California Community Colleges (4Cnet)

Sponsored by California Community Colleges and California State University (CSU) campuses, 4Cnet provisions access to high-speed, high-performance networking services and enables real-time IP multicasts. The 4Cnet telecommunications infrastructure features an ATM backbone network for enabling multimedia transmissions at rates of 155.52 Mbps (OC-3) and 622.08 Mbps (OC-12).

In addition to ATM, 4Cnet employs Frame Relay technology and dual self-healing SONET rings operating at 2.488 Gbps (OC-48) in San Francisco, Los Angeles, and San Diego. 4Cnet SONET rings support interconnectivity to Qwest PoPs (Points of Presence), CSU campuses, vBNS+, and CalREN-2 (California Research and Education Network, Phase 2). As noted, CalREN-2 employs an ATM-over-SONET infrastructure for enabling advanced teleservices.

3.12.2 Illinois

3.12.2.1 Northwestern University

Northwestern University utilizes an ATM-over-SONET ring topology for enabling information exchange at rates of 622.08 Mbps (OC-12) between its Chicago and Evanston campuses. This infrastructure is readily scalable to 2.488 Gbps (OC-48). Northwestern University also employs a SONET link to the SBC/AADS NAP in Chicago, also called the Chicago NAP, for enabling national and international peer-level information exchange.

3.12.3 North Carolina

3.12.3.1 North Carolina GigaNet (NCGN)

The North Carolina GigaNet (NCGN) links Duke University, North Carolina State University, and the University of North Carolina at Chapel Hill in a regional network configuration. Transmission rates at 622.08 Mbps (OC-12) are supported.

This ATM-over-SONET infrastructure facilitates access to academic services, commercial computing resources, and environmental programs.

3.12.3.2   North Carolina Research and Education Network (NC-REN)

A private communications network, the North Carolina Research and Education Network (NC-REN) operates a multichannel, multisite Interactive Video Network that links medical schools, universities, and research organizations statewide in point-to-point and point-to-multipoint configurations. The NC-REN Interactive Video Network enables videoconferences and provides access to tele-education courses, collaborative teleresearch, scientific teleworkshops, and teleseminars, and features an ATM-over-SONET infrastructure that supports information transport at rates reaching 155.52 Mbps (OC-3).

3.12.3.3   North Carolina Information Highway (NCIH)

The North Carolina Information Highway (NCIH) uses an ATM-over-SONET configuration that facilitates telecommuting, telemedicine research, online community activities, economic development, health education, and distance learning initiatives. This network provisions transmission services for state universities, schools, community colleges, hospitals, corporations, and government agencies at 155.52 Mbps (OC-3) and higher rates. Law enforcement personnel employ the NCIH to retrieve arrest records, photographs, and fingerprints. North Carolina residents access NCIH to obtain birth certificates, property records, and car titles.

Private-sector companies taking part in NCIH include local telephone exchange carriers such as BellSouth and interexchange carriers (IXCs) such as AT&T and Sprint. NCIH is part of the public switched network in North Carolina. This configuration is owned and operated by North Carolina communications providers.

3.12.3.3.1   NCIH Educational Initiatives

The NCIH ATM-over-SONET infrastructure delivers videoconferencing classes on subjects such as history, psychology, and Tai Chi to students throughout the state and AP (Advanced Placement) courses to students at isolated and remote locations. As an example, the Guilford County School System uses the NCIH platform to support a countywide network for rural and urban county high schools. This configuration fosters access to teletraining classes for instructors, PTA (Parent Teacher Association) telemeetings, and curricular enrichment tele-activities.

The North Carolina School of Science and Math uses the NCIH platform for delivering courses in calculus and statistics to students in historically underserved counties. Teacher training sessions and videoconferences are enabled by NCIH as well. North Carolina State University, the University of North Carolina at Wilmington, and Appalachian State University utilize the NCIH to provision access to distance education courses in law enforcement, pharmacy, nursing, emergency medicine, and library science.

The Northeast Telecommunications Network (NEAT-NET) interconnects Beaufort Community College with high schools in Beaufort, Tyrell, Washington, and

Hide Counties via the NCIH infrastructure. Moreover, North Carolina community colleges employ the NCIH platform to provide links to tele-education programs and advanced courses in radiation therapy, funeral training, criminology, history, psychology, accounting, and early childhood education.

3.12.3.3.2   NCIH E-Government (Electronic Government) Initiatives

The NCIH infrastructure supports diverse E-government applications and projects. As an example, Haywood County officials intend to use the NCIH platform for interconnecting libraries, hospitals, law enforcement agencies, schools, community colleges, public health groups, social service agencies, and county government departments in a virtual community network. Local agencies at the Catawba County Government Center intend to establish connections to the NCIH platform for enabling public-sector agencies to access Web resources.

3.12.3.3.3   NCIH Telemedicine Initiatives

Medical specialists at hospitals statewide that are linked to the NCIH conduct teleconsultations in real-time with patients and primary care physicians at remote healthcare clinics and review MRI (Magnetic Resource Imaging) and CAT (Computer Axial Tomography) scans. In addition, the University of North Carolina School of Public Health and the State Department of Environment, Health, and Natural Resources sponsor the Public Health Training and Information Network.

The Public Health Training and Information Network delivers teletraining sessions, undergraduate and graduate courses, public health updates, and professional certification programs. Public health agencies in Wilson, Cumberland, and Jackson Counties and healthcare providers in Hickory, Fayetteville, Elizabeth City, Wilson, and Chapel Hill access the Public Health Training and Information Network via NCIH connections.

3.12.3.3.3.1   North Carolina Healthcare Information and Communications Alliance (NCHICA)

The North Carolina Healthcare Information and Communications Alliance (NCHICA) utilizes the NCIH ATM-over-SONET infrastructure to support interactive telemedicine applications. Participants in NCHICA include the Duke University School of Medicine and Medical Center, the East Carolina University School of Medicine, and the Pittsburgh County Memorial Hospital.

The NCIH ATM-over-SONET platform supports improvements in the quality of healthcare available to rural and disadvantaged residents. For example, NCHICA initiatives enable rural patients to interact with medical specialists at distant locations and obtain specialized healthcare services and emergency treatment. General practitioners at rural and remote locations complete continuing medical education telecourses delivered via NCIH in programs sponsored by NCHICA.

In addition, NCHICA uses the NCIH infrastructure to facilitate interactive videoconferencing between general practitioners in rural communities and medical specialists on staff at major medical centers. Furthermore, NCHICA provisions links via NCIH to digitized medical records for enabling medical professionals to access clinical data.

3.12.3.3.3.2   NCIH TeleHome Care Program
An NCIH initiative, the North Carolina TeleHome Care Program provisions tele-medicine services to patients and their caregivers situated in remote and isolated locations. This initiative also supports interactive videoconferencing between uni-versity physicians, healthcare specialists, nutritionists, nurses, and/or social workers. Plans for installing additional NCIH links at rural hospitals, remote medical centers, local public schools, and SOHO (Small Office/Home Office) venues to support the NCIH TeleHome Care Program are in development.

3.12.3.3.3.3   Northwest Area Health Education Center
The Northwest Area Health Education Center employs the NCIH for interconnecting the Bowman Gray Medical School, the Catawba Memorial Hospital, and the Rowan County Hospital in a virtual regional network configuration. In addition to resource sharing, this virtual configuration enables medical specialists at the Bowman Gray Medical School to mentor interns and residents and enables primary care physicians and medical specialists to virtually participate in grand rounds at local community hospitals.

3.12.3.3.3.4   Eastern Carolina University School of Medicine
The Eastern Carolina University (ECU) School of Medicine utilizes an amalgam of technologies, including SONET, ISDN (Integrated Services Digital Network), ATM, T-1, POTS (Plain Old Telephone Service), and microwave, in an integrated network-ing configuration that supports diverse telemedicine applications and teleservices. In addition to serving as a component in the NCIH configuration, the ATM-over-SONET segment of the ECU School of Medicine Network facilitates connections to statewide networking implementations such as the North Carolina-Research and Education Network (NC-REN).

3.12.3.3.3.5   University of North Carolina at Chapel Hill (UNC-CH)
School of Medicine
The University of North Carolina at Chapel Hill (UNC-CH) School of Medicine supports videoconferencing between medical specialists at the UNC-CH School of Medicine and senior citizens and their caregivers in rural communities situated in North Hampton and Halifax Counties.

3.12.4   Wisconsin

3.12.4.1   BadgerNet

Sponsored by the State of Wisconsin Department of Administration, BadgerNet is a statewide network that employs an ATM-over-SONET infrastructure for enabling high-performance broadband network applications and services. BadgerNet partic-ipants include private and public schools, government agencies, technical colleges, and universities.

BadgerNet supports a broad spectrum of distance education activities, including delivery of college courses to high school students, interactive videoconferencing between high school students and their teachers with subject specialists and guest lecturers, teleseminars, teleworkshops, and at-work delivery of job training sessions.

In addition, BadgerNet enables connections to Web resources and digital library collections and archives.

3.13   SONET/SDH IMPLEMENTATION CONSIDERATIONS

As noted, Time-Division Multiplexing (TDM) supports SONET/SDH multimedia applications and services. SONET/SDH technologies define specifications for electrical and optical interfaces that operate at Layer 1 or the Physical Layer of the OSI Reference Model and support voice, video, and data transport at rates that reach 13.21 Gbps (OC-255). Optical distribution networks based on SONET/SDH technologies provision bandwidth on-demand for interactive tele-instructional applications and telemedicine programs. The decision to use SONET or SDH as a networking solution in the tele-education environment depends on the effectiveness of the SONET or SDH implementation in accommodating student and faculty requirements and institutional missions, goals, and objectives.

It is important to note that agreement exists on the potential value of SONET/SDH configurations in the academic arena. However, distribution of educational content via SONET/SDH configurations does not automatically guarantee learning or improvement in student performance. Future research involving optical network deployment in schools and universities must also focus on the practical design and deployment of pedagogical strategies and collaborative instructional activities for optimizing student skills and competencies in high-performance multimedia telelearning environments. Successful SONET/SDH implementation for supporting unique and important forms of telecollaboration and increasingly sophisticated computer applications for education and training requires an understanding of pedagogical issues and challenges as well as current and emergent technological advances in the optical networking domain.

Successors to proprietary optical configurations or first-generation optical networks, SONET/SDH technologies are second-generation optical network solutions. Despite their support of fast rates and scalable services, these technologies are limited in meeting the bandwidth requirements of complex volume-intensive distributed applications. As a consequence, third-generation optical network configurations such as POS (Packet-over-SONET/SDH), AONs (All-Optical Networks), PONs (Passive Optical Networks), APONs (Advanced Passive Optical Networks), and SuperPONs, and third-generation optical technologies and protocols such as WDM (Wavelength Division Multiplexing), and DWDM (Dense WDM), technologies are in development.

3.14   PACKET-OVER-SONET/SDH (POS) SOLUTIONS

3.14.1   POS FEATURES AND FUNCTIONS

Spiraling demand for fast networking services and high-capacity, high-performance networking solutions drives the development of Packet-over-SONET/SDH (POS) solutions. Endorsed by the Internet Engineering Task Force (IETF), Packet-over-SONET/SDH (POS) technology operates at Layer 2 or the Data-Link Layer of the OSI Reference Model and optimizes the transmission capabilities of the in-place

SONET/SDH infrastructure. POS solutions combine IP (Internet Protocol) utilization with SONET/SDH throughput, thereby improving network performance, ensuring seamless information transport, and enabling dependable and reliable delivery of bandwidth-intensive services and applications at multigigabit rates.

POS implementations map IP datagrams to SONET/SDH frames, support point-to-point connections, and streamline network operations. Moreover, POS installations optimize bandwidth availability by eliminating cellular overhead associated with ATM implementations. POS networks also support migration to IP (Internet Protocol) optical network configurations employing WDM and DWDM technologies.

POS implementations require the utilization of new networking equipment such as high-speed POS routers and POS switches to accommodate IP packet data requirements. Generally, POS installations support TDM and IP services on separate ring segments.

3.14.2   POS INITIATIVES IN THE ACADEMIC DOMAIN

3.14.2.1   ResearchChannel Consortium

Sponsored by the ResearchChannel Consortium, a coalition of research institutions and corporate research centers, the ResearchChannel initiative supports conversion of HDTV (High-Definition Television) signals into IP streams to enable distribution of high-quality audio and advanced video applications via Internet2 (I2) and the commodity Internet. The ResearchChannel Consortium also participates in the Internet2 Digital Video Network (I2-DVN) project and maintains a library of on-demand videos that demonstrate the capabilities of I2 in enabling high-performance, high-speed MPEG-2 (Moving Picture Experts Group-2) video distribution. It is interesting to note that the ResearchChannel Consortium also sponsored the first Internet2-to-cable television delivery experiment and pioneered the use of POS technologies for enabling video distribution.

3.14.2.2   Pacific/NorthWest GigaPoP (P/NWGP)

A next-generation IP testbed, the Pacific/NorthWest GigaPoP (P/NWGP) is the first GigaPoP in the United States to provision POS services at rates reaching 2.488 Gbps (OC-48). P/NWGP supports HDTV (High Definition Television) delivery, voice-over-Internet Protocol (VoIP) applications, advanced E-business (electronic business) projects, secure telehealthcare services, telesurgery, interactive videoconferencing, workforce training, and continuing distance education programs. P/NWGP participants include Montana State, Washington State, and Portland State Universities, and the Universities of Alaska, Montana, Washington, and Idaho.

3.14.3   POS VENDOR IMPLEMENTATIONS

3.14.3.1   BellSouth

BellSouth implements a Packet-over-SONET/SDH (POS) solution for the Florida Multimedia Internet Exchange (FloridaMIX) in South Florida. In addition to supporting

ATM services, the FloridaMIX is the first Network Access Point (NAP) in the United States to employ optical networking technology to transport IP traffic through multiple termination points. This NAP also provisions links to points throughout the United States, Latin America, Africa, Western Europe, and the Caribbean.

3.14.3.2 IXC Communications

IXC Communications supports implementation of Gemini 2000, an IP optical backbone network that employs POS technology. Gemini 2000 features 16 cross-country connections between San Francisco, California, and Washington, D.C., and provides services for NYSERNet 2000 (New York State Education and Research Network, Year 2000). Gemini 2000 enables network transmissions at rates reaching 2.488 Gbps (OC-48) and employs MPLS (MultiProtocol Label Switching) for providing QoS (Quality of Service) assurances and dependable transmission of time-sensitive traffic.

3.15   NEXT-GENERATION OPTICAL NETWORK SOLUTIONS

3.15.1   ALL-OPTICAL NETWORKS (AONs)

Configurations that support all-optical functions are called all-optical networks (AONs). Recent advances in optical networking technology also drive development and utilization of all-optical, next-generation, high-speed networking solutions to alleviate internetwork gridlock. All-optical networks (AONs) provision on-demand bandwidth and accommodate diverse communications requirements. AONs also support large-scale applications for dispersed user communities.

In all-optical circuit-switched networks, the process of converting optical signals to electronic signals and electronic signals to optical signals at intermediate network nodes is eliminated. It is interesting to note that although all-optical circuit-switched networks eliminate the need for optical-to-electronic and electronic-to-optical conversions, lightpaths in circuit-switched AONs can also be controlled electronically. In all-optical packet-switched networks, the process of signal conversion from optical-to-electronic formats and from electronic-to-optical formats results in latencies at intermediate network nodes and internode signal propagation delays, thereby compromising the speed and effectiveness of AON solutions.

All-optical networks foster development and utilization of numerous parallel broadband channels that enable interconnectivity of network interfaces and access nodes and use optical routing and switching services. In contrast to SONET/SDH and POS solutions, AONs enable significantly improved traffic throughput and network performance for telemedicine, tele-education, data warehousing, and E-commerce (electronic commerce) services and operations. AONs support multi-gigabit and multiterabit transmission rates over longer distances via the in-place optical fiber plant than SONET/SDH by dynamically assigning and reassigning diverse types of network traffic to optical channels for optimizing utilization of optical fiber bandwidth. (See Figure 3.4.)

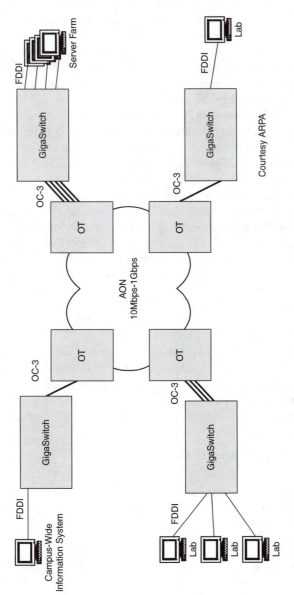

FIGURE 3.4 WDM dynamic reconfiguration in an AON (All-Optical Network) configuration.

3.15.2 PASSIVE OPTICAL NETWORKS (PONs)

PONs (Passive Optical Networks) are low-cost and low-maintenance high-speed last-mile solutions that support Fiber-in-the-Loop (FITL), Fiber-to-the-Home (FTTH), Fiber-to-the-Desk (FTTD), and Fiber-to-the-Curb (FTTC) network configurations, and enable high-speed, high-performance information transmission via point-to-multipoint optical fiber networks. PON (Passive Optical Network) equipment includes optical line termination devices (OLTDs) that are interlinked to optical network units (ONUs) or subscriber terminals. In addition, PONs employ passive optical network components such as directional or star couplers, wavelength routers, splitters, multiplexers, and filters that do not utilize electricity for signal transport.

PON implementations work in concert with Ethernet, IP, cable modem, and satellite technologies and feature greater bandwidth and fewer points of failure than conventional twisted copper pair, coaxial cable, and hybrid optical fiber and coaxial cable (HFC) solutions. In addition, PON operations are immune to lightning and other transient forms of electromagnetic interference.

PON transmissions are reliable, dependable, and robust over extended distances. Moreover, PON configurations are scalable and extendible and support always-on access to the commodity Internet and broadband network installations. Depending on user requirements, PONs enable rates reaching 622.08 Mbps (OC-12) in the downstream direction from the network to the customer premise and 155.52 Mbps (OC-3) in the upstream direction from the customer premise to the network.

3.15.2.1 PON Vendor Solutions

3.15.2.1.1 Alloptic GigaForce Solutions
Alloptic supports Ethernet PON GigaForce solutions for enabling FTTB (Fiber-to-the-Building), FTTH (Fiber-to-the-Home), and FTTC (Fiber-to-the-Curb) applications. GigaForce implementations dependably deliver high-speed data, full-screen broadcast television programs, and on-demand video to SOHO (Small Office/Home Office) venues and small- and medium-sized business establishments at rates ranging from 64 Kbps to 1 Gbps.

3.15.2.1.2 Mid-Atlantic Communications
In Washington, D.C., Mid-Atlantic Communications supports PON implementations that feature a scalable and extendible architecture for enabling integrated voice, video, and data transmission over the local loop to high-rise apartment complexes and multistory office buildings.

3.15.2.1.3 OnePath
OnePath plans to implement a PON solution for delivering cable television programming, direct broadcast television service, and fast Internet access in Fiber-to-the-Home (FTTH) configurations.

3.15.3 ATM PASSIVE OPTICAL NETWORKS (APONs)

APONs (ATM Passive Optical Networks) are flexible and scalable optical network solutions that support FTTH, FTTC, FTTB, and FTTCab (Fiber-to-the-Cabinet) topologies. APON configurations feature passive optical splitters that work in concert with the in-place optical fiber plant. Each APON enables the interconnection of up to 64 optical network units (ONUs) to a fiber optic segment. Bandwidth is shared by every ONU on the same fiber optic segment. APONs support transmission rates at 622.08 Mbps over distances that extend to 20 kilometers.

3.15.3.1 Full Service Active Network (FSAN) Consortium

According to the FSAN (Full Service Access Network) Consortium, an APON is a cost-effective architectural solution for enabling broadband fiber in the loop (FITL) implementations. Sponsored by the FSAN Consortium, the FSAN Optical Access Network Working Group developed the APON specification that was adopted by the ITU-T in the G.983.1 Recommendation in 1998. Bell Canada, Nippon Telegraph and Telephone (NTT), GTE, Korea Telecom, SingTel, and France Telecom participate in the FSAN Consortium.

3.15.4 SUPER PASSIVE OPTICAL NETWORKS (SUPERPONS)

SuperPONs provision communications services in areas of coverage that extend up to 100 kilometers, employ ATM-over-optical core infrastructures, and use optical amplifiers for signal regeneration. Downstream transmission rates at 2.488 Gbps (OC-48) and upstream transmission rates at 311 Mbps are supported.

3.15.4.1 Alcatel

Alcatel evaluates SuperPON capabilities and performance in testbed trials in Brussels.

3.16 WAVELENGTH DIVISION MULTIPLEXING (WDM)

3.16.1 WAVELENGTH DIVISION MULTIPLEXING (WDM) FUNDAMENTALS

Next-generation WDM (Wavelength Division Multiplexing) core optical networks exploit the tremendous capabilities of optical fiber by dividing optical spectrum and available bandwidth into two or four non-overlapping channels or wavelengths on each fiber optic strand. Every wavelength or channel enables operations at a maximum rate of 40 Gbps. WDM solutions interwork with IP, ATM, SONET/SDH, Gigabit Ethernet, and DWDM technologies.

In WDM implementations, optical amplifiers replace electrical repeaters in enabling signal regeneration. In addition, optical amplifiers employ erbium to enhance the power of optical wavelengths or colors of light and enable trouble-free transmission of optical signals over extended distances. WDM signal robustness depends on the number of optical amplifiers used in the system. Generally, multiple optical amplifiers degrade network performance by generating noise and crosstalk on fiber optic links.

Optical cross-connects support multicasting functions and dynamic network reconfiguration at the wavelength level in order to accommodate changes in bandwidth demand in WDM installations. In addition, optical cross-connects support wavelength grooming and conditioning to optimize utilization of the in-place fiber optic plant and provision transparent and scalable network operations. In the present-day WDM environment, optical cross-connects support electronic operations at the fiber optic core because optical signals entering and exiting optical cross-connects transit optical and electronic interfaces.

With the integration of an all-optical core in an optical cross-connect, future network infrastructures are expected to support increased transmission rates and sophisticated services such as transparent optical switching and optical routing functions in the Optical Layer. The Optical Layer enables robust transmissions and fundamental network operations at a sublayer of the Physical Layer or Layer 1 of the OSI (Open Systems Interconnection) Reference Model.

Effective WDM implementation involves precise alignment of WDM lasers with standardized wavelengths. The wavelength setting must also accommodate application requirements.

WDM configurations also include optical multiplexers and optical demultiplexers at endpoints of the WDM light path for multiplexing signals onto the optical fiber physical medium and then separating signals and sending them to receivers. Optical gateways perform bridging functions for enabling WDM network configurations to support TDM (Time-Division Multiplexing) service and transmission of IP packets and ATM cells. Optical wavelength converters, optical filters, flat gain optical amplifiers, and optical add-and-drop multiplexers (OADMs) also contribute to seamless WDM operations.

Constant vigilance and systematic network management, administration, and maintenance operations are required to ensure WDM system reliability, dependability, and availability. A spare optical fiber route or redundant optical fiber pathway ensures communications operations in case of network outages resulting from cuts in the optical fiber plant.

3.16.2 WDM and Frequency-Division Multiplexing (FDM)

As noted, WDM is regarded as the optical equivalent of FDM (Frequency-Division Multiplexing). As with WDM, FDM enables numerous signals to be combined for transmission via the main communications channel. WDM and FDM also assign each signal to a different frequency within the main channel. With WDM and FDM, each channel uses pre-assigned bandwidth. Like FDM, WDM employs different frequencies for each bitstream and combines modulated frequencies for transmission. In contrast to FDM, WDM only works in concert with fiber optic operations.

3.16.3 WDM and Time-Division Multiplexing (TDM)

WDM and TDM are key protocols for optimizing communications transport via fiber optic cabling. With TDM utilization, each signal is assigned to a fixed timeslot in a pre-assigned rotation for transiting the network and each channel corresponds to a unique timeslot allocation. By contrast, in WDM implementations, each channel corresponds to a unique wavelength allocation and bitstreams are concurrently transmitted across the optical fiber link via distinct wavelength channels. In addition,

TDM supports dynamic channel assignment by altering timeslot assignments. WDM enables dynamic channel reassignment by altering wavelengths.

WDM and TDM support reconfiguration in response to variations in traffic patterns, demographic changes, and real-time demand. WDM and TDM also optimize optical fiber bandwidth capabilities and network capacity. In addition, WDM and TDM enable broadband channel creation and deployment and support voice, video, and data transmission at 2 Tbps (Terabits per seconds) and higher rates over a single optical fiber strand.

3.17   DENSE WAVELENGTH DIVISION MULTIPLEXING (DWDM)

An innovative optical transmission technique, DWDM (Dense WDM) technology enables multiple optical signals on different wavelengths or channels to be transmitted on a single optical fiber strand. The term "dense" refers to high-wavelength or high-channel count per optical fiber strand. The DWDM Optical Layer routes multiple optical signals over multiple wavelengths or channels via a direct path between two endpoints in a DWDM network. In addition, the Optical Layer enables fast and reliable optical signaling restoration. Generally, DWDM implementations employ all-optical ring topologies.

3.17.1   DWDM TRANSMISSION FUNDAMENTALS

DWDM supports high-speed network transmission at the Optical Layer. For example, DWDM fosters transmission at 400 Gbps by using 40 wavelengths. Each wavelength operates at 10 Gbps (OC-192). Typically, a DWDM network supports transmission of one or more terabits of information.

DWDM technology optimizes optical fiber networking capabilities and bandwidth availability by dividing the available light spectrum into wavelengths or channels, with each wavelength or channel carrying a different color of light. DWDM networks transmit multiple optical signals via multiple wavelengths or channels on the same optical fiber strand. Optical signals on each DWDM wavelength are amplified as a group to ensure fast and dependable transmission.

DWDM networks employ a multiservice infrastructure that enables the Optical Layer to directly interoperate with Gigabit Ethernet, ATM, and IP deployments, thereby eliminating the need to use SONET/SDH overlays in long-haul optical network implementations. In contrast to WDM and DWDM operations, SONET/SDH technologies support information transport via a single channel or wavelength of light via each fiber optic strand.

Optical fiber installations represent a significant investment. However, both the WDM and DWDM transmission processes enable communications providers to significantly expand the bandwidth capacity of the fiber optic plant.

3.17.2   DWDM NETWORK ELEMENTS AND COMPONENTS

As with WDM, DWDM employs a mix of optical devices that include optical add-and-drop multiplexers (ADMs), optical gateways, optical cross-connects, optical switches, optical routers, optical wavelength converters, and optical terminal equipment. Like WDM equipment, DWDM devices support diverse optical interfaces; work in concert with technologies such as IP, SONET/SDH, WDM, Gigabit Ethernet, and ATM; and enable high-capacity, high-speed transmission rates.

DWDM implementations employ Optical Bi-directional Line Switched Ring (OBLSR) topologies to optimize bandwidth capacity of the in-place fiber optic plant and the amount of traffic transported via the Optical Layer. In DWDM implementations, an EDFA (Erbium Doped Fiber Optic Amplifier) with advanced filtering techniques extends the reach of optical signals and streamlines the transmission process by directly increasing signal power and eliminating the need for electronic conversion.

By using state-of-the-art wavelength routing protocols and gigabit routers, DWDM technology allows for dynamic provisioning of wavelength or channel capacity to support bandwidth-intensive applications as well. In order to accommodate DWDM requirements, a high-power tunable laser transmits light at specific wavelengths or channels with very narrow spectral widths. This procedure enables development of multiple, independent, narrowly spaced transmission channels or wavelengths on a single fiber optic strand.

DWDM system components such as tunable lasers optimize DWDM performance by eliminating adjacent channel interference, fostering flexible network operations, and supporting expanded bandwidth capacity. Tunable lasers also enable communications carriers to readily adapt to capacity changes by automatically adding or dropping channels or wavelengths in response to application requirements.

3.17.3 DWDM Implementation Considerations

In planning a DWDM network implementation, the goals and objectives for DWDM deployment, the current state of the in-place network infrastructure, and bandwidth requirements for accommodating diverse applications must be clarified. Costs for network installation, operations, management, administration, and maintenance must also be determined. In the initial phase of implementation, DWDM systems are typically connected to SONET/SDH transport equipment, support SONET/SDH applications, and employ the in-place fiber optic plant for long-haul optical network transmission.

It is important to note that DWDM standards are not yet fully defined. As a consequence, DWDM systems generally use proprietary management and multiplexing procedures.

Moreover, DWDM implementations are also subject to operational constraints. As an example, DWDM service is adversely affected by sudden temperature changes in optical-cross connects and manufacturing imperfections in the fiber optic plant. Signal attenuation, optical signal-to-noise ratio (OSNR), chromatic dispersion, and crosstalk at the Optical Layer also disrupt DWDM network performance.

3.17.4 WDM and DWDM Network Services

DWDM solutions streamline network management and maintenance services, eliminate the need for amplifiers in long-haul implementations, and provide increased bandwidth and channel capacity. DWDM solutions also support operations in short- and long-haul networks.

In DWDM deployments, densely packed parallel wavelengths significantly increase the bandwidth capacity of fiber optic networks. As with WDM, DWDM is a multiplexing technique that works in concert with optical fiber. In parallel with WDM, DWDM combines multiple optical signals that are amplified and transmitted as a group via a single fiber optic strand. WDM and DWDM solutions support an array of OC levels; work in concert with a mix of technologies, architectures, and protocols; and interoperate with standards-compliant Gigabit Ethernet, SONET/SDH, ATM, and IP networks.

WDM and DWDM standards are not fully defined or implemented. WDM and DWDM solutions are deployed primarily in testbeds environments. Capabilities of WDM and DWDM configurations are still under investigation.

3.18 WDM AND DWDM STANDARDS ACTIVITIES

Standards organizations that define specifications for WDM and DWDM technologies and equipment include the Internet Engineering Task Force (IETF), the American National Standards Institute (ANSI), the International Telecommunications Union-Telecommunications Standards Sector (ITU-T), the European Telecommunications Standards Institute (ETSI), and the Institute of Electrical and Electronics Engineers (IEEE). These organizations develop specifications that support optical network scalability, extendibility, and interoperability, and clarify approaches for enabling ATM-over-WDM, ATM-over-DWDM, IP-over-WDM, IP-over-DWDM, Gigabit Ethernet-over-WDM, Gigabit Ethernet-over-DWDM, SONET/SDH-over-WDM, and SONET/SDH-over-DWDM implementations in multivendor environments.

3.18.1 INTERNATIONAL TELECOMMUNICATIONS UNION-TELECOMMUNICATIONS STANDARDS SECTOR (ITU-T)

The ITU-T G.691 and the ITU-T G.692 Recommendations establish parameters for photonic network implementations. In addition, the ITU-T G.692 Recommendation clarifies procedures for implementation of a 100 GHz (Gigahertz) channel spacing plan for DWDM systems. ITU-T Recommendations also facilitate development of standardized point-to-point optical networks; enable standards-compliant SONET/SDH, WDM, and DWDM deployments; and describe Optical Layer or Layer 1 functions and capabilities.

3.19 WDM AND DWDM MARKETPLACE

3.19.1 CAMBRIAN COMMUNICATIONS

Cambrian Communications supports mixed-mode local and long-haul IP-over-mDWDM (Metropolitan DWDM) next-generation networks that facilitate scalable bandwidth, media-rich applications, and rapid on-time information delivery in cities along the East Coast of the United States. For example, Cambrian mDWDM networks are implemented in Baltimore, Philadelphia, New York City, and Washington, D.C. In addition, Cambrian Communications develops DWDM solutions for NSPs (Network Service Providers) that cannot afford dark fiber installation and the necessary optical equipment for initiating information transport. Dark fiber refers to an unused and unmetered in-place fiber optic plant with unlimited capacity that has not yet supported light pulse transmissions.

3.19.2   Cisco Systems and GNG Networks

Cisco Systems and GNG Networks (a Korean IP infrastructure carrier) support implementation of IP-over-SONET/SDH and IP-over-DWDM networks for enabling interconnectivity between major cities in Korea in a configuration that extends to approximately 8,400 kilometers.

3.19.3   Corvis Corporation

Corvis Corporation supports an all-optical IP-over-DWDM solution that operates at rates between 2.488 Gbps (OC-48) and 10 Gbps (OC-192), and supports lightwave transmission without signal regeneration over a distance that extends to 3,200 kilometers. The Corvis optical solution also enables transmissions between 400 Gbps and 2.4 Tbps (Terabits per second) to support disaster recovery operations and real-time coverage of televised news events and facilitate implementation of multigigabit and multiterabit ATM-over DWDM and IP-over-DWDM network configurations.

3.19.4   ECI Telecom

Developed by ECI Telecom, the LumiNet solution employs DWDM technology for enabling multimedia transmissions at rates reaching 400 Gbps in long-haul networks over distances that extend to 700 kilometers. In addition, ECI Telecom employs a DWDM bridge for interlinking an undersea optical fiber network to a terrestrial LumiNet multichannel DWDM system. LumiNet DWDM solutions interoperate with SONET/SDH, ATM, IP, and Gigabit Ethernet technologies. LumiNet also enables transmission of video-over-optical fiber (VOO) applications. With VOO implementations, vast amounts of high-quality uncompressed video signals are transmitted via multiple video channels over long distances. VOO deployments eliminate the need for MPEG-2 compression, optical repeaters, and optical amplifiers.

3.19.4.1   ECI Telecom and Bankveret Telecom

ECI Telecom and Bankveret Telecom, the Telecommunications Division of the National Railway Operator in Sweden, support a DWDM dual self-healing fiber optic ring network that provisions networking services for Stockholm, Malmo, and Goteborg.

3.19.5   Global Crossing

Global Crossing operates a terrestrial optical fiber network across the United States in a configuration called North American Crossing. This network employs a mix of technologies and services, including high-density WDM (HDWDM), SONET, Frame Relay, and IP, and supports rates ranging from 2.488 Gbps (OC-48) to 10 Gbps (OC-192). Approaches for enabling transmission at 1.28 Tbps via a single optical fiber strand are in development.

3.19.6   GTS (Global TeleSystems)

GTS (Global TeleSystems) supports implementation of an all-optical Trans-European DWDM Network for enabling information transport at 10 Gbps (OC-192). The GTS Trans-European Network also includes an SDH protected ring network overlay that operates directly on top of the DWDM infrastructure at rates of 155.52 Mbps (OC-3).

In addition, the GTS Trans-European DWDM Network features Points of Presence (PoPs) in key European cities and provisions communications coverage in an area that extends to 25,000 kilometers. The GTS DWDM optical configuration also supports bandwidth-intensive operations for 50 broadband City Enterprise Networks (CENs). CENs are municipal optical networks that enable networking operations in metropolitan areas that include Prague, Budapest, Berlin, Paris, London, Amsterdam, Frankfurt, Madrid, and Stockholm. For the Vienna CEN, GTS supports rates at 100 Gbps via an IP-over-DWDM infrastructure. CENs in Moscow, Kiev, and St. Petersburg enable applications that include E-business transactions, Internet connectivity, and high-speed multimedia transmission.

The GTS Trans-European DWDM Network consists of four subnetworks. Each subnetwork employs an IP-over-DWDM platform. Subnetworks are situated in Central, Southern, Northern, and Eastern Europe. The Eastern European Subnetwork interconnects CENs in metropolitan areas that include Nuremburg, Munich, Salzburg, and Dresden. With the addition of newly lit fiber optic pairs to the GTS DWDM configuration in 2000, this DWDM network currently facilitates trans-European networking operations and transmissions at rates of 1 Tbps.

3.19.7   Iaxis

Iaxis operates a multiterabit wireline trans-European network that became operational in 1999. The Iaxis trans-European WDM network extends over 3,000 kilometers. For Phase 1, network services were provisioned in Brussels, Antwerp, Rotterdam, Amsterdam, Frankfurt, Dusseldorf, Strasbourg, Paris, and London. Phase 2 marks the expansion of this network to cities in Germany and Switzerland. Upon completion, the Iaxis trans-European WDM network will provision communications coverage over an area that spans 12,000 kilometers.

3.19.8   Lucent Technologies

3.19.8.1   Lucent Technologies WaveWrapper Solution

Developed by Lucent Technologies, WaveWrapper solutions provision network management and administrative services, including error correction, performance monitoring, and ring-protection for safeguarding information transport in DWDM networks.

WaveWrappers are small digital wrappers that are placed around each wavelength that transits the optical network. In WaveWrapper implementations, IP, ATM, and Gigabit Ethernet traffic is transported directly over the DWDM infrastructure.

3.19.8.2 Lucent Technologies and Canoga Perkins

In 2000, Lucent Technologies and Canoga Perkins supported implementation of enhanced Wideband WDM (WWDM) technology in a trial network between the University of Washington and the Pacific/Northwest GigaPoP (P/NWGP) that enabled ultrafast transmissions. As noted, P/NWGP is a next-generation optical network testbed that interlinks research institutions and universities throughout Alaska, Idaho, Washington, Montana, and Oregon to one another and to advanced research networks such as I2.

3.19.8.3 Lucent Technologies and British Telecom

Lucent Technologies also supports a DWDM installation that enables terabit transmission rates via the Global Transport Network developed by British Telecom. Based on Lucent WaveStar technology, this network also supports ATM, Gigabit Ethernet, and IP operations directly over a DWDM infrastructure.

3.19.9 METROMEDIA FIBER NETWORK (MFN)

The Metromedia Fiber Network (MFN) is a metropolitan SONET-over-DWDM configuration that works in concert with POS, IP, ATM, Frame Relay, and Gigabit Ethernet technologies. The MFN platform enables seamless transmission at fixed costs. Currently, rates ranging from 155.52 Mbps (OC-3) to 10 Gbps (OC-192) are enabled. In addition, the MFN configuration supports optimal transmission rates reaching 2 Tbps.

Government agencies, energy and oil companies, banks, hotels, medical centers, and high-technology firms utilize MFN metropolitan implementations to access advanced Internet, intranet, and extranet applications. In addition, research centers, universities, scientific laboratories, and libraries employ MFN configurations for enabling telecollaborative research, tele-education services, and digital library initiatives.

3.19.9.1 MFN Implementations

The MFN dark fiber network is installed in New York City, Boston, Chicago, Philadelphia, Los Angeles, San Jose, Dallas/Fort Worth, Houston, and Washington, D.C. In San Francisco, MFN operates multiple overlapping and interlinked optical fiber rings for providing redundant, reliable, and secure services to corporations in urban business districts.

In 2000, MFN implemented an IP-over-DWDM network at George Mason University (GMU). This private optical network supports distance learning programs, LAN-to-LAN services, and advanced Web applications, and interlinks the main

GMU campus with satellite GMU campuses in the Northern Virginia cities of Arlington and Fairfax.

3.19.10 Qwest Communications Network

Qwest Communications supports implementation of a fiber optic network configuration that features a self-healing, dual-ring architecture. This network employs an IP-over-SONET-over-DWDM infrastructure to enable transmissions at rates of 10 Gbps (OC-192) across the United States. Designed for bandwidth-intensive and time-sensitive applications, this high-capacity network supports Web hosting services, IP VPNs (Virtual Private Networks), high-definition television (HDTV) programming, video streaming, and high-quality voice applications.

3.19.11 Teleglobe Communications

The Teleglobe Communications backbone network in the United States employs gigabit switch routers at Points of Presence (PoPs) that support network services and transmissions via a POS-over-DWDM infrastructure. PoPs are situated in New York City, Atlanta, Dallas, Chicago, Denver, San Francisco, and Washington, D.C. In the European Union, Teleglobe implements a DWDM 40-channel network that features a dual fiber-optic ring topology; interconnects London, Brussels, Paris, Amsterdam, and Frankfurt in a wide area networking configuration; and facilitates transmissions at rates of 10 Gbps (OC-192).

3.20 DARPA BROADBAND INFORMATION TECHNOLOGY (BIT) AND NEXT-GENERATION INTERNET (NGI) PROGRAMS

Research initiatives in the optical network domain are funded by the U.S. Department of Defense Advanced Research Projects Agency (DARPA) Information Technology Office (ITO). DARPA initiatives foster development of advanced all-optical network architectures, technologies, protocols, and multiplexing techniques for optimizing available bandwidth on fiber-optic strands. These initiatives enable technical innovations in optical network services, optical signaling and switching techniques, and optical devices, and support rapid and transparent network reconfiguration following sudden network failures. DARPA research enables real-time applications in tele-education and telemedicine, ultrafast access to and retrieval of multigigabits of data in massive datasets, and 3-D simulations.

The DARPA Next-Generation Internet (NGI) and the DARPA Broadband Information Technology (BIT) Programs sponsor implementation of all-optical technologies in global optical networks that support terabit transmission rates. Findings contribute to the enhancement of the NGI infrastructure and new generations of optical services. WDM technologies developed under the BIT Program are implemented by Sprint, WorldCom, AT&T, and Qwest. Participants in the BIT Program include the Multiwavelength Optical Networking Consortium (MONETC), the National Transparent All Optical Network Consortium (NTONC), ICON

(IBM/Corning Optical Network), and the WDM with Electronic Switching Technology (WEST).

3.20.1 ALL-OPTICAL NETWORKING CONSORTIUM

AT&T, Digital Equipment Corporation (DEC), and the Massachusetts Institute of Technology (MIT) participate in the All-Optical Networking Consortium. This consortium supports implementation of an optical networking testbed in the Boston metropolitan area that uses a 20-channel WDM system. Data rates per wavelength range from 10 Mbps to 10 Gbps. Consortium participants develop next-generation ultrafast optical networks that employ TDM technology for enabling transmissions at terabit rates and identify methods and approaches to facilitate shared utilization of supercomputers, media banks, and high-speed video servers.

3.20.2 BOSTON TO WASHINGTON, D.C., FIBER-OPTIC NETWORK (BoSSNET)

A SuperNet regional testbed initiative, BoSSNET (Boston-South Optical Network) is a high-capacity, dark fiber optic network that upon completion will extend from Boston, Massachusetts, to Washington, D.C. BoSSNET features a fiber optic ring topology that enables access to high-capacity services via a multichannel WDM infrastructure. Transmission rates at 10 Gbps (OC-192) are supported. The Massachusetts Institute of Technology (MIT) participates in this project.

3.20.3 HIGH-SPEED CONNECTIVITY CONSORTIUM (HSCC)

The High-Speed Connectivity Consortium (HSCC) supports implementation of a nationwide multigigabit WDM network testbed that supports voice, video, and data transmission at 2.488 Gbps (OC-48). This all-optical network provisions advanced services with QoS assurances and supports applications that include data mining, electronic commerce (E-commerce), telemedicine, distance learning, video telephony, and collaborative teleworking. The University of Washington, Carnegie Mellon University (CMU), Qwest Communications, and Ciena Corporation participate in the HSCC testbed project.

3.20.4 MULTIWAVELENGTH OPTICAL NETWORK (MONET) AND THE MONET CONSORTIUM (MONETC)

The Multiwavelength Optical Networking Consortium (MONETC) defines methods and techniques for developing and implementing the Multiwavelength Optical Network (MONET). MONET is a nationwide optical network that employs a WDM backbone for enabling bandwidth-intensive government and business applications. Currently in development, this testbed employs a dual-ring topology that supports ATM and SONET overlays and provides metropolitan and regional connections.

The MONET initiative enables an assessment of procedures for effectively deploying an all-optical network infrastructure that enables trouble-free operations and an evaluation of the capabilities of multiwavelength lasers. MONETC participants include the University of California at Davis (UC Davis), AT&T, Lucent Technologies, Bellcore, BellSouth, and SBC.

A long-haul MONET segment interlinks optical testbeds in New Jersey and Washington, D.C. This segment supports transmissions at rates of 2.488 Gbps (OC-48) over a distance that extends to 2,000 kilometers. In New Jersey, an eight-channel short-haul MONET WDM testbed segment also supports information transmission at rates reaching 2.488 Gbps (OC-48).

The Washington, D.C., Area Network, a MONET WDM testbed segment, interlinks the National Security Agency (NSA), DARPA, the Naval Research Laboratory (NRL), the U.S. Department of Defense Intelligence Agency (DIA), NASA, and the U.S. Department of Defense Information Systems Agency (DISA).

3.20.5 NATIONAL TRANSPARENT OPTICAL NETWORK, PHASE II (NTON, PHASE II) AND THE NTON CONSORTIUM (NTONC)

Currently in development, the National Transparent All Optical Network, Phase II (NTON II) is a 2,500-kilometer, high-performance, high-speed, all-optical WDM network that interconnects academic institutions, government agencies, research centers, and private-sector laboratories on the West Coast of the United States. NTON II provides services in an area of coverage that extends from San Diego, California to Seattle, Washington, and enables transmission rates at 20 Gbps. NTON II PoPs (Points of Presence) are situated in San Francisco, Portland, and Los Angeles.

The National Transparent All Optical Network, Phase II (NTON II) employs a WDM platform with each optical fiber strand carrying four wavelengths or channels. Currently, the NTON II installation provides direct connections to major universities on the West Coast at 2.488 Gbps. The NTON II segments from San Francisco to Los Angeles, Seattle to Portland, and Los Angeles to San Diego provision access to high-capacity broadband applications and services.

In the San Francisco Bay area, the NTON II segments include an optically switched ring configuration and a regional network that interlinks the University of California at Berkeley and Silicon Valley to the NTON II platform. In addition, the NTON II maintains interconnections to Internet2 GigaPoPs and broadband research networks such as vBNS+, Abilene, the NASA Research and Education Network (NASA-NREN), the U.S. Department of Defense Research and Engineering Network (DREN), and SuperNet.

The NTON II testbed supports network experiments and pilot projects to determine the capabilities of optical network equipment, all-optical network architectures, and optical protocols in supporting reliable network operations and QoS (Quality of Service) guarantees. Moreover, the NTON II testbed enables high-energy physics and nuclear physics projects, digital library initiatives, implementation of interactive terrain visualization systems, distributed simulation testing of network weapons systems, and distributed healthcare imaging applications. In 1999, the NASA Jet Propulsion Laboratory, the first entity to utilize the NTON II segment between Los Angeles and San Francisco, transmitted multiple uncompressed streams of high-definition video to designated locations.

The National Transparent All Optical Network Consortium (NTONC) facilitates implementation of telecollaborative all-optical network design projects and supports telemedicine projects in partnership with the Kaiser Permanente Division of Research. In addition, the NTONC supports development of optical networking components and optical technologies for the Next-Generation Internet (NGI) segment developed by the WEST (WDM with Electronic Switching Technology) Coast Team.

Participants in the NTON II Consortium include the University of Southern California (USC), the University of California at San Diego (UC San Diego), and the University of California at Berkeley (UC Berkeley). Moreover, Case Western Reserve, Columbia and Stanford Universities; the NASA Jet Propulsion Laboratory; and the Lawrence Livermore National Laboratory also participate in NTON II research. NTONC vendors and communications carriers include Nortel Networks, Sprint, United Technologies, Microsoft, and the Boeing Information Space and Defense System Group. (See Figure 3.5.)

3.20.6 SuperNet

The SuperNet configuration provisions network services via an IP-over-WDM infrastructure that serves as a prototype for next-generation all-optical networks. SuperNet consists of several interlinked and interoperable regional testbeds, including NTON II, the High-Speed Connectivity Consortium (HSCC) Network, BoSSNET, and ATDNet. Additional SuperNet testbeds include ONRAMP (Research and Demonstration of a Next-Generation Internet Optical Network for Regional Access using Multiwavelength Protocols), MONET, and ABone (Active Network Backbone).

The SuperNet infrastructure employs gigabit and terabit IP routers and enables transmission rates in excess of 100 Gbps. The SuperNet configuration supports innovations in OTDM (Optical Time-Division Multiplexing) and implementation of reconfigurable optical network technologies.

3.20.7 WDM with Electronic Switching Technology (WEST) Coast Team

The WEST (WDM with Electronic Switching Technology) Coast Team supports implementation of a next-generation optical cross-connect that facilitates transmissions at 120 Gbps via a WDM infrastructure. This cross-connect enables signal regeneration, data switching services, fast circuit reconfiguration for network protection, and multiplexing and demultiplexing of WDM optical signals, and streamlines network management operations to enable fast, reliable, and dependable multimedia transport.

WEST Coast Team participants include the Rockwell Science Center, Ortel Corporation, the California Institute of Technology (Cal Tech), and the NASA Jet Propulsion Laboratory. In addition, the University of California at Los Angeles (UCLA), the University of California at San Diego (UC San Diego), and the University of California at Santa Barbara (UC Santa Barbara) participate in WEST Coast Team research initiatives.

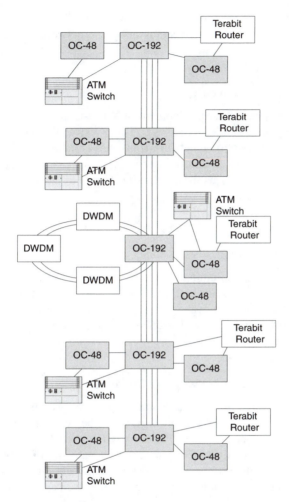

FIGURE 3.5 DWDM implementation in the National Transparent Optical Network, Phase II (NTON II).

3.21 U.S. WDM (WAVELENGTH DIVISION MULTIPLEXING) AND DWDM (DENSE WDM) UNIVERSITY INITIATIVES

3.21.1 MASSACHUSETTS

3.21.1.1 Research and Demonstration of a Next-Generation Internet Optical Network for Regional Access using Multiwavelength Protocols (ONRAMP)

Sponsored by the Massachusetts Institute of Technology (MIT) Lincoln Laboratory as part of the Next-Generation Internet (NGI) SuperNet testbed program, the

ONRAMP project identifies approaches for optimizing WDM bandwidth utilization and supports development of next-generation optical networking architecture.

3.21.2 NEW JERSEY

3.21.2.1 Princeton University

Princeton University supports development of a 100-Gbps scalable multiprocessor interconnection system. This system uses OTDM (Optical Time-Division Multiplexing) and conventional electronic network solutions. Capabilities of an ultrafast Terahertz Optical Asymmetric Demultiplexer (TOAD) in enabling rapid information transport are assessed. Methods for interlinking TOAD to the MONET WDM backbone network are explored.

3.21.3 NEW YORK

3.21.3.1 Columbia University

The Lightwave Group at Columbia University examines WDM capabilities in provisioning LAN interconnectivity, high-speed ATM applications, and interactive video services. In addition, the Lightwave Group also conducts research that contributes to innovations in all-optical network architectures, operations, and services.

3.21.4 NORTH CAROLINA

3.21.4.1 North Carolina Networking Initiative (NCNI)

A statewide collaborative project, the North Carolina Networking Initiative (NCNI) supports deployment of DWDM technology in a fiber optic testbed that links the University of North Carolina at Chapel Hill (UNC-CH), North Carolina State University (NCSU), and Duke University. NCNI evaluates the capabilities of a DWDM dark fiber network testbed in handling campus and research traffic, and supports development of network monitoring devices for enabling seamless management operations.

3.21.5.2 University of North Carolina at Chapel Hill (UNC-CH)

In 1998, the Center for Advanced Video Network Engineering Research (CAVNER) and the Networking and Communications Group at the University of North Carolina at Chapel Hill (UNC-CH) deployed a first-generation DWDM network for supporting real-time transmission of uncompressed and compressed full-motion video across campus. This implementation enables tele-education services, scientific imaging, and telemedicine applications at multigigabit rates over the in-place fiber optic plant. SONET-over-DWDM and Gigabit Ethernet-over-DWDM technical services are supported.

3.22 INTERNATIONAL OPTICAL RESEARCH NETWORK INITIATIVES

International optical research network initiatives foster development of sophisticated techniques and methods to enable implementation of next-generation extendible and scalable optical networks that robustly support multimedia transmission at multigigabit rates. Representative WDM and DWDM implementations in the international area are highlighted in this section.

3.22.1 CANADA

3.22.1.1 CA*net3 (Canadian Network for the Advancement of Research, Industry, and Education, Phase 3)

*3.22.1.1.1 CA*net3 Technical Features and Functions*

Operational since 1998, CA*net3 employs a DWDM Optical Layer that works in concert with IP network overlays in enabling high-speed networking services and applications. The CA*net3 infrastructure supports ATM-over-DWDM, SONET-over-DWDM, Gigabit Ethernet-over-DWDM, and Gigabit Ethernet-over-Coarse Wavelength Division Multiplexing (CWDM) technologies. CA*net3 network nodes are situated at Calgary, Regina, Vancouver, Winnipeg, Toronto, Ottawa, Fredericton, and Halifax. At Chicago, CA*net3 supports connections to the STAR TAP NAP to enable peer-level traffic exchange.

CA*net3 utilizes fiber-optic dual-ring configurations for doubling network capacity and ensuring a backup optical fiber solution in the event of a fiber cut. Currently, CA*net3 supports active network connections between Toronto and Winnipeg. Upon completion, CA*net3 will enable information transport at rates of 40 Gbps across Canada.

*3.22.1.1.2 CA*net3 Applications and Services*

CA*net3 supports implementation of Optical Regional Advanced Networks (ORANs) as replacements for RANs (Regional Area Networks). CA*net3 GigaPoPs enable access to and distribution of broadband resources via ORANs. Schools, universities, and research centers that require on-demand bandwidth for supporting volume-intensive transmissions are among the initial users of CA*net3.

CA*net3 is not intended to replace CA*net II or compete with the commercial Internet. Rather, CA*net3 serves as an advanced academic and research testbed for enabling telecollaborative initiatives in fields that include E-commerce, E-government, telemedicine, and tele-education.

*3.22.1.1.3 CA*net3 Initiatives in the Educational Domain*

CA*net3 (Canadian Network for the Advancement of Research, Industry, and Education, Phase 3) is an enabler of diverse initiatives in the tele-education domain that optimize utilization of the high-performance, bandwidth-intensive DWDM networking infrastructure. These initiatives explore the role of videoconferencing in enabling telelearning activities in distributed educational environments. CA*net3 projects also facilitate utilization of interactive tele-education modules by students attending Canadian veterinary colleges and the implementation of telementoring and telecoaching sessions for adult learners.

In addition, CA*net3 initiatives support integration of multimedia streaming technologies into classroom activities and deployment of a policy-caching service that enables specified Web content to be distributed automatically to the cache in anticipation of student requests. In 1999, Canada and Mexico signed an agreement to support advanced collaborative tele-education programs and shared use of remote laboratories and digital library resources available on CA*net3 and CUDI, the NREN (National Research and Education Network) of Mexico.

3.22.1.1.4   CANARIE and Bell Canada Consortium

The Canadian Network for the Advancement of Research, Industry, and Education (CANARIE) and the Bell Canada Consortium sponsor CA*net3 deployment. Participants in the Bell Canada Consortium include Nortel Networks and Cisco Systems. Approaches to extend gigabit Internet connectivity to every Canadian residence by 2005 are under consideration.

3.22.1.1.5   Photonic Systems Group

Sponsored by the National Resarch Council of Canada and the Institute of National Measurement Standards, the Photonic Systems Group supports implementation of optical solutions to enable massive data storage, long-haul communications, and network interconnectivity. The Photonic Systems Group also promotes implementation of all-optical networks that transport voice, video, and data at 100 Gbps and higher rates, and fosters development of fiber-optic network services and applications.

In a joint initiative, the Photonic Systems Group, Nortel Technology, and the Georgia Institute of Technology implemented an all-optical gigabit network enabling transmission rates at 100 Gbps. In another collaborative project, the Photonic Systems Group and McGill University benchmarked capabilities of optical devices such as tunable lasers and optical amplifiers in supporting high-capacity operations in an all-optical network testbed installation.

3.22.1.1.6   Optical Processing and Computing Consortium of Canada (OPCOM)

Sponsored by the National Research Council of Canada and the Ottawa Carleton Resarch Institute, the Optical Processing and Computing Consortium of Canada (OPCOM) promotes development of optical-to-electronic and electronic-to-optical solutions and implementation of advanced photonic systems.

3.22.2   United Kingdom

3.22.2.1   Centre for Communications Systems Research (CCSR)

The Centre for Communications Systems Research (CCSR) works with the University of London, Cambridge University, Imperial College, King's College, University College London (UCL), and Queen Mary Westfield College in supporting the design and deployment of next-generation, wavelength-routed optical networks (WRONs). The CCSR also develops methods for managing ATM-over-SDH configurations that deliver next-generation applications and procedures for effectively utilizing WDM technology in trans-continental networks.

3.23 EUROPEAN COMMISSION ADVANCED COMMUNICATIONS TECHNOLOGY AND SERVICE (EC-ACTS) PROGRAM

Photonics initiatives sponsored by the European Commission Advanced Communications Technology and Services (EC-ACTS) Program fostered development of Optical Layer technologies, optical equipment, and all-optical network architectures. Representative EC-ACTS optical initiatives are highlighted in this section.

3.23.1 BROADBAND OPTICAL NETWORK USING ATM PASSIVE OPTICAL NETWORK ACCESS FACILITIES IN REALISTIC TELECOMMUNICATIONS ENVIRONMENTS (BONAPARTE)

The BONAPARTE initiative supported development of a reliable, extendible, and affordable broadband ATM Passive Optical Network (APON). This APON solution fostered delivery of vast amounts of data over a distance of 10 kilometers. Aggregated transmission rates of 622.08 Mbps upstream and downstream were enabled. APON pilot tests were carried out in Madrid, Turin, Hamburg, and Basel at sites that were interconnected via a pan-European ATM network. BONAPARTE teleteaching trials were conducted in university environments. Hospital trials supported teleradiology teleconsultations and enabled real-time interactions between off-site physicians and patients undergoing radiological examinations.

3.23.2 HIGHWAY

The HIGHWAY initiative explored the functionality of ultra-high-speed optical-electronic components and subsystem technologies for enabling reliable operations in 40 Gbps TDM optical configurations. Field trials demonstrated the effectiveness of ultrafast, high-capacity optical networks in supporting bandwidth-intensive applications with an aggregate capacity above 1 Tbps.

3.23.3 METROPOLITAN OPTICAL NETWORK (METON)

The METON initiative supported development of a next-generation WDM Transport Layer to enable deployment of a hybrid ATM and SDH multiservice Metropolitan Area Network (MAN). METON trials were carried out in the Stockholm segment of the Scandinavian Gigabit Network.

3.23.4 OPTICAL PAN-EUROPEAN NETWORK (OPEN)

The OPEN (Optical Pan-European Network) initiative explored the suitability of using an optical pan-European overlay network employing WDM technology for interlinking major European municipalities. Capabilities of a multiwavelength pan-European optical networking configuration based on a mesh topology were evaluated. The performance of networking operations and services were examined in field trials between Paris and Brussels, and between Oslo and Thisted.

3.23.5 PAN-EUROPEAN LIGHTWAVE CORE AND ACCESS NETWORK (PELICAN)

The PELICAN project contributed to the implementation of a pan-European WDM optical network. Field trials identified approaches that enabled implementation of an optical transport WDM network built on an in-place optical fiber infrastructure that interconnected cities in France and Belgium. This broadband, full-service optical network also supported ATM-over-WDM and IP-over-WDM technologies and enabled distance learning and collaborative telecommuting applications.

3.24 EUROPEAN COMMISSION INFORMATION SOCIETY TECHNOLOGIES (EC-IST) PROGRAM

A successor to the EC-ACTS Program, the European Commission Information Society Technologies (EC-IST) Program initiated operations in 1999. The EC-IST Program sponsors network trials to benchmark the performance of multimedia applications and services enabled by new technologies and supports assessment of advanced wireline and wireless communications protocols, architectures, standards, and technologies in network testbeds. In addition, the EC-IST Program facilitates development and deployment of innovative high-performance networking services such as IPv6 (Internet Protocol version 6); advanced third-generation (3G) cellular networks; and high-speed, next-generation optical networking configurations that employ ATM-over-WDM (Wavelength Division Multiplexing) and ATM-over-DWDM (Dense WDM) solutions.

The EC-IST program also supports the establishment of a universal network infrastructure that provisions access to information resources at any time and from any place on the planet, regardless of user location or mobility. Representative EC-IST initiatives in the optical networking domain are highlighted in this section.

3.24.1 ATRIUM

The ATRIUM testbed enables development of advanced IP routers to support MPLS-over-WDM (MultiProtocol Label Switching-over-WDM) services and applications. The capabilities of MPLS-over-WDM solutions in provisioning reliable QoS (Quality of Service) transmissions and rapid service restoration in the event of network node or loop failures are evaluated.

3.24.2 DAVID

The DAVID initiative explores approaches for developing a Packet-over-DWDM network that integrates optical and electronic technologies and components and facilitates high-performance, high-speed data and voice integration. This multiservice network supports bandwidth-intensive multimedia transmissions at rates greater than 1 Tbps (Terabits per second) and provides metropolitan and long-haul backbone network services.

3.24.3 METEOR

The METEOR initiative fosters development of a DWDM optical metropolitan network that employs an optical ring topology. Designed to support transmission rates at 40 Gbps — and ultimately at 1 Tbps — this DWDM infrastructure enables seamless access to bandwidth-intensive multimedia applications. Capabilities of optical network elements such as optical add-and-drop multiplexers (OADMs) in provisioning transparent optical transmission functions and network management services are also examined.

3.25   UNDERSEA OPTICAL NETWORK SOLUTIONS

Spiraling demand for high-capacity networks to support multimedia services and applications contributes to deployment of undersea optical networks. Initially implemented in the mid-1990s, undersea optical networks originally employed SONET/SDH infrastructures. Present-day undersea initiatives typically employ WDM and DWDM technologies for enabling oceanic transmissions at multigigabit and multiterabit rates over long distances. Representative initiatives in the underwater optical network domain are now explored.

3.25.1   ALCATEL

3.25.1.1   SEA-ME-WE 3

Alcatel provisions network management operations for undersea network configurations such as SEA-ME-WE 3. Featuring ATM-over-WDM and SDH-over-WDM technologies, the SEA-ME-WE 3 configuration upon completion will extend over 38,000 kilometers and feature 40 landing points in 34 countries between the European Union and Asia. Transmission rates ranging from 2.488 Gbps to 50 Gbps over two pairs of undersea fiber optic cabling will be enabled. Each optical fiber strand supports eight wavelengths or fiber optic channels. Each wavelength or channel supports multimedia transmission at 2.488 Gbps (OC-48 or STM-16). Network segments between Hong Kong and Korea are currently in operation.

3.25.2   AT&T

AT&T employs WDM technology to provide communications services via a trans-Atlantic fiber optic ring network to landing points in the United States, Great Britain, and France. This undersea network also utilizes forward error correction (FEC) control mechanisms for enabling reliable transmission of multimedia applications at 10 Gbps. Infrastructure upgrades supporting transmission rates at 20 Gbps are in development.

3.25.3   FLAG (FIBER OPTIC LINK AROUND THE GLOBE) TELECOM

FLAG (Fiber Optic Link Around the Globe) Telecom operates underwater networks that support SDH-over-DWDM technologies and administers network landing points in Italy, Egypt, Spain, Jordan, the United Arab Emirates, Saudi Arabia, Malaysia, India, Korea, Thailand, and China. FLAG Telecom gateways situated in Eastern Africa, the Asia-Pacific region, the United Kingdom, South America, the Middle East, Latin America, the European Union, and the United States are designed to enable worldwide connectivity. FLAG Telecom also supports interconnections of undersea networks to terrestrial broadband networks such as the FLAG Europe-Asia fiber optic network and the GTS Trans-European DWDM Network.

FLAG Telecom buries optical fiber cable below the seabed to eliminate disruption to undersea habitats and to safeguard undersea networks from damage caused by strong harbor currents, adverse weather conditions, electromagnetic interference, and disruptions associated with congested shipping lanes. The FLAG Telecom undersea fiber optic system supports rates at 2.488 Gbps (OC-48 or STM-16) for enabling high-speed multimedia transmission, fully protected voice services, and Internet broadcasts. Upon completion, the FLAG undersea DWDM network will interlink landing points in the United Kingdom and Japan via the Pacific Ocean, the South China Sea, the Indian Ocean, the Red Sea, the Mediterranean Sea, and the Atlantic Ocean.

3.25.3.1 FA-1 (Flag Atlantic-1) Trans-Atlantic Network

As with other FLAG undersea initiatives, the FA-1 (Flag Atlantic-1) Trans-Atlantic Network employs laser-generated lightwaves for transmitting digital information via WDM and DWDM technologies. The FA-1 Trans-Atlantic Network features a dual terrestrial and undersea optical fiber infrastructure for interconnecting landing points in the United States, the European Union, the Asia-Pacific region, and the Middle East. In addition, the FA-1 Trans-Atlantic Network provisions 5 Tbps of total raw capacity and supports secure digital transmissions via an SDH-over-DWDM backbone network at rates reaching 2.5 Tbps.

3.25.3.2 FL-A1 (FLAG ASIA) Segment

Currently in development by FLAG Telecom, the FL-A1 (Flag Asia) segment is an undersea DWDM network that provisions services for international communications carriers. GTS expects to lease protected capacity on FL-A1 for enabling seamless connections at 400 Gbps and unprotected capacity at 800 Gbps in an area that extends from landing points in New York to the GTS Trans-European DWDM Network.

3.25.4 FRANCE TELECOM

France Telecom employs an SDH-over-WDM optical infrastructure for fiber optic undersea networks that support high-speed, full-duplex voice, video, and data transmission, and provision on-demand bandwidth. France Telecom participates in the Americas II and Atlantis-2 undersea fiber optic initiatives.

3.25.4.1 Americas II Undersea Network

The Americas II fiber optic undersea network links North America, South America, and the Caribbean region in a WDM configuration that extends over 8,000 kilometers. This system features four pairs of fiber optic cabling and a landing point in Martinique. Information transport at 40 Gbps is supported.

3.25.4.2 Atlantis-2 Undersea Network

The Atlantis-2 WDM undersea fiber optic network fosters interconnections between landing points in Portugal, the Canary Islands, the Cape Verde Islands, Senegal,

Brazil, and Argentina. This configuration employs two pairs of fiber optic cabling. Each fiber optic strand supports rates ranging between 5 Gbps and 20 Gbps. The network extends to 12,000 kilometers.

3.25.5 GLOBAL CROSSING

3.25.5.1 Atlantic Crossing 1 (AC-1) and Atlantic Crossing 2 (AC-2) Undersea Networks

Sponsored by Global Crossing, the Atlantic Crossing 1 (AC-1) trans-Atlantic DWDM undersea fiber optic network links landing points in Germany, the Netherlands, the United Kingdom, and New York City in a fiber optic configuration extending to 14,000 kilometers. In a related initiative, Global Crossing also supports a 2.5 Tbps DWDM undersea network configuration called Atlantic Crossing 2 (AC-2). The AC-2 undersea network interoperates with the AC-1 installation.

3.25.5.2 Mid-Atlantic Crossing and Pan American Crossing Undersea Networks

Developed by Global Crossing, the Mid-Atlantic Crossing undersea fiber optic network interconnects landing points in the Caribbean region and the Eastern United States in a WDM configuration that extends to 7,500 kilometers. Also a Global Crossing project, the Pan American Crossing undersea WDM fiber optic network interconnects landing points in the Caribbean region, Central America, Mexico, and the Western United States in a configuration covering 8,900 kilometers.

3.25.5.3 Global Crossing Undersea Fiber Optic Network in Japan

In Japan, Global Crossing supports implementation of a high-speed undersea fiber optic network equipped with DWDM technology that interlinks landing points in Tokyo, Nagoya, and Osaka in a configuration that extends to 1,200 kilometers.

3.25.6 IAXIS

Iaxis supports construction of a 10,000-kilometer WDM network in the Mediterranean Sea that interlinks landing points in Africa, the European Union, and the Asia-Pacific region.

3.25.7 SOUTH ATLANTIC TELEPHONE/WESTERN CABLE/SOUTHERN AFRICA FAR EAST, PHASE 3 (SAT-3/WASC/SAFE)

Currently in development, SAT-3/WASC/SAFE is an undersea fiber optic network that accommodates Africa's expanding telecommunications requirements. SAT-3/WASC segments support connections between landing points at sites in Senegal, Portugal, the Ivory Coast, Ghana, Nigeria, and Angola. A WDM Optical Layer enables transmission rates at 20 Gbps, with upgrades to 40 Gbps available. SAFE segments interlink landing points in South Africa, Mauritius, and Malaysia.

As with other undersea networks, SAT-3/WASC/SAFE implementation involves mapping the topography of the ocean bed with sonar prior to fiber optic installation. Aluminum sheaths protect optical fiber from shark bites. These sheaths are situated between depths of 1,000 meters and 3,000 meters. Moreover, fiber optic cabling at shallower depths is also encased in aluminum sheaths to prevent damage from rocks, fishing nets, and anchors. A satellite remote control and surveillance system for SAT-3/WASC/SAFE aids in the identification of network faults. Upon completion, this network will support transmission over a maximum of eight fiber optic pairs for enabling a total rate of 80 Gbps.

3.25.8   TELECOM NEW ZEALAND, OPTUS, SOUTHERN CROSS CABLES, AND WORLDCOM

3.25.8.1   Trans-Pacific Southern Cross Network

Telecom New Zealand, Optus, Southern Cross Cables, and WorldCom sponsor the Trans-Pacific Southern Cross Network. Presently in development, this self-restoring, undersea fiber optic, long-haul configuration interlinks landing points in Australia, New Zealand, Fiji, Hawaii, and the West Coast of the United States in a terrestrial and undersea fiber optic network configuration. A high-speed, high-capacity config-uration, the Trans-Pacific Southern Cross Network employs SONET/SDH-over-WDM/DWDM technologies for provisioning an initial capacity of 40 Gbps on each fiber optic pair. Upon completion, the Trans-Pacific Southern Cross Network will enable transmission rates reaching 480 Gbps and transport high-quality, high-per-formance voice, video, and data signals in an area of coverage that extends to 30,500 kilometers.

3.26   SUMMARY

The spiraling increase in the volume of information traffic transmitted across com-munications networks contributes to the development of terrestrial and undersea optical fiber network solutions for provisioning additional network capacity. The communications capacity of fiber optic cabling is immense. According to the NSF (National Science Foundation), a single strand of optical fiber possesses an accessible bandwidth approaching 100,000 Gigahertz (GHz). By comparison, the entire radio spectrum managed by the Federal Communications Commission (FCC) for enabling FM and AM radio and television programming, cellular telephony, and satellite services consists of approximately 60 GHz.

In this chapter, representative SONET/SDH solutions in enabling dependable and reliable voice, video, data, and still-image transmission, and rapid access to bandwidth-intensive applications are described. Capabilities of SONET/SDH imple-mentations in fostering reliable tele-education and telemedicine programs, telecol-laborative research, and E-government activities are examined.

Optical networks are evolving at a remarkable pace. Unprecedented demand for high-speed, high-capacity optical solutions contributes to the popularity of optical fiber networks that employ third-generation WDM and DWDM optical technologies and the emergence of optical network configurations such as PONs, SuperPONs, APONs, and AONs. Distinguishing characteristics of these optical network configurations are highlighted.

Distinctive attributes of WDM and DWDM technologies, architectures, components, and solutions are introduced. The role of WDM and DWDM technologies in contributing to the development of a new Optical Layer that serves as a Sublayer of the Physical Layer or Layer 1 of the OSI (Open Systems Interconnection) Reference Model is highlighted. Network services provided by third-generation WDM and DWDM technologies in supporting extendible, reliable, and dependable next-generation network configurations and provisioning access to bandwidth-intensive multimedia applications with differentiated QoS guarantees in the educational, corporate, medical, and governmental sectors are reviewed. Representative examples of terrestrial and undersea WDM and DWDM configurations in supporting ultrafast information transport and providing unprecedented bandwidth for next-generation Internet innovations and services are also examined.

3.27 SELECTED WEB SITES

Advanced Research and Development Network Operations Center (ARDNOC). CA*net3 Optical Internet Backbone. Last modified on April 23, 2001.
Available: http://www.canet3.net/optical/optical.html
Advanced Technology Demonstration Network. ATDNet. Last modified on August 8, 2001.
Available: http://www.atd.net/
Alcatel. Submarine Networks.
Available: http://www.alcatel.com/submarine/
All-Optical Networking Consortium. Homepage. Last modified on July 30, 1997.
Available: http://www.ll.mit.edu/aon/index.html
Canadian Network for the Advancement of Research, Industry, and Education (CANARIE). CA*net3. Last modified on September 25, 2001.
Available: http://www.canarie.ca/advnet/canet3.html
European Commission. The ACTS Information Window.
Available: http://www.de.infowin.org/
FLAG (Fiber Optic Link Around the Globe) Telecom. Overview.
Available: http://www.flag.bm/index_e1.htm
France Telecom. Welcome to France Telecom's Marine Activities Website.
Available: http://www.marine.francetelecom.fr/english/home.htm
Global Crossing. The Expanding Network.
Available: http://www.globalcrossing.com/network.html?bc = Network
High-Speed Connectivity Consortium (HSCC). Home Page. Last modified on July 12, 2001.
Available: http://www.hscc.net/
LightReading: The Global Site for Optical Networking.
Available: http://www.lightreading.com/

Massachusetts Institute of Technology Lincoln Laboratory Advanced
    Networks Group. BoSSNET (Boston-South Network).
    Available: http://www.ll.mit.edu/AdvancedNetworks/bossnet.html
Metropolitan Research and Education Network. MREN: Advanced networking
    for advanced applications. Last modified on February 21, 2001.
    Available: http://www.mren.org/
Optical Domain Service Interconnect Coalition.
    Available: http://www.odsicoalition.com/index.asp
PennWell Corporation. LIGHTWAVE. Fiber Optic Communications,
    Bandwidth Access, and Telecommunications.
    Available: http://lw.pennwellnet.com/home.cfm
Wisconsin Bureau of Network Services. BadgerNet Overview. Last modified
    on August 20, 2001.
    Available: http://www.doa.state.wi.us/dtm/bns/overview/history.htm
University Corporation for Advanced Internet Development (UCAID). About
    Abilene.
    Available: http://www.ucaid.edu/abilene/html/about.html
U.S. Department of Defense Advanced Research Projects Agency (DARPA)
    Information Technology Office (ITO). Next-Generation Internet (NGI).
    Project List.
    Available: http://www.arpa.mil/ito/research/ngi/projlist.html
U.S. Department of Defense Advanced Research Projects Agency (DARPA)
    Information Technology Office (ITO). Next-Generation Internet (NGI).
    SuperNet Testbed.
    Available: http://www.arpa.mil/ito/research/ngi/supernet.html

4 Ethernet Networks

4.1 INTRODUCTION

The popularity of Web entertainment, E-commerce, and distance learning, and the concurrent demand for ready access to Web applications in the local networking environment, contribute to the widespread implementation of Ethernet communications networks. A flexible, reliable, scalable, and dependable technology, the Ethernet technology suite sustains intranet, extranet, and Internet connectivity in LAN (Local Area Network) environments and interoperates with wireline and wireless solutions.

The Ethernet technology suite is defined by a series of specifications and supplements associated with the IEEE (Institute of Electrical and Electronics Engineers) 802.3 family of LAN standards. An Ethernet LAN supports operations in a delimited geographic area such as a classroom, a multistory office building, a factory, or a cluster of buildings on a university campus, and enables information exchange and shared workgroup applications over a common communications medium. The multiservice Ethernet platform interworks with technologies that include Frame Relay (FR), FDDI (Fiber Data Distributed Interface), Fibre Channel (FC), DSL (Digital Subscriber Line), cable modem, WDM (Wavelength Division Multiplexing), and DWDM (Dense WDM).

A conventional, in-place, 10 Mbps (Megabits per second) Ethernet platform can be readily upgraded to Fast Ethernet, Gigabit Ethernet, and ultimately 10 Gigabit (Gigabits per second or Gbps) Ethernet. As a result, institutions and organizations with in-place 10 Mbps Ethernet implementations that migrate to faster and more powerful Ethernet solutions preserve their investments in Ethernet equipment while gaining additional bandwidth capacity and infrastructure services.

A ubiquitous LAN solution, Ethernet is the technology of choice for local network implementations in schools, colleges, universities, libraries, research centers, scientific organizations, government agencies, corporations, and hospitals. As noted in Chapter 3, Ethernet is also an enabler of high-performance high-capacity metropolitan optical network solutions. (See Figure 4.1.)

4.2 PURPOSE

Demand for scalable, reliable, dependable, and affordable networks in all types of environments contributes to Ethernet's dominance in the LAN (Local Area Network) arena. This chapter provides an introduction to the distinctive attributes of Ethernet, Fast Ethernet. Gigabit Ethernet, and 10 Gigabit Ethernet technical solutions. Representative initiatives supported by the four Ethernets in tele-education, E-government

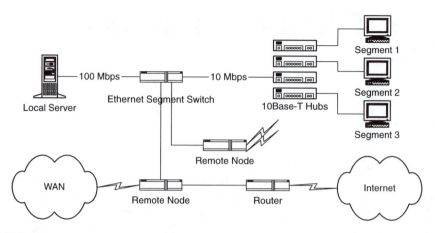

FIGURE 4.1 An Ethernet/Fast Ethernet LAN configuration with multiple access points for enabling remote connections via dial-up service, the Internet, and WAN links.

(electronic government), telehealthcare, and E-business (electronic business) environments are highlighted. In addition, capabilities of the Ethernet technology suite in enabling shared use of network resources and VPN (Virtual Private Network) implementations, advanced applications and Class of Service (CoS) assurances, and home phoneline network solutions are described.

4.3 FOUNDATIONS

Robert Metcalf, then affiliated with the Xerox Palo Alto Research Center, is credited with originally demonstrating Ethernet capabilities in 1973. The term "Ethernet" was the name of the initial LAN product developed by Digital Equipment Corporation (DEC), Intel, and Xerox (DIX) that Metcalf used. The word "Ether" referred to the wireline cabling that was employed. With the sponsorship of DIX, 10 Mbps Ethernet was adopted as the foundation for the IEEE 802.3 LAN standard in 1984.

In 1995, Fast Ethernet was adopted as an extension to the Ethernet specification and officially called the IEEE 802.3u standard. Fast Ethernet sustains data transmission via twisted copper pair, coaxial cable, and optical fiber in the local area at rates reaching 100 Mbps.

The inability of Fast Ethernet to effectively handle the expanding volume of traffic carried by the LAN backbone and the ongoing need for increased capacity contributed to the emergence and standardization of Gigabit Ethernet. The Gigabit Ethernet standard facilitates transmission rates at 1000 Mbps or 1 Gbps.

Gigabit Ethernet retains many of the same features as Ethernet and Fast Ethernet and therefore interworks with the already installed base of Ethernet and Fast Ethernet networks in diverse communications environments. A recent Ethernet innovation, 10 Gigabit Ethernet offers substantial performance enhancements in comparison to its Ethernet predecessors.

The four Ethernets support operations, applications, and topologies that follow traditional Ethernet guidelines. Regardless of rates enabled, Ethernet networks effectively transmit traffic in localized environments. In addition, Ethernet configurations work in concert with other protocols, technologies, and network architectures in supporting sophisticated levels of service and complex voice, video, and data applications that feature Class of Service (CoS) assurances.

4.4 ETHERNET TECHNICAL BASICS

In the present-day environment, the IEEE 802.3 specification and its Annexes describe a suite of Ethernet systems that are capable of running at speeds of 10 Mbps, 100 Mbps, 1000 Mbps or 1 Gbps, and 10000 Mbps or 10 Gbps via diverse media. Ethernet technology is widely employed in present-day corporate, medical, academic, and home networking environments.

In wireline configurations, Ethernet technologies transmit information via optical fiber, coaxial cable, and twisted copper pair or the same cabling used for the Public Switched Telephone Network (PSTN). Ethernet is also a popular enabler of hybrid wireline and wireless configurations employing satellite and microwave technologies. The capabilities of wireless Ethernet configurations are examined in Chapter 9.

4.4.1 Ethernet Frame Format

According to the IEEE 802.3 LAN standard, the Ethernet frame is the basic unit of transport in an Ethernet network. This frame functions as an envelope for carrying data through the networking configuration.

The Ethernet frame consists of several different fields. The frame begins with an eight-byte preamble for synchronizing operations and alerting the Ethernet NIC (Network Interface Card) to accept incoming data. The frame next features a six-byte destination address field, six-byte source address field, a two-byte type field, and a data field that contains a maximum of 1500 bytes. The Ethernet frame concludes with a four-byte frame check sequence field for verifying data integrity. Ethernet frames vary in length and size from 64-byte packets to 1514-byte packets. The Ethernet frame format is used consistently across Ethernet, Fast Ethernet, Gigabit Ethernet, and 10 Gigabit Ethernet platforms.

4.4.2 Carrier Sense Multiple Access with Collision Detection (CSMA/CD) Protocol

The Ethernet LAN specification describes a contention Media Access Control (MAC) protocol called Carrier Sense Multiple Access with Collision Detection (CSMA/CD). The CSMA/CD protocol defines the process by which the network allocates transmission rights among network stations or nodes that share a common media. In the present-day environment, CSMA/CD is a commonly used protocol in Ethernet networks with bus, tree, and star topologies.

In moving forward with the Carrier Sense Multiple Access (CSMA) process, a network station or node seeking to transmit a packet first listens to the Ethernet medium in order to sense if the medium is busy. The Multiple Access process ensures that every network station is linked to the shared communications medium. In the absence of signaling on the Ethernet medium, any network station can initiate the transmission process. Conversely, if a signal is detected, a network station withholds transmission until the medium is idle.

Collisions occur when two or more network stations simultaneously initiate transmissions upon sensing a clear channel. Upon detecting a collision, the transmitting stations execute an algorithm that supports a back-off procedure. These network stations then wait until the medium is clear prior to initiating the retransmission process. In an Ethernet network, packet delivery is not guaranteed. As a consequence, Ethernet transport of packets to destination network nodes is categorized as "best-effort delivery" or "ordered chaos."

An Ethernet local network is also known as a CSMA/CD LAN. CSMA/CD originated with the implementation of the Aloha protocol in the Aloha radio network in the late 1960s. The Aloha protocol is a random access control method that led to the development of CSMA/CA (Carrier Sense Multiple Access/Collision Access) and subsequently to CSMA/CD.

With CSMA/CA, information transmission occurs whenever a network station has a frame or information packet to transmit. If a frame does not reach its destination, the frame is re-transmitted after a random period of time. If two frames collide, the data encapsulated in both frames are damaged or lost. The slotted Aloha version of the Aloha protocol minimizes the number of network collisions by assigning timeslots for each frame or packet to be transported. Developed by the University of Hawaii and the U.S. Department of Defense Advanced Research Projects Agency (DARPA), the Aloha radio network was a contention system that featured a shared channel, supported satellite transmissions, and enabled information transport in networks with ring topologies.

4.4.2.1   IEEE 802.3 CSMA/CD Working Group of the IEEE LAN and MAN Standards Committee

The IEEE 802.3 Working Group of the IEEE LAN (Local Area Network) and MAN (Metropolitan Area Network) Standards Committee defines CSMA/CD specifications for Ethernet installations. This Working Group also coordinates standards activities with the Fast Ethernet Alliance and the Gigabit Ethernet Alliance now known as the I0 Gigabit Ethernet Alliance (I0GEA). Vendors, computer manufacturers, and network providers that promote the publication and adoption of IEEE 802.3 Ethernet standards typically participate in the Fast Ethernet Alliance and the Gigabit Ethernet Alliance

4.5   TECHNICAL FUNDAMENTALS

Regardless of the speed supported, a typical Ethernet system consists of physical components that include network stations or network nodes; internetworking devices such as hubs, bridges, routers, and switches for interlinking network segments; repeaters for regenerating Ethernet signals; and NICs (Network Interface Cards). Network stations or nodes are linked to an Ethernet network with an Ethernet adapter or interface. This adapter performs MAC (Media Access Control) functions for transporting and receiving Ethernet frames over the communications channel. The communications channel or medium provides a transmission pathway for signal transport between the Ethernet adapter and the Ethernet transceiver.

Ethernet local networks feature network nodes or end-stations and interconnecting media. Depending on their functions, network nodes are categorized as Data Terminal Equipment (DTE) and Data Communications Equipment (DCE). PCs, print servers, and file servers are examples of DTE devices. DCE includes communications interface units such as NICs (Network Interface Cards) and modems and stand-alone devices such as routers and switches that receive and forward Ethernet frames or packets across the network.

Ethernet transceivers connect network stations or nodes to the Ethernet transmission medium. Typically, transceivers are built into network interface cards (NICs), work with diverse Ethernet platforms, and support operations that are compliant with the IEEE 802.3 specification.

Ethernet transceivers receive and transmit signals over the connecting medium and use electronic capabilities for handling collision detection. When a collision occurs, the Ethernet transceiver puts a jamming signal on the communications link to alert other transceivers in the configuration that a collision has taken place.

Although the IEEE 802.3 standard is based on Ethernet configurations and capabilities, some minor differences between the wording in this specification and the terms used in actual Ethernet implementations remain. For example, the phrase "connecting cable" in practical Ethernet LANs is called the "Information Attachment Unit" in the IEEE 802.3 specification and the term "transceiver " in localized network deployments is described as the "Medium Attachment Unit" (MAU) in the IEEE 802.3 standard.

Ethernet technology supports data encapsulation, collision detection, and data encoding functions. In addition, Ethernet technology enables administrative operations such as media access management. New generations of Ethernet switches enable error detection and recovery and support utilization of the Simple Network Management Protocol (SNMP) for monitoring the network and maintaining trouble-free network performance.

4.5.1 ETHERNET PROTOCOL STACK

Ethernet, Fast Ethernet, Gigabit Ethernet, and 10 Gigabit Ethernet support operations at Layer 1 or the Physical Layer and Layer 2 or the Data-Link Layer of the OSI (Open Systems Interconnection) Reference Model. The Data-Link Layer or Layer 2 is responsible for transmission and reception of Ethernet frames, specifies procedures for accessing the Physical Layer, and employs multiplexing techniques such as CSMA/CD for effective transmission. The Data-Link Layer includes the Media Access Control (MAC) Sublayer and the Logical Link Control (LLC) Sublayer. At the Data-Link Layer, the four Ethernets employ the MAC Sublayer for translating data into Ethernet frame formats.

Initially designed to support half-duplex transmission, the Ethernet MAC protocol also supports burst mode operations for enabling concurrent full-duplex services that feature bi-directional transmission via point-to-point links with gigabit capacity. The Ethernet MAC (Media Access Control) protocol address is a hardware address that indicates the identification of every node on the network. Originally designed to support half-duplex transmission, the Ethernet MAC protocol also supports burst mode operations for enabling concurrent full-duplex services that feature bi-directional packet transport via point-to-point links with gigabit capacity.

4.5.2 ETHERNET TOPOLOGIES

A network topology refers to the arrangement of network stations or nodes on the communications medium. A network station or node is an active device linked to the network. Network stations include computers, laptops, notebooks, scanners, and printers.

In a bus topology, network stations or nodes are connected to the common communications medium and arranged linearly along this shared medium between network termination points situated at the two endpoints of the Ethernet segment. Centralized hubs and switches dependably move traffic through the Ethernet network. These devices can also transform an Ethernet bus configuration into a star format. In a bus configuration, malfunctioning Ethernet nodes or cuts anywhere in the wireline infrastructure bring network operations to a halt until transmission problems are resolved.

In a star configuration, network stations are arranged around a network core and linked to the centralized hub or switch by the shared communications channel. Each network station in a star configuration features a dedicated point-to-point link to the centralized hub or switch. This link acts as a connection point for enabling network nodes attached to the network to communicate with each other. Depending on their locations in the star configuration, cuts in the common communication medium or malfunctioning Ethernet nodes do not necessarily interfere with the functions of other nodes attached to the common communications channel and network operations.

A centralized hub or switch also enables several network stations to transmit information simultaneously as long as the transmissions are sent to different network stations. However, if some stations attempt to transmit data to the same recipient concurrently, the switch stores and then forwards this data at a later time to avoid collisions. Distances between network stations are extended through the use of internetworking devices such as gateways and routers. (See Figure 4.2.)

4.5.3 ETHERNET TRANSMISSION CAPABILITIES

Ethernet networks support half-duplex or uni-directional transmissions and full-duplex or bi-directional transmissions. By default, a standard Ethernet NIC enables communications in half-duplex mode. Full-duplex mode doubles the carrying capacity of a connection. For example, in a conventional full-duplex 10 Mbps Ethernet network, rates reaching 20 Mbps are supported.

All Ethernet topologies have segment length limitations. The maximum transmission distance varies between different types of Ethernet LANs. For example, 10BASE-T networks support transmissions over distances that reach 100 meters per segment via twisted copper wire pairs. By contrast, 10BASE-F optical fiber networks support transmissions over distances that reach 2,000 meters per segment via fiber optic cabling.

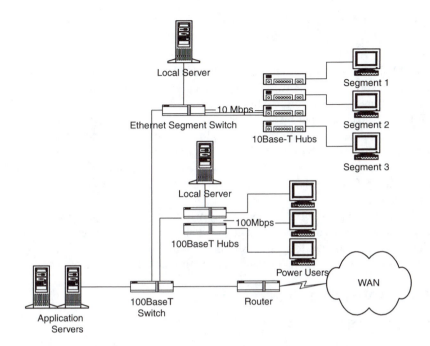

FIGURE 4.2 An Ethernet/Fast Ethernet LAN configuration provisioning router-based WAN connections.

4.6   10 Mbps ETHERNET

4.6.1   MULTIPLEXING FUNDAMENTALS

10 Mbps Ethernet implementations enable baseband transmission so that the communications link functions as a single channel capable of supporting a single transmission at any point in time. Because signals on baseband wiring travel at rates ranging from 1 Mbps to 10 Mbps, sending network stations employ TDM (Time-Division Multiplexing) for dividing the medium into timeslots to enable information transport at predetermined times. Multiple signals are combined into a composite signal by the TDM multiplexer for uni-directional or bi-directional transmission via the shared communications medium. At the termination point, a TDM demultiplexer separates out individual signals from the composite signal for distribution to recipient stations.

By contrast, the Frequency-Division Multiplexing (FDM) multiplexer divides the shared communications link into numerous channels so that more than one transmission occupies the communications link concurrently. Multiple signals are intermixed for one-way or two-way transmission via these channels to the termination point. FDM demultiplexers then separate the signals and distribute them to the designated recipient nodes.

4.6.2   ETHERNET INSTALLATIONS

Four types of media are used in 10 Mbps Ethernet installations. These media include thick coaxial cabling for 10BASE-5; thin coaxial cable for 10BASE-2; Unshielded

Twisted Pair (UTP) copper wiring for 10BASE-T; and optical fiber for 10BASE-F and 10BASE-F sub-categories, specifically, 10BASE-FL (Fiber Optic Inter Repeater Link) and 10 BASE-FB (Fiber Backbone).

10BASE-5, 10BASE-2, and 10BASE-F installations employ bus or linear topologies. By comparison, 10BASE-T installations employ a star topology for linking network segments to a centralized hub.

Ethernet coaxial cable installations employ standard cable system technology and transmit data in an analog format. Amplifiers support signal strength. Originally designed to work with thick coaxial cable resembling yellow garden hose, Ethernet now facilitates data transport via an array of wireline options including Shielded Twisted Pair (STP), Unshielded Twisted Pair (UTP), and single-mode and multimode fiber optic cable.

Single-mode optical fiber supports transmission via a single lightpath or channel over longer network segments at higher bandwidths and at higher costs than multimode optical fiber. Multimode fiber optic cabling uses multiple lightpaths or channels for transporting multiple streams of voice, video, and data over the network.

4.6.3   10 Mbps Ethernet Specifications

4.6.3.1   10BASE-5

The initial Ethernet specification was 10BASE-5 or thick-wire Ethernet. 10BASE-5 supported a 10 Mbps data rate and baseband transmission on thick coaxial cable. In parallel with other Ethernet variants, the descriptor 10BASE-5 highlights essential Ethernet capabilities. Specifically, 10BASE-5 indicates that data rates at 10 Mbps are enabled, baseband signaling or the insertion of digital signals directly into the transmission medium is employed, and the maximum segment length is 500 meters. (Meters are determined by multiplying the final number in the descriptor by 100.)

4.6.3.2   10BASE-2

Subsequent to 10BASE-5, the 10BASE-2 specification, also known as thin wire Ethernet and Cheapernet, was officially recognized. With 10BASE-2, the maximum segment length is 200 meters.

4.6.3.3   10BASE-T

Until the adoption of the Ethernet 10BASE-T specification, coaxial cable was the preferred communications medium for LANs. Network problems associated with 10BASE-2 cabling contributed to the development of 10BASE-T. The letter "T" refers to 10BASE-T usage of UTP (Unshielded Twisted Pair). UTP is generally employed in the PSTN (Public Switched Telephone Network) and commonly serves as the foundation for present-day Ethernet deployments.

10BASE-T employs two pairs of voice grade Category 3 or Category 5 UTP wiring for enabling baseband transmission at 10 Mbps over distances of 100 meters. Also called twisted pair Ethernet, 10 BASE-T supports half-duplex and full-duplex operations.

Distinguished by its support of interoperability, ease of use, and affordability, 10BASE-T solutions are widely used for local area network (LAN) implementations. Basic 10BASE-T network components include cabling, repeaters, adapter cards, and centralized hubs or switches for supporting star topologies.

In a 10BASE-T star topology, a transmission from a network station or node is received by the hub and then retransmitted to its destination address. In 10BASE-T implementations, the term "switches" also refers to "switching hubs." A 10BASE-T switching hub is a centralized hub with advanced switching capabilities for provisioning each network station or node with a virtual private Ethernet connection. 10BASE-T configurations that use switching hubs are also called switched 10BASE-T LANs or switched LANs (SLANs).

4.6.3.4 10BASE-F

The letter "F" in 10BASE-F Ethernet specifications signifies fiber optic cabling. 10BASE-F implementations support baseband transmission rates at 10 Mbps and employ optical fiber cabling to facilitate in-building point-to-point connections and SLAN (switched LAN) solutions. Ethernet-over-fiber optic configurations employ 10BASE-FL, 10BASE-FP, and 10BASE-FP technologies.

4.6.3.4.1 10BASE-FL

Adopted in 1993, the IEEE 802.3j standard includes a description of 10BASE-FL. The letters "FL" signify the capability of 10BASE-FL solutions in interoperating with FOIRL (Fiber Optic Inter Repeater Link) deployments. Designed to replace FOIRL solutions, 10BASE-FL configurations employ interlinked repeater hubs for enabling baseband transmissions at 10 Mbps over distances that extend to 2,000 meters.

The number of network stations that are interlinked in a 10 Mbps baseband Ethernet LAN depends on types of cabling that are employed. As an example, 10BASE-2 thick wire installations support up to 30 nodes per segment. By contrast, 10BASE-FL optical fiber configurations support a maximum of 1,024 nodes per segment. Available from Cisco Systems, an Ethernet 10 BASE-FL port adapter supports five IEEE 802.3 Ethernet FL interfaces that collectively enable half-duplex transmission rates reaching 50 Mbps.

4.6.3.4.2 10BASE-FB

Delineated in the IEEE 802.3j specification, 10BASE-FB involves the use of a fiber optic backbone network that employs synchronous signaling and enables the integration of additional repeaters to the network segment. As with 10BASE-FL, 10BASE-FB configurations support baseband transmission at 10 Mbps over distances that extend to 2,000 meters.

4.6.3.4.3 10BASE-FP

First released in the year 2000, 10BASE-FP describes an optical fiber passive star network configuration that supports half-duplex transmission rates at 10 Mbps via loop lengths that extend to 500 meters. 10BASE-FP refers to the Fiber Passive System that interlinks multiple information appliances to fiber optic cabling without repeaters. Point-to-point transmissions are supported.

4.6.3.5 10BROAD-36

10BROAD-36 supports broadband Ethernet transmissions at rates of 10 Mbps via coaxial cable network segments that extend to 3,600 meters.

4.6.4 10 MBPS ETHERNET MARKETPLACE

10 Mbps Ethernet interoperates with diverse technologies. Vendors such as 3Com and Extreme Networks support Ethernet-over-VDSL (Very High-Speed Digital Subscriber Line) implementations, Ethernet-over-cable modem configurations, and Ethernet-over-T-1 installations.

4.6.4.1 3Com Solutions

3Com supports a Visitor and Community Network (VCN) System for provisioning 10 Mbps Ethernet services in condominiums, small- and medium-sized business establishments, hotels, and office buildings. This VCN solution enables video-on-demand (VOD), high-speed, always-on Internet connections for Web browsing, and IP telephony.

4.6.4.2 Extreme Networks

Extreme Networks provisions an advanced platform for enabling Ethernet to operate over multiple transmission systems, including T-1 networks for extending broadband metropolitan (metro) services to commercial buildings and SOHO (Small Office/Home Office) venues. Extreme Networks also supports an Ethernet-over-T-1 installation that employs multilink, point-to-point connections for enabling multiple channel consolidation and supporting transmissions ranging from 1.544 Mbps (T-1) to 6 Mbps in wider area installations

4.6.5 10 MBPS ETHERNET IMPLEMENTATION CONSIDERATIONS

The 10 Mbps Ethernet network configuration is an affordable, flexible, reliable, and dependable network solution, particularly for those institutions with limited budgets. However, despite technical innovations, realization of the full 10 Mbps rate in conventional Ethernet implementations is rarely attained. Generally, rates ranging between 3 and 4 Mbps are supported. Because conventional 10 Mbps Ethernet LANs lack adequate bandwidth for implementing multimedia applications, shared Ethernet deployments featuring a mix of shared Ethernet hubs and 10/100 (10 Mbps Ethernet and 100 Mbps Fast Ethernet) Ethernet switches for supporting increased bandwidth are typically implemented. (See Figure 4.3.)

4.6.6 10 MBPS ETHERNET TO 100 MBPS FAST ETHERNET MIGRATION

Migration from one Ethernet variant to another is relatively straightforward. Fast Ethernet products such as adapters, repeaters, stackable repeaters, and switches can be integrated into a conventional 10 Mbps Ethernet network incrementally or all at once to provide additional bandwidth and capacity for enhancing network performance.

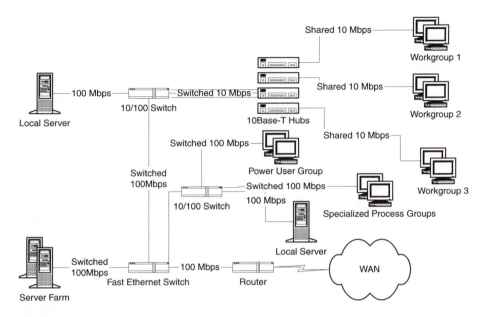

FIGURE 4.3 A switched Ethernet/Fast Ethernet network featuring router-supported WAN links. Ethernet subnetworks provide connections to workgroups and power users.

4.7   FAST ETHERNET

4.7.1   IEEE 802.3u or Fast Ethernet Specification

Demand for higher transmission speeds culminated in the development of the Fast Ethernet specification. Also called the IEEE 802.3u specification, Fast Ethernet, or 100 Mbps Ethernet, the Fast Ethernet specification employs the CSMA/CD process for enabling transmission rates at 100 Mbps. The Fast Ethernet specification also defines several transmission methods that vary in the cabling employed. Popular 100 Mbps configurations include 100BASE-T, 100BASE-T2, 100BASE-T4, 100BASE-TX, and 100BASE-FX.

4.7.2   Fast Ethernet Transmission Methods

4.7.2.1   100BASE-T

100BASE-T or Fast Ethernet employs four pairs of UTP (Unshielded Twisted Pair) Categories 3, 4, or 5 copper wires to facilitate baseband operations and transmissions at 100 Mbps over network segments that extend to 100 meters. 100BASE-T subcategories include 100BASE-T2, 100BASE-T4, and 100BASE-TX.

4.7.2.1.1   100BASE-T2

100BASE-T2 supports baseband operations and digital signal processing techniques for enabling transmission via two pairs of Categories 3, 4, or 5 UTP copper wires over network segments that extend to 100 meters. 100BASE-T configurations support half-duplex and dual-duplex transmission by using two simplex or one-way links to enable bi-directional services.

4.7.2.1.2 100BASE-T4

100BASE-T4 splits the data stream for enabling half-duplex baseband transmissions via four pairs of Categories 3, 4, or 5 UTP copper wires in 100-meter segments. 100BASE-T4 is the half-duplex version of 100BASE-T.

4.7.2.1.3 100BASE-TX

100BASE-TX employs two pairs of STP or UTP Category 5 copper wires for linking centralized hubs, switches, and network stations or nodes together in 100-meter segments. It is important to also note that the maximum distance between two end-stations over which 100BASE-TX transmissions are enabled is 400 meters. 100BASE-TX implementations support baseband operations and employ one pair of copper wires to transmit data and the other pair of copper wires to receive data at rates reaching 100 Mbps.

4.7.2.2 100BASE-FX

100BASE-FX installations employ two strands of multimode fiber optic cabling for enabling full-duplex transmission via network segments that extend to 400 meters. By using a version of the Fiber Data Distributed Interface (FDDI) Physical Layer standard developed by the American National Standards Institute (ANSI), 100BASE-FX also supports single-mode optical fiber solutions for enabling information transport via network segments that extend to 5,000 meters.

4.7.3 FAST ETHERNET SWITCHES

10/100 Ethernet switches enable the development of a dual-mode network by supporting 10 Mbps Ethernet and 100 Mbps Fast Ethernet operations. With the use of 10/100 Ethernet switches, only minor changes are required in the existing infrastructure to effectively migrate from 10 Mbps Ethernet to 100 Mbps Ethernet.

10/100 Ethernet switches work in concert with the installed Ethernet infrastructure. As a consequence, migration to a 100 Mbps Ethernet solution is readily accomplished by replacing the 10BASE-T switch with a 100BASE-T switch. Fast Ethernet switches with speed translation ports also support 10 Mbps Ethernet and 100 Mbps Ethernet implementations. Because the costs for 10 Mbps Ethernet switches and 10/100 Ethernet switches are nearly identical, enterprises and organizations typically employ 10/100 Ethernet switches to accommodate current and projected applications.

Fast Ethernet switches interlink individual LANs between buildings into larger configurations such as switched LANs (SLANs). These devices provide centralized connection points for network servers and consolidate traffic in LAN installations for improved network performance. Fast Ethernet backbone switches work in concert with legacy equipment, support full-duplex transmissions, facilitate network management and advanced packet filtering functions, and enable interconnectivity

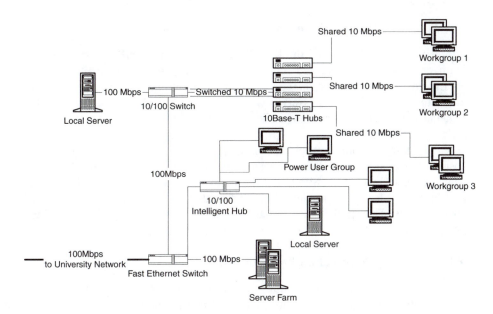

FIGURE 4.4 A switched Ethernet departmental network that interworks with a university Fast Ethernet backbone network.

between Fast Ethernet LANs and Token Ring LANs to alleviate network congestion. (See Figure 4.4.)

4.8   FAST ETHERNET STANDARDS ORGANIZATIONS AND ACTIVITIES

4.8.1   FAST ETHERNET ALLIANCE

The Fast Ethernet Alliance promotes development of Fast Ethernet solutions that are compatible with the CSMA/CD protocol endorsed in the original Ethernet specification. This Alliance also supports the implementation of interoperable Ethernet and Fast Ethernet products such as 10/100 Ethernet switches.

4.8.2   FAST ETHERNET CONSORTIUM

Established in 1993, the Fast Ethernet Consortium performs scheduled testing of IEEE 802.3u equipment and software to verify interoperability support, standards conformance, and product quality. A vendor group, the Fast Ethernet Consortium also benchmarks the capabilities of Fast Ethernet switches, hubs, routers, and other internetworking products at the University of New Hampshire InterOperability Lab (IOL). In addition, the Fast Ethernet Consortium evaluates Fast Ethernet functions in research testbeds and actual networking environments.

4.8.3   ATM FORUM

4.8.3.1   MultiProtocol-over-ATM (MPOA) Specification

Endorsed by the ATM Forum, the MultiProtocol-over-ATM (MPOA) specification optimizes traffic throughput and supports ATM interoperability with Ethernet and Fast Ethernet configurations, thereby enabling Ethernet and Fast Ethernet configurations to benefit from ATM services. As an example, MPOA supports packet exchange between Ethernet, Fast Ethernet, and Token Ring LANs via an ATM backbone network.

MPOA also supports IP applications; streamlines the transmission process; and transports voice, video, and audio signals to destination sites via ATM SVCs (Switched Virtual Circuits). In addition, MPOA enables network traffic at Layer 3 or the Network Layer of the OSI Reference Model to be switched at the Media Access Control (MAC) Sublayer of the Data-Link Layer of the Ethernet protocol stack. The ATM Forum MultiProtocol-over-ATM Working Group clarifies MPOA operations.

4.8.3.2   Routing Information Protocol (RIP)

The ATM Forum endorses the Routing Information Protocol (RIP). This protocol establishes links between ATM subnetworks and Ethernet and Fast Ethernet networks to support fast and dependable IP services.

4.8.4   FAST ETHERNET COMPETITOR TECHNOLOGIES

4.8.4.1   100VG-AnyLAN

4.8.4.1.1   100VG-AnyLAN Transmission

The 100VG-AnyLAN is an IEEE 802.5 specification for 100 Mbps LAN installations that utilize either Fast Ethernet or Token Ring frames for transmission. The 100VG-AnyLAN supports information transport via four pairs of Category 1, 3, 4, or 5 Unshielded Twisted Pair (UTP) copper cabling. In contrast to Ethernet and Fast Ethernet utilization of CSMA/CD, 100VG-AnyLAN uses a demand-priority access technique that eliminates the need to monitor the network for collisions.

4.8.4.1.2   100VG-AnyLAN Operations

In a 100VG-AnyLAN configuration, the hub or root controls network access. With 100VG-AnyLAN demand-priority operations, a network station or node seeking to transmit signals submits a request to the hub. The hub immediately acknowledges the request if the network is idle. Upon receiving the acknowledgment, the network station initiates the transmission process by sending voice, video, and data to the hub. When the hub receives more than one request concurrently, the hub uses a round-robin technique to acknowledge each request in turn. High-priority requests for transmission involving time-sensitive applications such as videoconferencing are serviced ahead of normal-priority requests for transport of time-insensitive data such as e-mail and file transfers. Generally, the hub does not grant priority access to a port more than twice in a row, thereby enabling the fair participation of all network stations in LAN activities.

100VG-AnyLAN solutions employ hubs that are more costly than centralized hubs used with 100BASE-T installations. In addition, 100VG-AnyLAN is limited to provisioning rates at 100 Mbps because this technology does not support full-duplex transmissions. By contrast, 100BASE-T supports full-duplex transmissions at 200 Mbps and enables straightforward migration to gigabit rates enabled by Gigabit Ethernet technology.

Developed by Hewlett-Packard, 100VG-AnyLAN was expected to be a strong contender to Fast Ethernet in the LAN domain. In comparison to 100VG-AnyLAN, Fast Ethernet dominates the present-day marketplace.

4.8.4.2 Fiber Data Distributed Interface (FDDI)

4.8.4.2.1 FDDI Capabilities

Initially standardized in the ANSI (American National Standards Institute) X3T9.5 specification released in 1989, the international FDDI specification is also known as ISO (International Standards Organization) 9314. FDDI supports transmissions via fiber optic cabling. In addition, FDDI also works with standard twisted pair copper cabling in enabling interconnectivity. The copper version of FDDI is called Copper Distributed Data Interface (CDDI).

Fiber Distributed Data Interface (FDDI) is an ANSI standards-compliant technology for 100 Mbps Token Ring LANs. Internetworking devices such as routers, bridges, and switches foster FDDI interoperability with star and bus LAN topologies. In addition, FDDI serves as a fiber optic backbone for LAN installations.

4.8.4.2.2 FDDI Transmission

The maximum frame size supported by FDDI includes 4500 bytes, with 4478 bytes allocated for the payload or information field. The remaining bytes, including a header field and a trailer field, ensure dependable information delivery. FDDI utilizes a timed token passing scheme for network access. A token consists of a three-octet FDDI frame. By contrast, the IEEE 802.5 Token Ring LAN specification supports utilization of a priority reservation token access methodology.

4.8.4.2.3 FDDI Operations

FDDI LAN installations employ a dual-ring topology. The main ring utilizes dual-attachment stations to support a secondary ring that enables operations in case of a failed station or fiber cut. Each ring provisions network service for up to 500 nodes. By using repeaters, each ring supports network operations for 1000 nodes. In an FDDI LAN, FDDI protocol analyzers monitor networking activities and the Simple Network Management Protocol (SNMP) enables seamless networking operations.

FDDI WANs (Wide Area Networks) facilitate information transport over distances that extend to two kilometers via multimode optical fiber. With single-mode optical fiber, FDDI WAN configurations enable transmissions at distances that extend to 10 kilometers.

FDDI is a flexible internetworking technology that interoperates with Ethernet, Fast Ethernet, ATM, and satellite networking solutions. Translational and encapsulating bridges that foster interconnectivity between LAN media featuring dissimilar MAC (Medium Access Control) protocol sublayers enable FDDI and Gigabit Ethernet interoperability. FDDI-to-Gigabit Ethernet migration is accomplished by replacing FDDI hubs, concentrators, or routers with Gigabit Ethernet switches and the installation of new Gigabit Ethernet interfaces. This process enables enterprises to retain investments in the optical fiber infrastructure while also increasing the aggregate bandwidth substantially for the upgraded network segments.

FDDI was expected to be the enabler of next-generation LAN configurations. However, in contrast to Fast Ethernet solutions, FDDI installations are considerably more expensive to implement and maintain. Moreover, FDDI installations are limited in supporting high-performance multimedia applications.

In contrast to FDDI, Fast Ethernet implementations are affordable and extendible, operate over twisted copper wire pairs, and support backward compatibility with in-place 10 Mbps Ethernet LANs. These factors also contributed to the selection of the Fast Ethernet specification or IEEE 802.3u as an extension to the IEEE 802.3 standard.

4.8.4.2.4   FDDI and FDDI II (FDDI, Phase 2)

FDDI II is based on the original FDDI format. As with FDDI, FDDI II technology supports network operations at rates reaching 100 Mbps. FDDI II was designed specifically to support multimedia applications and enable voice and video transport. FDDI II solutions use a synchronous bandwidth allocation scheme that enables network nodes to reserve bandwidth for regularly transmitted, delay-sensitive network traffic. Because FDDI II technology is ineffective in supporting broadcast-quality video, FDDI II solutions are used infrequently.

4.9   GIGABIT ETHERNET TECHNICAL FUNDAMENTALS

As with its Ethernet and Fast Ethernet predecessors, Gigabit Ethernet technology works in wireless and wireline environments. A multifunctional communications technology, Gigabit Ethernet significantly expands available bandwidth for enabling interactive, high-performance multimedia applications. In comparison to Fast Ethernet, Gigabit Ethernet increases the rates at which data are transmitted by a factor of ten. In contrast to its predecessor, Gigabit Ethernet employs the physical signaling scheme described in the Fibre Channel specification for enabling packet transmission.

4.9.1   GIGABIT ETHERNET FUNCTIONS

Gigabit Ethernet networks utilize the underlying infrastructure established with Ethernet and Fast Ethernet specifications. As a consequence, Gigabit Ethernet enables a straightforward migration path to higher performance levels without disrupting in-place networking operations. Generally viewed as an extension of Ethernet and Fast Ethernet, Gigabit Ethernet technology ensures seamless interworking with Ethernet and Fast Ethernet services and operations.

Gigabit Ethernet also employs the same protocols, transmission schemes, frame size, frame format, flow control procedures, and access methods as Ethernet and Fast Ethernet. Therefore, the need for complex emulation and translation techniques to support migration from Ethernet and Fast Ethernet implementations to Gigabit Ethernet solutions is eliminated.

4.9.2 GIGABIT ETHERNET ARCHITECTURE

Gigabit Ethernet Physical Layer specifications support half-duplex and full-duplex operations; gigabit transmissions over two strands of optical fiber or two pairs of shielded twisted copper wires; and encoding schemes defined by the ANSI Fibre Channel standard. The network topology for Gigabit Ethernet conforms to the conventional specifications for Ethernet and Fast Ethernet. In conjunction with Ethernet and Fast Ethernet, Gigabit Ethernet employs Layer 1 or the Physical Layer and the Media Access Control (MAC) Sublayer of Layer 2 or the Data-Link Layer for information transport. Additionally, Gigabit Ethernet works in conjunction with upper layer protocols such as IP (Internet Protocol) and TCP (Transmission Control Protocol). In terms of the OSI Reference Model, IP and TCP utilize the Transport Layer or Layer 4 and the Network Layer or Layer 3 for enabling communications services between applications.

4.9.3 GIGABIT ETHERNET OPERATIONS

Gigabit Ethernet implementation is basically a replication of deployment procedures associated with its predecessors. Moreover, in-place Ethernet and Fast Ethernet management systems and equipment function in concert with Gigabit Ethernet management systems, services, and applications, thereby enabling cost-effective migration to advanced Gigabit Ethernet solutions.

Early Gigabit Ethernet implementations employed optical fiber for supporting full-duplex transmissions and facilitating building-to-building LAN interconnections. As evidenced by the ratification of the Gigabit Ethernet over copper cabling standard by the IEEE, Gigabit Ethernet also works effectively with Unshielded Twisted Pair (UTP) in delivering services to the desktop.

As with previous Ethernet generations, Gigabit Ethernet is scalable and extendible, augments functions of in-place networks, and enables a broad array of applications. In the LAN arena, Gigabit Ethernet supports links between servers and switches and interconnects clusters of servers, server farms, and network segments. In addition, Gigabit Ethernet provisions high-speed connections between buildings and supports transmissions ranging from 10 Mbps to 1000 Mbps or 1 Gbps and higher rates. Moreover, Gigabit Ethernet installations are less costly than ATM in terms of implementation, operations, administration, and maintenance.

It is important to note that Gigabit Ethernet and its predecessors are subject to distance and signal constraints. Transmission impairments result in attenuation or signal power loss, near-end crosstalk (NEXT), and latencies in signal transport.

4.9.4 IEEE 802.3z OR FIBER OPTIC GIGABIT ETHERNET IMPLEMENTATIONS

The IEEE 802.3z specification for extending Gigabit Ethernet functions into the optical fiber environment was ratified in 1998. Endorsed by the IEEE 802.3z Gigabit

Ethernet Task Force, the Gigabit Ethernet Alliance, and the IEEE Standards Committee, the IEEE 802.3z standard defines Ethernet operations at rates of 1000 Mbps or 1 Gbps for half-duplex transmissions and Ethernet operations at 2000 Mbps or 2 Gbps for full-duplex transmissions.

The Gigabit Ethernet IEEE 802.3z standard also clarifies capabilities of transceivers that operate in conjunction with single-mode and multimode optical fiber plants and supports ongoing utilization of in-place optical fiber links that interconnect multiple buildings in campus LANs. The role of optical components such as optical lasers in transporting data-, voice-, and video-over-optical fiber is also indicated.

4.9.4.1   Gigabit Media Independent Interface (GMII)

The IEEE 802.3z specification also describes Gigabit Media Independent Interface (GMII) services. The GMII enables interconnectivity of MAC Sublayer protocol devices with the Physical Layer or Layer 1 of the OSI Reference Model. Furthermore, the GMII supports half-duplex and full-duplex transmissions, multivendor interoperability, and backward compatibility with Ethernet and Fast Ethernet installations. In addition, the GMII enables virtual independent pathways or channels for data transmission and reception.

4.10   GIGABIT ETHERNET SOLUTIONS

Based on the work of the IEEE 802.3z Gigabit Ethernet Task Force and the Gigabit Ethernet Alliance and ratified by the IEEE, the Gigabit Ethernet standard defines network interfaces, repeater operations, topologies, and capabilities of 1000BASE-SX and 1000BASE-LX configurations. The IEEE 802.3z Gigabit Ethernet Task Force and the Gigabit Ethernet Alliance also clarify features and functions of 1000BASE-LH and 1000BASE-CX networks.

4.10.1   1000BASE-SX

Based on the Fibre Channel signaling standard, 1000BASE-SX describes Gigabit Ethernet short wavelength solutions. In supporting full-duplex mode transmissions, each 1000BASE-SX network segment enables transmission via multimode optical fiber at distances that extend to 275 meters. In addition, each 1000BASE-SX network segment supports full-duplex information transport over single-mode optical fiber at distances that extend to 550 meters.

4.10.2   1000BASE-LX

1000BASE-LX describes long wavelength Gigabit Ethernet solutions. With full-duplex mode operations, each 1000BASE-LX network segment facilitates transmissions via single-mode optical fiber at distances that reach 5000 meters. In enabling full-duplex operations over multimode optical fiber, each 1000BASE-LX network segment enables voice, video, and data transport at distances that extend to 550 meters.

4.10.2.1   Differential Mode Delay (DMD)

The IEEE 802.3z Gigabit Ethernet Task Force also developed a solution for reducing the adverse impact of Differential Mode Delay (DMD), a condition that generates jitter in LED (Light Emitting Diode) installations supporting transmissions via multimode optical fiber. According to this Task Force, the utilization of conditioners with 1000BASE-LX and 1000BASE-SX configurations and dispersal of light pulses evenly through every lightpath or channel in each network segment resolves DMD transmission disruptions.

4.10.3   1000BASE-LH

Supported by the IEEE 802.3z Gigabit Ethernet Task Force and the Gigabit Ethernet Alliance, 1000BASE-LH defines long-haul fiber optic Gigabit Ethernet Metropolitan Area Network (MAN) solutions that operate in multivendor urban environments.

4.10.4   1000BASE-CX

Developed by the IEEE 802.3z Gigabit Ethernet Task Force and the Gigabit Ethernet Alliance, 1000BASE-CX refers to Gigabit Ethernet transmission over Category 5 Unshielded Twisted Pair (UTP). Each 1000BASE-CX network segment enables full-duplex operations at distances that extend to 25 meters.

4.10.5   IEEE 802.3AB OR 1000BASE-T

The IEEE ratified the IEEE 802.3ab standard, also known as 1000BASE-T, in 1999. This standard is based on research efforts sponsored by the Gigabit Ethernet Alliance. To verify 1000BASE-T capabilities, tests benchmarking the performance of Gigabit Ethernet over copper wiring were conducted by Gigabit Ethernet Alliance participants, including Alteon Web System, Extreme Networks, 3Com, and Sun Microsystems, at the Silicon Valley Networking Lab (SVNL). These tests evaluated Gigabit Ethernet performance in seamlessly enabling interoperability, full-motion video, groupware applications, and remote file transfers over Category 5 UTP (Unshielded Twisted Pair). Findings contributed to IEEE endorsement of the 1000BASE-T Gigabit Ethernet-over-copper wire specification.

1000BASE-T solutions utilize four pairs of Category 5 UTP copper wires for enabling transmissions up to 100 meters over a single network segment. Each copper pair supports throughput at 250 Mbps for enabling a total transmission of 1000 Mbps or 1 Gbps in half-duplex mode and 2000 Mbps or 2 Gbps in full-duplex mode. Because 1000BASE-T installations work in conjunction with the installed Category 5 wireline infrastructure, the need to rewire ceilings, walls, or raised floors is eliminated.

1000BASE-T is compatible with Ethernet and Fast Ethernet technologies and works in concert with 56 Kbps (Kilobits per second) modems. A high-performance networking solution, 1000BASE-T also enables innovative desktop applications and high-speed server connectivity. Techniques employed by 1000BASE-T for eliminating signal degradation resulting from signal attenuation, echo, impulse noise, and near-end crosstalk associated with UTP transmission include line coding, pulse shaping, and forward error correction (FEC). Specifications for enabling 1000BASE-T implementations to support traffic over Category 6 and Category 7 cabling systems are in development.

4.10.6   1000BASE-2

The Gigabit Ethernet Alliance endorses the utilization of Category 5e (Enhanced Category 5) cabling for 1000BASE-2 installations.

4.11   GIGABIT ETHERNET PROTOCOLS

Gigabit Ethernet provisions higher level services than its predecessors by leveraging capabilities of technologies, standards, and protocols such as the Resource Reservation Protocol (RSVP), the Real-Time Transit Protocol (RTP), the Real-Time Control Protocol (RTCP), and the Real-Time Streaming Protocol (RTSP). In addition, Gigabit Ethernet also supports operations in concert with the IEEE 802.1p and the IEEE 802.1Q specifications. To enable multimedia transmission, Gigabit Ethernet reserves bandwidth for voice, video, and data transmission; assigns priority CoS (Class of Service) tags to packets for conveying CoS levels to internetworking devices; and provisions priority queues to transport real-time data, voice, and video traffic over IP.

4.11.1   RESOURCE RESERVATION PROTOCOL (RSVP)

Also called the Resource Reservation Setup Protocol, the RSVP (Resource Reservation Protocol) dynamically allocates bandwidth to network applications in conventional packet networks. With RSVP implementation, multimedia applications running on Gigabit Ethernet implementations can request a specified CoS (Class of Service) assurance. Internetworking devices such as routers and switches respond to an RSVP request by creating a virtual connection through the network for enabling the specified CoS assurance.

Vendors such as 3Com and Cisco Systems employ RSVP services to facilitate dependable and reliable delivery of multimedia network applications and support distributed broadband connections. In addition to providing CoS assurances, the Resource Reservation Protocol supports QoS (Quality of Service) functions such as guaranteed bandwidth reservation for enabling real-time multimedia transmissions via LANs and VLANs (Virtual LANs). RSVP facilitates operations at Layer 3 or the Network Layer of the OSI Reference Model.

4.11.2   REAL-TIME TRANSIT PROTOCOL (RTP)

Gigabit Ethernet also works in concert with the Real-Time Transit Protocol (RTP). RTP is a transport protocol that supports delivery of continuous media such as real-time audio, video, or simulation data over the Internet. RTP supports time-stamping, sequence numbering, and payload identification for ensuring dependable packet delivery in real-time. In addition, RTP provides a framework for end-to-end network delivery of IP multicast traffic, works in concert with RSVP operations, and enables resource reservation for transporting bandwidth-intensive applications.

RTP facilitates Gigabit Ethernet applications that include Web-based videoconferences, voice-over-IP (VoIP), and real-time transport of multimedia on-demand. In ATM networks, RTP provisions QoS guarantees.

4.11.3 REAL-TIME CONTROL PROTOCOL (RTCP)

The Real-Time Control Protocol (RTCP) works in conjunction with the RTP to provide data on distribution faults that support network management operations. By controlling transmission intervals and preventing IP multicasts from inundating network resources, RTCP enables extendible and scalable RTP operations.

4.11.4 REAL-TIME STREAMING PROTOCOL (RTSP)

The Real-Time Streaming Protocol (RTSP) supports streaming voice, video, and data for enabling point-to-multipoint IP multicast distributions. Streaming operations involve dividing voice, video, and data into numerous packets for network transmission and reconstituting the packets at the destination address.

4.12 GIGABIT ETHERNET CLASS OF SERVICE (COS) ASSURANCES

Gigabit Ethernet provisions Class of Service, also referred to as QoS assurances, rather than QoS guarantees. In provisioning higher-level services, Gigabit Ethernet leverages capabilities of technologies and protocols such as MPLA (MultiProtocol Link Aggregation) and MPLS (MultiProtocol Label Switching) to support network extendibility, scalability, and reliable packet transmission.

4.12.1 MULTIPROTOCOL LINK AGGREGATION (MPLA)

The IETF MultiProtocol Link Aggregation (MPLA) Working Group develops MPLA technology to support the integration of multiple physical connections into one virtual connection for enabling fast and dependable information transport in point-to-point and point-to-multipoint implementations. MPLA technology optimizes network performance and enables cost-effective and scalable operations for accommodating current and next-generation transmission requirements for applications such as videoconferencing, telecollaborative engineering, and interactive VR (Virtual Reality) simulations.

4.12.2 MULTIPROTOCOL LABEL SWITCHING (MPLS)

The IETF (Internet Engineering Task Force) MultiProtocol Label Switching (MPLS) Working Group establishes criteria for utilization of an MPLS solution. Based on a label-switching technology, MPLS supports the attachment of a label to every network packet that establishes the packet path through the network. Labels such as CoS and QoS identify packets that warrant priority services. In addition, the MPLS solution defines procedures for distributing IP multicasts and label packets to optimize network performance. Gigabit Ethernet supports MPLS functions. In addition, Packet-over-SONET/SDH (POS), Frame Relay, and ATM also benefit from MPLS operations.

4.13 GIGBIT ETHERNET STANDARDS ORGANIZATIONS AND ACTIVITIES

4.13.1 CELLS IN FRAME (CIF) ALLIANCE

The Cells in Frame (CIF) Alliance identifies approaches for deploying ATM in conjunction with Fast Ethernet and/or Gigabit Ethernet technologies in mixed-mode network configurations.

4.13.2 GIGABIT ETHERNET ALLIANCE

The Gigabit Ethernet Alliance is a multivendor consortium that supports implementation of affordable and interoperable Gigabit Ethernet solutions. This group encourages the seamless migration from in-place Ethernet and Fast Ethernet installations to Gigabit Ethernet deployments in response to the demand for increased network bandwidth. The Gigabit Ethernet Alliance also endorses the continued use of the conventional Ethernet frame format and traditional network management guidelines.

In addition, the Gigabit Ethernet Alliance supports backward compatibility of Gigabit Ethernet installations with on-site Ethernet and Fast Ethernet networks for leveraging infrastructure investments. The Gigabit Ethernet Alliance also conducts interoperability tests for verifying capabilities of Gigabit Ethernet products. Founding members of the Gigabit Ethernet Alliance include Cisco Systems, Sun Microsystems, VLSI Technology, Compaq, LSI Logic, Intel, and UB Networks. The Gigabit Ethernet Alliance is now part of the 10 GEA.

4.13.3 GIGABIT ETHERNET CONSORTIUM

Formed in 1997, the Gigabit Ethernet Consortium evaluates Gigabit Ethernet product quality and the performance, reliability, and availability of Gigabit Ethernet implementations in research and actual environments. The Gigabit Ethernet Consortium employs a series of test suites for ensuring standards compliance of IEEE 802.3z devices such as switches, routers, and transceivers in research trials conducted at the University of New Hampshire InterOperability Lab (IOL). As with the Fast Ethernet Consortium, the Gigabit Ethernet Consortium is a vendor coalition.

4.13.4 IETF (INTERNET ENGINEERING TASK FORCE) IPNG (IP NEXT-GENERATION) WORKING GROUP

The IETF (Internet Engineering Task Force) IPng (IP Next-Generation) Working Group defines approaches for migrating from IPv4 (Internet Protocol version 4) to IPv6 (IP version 6). Endorsed by the IETF, RFC 1752 clarifies procedures for transmitting IPv6 packets via Gigabit Ethernet configurations.

4.14 ETHERNET VLANS (VIRTUAL LANS)

4.14.1 VLAN CAPABILITIES

Ethernet, Fast Ethernet, and Gigabit Ethernet switches enhance network performance and security and facilitate access to bandwidth-intensive multimedia applications. These devices also enable network segmentation to support development and implementation of VLANs (Virtual Local Area Networks). An Ethernet, Fast Ethernet, or Gigabit Ethernet VLAN refers to a logical grouping of network stations or nodes that function as if they are connected to a single shared medium regardless of their physical network locations. Switches are sophisticated devices that foster interconnections of network stations and resources situated on physically different network segments into logical VLAN configurations.

Ethernet, Fast Ethernet, and Gigabit Ethernet VLANs are typically deployed for enabling interactive communications among individuals participating in common workgroup activities. Generally found in larger enterprises such as universities and corporations, VLANs enable constituent workgroups to communicate with each other, share multimedia resources and available bandwidth, and jointly participate in online activities.

4.14.2 VLAN OPERATIONS

An Ethernet, Fast Ethernet, or Gigabit Ethernet VLAN is a flexible configuration that is easy to manage, administer, and maintain. Network administrators readily identify network stations in trouble in VLAN implementations, thereby eliminating the need to shut down the network. Because network stations can be quickly added to and deleted from VLAN configurations, the need to rewire the on-site installation for enabling diverse workgroup tele-applications is eliminated. Inasmuch as network routers move traffic between an enterprisewide network for supporting intra-workgroup VLAN capabilities, information flow is predictable and on-demand bandwidth can be made available for multimedia applications. In addition, network servers that are configured to provision applications for multiple VLAN workgroups generally can optimize network response time by eliminating excessive router traffic and reducing network bottlenecks.

4.14.3 IEEE 802.1Q VLAN SPECIFICATION

VLAN implementations typically conform to requirements described in the IEEE 802.1Q specification. Designed for Ethernet, Fast Ethernet, and Gigabit Ethernet VLANs, the IEEE 802.1Q specification defines VLAN attributes and capabilities and indicates approaches for resource sharing and information exchange in VLAN configurations.

Moreover, the IEEE 802.1Q specification describes VLAN architecture, clarifies network administrative functions, and streamlines the VLAN implementation process. In addition, this specification provides a foundation for enabling the interoperability of VLAN installations in multivendor environments.

4.14.3.1   IP Multicasts

Protocols defined by the IEEE 802.1Q specification support IP multicasts in Ethernet, Fast Ethernet, and Gigabit Ethernet VLAN configurations. IP multicasts enable multimedia distribution via the Internet. Traditional unicast approaches for transmitting voice, video, and data via point-to-point links are no longer effective in provisioning multimedia delivery. As a consequence, the Multicast Backbone (MBone) provisions point-to-multipoint and multipoint-to-multipoint multicast distributions to specified groups of recipients.

4.14.3.2   Multicast Protocols

In Ethernet, Fast Ethernet, and Gigabit Ethernet VLANs, the Generic Attribute Registration Protocol (GARP), the GARP VLAN Registration Protocol (GVRP), and the GARP Multicast Registration Protocol (GMRP) support dynamic creation of multicast groups. Multicast routing protocols such as the Distance Vector Multicast Routing Protocol (DVMRP) foster development of multicast delivery paths for transmission of multicast packets through the network to specified multicast groups. The GARP Multicast Registration Protocol (GMRP) and the Internet Group Management Protocol (IGMP) identify problems with multicast transmission and facilitate implementation of procedures for multicast delivery that do not compromise network operations.

4.15   IEEE 802.1P SPECIFICATION

4.15.1   IEEE 802.1P CAPABILITIES

Approved in 1998 by the IEEE 802.1p Working Group as part of the IEEE 802.1 standard for bridges, the IEEE 802.1p specification supports development of next-generation converged Ethernet configurations. These configurations deliver scalable applications with differentiated service requirements such as desktop videoconferencing, televideo training, and voice-over-IP (VoIP). The IEEE 802.1p specification optimizes network throughput, transmission efficiencies, and network response times. Moreover, the IEEE 802.1p specification enables seamless operations across multivendor environments, equipment monitoring, and network administrative activities.

The IEEE 802.1p specification also fosters establishment of multicast groups to reduce inundation of network resources when IP multicasts are used for multimedia distribution. In addition, this specification supports priority queuing mechanisms for handling high-priority traffic.

4.15.2   IEEE 802.1P AND IEEE 802.1Q OPERATIONS

As with the IEEE 802.1Q standard, the IEEE 802.1p specification defines the purpose of the Generic Attribute Registration Protocol (GARP). GARP enables specific applications and protocols such as the GARP Multicast Registration Protocol (GMRP) and the GARP VLAN Registration Protocol (GVRP). With GMRP, a network station requests VLAN admission as opposed to requesting admission to a multicast domain.

In parallel with IEEE 802.1Q operations, IEEE 802.1p GMRP service also delivers IP multicasts by using GARP for registering multicast membership. GMRP works in concert with IP multicast protocols operating at the Network Layer or Layer 3 of the OSI Reference Model in order to provide efficient multicast operations, administrative control of multimedia traffic, bandwidth conservation, and management of multicast addresses.

4.16   GIGABIT ETHERNET AND FIBRE CHANNEL

The Gigabit Ethernet approach for information transport is based on the ANSI (American National Standards Institute) Fibre Channel (FC) specification. The ANSI Fibre Channel specification defines transmission capabilities and optical components for enabling gigabit speeds at the Physical Layer or Layer 1 of the OSI Reference Model. Fibre Channel technology fosters transmission at rates reaching 1.063 Gbps. With enhancements, Fibre Channel technology supports information transport at 1.250 Gbps and ensures a 1 Gbps rate of transmission for voice, video, and data applications.

4.17   GIGABIT ETHERNET AND ATM

4.17.1   GIGABIT ETHERNET PACKETS AND ATM CELLS

As with Ethernet and Fast Ethernet, Gigabit Ethernet supports transmission of packets ranging in size from 64 to 1514 bytes. The unpredictability of packet size restricts the suite of Ethernet technologies in enabling explicit QoS guarantees, particularly in wider area networks. As a consequence, transmissions in Gigabit Ethernet installations are based on CoS assurances.

In comparison to Gigabit Ethernet, ATM networks employ a 53-byte fixed-sized cell for accommodating on-demand bandwidth requests from network stations or nodes in local area and wider area environments. A multipurpose communications solution, ATM networks integrate video, data, and voice traffic for transport via a consolidated high-speed network platform.

ATM technology controls jitter and limits latency, implements flow control and congestion control mechanisms, and explicitly supports QoS guarantees for ensuring real-time traffic delivery at 1.544 Mbps (T-1) and higher rates. ATM architecture is flexible and extendible and fosters LAN-to-WAN consolidation. ATM protocols work in concert with technologies that include IP, FDDI, Ethernet, Fast Ethernet, Frame Relay, ISDN, cable modem, and DSL (Digital Subscriber Line).

4.17.2   GIGABIT ETHERNET CLASS OF SERVICE (CoS) ASSURANCES AND ATM QUALITY OF SERVICE (QoS) GUARANTEES

Gigabit Ethernet supports high-speed connectivity for enabling bandwidth-intensive asynchronous data, video, and voice transport with predictable response times. In addition, Gigabit Ethernet provisions very high bandwidth to Fast Ethernet segments for supporting switched intra-building backbone services and inter-switch operations. Gigabit Ethernet also leverages Ethernet and Fast Ethernet technologies and in-place equipment for supporting gigabit transmission rates.

Gigabit Ethernet supports VLAN configurations, Web activities, network management, congestion control services, and multimedia applications that are based on differentiated Class of Service (CoS) assurances. To provision differentiated CoS assurances, Gigabit Ethernet switches and routers prioritize transmission of delay-sensitive voice, video, and data traffic. For example, Gigabit Ethernet implementations supported by 3Com and Cisco Systems prioritize CoS traffic levels in accordance with the IEEE 802.1p and the IEEE 802.1Q specifications to ensure multimedia delivery across Ethernet, Fast Ethernet, and Gigabit Ethernet platforms.

Gigabit Ethernet competes with ATM in enabling broadband services. However, at present, Gigabit Ethernet is not a proven solution for reliably transporting real-time voice, data, and video. In contrast to Gigabit Ethernet, ATM enables effective multimedia transmission in local, regional, and wide area implementations with strict QoS guarantees.

To resolve QoS and CoS conflicts, the ATM Forum and the IETF (Internet Engineering Task Force) standardize methods and procedures for mapping Ethernet, Fast Ethernet, and Gigabit Ethernet CoS assurances to ATM QoS guarantees across ATM networks. Moreover, procedures for mapping ATM QoS levels to Ethernet CoS assurances across Ethernet networks are also under consideration.

In addition, the ATM Forum supports utilization of PVCs (Permanent Virtual Circuits) for interworking CoS-to-QoS operations. Developed by the IETF, the IP Precedence and Type of Service specification enables Ethernet CoS priorities to be maintained in wide area ATM configurations by indicating the priority or service level required in each IP packet header.

4.17.3 GIGABIT ETHERNET LANS AND ATM LANES (LAN EMULATIONS)

Ethernet, Fast Ethernet, and Gigabit Ethernet solutions support multimedia services in DANs (Desktop Area Networks) and LANs. Although ATM was designed as a wider area network, the ATM Forum adopted the LANE (Local Area Network Emulation) protocol for enabling ATM operations in LAN configurations. LANE operates as a network overlay on top of ATM Adaptation Layer 5 (AAL5) of the ATM protocol stack. The ATM Forum also supports the encapsulation of Ethernet, Fast Ethernet, and Gigabit Ethernet packets into ATM cells for enabling internetwork interoperability.

ATM LANs that enable LANE (LAN Emulation) functions are classified as IEEE 802.3 Ethernet LANs or IEEE 802.5 Token Ring LANs. The LAN emulation (LANE) protocol also enables Ethernet and Fast Ethernet local networks to interoperate with ATM local networks. Specifications approved by the ATM Forum such as FATE (Frame ATM Transport over Ethernet) define parameters for mapping and encapsulating Ethernet packets into ATM cells for transmission via SVCs (Switched Virtual Connections) across an ATM platform. In parallel with Ethernet, Fast Ethernet, and Gigabit Ethernet technologies, LANE also enables the re-use of legacy networks.

Switches, routers, multiplexers, and hubs for interworking ATM and Gigabit Ethernet technologies in the same local and wider area multivendor environment are in development. Vendors active in the Gigabit Ethernet and ATM arena include Asante Technologies, Cabletron, Cisco, Digital Equipment Corporation (DEC), and Newbridge Networks. In 1998, Intel and Fore Systems formed an alliance to develop integrated ATM and Gigabit Ethernet implementations for enabling multimedia services and applications.

4.17.4 GIGABIT ETHERNET VERSUS ATM

The decision to use a Gigabit Ethernet infrastructure or an ATM platform depends on such variables as costs, application requirements, enterprisewide goals and objectives, and the in-place wiring structure. Gigabit Ethernet offers viable options for improving network availability and performance, particularly in situations where network services repeatedly come to a standstill as a consequence of traffic congestion. Gigabit Ethernet also provisions a migration path to higher-speed 10 Gigabit Ethernet configurations without large-scale infrastructure investments. However, as noted, Gigabit Ethernet provisions CoS assurances as opposed to QoS guarantees. The intent to safeguard investments in the Ethernet infrastructure must also be weighed against potential risks of limited network performance.

4.18 GIGABIT ETHERNET MARKETPLACE

4.18.1 CISCO SYSTEMS

Cisco Systems supports implementation of a high-performance LAN routing and switching solution that works in concert with Gigabit Ethernet technology. Based on the Cisco Catalyst product series, this solution enables voice, video, and data delivery to the desktop with CoS assurances; accommodates electronic commerce (E-commerce) applications; and, in addition to Gigabit Ethernet, supports integration with in-place Ethernet and Fast Ethernet systems and next-generation 10 Gigabit Ethernet installations.

4.19 GIGABIT ETHERNET IMPLEMENTATION CONSIDERATIONS

Fast Ethernet and Gigabit Ethernet share common network components and support joint topologies and network architectures. These technologies also facilitate connectivity to legacy networks and provide a scalable approach for achieving very high data rates. The transition to higher-speed Ethernet networks can be readily accomplished without large-scale reinvestment.

Fast Ethernet and Gigabit Ethernet networks provide high-capacity advanced network services for new classes of applications that cannot function effectively in the current network infrastructure. The appeal of Fast Ethernet and Gigabit Ethernet stems from their structural and functional compatibility with each other and with conventional Ethernet solutions.

In parallel with Ethernet and Fast Ethernet, Gigabit Ethernet can be implemented in incremental phases, employs CSMA/CD to resolve contention for the shared medium, and supports operations in full-duplex and half-duplex modes. In full-duplex operations, Gigabit Ethernet enables rates at 2 Gbps via point-to-point and switch-to-server links. Gigabit Ethernet also employs frame flow control as a CSMA/CD enhancement for enabling switch-to-node and switch-to-switch connections. In instances of shared port connections, Gigabit Ethernet supports transmissions in half-duplex mode at rates reaching 1 Gbps and transmissions in full-duplex mode at rates reaching 2 Gbps.

Typically, Gigabit Ethernet is deployed in backbone network environments that require increased bandwidth between routers, centralized hubs, repeaters, servers, and switches. As with Fast Ethernet, Gigabit Ethernet enables advanced applications and services in fields that include distance learning, telemedicine, E-government, and E-commerce. Gigabit Ethernet also works in conjunction with established applications, platforms, operating systems, protocols, and technologies.

In addition, Gigabit Ethernet interoperates with Ethernet and Fast Ethernet technologies and on-site equipment supporting earlier Ethernet deployments. As a consequence, migration to a Gigabit Ethernet installation is a straightforward process. Gigabit Ethernet is a robust multivendor solution that can be built with Gigabit Ethernet devices such as adapters and 100/1000 Mbps dual-mode Gigabit Ethernet switches from diverse manufacturers.

Gigabit Ethernet is distinguished from its Ethernet predecessors by its support of higher bandwidths, faster response times, and substantial improvements in backbone and server performance. A versatile and flexible technology, Gigabit Ethernet supports Layer 3 or Network Layer switching and routing operations and employs upper-layer protocols for enabling IP multicasts and CoS assurances. In addition, Gigabit Ethernet facilitates seamless network extendibility and scalability at those enterprises with plans to integrate additional servers and Web caches into the network infrastructure for supporting data-intensive applications such as bulk file transfers and on-demand publishing via in-place configurations.

Nonetheless, there are challenges associated with effective Gigabit Ethernet deployment. Gigabit Ethernet implementations require new tools and components for network monitoring, troubleshooting, and management operations. In addition, Gigabit Ethernet installations require a flexible architecture that supports switched data streams, throughput at gigabit rates, and access to intranets, extranets, and next-generation applications and services. Gigabit switching and interface products are necessary to provision a smooth and scalable migration path from current Ethernet and Fast Ethernet configurations to Gigabit Ethernet network installations. Moreover, approaches for enabling the co-existence and interoperations of Gigabit Ethernet and ATM services must also be defined.

Gigabit Ethernet operations are subject to functional constraints. Gigabit Ethernet standards for transporting traffic over copper cabling are relatively recent and not fully tested in actual networking environments. Strategies for transporting multimedia via Gigabit Ethernet links are still in development. Further, Gigabit Ethernet is not currently capable of fully guaranteeing congestion control and enabling automatic rerouting of network traffic in case of network gridlock.

Despite expectations associated with Gigabit Ethernet implementations, early field trials of Gigabit Ethernet operations indicate the installed base of workstations and servers may not be able to support gigabit links and take advantage of higher throughput. As a consequence, although Gigabit Ethernet is perceived as yet another Ethernet incarnation, its implementation involves establishing staff responsibilities for Gigabit Ethernet services and a budget for acquisition of standards-compliant Gigabit Ethernet switches, repeaters, hubs, and routers to support substantial improvements in network operations and functions.

New devices such as buffered distributors and full-duplex repeaters for enabling interconnectivity between two or more Gigabit Ethernet configurations operating at 1 Gbps and faster rates are in development. In the absence of auxiliary equipment such as buffered distributors, Gigabit Ethernet is limited in provisioning CoS assurances for multimedia traffic and sustaining error-free multimedia transport. The provision of sustained support and seamless interoperability is critical to the acceptance and deployment of Gigabit Ethernet technology.

Future Gigabit Ethernet implementations are expected to routinely interwork with Ethernet and Fast Ethernet switches featuring speed translation capabilities for enabling interoperable services. In addition, approaches for provisioning full-scale interoperability between Gigabit Ethernet and ATM technologies are also in development.

A Gigabit Ethernet solution enables an enterprise to leverage its investment in the in-place Ethernet and Fast Ethernet infrastructure while boosting network performance and realizing the benefits of high-speed networking. With Gigabit Ethernet, research centers, educational institutions, government agencies, and hospitals can accommodate accelerating bandwidth requirements for enabling such network applications as videoconferencing, data warehousing, medical imaging, and scientific simulations, and fostering dependable access to intranets, extranets, and Web resources.

As a consequence of its capabilities in interworking with diverse narrowband and broadband technical solutions, Gigabit Ethernet technology is regarded as a key enabler of next-generation network implementations. The popularity of Gigabit Ethernet solutions is reflected in the wide range of initiatives sponsored by healthcare centers, universities, corporations, and government organizations. Representative initiatives are examined later in this chapter.

4.20   10 GIGABIT ETHERNET

4.20.1   10 GIGABIT ETHERNET TECHNICAL FUNDAMENTALS

The remarkable growth of Internet traffic and the widespread popularity of Ethernet technologies contribute to development and implementation of the 10 Gigabit Ethernet specification by the IEEE 802.3 High-Speed Study Group and the formation of the IEEE 802.3ae Task Force for coordinating standards activities in the 10 Gigabit Ethernet domain.

4.20.2   10 GIGABIT ETHERNET OPERATIONS

Gigabit Ethernet technology is a logical extension to the Ethernet family of standards and retains the IEEE 802.3 frame format and packet size of its predecessors. Moreover,

10 Gigabit Ethernet solutions are designed to work in concert with Ethernet, Fast Ethernet, and Gigabit Ethernet technologies, architectures, and protocols.

10 Gigabit Ethernet networks support operations at Layer 1 or the Physical Layer and Layer 2 or the Data-Link Layer of the OSI Reference Model. By working in conjunction with the IEEE 802.3 MAC (Media Access Control) protocol, 10 Gigabit Ethernet enables full-duplex operations and utilizes bridging equipment already employed for Ethernet, Fast Ethernet, and Gigabit Ethernet networks. In addition, by conforming to IEEE 802.3 protocols, 10 Gigabit Ethernet ensures operational compatibility with the installed base of Ethernet networks.

In parallel with the three Ethernets, 10 Gigabit Ethernet enables upper layer functions and provisions QoS assurances. 10 Gigabit Ethernet applications include IP telephony, video-on-demand (VOD), Web caching, server load balancing, and security and policy enforcement. 10 Gigabit Ethernet technology also operates with IETF (Internet Engineering Task Force) protocols such as MPLS (MultiProtocol Label Switching) and Simple Network Management Protocol (SNMP).

Moreover, the Management Information Base (MIB) for 10 Gigabit Ethernet works in concert with the IEEE 802.3 MIBs (Management Information Bases) for Ethernet, Fast Ethernet, and Gigabit Ethernet installations. Additionally, 10 Gigabit Ethernet specifications support Ethernet, Fast Ethernet, and Gigabit Ethernet flow control procedures. However, in contrast to its predecessors, 10 Gigabit Ethernet is not expected to support CSMA/CD operations at half-duplex mode.

In the LAN arena, 10 Gigabit Ethernet fosters high-speed interconnectivity between large-scale switches and bandwidth-intensive backbone networks, and supports seamless high-speed Internet, intranet, and extranet connections. In addition to provisioning LAN services, 10 Gigabit Ethernet supports operations performed in metropolitan and wider area networks. Gigabit Ethernet technology also enables significantly higher bandwidth capabilities than its predecessors for supporting current and next-generation voice, video, and data applications.

4.21   10 GIGABIT ETHERNET STANDARDS ORGANIZATIONS AND ACTIVITIES

4.21.1   10 GIGABIT ETHERNET ALLIANCE (10GEA)

The 10 Gigabit Ethernet Alliance (10GEA) is an industry forum that supports wide-scale implementation of the 10 Gigabit Ethernet specification in LANs, MANs, and WANs and works with the IEEE 802.3ae Task Force to facilitate 10 Gigabit Ethernet standards development. In addition, the 10GEA promotes implementation of 10 Gigabit Ethernet solutions that work in concert with SONET/SDH (Synchronous Optical Network and Synchronous Digital Hierarchy), WDM (Wavelength Division Multiplexing), and DWDM (Dense WDM) technologies. Gigabit Ethernet Alliance members include Cisco, 3Com, Extreme Networks, Sun Microsystems, Intel, Nortel Networks, Ascend Communications, and Lucent Technologies. In addition, Apple Computer, 3M, Hewlett Packard, and Compaq Computer participate in the 10GEA as well.

4.21.2   IEEE P802.3A 10 GIGABIT ETHERNET TASK FORCE

Following approval by the IEEE 802.3 Higher Speed Study Group, 10 Gigabit Ethernet specifications were sent to the IEEE Standards Board. In 2000, this Board authorized moving forward with 10 Gigabit Ethernet specifications within the IEEE P802.3ae 10 Gigabit Ethernet Task Force. The IEEE P802.3ae 10 Gigabit Ethernet Task Force supports development of 10 Gigabit Ethernet protocols that function in all-optical LAN, MAN, and WAN networking environments; clarifies guidelines for maintaining the Ethernet, Fast Ethernet, and Gigabit Ethernet frame size and frame format; and supports full-duplex 10 Gigabit Ethernet operations. This Task Force also facilitates development of Physical Layer or Layer 1 specifications for enabling 10 Gigabit Ethernet transmissions over multimode optical fiber at distances between 100 meters and 300 meters and single-mode optical fiber at distances of 2,000, 10,000, and 40,000 meters or 2, 10, and 40 kilometers.

4.21.3   IEEE 802.AH ETHERNET FIRST MILE TASK FORCE

In 2001, the IEEE 802.ah Ethernet in the First Mile Task (EFM) Force began work on development of a wireline 10 Mbps Ethernet local loop standard that provisions communications services over the first mile from the customer premise to the local telephone exchange. This Task Force clarifies network operations, management services, and architectures for first mile solutions. In addition, specifications to support 10 Mbps transmissions via point-to-point links over twisted copper pair wiring at distances that extend to 750 meters are in development. Moreover, standards for enabling 1000 Mbps or 1 Gbps transmissions via point-to-point and point-to-multipoint connections via optical fiber connections over distances that extend to 10,000 meters or 10 kilometers are under consideration.

4.22   10 GIGABIT ETHERNET SOLUTIONS

Gigabit Ethernet technology supports interconnectivity between geographically distributed local networks; enables integrated LAN, MAN, and WAN implementations; and provisions affordable fast access to bandwidth-intensive multimedia applications over single-mode and multimode optical fiber. 10 Gigabit Ethernet services are carried directly over SONET/SDH, WDM, and DWDM infrastructures. For example, 10 Gigabit Ethernet-over-DWDM supports multimedia transmissions via a 50-micron multimode optical fiber core at distances that extend to 65 meters. With a 62.5-micron multimode optical fiber core, each 10 Gigabit Ethernet-over-DWDM network segment enables transmissions at a distance that extends to 300 meters. With a 9.0-micron single-mode optical fiber core, each 10 Gigabit Ethernet-over-DWDM network segment supports multimedia transport over distances extending from 10 kilometers to 40 kilometers.

4.22.1   10 GIGABIT OPTICAL ETHERNET IN ACTION

Accelerating demand for fast Internet access and higher bandwidth services contributes to the development of optical Ethernet solutions in the vendor marketplace that are designed to work in conjunction with SONET/SDH, WDM, and DWDM technologies. Network configurations based on optical Ethernet solutions are scalable, extendible, and reliable, and support multigigabit interactive video, voice, and data applications.

Implementation of 10 Gigabit Ethernet-over-DWDM installations in a metropolitan area network (MAN or metro) configuration enables service providers to support end-to-end IP information transport, use ubiquitous Ethernet interfaces, and extend the reach of VLANs across metropolitan and long-haul networks. 10 Gigabit Ethernet-over-DWDM implementations dependably deliver next-generation Ethernet services to multiple sites. By eliminating the relatively time-consuming process of electronic-to-optical and optical-to-electronic conversion and encapsulating Ethernet frames into other packet formats, 10 Gigabit Ethernet-over-DWDM installations streamline network operations.

4.23   10 GIGABIT ETHERNET MARKETPLACE

4.23.1   CISCO SYSTEMS

A founding member of the 10 Gigabit Ethernet Alliance, Cisco Systems supports 10 Gigabit Ethernet solutions that enable operations over Category 6 twisted copper pair.

4.23.2   COGENT COMMUNICATIONS

A specialized ISP (Internet Service Provider), Cogent Communications enables implementation of an all-optical, long-haul backbone network based on 10 Gigabit Ethernet technology. This sophisticated backbone network supports high-speed access to Internet services and transparent VLAN, intranet, and extranet applications. In addition, Cogent Communications solutions employ Gigabit Ethernet technology for enabling broadband metropolitan networking applications at 2.488 Gbps (OC-48) within cities. Each 10 Gigabit Ethernet metropolitan network is also interlinked to the long-haul 10 Gigabit Ethernet network backbone that is also supported by Cogent Communications.

4.23.3   EXTREME NETWORKS

Extreme Networks enables implementation of VMAN (Virtual MAN) solutions that support VLAN aggregation in co-located environments and operate in conjunction with 10 Gigabit Ethernet-over-WDM technology. This infrastructure uses eight full-duplex channels for supporting operations at multigigabit rates and provisions CoS assurances for delay-sensitive voice, data, and video applications.

4.23.4   LUCENT TECHNOLOGIES

In 1999, Lucent Technologies implemented a prototype 10 Gbps GigaChannel Ethernet multiplexer that works in concert with DWDM technologies and supports optical

Ethernet implementations. The network infrastructure also includes inline EDFAs (Erbium-Doped Fiber Amplifiers) and single-mode and multimode optical fiber. This Lucent optical Ethernet solution facilitates ultrafast Ethernet frame transmission, supports standardized link aggregation protocols, and provisions network services that operate at the Data-Link Layer or Layer 2 and the Network Layer or Layer 3 of the OSI Reference Model. In addition, VLAN functions and QoS transmissions over distances of 40 kilometers are supported.

4.23.5 NORTEL NETWORKS

Nortel Networks supports multiservice SONET/SDH and DWDM optical platforms that work in concert with 10 Gigabit Ethernet for enabling high-bandwidth transmission. In addition, Nortel 10 Gigabit Ethernet solutions enable transmissions from the core network to the desktop in DAN (Desktop Area Network) configurations. 10 Gigabit Ethernet solutions available from Nortel Networks also support transparent VPN (Virtual Private Network) operations across public or shared metropolitan networks, enable long-haul deployments, and conform to the IEEE 802.1Q and the IEEE 802.1p specifications.

4.23.5.1 Nortel Networks and Korea Telecom

Korea Telecom implements optical 10 Gigabit Ethernet solutions developed by Nortel Networks for supporting high-speed, high-performance Internet services; CoS assurances; VPN implementations; and secure intranet and extranet interconnections. These multiservice, multifunctional broadband optical networks work in concert with DSL (Digital Subscriber Line) technology for transporting voice, video, and data to SOHO (Small Office/Home Office) venues and delivering customized IP and multimedia services to corporate entities.

4.23.6 TELESON COMMUNICATIONS

A high-speed Internet access provider, Telseon Communications supports optical 10 Gigabit Ethernet solutions that enable multimedia applications, IP services, Storage Area Networks (SANs), and LAN-to-LAN interconnectivity in Denver, St. Louis, Chicago, Seattle, and Atlanta.

4.24 WIRELESS ETHERNET SOLUTIONS

As with wireline Ethernet networks, wireless Ethernet solutions are reliable, flexible, extendible, and scalable. Moreover, wireless Ethernet solutions are affordable and easily implemented. Wireless Ethernet implementations support applications in homes, airports, hotels, parks, museums, schools, hospitals, office buildings, banks, cruise ships, casinos, restaurants, and rental car agencies. These configurations also facilitate inventory tracking in stores and libraries, telemedicine applications in rural locations, and roadside emergency assistance. Wireless Ethernet features and functions are examined in Chapter 9.

4.25 ETHERNET HOME PHONELINE NETWORKS

Demand for Internet access and multimedia services drives network deployment in the education, government, telemedicine, and commercial sectors. This demand also fosters development of home networks for provisioning low-cost shared interconnectivity to Web resources and transporting data between information appliances within the home environment.

4.25.1 HOME PHONELINE NETWORK FUNDAMENTALS

A recent innovation in the home networking domain, home phoneline networks are typically based on 10 Mbps Ethernet technology. Developed by the HomePNA (Home Phoneline Network Alliance), these networks support high-speed, affordable, and reliable data transport; file and application sharing; home automation; telephone services; and fax transmissions. Ethernet components that support realization of a home phoneline networking solution include 10 Mbps Ethernet NICs (Network Interface Cards) and Ethernet-to-home network adapters that enable any in-home device such as a laptop or a desktop PC (Personal Computer) with an Ethernet port to interoperate with the home phoneline network. Approximately 50 network devices and peripherals can be interlinked in a home phoneline network configuration. A HomePNA solution supports operations with devices and PCs that are up to 1,000 feet apart. In addition to 10 Mbps Ethernet, home phoneline networking solutions also work in concert with POTS (Plain Old Telephone Service), DSL (Digital Subscriber Line), satellite, and cable modem technologies.

4.25.2 HOMEPNA SPECIFICATIONS

Developed by the HomePNA in 1998, the HomePNA solution is based on core 10 Mbps Ethernet networking capabilities that are adapted to home phoneline utilization. A *de facto* industry standard, the HomePNA 1.0 specification supports the Ethernet CSMA/CD protocol described in the IEEE 802.3 standard to enable multiple device connectivity to a common communications medium.

In addition, home phoneline networks are also compatible with the Ethernet MAC (Media Access Control) specification and comply with U.S. FCC (Federal Communications Commission) regulatory requirements. HomePNA 1.0 installations use Ethernet software and hardware, enable rates at 1 Mbps, and support operations via twisted pair, coaxial cable, and optical fiber wiring commonly found in homes and high-rise apartment buildings.

4.25.2.1 HomePNA 1.0 Specification

Information appliances and peripherals that are compliant with the HomePNA 1.0 specification interoperate with each other and the Internet without interruption of conventional telephony and fax services. Because HomePNA installations utilize the in-place wiring structure, the need for additional cabling is eliminated. Home phoneline networks enable seamless transport in home and small office environments by accommodating varying levels of signal noise generated by heaters, air conditioners, and consumer appliances such as dishwashers, washing machines, dryers, microwave ovens, and refrigerators.

Home phoneline networks that are compliant with the HomePNA 1.0 specification support operations in the spectrum between the 5.5 MHz (Megahertz) and the 9.5 MHz frequency bands. The HomePNA solution employs signals with low energy levels to comply with FCC requirements. In addition, frequency selective filters and FDM (Frequency-Division Multiplexing) techniques enable standards-compliant home phoneline networks to reduce signal interference generated by traditional POTS (Plain Old Telephone Service) applications that operate in adjacent frequencies.

In contrast to residential broadband access solutions employing cable modem and DSL technologies, the HomePNA 1.0 specification eliminates the need for special devices such as splitters and additional wireline installation at the customer premise. Data throughput is achieved by encoding multiple data bits into each signal. Developed by Tut Systems, HomeRUN technology serves as the foundation for the HomePNA 1.0 standard.

4.25.2.2 HomePNA 2.0 Specification

Endorsed in 1999, the HomePNA 2.0 *de facto* specification fosters migration to standardized 10 Mbps Ethernet solutions in residential venues. Vendors such as Epigram and Lucent Technologies developed the framework for the HomePNA 2.0 installation.

Home phoneline networks that are compliant with the HomePNA 2.0 specification employ selective frequencies in spectrum between the 2 MHz and 30 MHz RF (Radio Frequency) spectrum. The HomePNA 2.0 specification includes a time modulation line coding technique for fostering data transport via the in-place wireline infrastructure.

The HomePNA 2.0 specification supports backward compatibility with the HomePNA 1.0 specification, co-existence with other technologies and communications services at SOHO (Small Office/Home Office) venues, and rapid transmission of high-bandwidth applications such as large graphics files. Moreover, the HomePNA 2.0 specification enables home phoneline networks to interoperate with clustered devices that work in concert with USB (Universal Serial Bus) technology. The HomePNA 2.0 specification also supports the IEEE 1394 High-Performance Serial Bus specification.

4.25.3 HomePNA Installations

HomePNA specifications foster access to fixed wireline broadband networks via a standardized high-speed home network interface. This plug-and-play solution enables peripheral, file, and application sharing; automatic software updates and backups; multi-user entertainment; home appliance automation; and home security services. With the installation of the home phoneline network, devices such as desktop PCs (Personal Computers), digital cameras, printers, scanners, cordless phones, HDTV (High Definition Television) sets, and palmtop computers support links to voice, video, and data applications throughout the home or small office.

HomePNA 10 Mbps Ethernet installations are scalable, flexible, and extendible. Each RJ-11 telephone jack in a home or small office enables the addition of a network terminal or device to the home phoneline network infrastructure. In addition to 10 Mbps Ethernet solutions, standardized home phoneline networks operate in conjunction with 28.8 Kbps, 33.6 Kbps, and 56 Kbps dial-up modem connections; high-speed always-on powerline networks; and ISDN (Integrated Services Digital Network) installations.

4.25.4 HOME PHONELINE NETWORK ALLIANCE (HOMEPNA)

The rapid growth of households with multiple computers and devices contributed to the development of the Home Phoneline Network Alliance (HomePNA) in 1998. HomePNA supports the design, deployment, and adoption of integrated residential phoneline networking solutions.

To promote widespread utilization of home phoneline networking solutions, the HomePNA develops certification standards for ensuring interoperability among HomePNA products, devices, and components. The HomePNA validates capabilities of HomePNA certified products at the CEBus Industry Council (CiC) Laboratory at Purdue University. Tests for verifying functions of HomePNA services are also conducted at the Silicon Valley National Laboratory (SVNL). In addition, trials benchmarking the performance of HomePNA installations are carried out in single-family homes, apartment buildings, duplexes, and condominiums throughout the United States.

4.25.5 HOME PHONELINE MARKETPLACE

Products that are compatible with the HomePNA 2.0 specifications are available from IBM, Lucent Technologies, 3Com, NetGear, AT&T, Compaq, Epigram, Hewlett Packard, Tut Systems, and Rockwell Semiconductor Systems.

4.25.5.1 Intel Corporation

Intel Corporation supports a home phoneline network solution that delivers transmissions at rates up to 10 Mbps over existing telephony lines to a residential location. In addition to supporting compliance with the HomePNA 2.0 specification, the Intel home phoneline network solution enables every user in a residence to share Internet connectivity via one modem, one phoneline, and one Internet Service Provider (ISP) account.

This solution requires the use of Intel Web sharing and installation software for supporting network management operations and the Intel 10 Mbps Ethernet Network Interface Card (NIC) to enable transfer of high-resolution graphics and full-motion video between information appliances within the customer premise. An advanced encoding method fosters dependable data transmission and access to broadband services via ordinary copper phonelines. Separate channels transport voice, video, and data signals on different frequencies.

4.25.6   ADDITIONAL HOME NETWORKING OPTIONS

4.25.6.1   European Telecommunications Standards Institute (ETSI) Multimedia Home Platform (MHP)

In contrast to home phoneline networks and their utilization of Ethernet technology, the Multimedia Home Platform (MHP) for home access networks (HANs) developed by the ETSI (European Telecommunications Standards Institute) is based on an ATM interface. This interface provisions connections to external networks, supports operations at 25 Mbps or 51.84 Mbps, and works in conjunction with cable modem and DSL technologies.

4.25.6.2   Open Services Gateway Initiative (OSGi)

The OSGi Consortium develops open standards for interlinking information appliances such as Web phones, Personal Digital Assistants (PDAs), and gaming devices via a single gateway. This gateway serves as a single point of contact for managing smart appliances, multimedia entertainment applications, smart home security systems, and IP telephony services in home networks. Developed by a vendor consortium that includes Motorola, IBM, Nortel Networks, Oracle, Deutsche Telekom, and Alcatel, this initiative employs an Application Layer framework for supporting home networking platforms and multivendor solutions. The OSGi solution is also compliant with HomePNA specifications.

In residential venues, the OSGi gateway routes incoming signals to the appropriate in-home termination point and enables intelligent information appliances to access networking applications concurrently. OSGi architecture is an enabler of energy management functions and home security and home entertainment systems. OSGi home phoneline networking implementations support transmission of multimedia from trusted service providers on the external network to the home or internal client. Based on the Java programming language, OSGi technology interworks with Microsoft Windows, Bluetooth, and Jini solutions.

4.25.6.3   Jini Network Technology

Available from Sun Microsystems, Jini network technology supports current and emergent network applications that are constructed from networks and objects; enables device operations that are independent of in-place hardware and software systems; and optimizes capabilities of in-place network resources. Jini architecture facilitates development of a straightforward network infrastructure for network service delivery that is flexible, scalable, and extendible. Administrative support and supervision are not required for adding or deleting services to Jini networks. As with OSGi, Jini network technology employs the Java programming language.

In the home network arena, Jini network technology works in conjunction with Bluetooth technology for enabling wireless network connectivity via short-range radio links between portable devices such as laptops, notebooks, and PDAs and employs FireWire technology for interlinking portable devices and computer peripherals. In home networking installations, Jini network technology also interoperates with the Universal Plug and Play Protocol for provisioning seamless information transmission and the Service Location Protocol for enabling network nodes to find and identify network services.

4.26   U.S. TELE-EDUCATION INITIATIVES

4.26.1   ALASKA

4.26.1.1   Arctic Region Supercomputing Center (ARSC)

The Arctic Region Supercomputing Center (ARSC) supports Gigabit Ethernet connections to the University of Alaska Science Research Institute Network and the U.S. Department of Defense Research and Education Network (DREN) for enabling interactive participation in broadband telecollaborative research projects and teleresearch programs.

4.26.1.2   AuroraNet

Situated on the North Slope of Alaska, AuroraNet employs a combination of wireless and wireline satellite, Ethernet, and Fast Ethernet technologies for delivering voice, video, and data services to remote communities in Nuiqsut, Atqasuk, Point Hope, Point Lay, and Kaktovik. Satellite videoconferencing available via AuroraNet supports virtual learning communities in public schools and a regional college. Ethernet and Fast Ethernet technologies enable teleconsultations between primary care physicians treating patients in rural clinics and medical specialists at urban hospitals and provision a framework for conducting E-government (electronic government) activities in Barrow.

4.26.2   CALIFORNIA

4.26.2.1   Santa Clara University (SCU)

Santa Clara University (SCU) utilizes a Gigabit Ethernet backbone network for provisioning access to Web services and digital library resources, links to educational networks such as CREN (Corporation for Research and Educational Networking), and connectivity to SCU intranets. The SCU Gigabit Ethernet backbone network also enables videoconferencing and utilization of an Electronic Reservation (E-Res) system featuring course notes, syllabi, and assignments. In addition, this backbone network includes Ethernet and Fast Ethernet segments for enabling access to advanced applications in graduate-level engineering courses and business programs.

4.26.2.2   United States (U.S.) Navy Postgraduate School (NPS)

Situated in Monterey, the United States Navy Postgraduate School (NPS) employs a networking configuration consisting of ATM and Ethernet technologies. The ATM backbone network supports rates at 622.08 Mbps (OC-12) and features a star topology for optimizing traffic throughput. In addition, ATM edge devices operating at 155.52 Mbps (OC-3) connect 18 buildings on the school campus and interlink NPS Ethernet LANs to the NPS enterprisewide ATM network. The NPS network supports multimedia services, tele-education projects, videoconferencing, and dissemination of classroom lectures directly to students at their desktop PCs.

4.26.2.3   University of California at Berkeley (UC Berkeley)

The University of California at Berkeley (UC Berkeley) employs a high-speed, high-capacity experimental backbone network called BMRCnet (Berkeley Multimedia Research Center Network). BMRCnet utilizes Ethernet and ATM technologies for interlinking multimedia laboratories, electronic classrooms, computer storage servers, and network systems across campus. BMRCnet also supports development of high-volume multimedia applications such as interactive television programming and video-on-demand. A distributed storage network for maintaining terabits of digital resources is also in development. The BMRCnet backbone supports server and desktop connections at 155.52 Mbps (OC-3).

4.26.2.4   University of California at Riverside (UC Riverside)

The University of California at Riverside replaced an FDDI (Fiber Data Distributed Interface) network with a Gigabit Ethernet installation at the College of Engineering. This scalable Gigabit Ethernet network interlinks buildings on the UC Riverside campus; supports high-speed access to the Internet, I2 (Internet2), and other Cal-REN-2 (California Research and Education Network, Phase 2) networks; and enables a wide array of bandwidth-intensive applications at rates ranging from 10 Mbps to 1 Gbps.

4.26.3   GEORGIA

4.26.3.1   Emory University

Emory University utilizes an ATM network backbone that works in concert with departmental Fast Ethernet networks at rates reaching 155.52 Mbps (OC-3). This hybrid infrastructure supports distance learning programs, Internet connectivity, and videoconferencing applications. In addition, this mixed-mode network also facilitates voice, video, and data transmission; video production initiatives; telecollaborative research projects; and interactive workgroup teletraining sessions.

4.26.4   INDIANA

4.26.4.1   Butler University

Situated in Indianapolis, Butler University employs an integrated Ethernet, Fast Ethernet, and Gigabit Ethernet configuration for enabling data-intensive file transmissions, distance education services, interactive Web classes, and access to Web resources.

4.26.5 KENTUCKY

4.26.5.1 Morehead State University

Morehead State University uses an ATM networking configuration that works in concert with 10 Mbps Ethernet links to campus dormitories for enabling students to access tele-education applications.

4.26.6 MASSACHUSETTS

4.26.6.1 Boston's Kids Compute 2001

Boston's Kids Compute 2001 is a multiservice, multifunctional initiative that promotes implementation of Ethernet and Fast Ethernet technologies in Boston public schools for supporting tele-education and teletraining. Through Kids Compute 2001, high school students enroll in tele-education courses in composition, calculus, and criminal justice to earn college credits. Kids Compute 2001 also supports videoconferencing, video-over-IP, and a faculty intranet that provisions access to national educational guidelines and standards, lesson plans, student projects, and procedures for introducing new technologies into the classroom. An online grading system and attendance roster facilitates completion of administrative tasks associated with Kids Compute 2001 implementation. Moreover, a Frame Relay metropolitan network interlinks each public school in the Boston area to district offices and to the Web.

4.26.6.2 Suffolk University

Situated in downtown Boston, Suffolk University employs a Gigabit Ethernet solution in its law school facility for supporting connections to legal resources, digital publications, databases, electronic casebooks for in-class citations, and multimedia applications. This network configuration also supports online class registration and access to course grades and transcripts.

4.26.7 MICHIGAN

4.26.7.1 Rockford Public School System

Located in Grand Rapids, the Rockford Public School System employs a combination of Ethernet, Fast Ethernet, and Gigabit Ethernet technologies for provisioning data, voice, and video transmission and enabling applications such as interactive videoconferencing and IP telephony at school sites and administrative facilities.

4.26.8 MISSOURI

4.26.8.1 Washington University School of Law

The Washington University School of Law in St. Louis utilizes a mix of Ethernet, Fast Ethernet, and FDDI technologies to enable video moot court competitions, on-demand Internet access, and interactive videoconferences between students, faculty, and judges.

4.26.9 NEVADA

4.26.9.1 University of Nevada at Las Vegas (UNLV)

The University of Nevada at Las Vegas (UNLV) employs an Ethernet and Fast Ethernet solution for provisioning access to campus VLANs, preserving network uptime, and ensuring network reliability and availability. Operated by the Campus Housing ResNet (Residential Network) Office, the UNLV ResNet employs 10BASE-T Ethernet connections for enabling on-campus students in dormitories to access campus networks and Web services. In addition to the Ethernet and Fast Ethernet solution, the UNLV School of Education supports satellite telecourse delivery to K–12 teachers unable to attend classes on campus. Approaches for upgrading the campus backbone network to Gigabit Ethernet as a replacement for the in-place ATM network are under consideration.

4.26.10 NEW JERSEY

4.26.10.1 Sewell School District

The Sewell School District employs Ethernet, Fast Ethernet, and ATM technologies in a sophisticated configuration the supports voice, video, and data delivery. This scalable network supports rates ranging from OC-3 (155.52 Mbps) to OC-12 (622.08 Mbps) and enables applications such as distance education, videoconferencing, electronic testing with immediate feedback, student Web page development, and telecollaborative activities. This network also provisions access to Web services, digital library resources, and faculty and administrative intranets.

4.26.10.2 Vineland Public School District

The Vineland Public School District utilizes a Fast Ethernet and ATM metropolitan area configuration for fostering voice, video, and data transmissions to city government agencies and school buildings at rates reaching 622.08 Mbps (OC-12). This configuration also supports E-government and tele-education services and provisions access to Web resources for courses in science, mathematics, and foreign languages.

4.26.11 NEW YORK

4.26.11.1 LAKENET

The Wayne-Finger Lakes Board of Cooperative Educational Services (BOCES) in New York State sponsors LAKENET, a WAN employing Ethernet, Fast Ethernet, FDDI, and ATM technologies, for expanding curricular options and educational opportunities at public schools. LAKENET interlinks approximately 40 school districts, delivers distance learning programs and teletraining sessions, and supports connectivity to a vast array of online resources.

4.26.11.2 New York City School System

The New York City Board of Education employs Ethernet and Fast Ethernet in a scalable network configuration supported by the New York City School System. This network facilitates interdisciplinary instruction, workgroup computing, Web browsing, and interactive tele-education programs such as the Virtual Enterprise, a vocational telelearning initiative that prepares high school juniors and seniors for employment.

4.26.11.3 Northeast Parallel Architecture Center (NPAC) at Syracuse University

The network infrastructure developed by NPAC (Northeast Parallel Architecture Center) at Syracuse University uses Ethernet, Fast Ethernet, and Gigabit Ethernet technologies for enabling access to Web resources, multimedia applications, high-performance distributed simulations, and connections to advanced federal network configurations such as DREN.

4.26.11.4 Rochester Institute of Technology (RIT)

At the Rochester Institute of Technology (RIT), a 10 Mbps Ethernet implementation enables students to access the RIT intranet from libraries, classrooms, the cafeteria, and dormitories, and register online for courses, access digital account records maintained by the Bursar's Office, and reserve campus apartments for the next school year.

4.26.12 OHIO

4.26.12.1 Ohio State University

The Fisher College of Business at Ohio State University in Columbus employs Gigabit Ethernet as its core networking technology. This high-speed, high-performance network supports administrative services and academic applications such as video instruction, videoconferencing, and VLAN workgroup telecollaborative projects.

4.26.13 OKLAHOMA

4.26.13.1 Oklahoma State University (OSU)

Oklahoma State University (OSU) employs an ATM backbone network that works in concert with Ethernet and Fast Ethernet segments for enabling delivery of undergraduate telecourses to high school students in metropolitan areas and remote communities.

4.26.13.2 Western Heights School District

The Western Heights School District implements Fast Ethernet over fiber-optic cabling for interlinking schools throughout the district. Approaches are in development for upgrading the districtwide Fast Ethernet network backbone to Gigabit Ethernet for supporting real-time videoconferencing, faculty and student telecollaboration, and distribution of broadcast video programs.

4.26.14 PENNSYLVANIA

4.26.14.1 Widener University

A private regional institution with two campuses in Pennsylvania and one campus in Delaware, Widener University employs an ATM backbone network that supports rates at 155.52 Mbps (OC-3) for delivering voice, video, and data to classrooms, cafeterias, staff offices, and desktop PCs via Ethernet and Fast Ethernet segments. This configuration enables videoconferencing, telelectures, tele-education applications, and Web browsing.

4.26.15 TEXAS

4.26.15.1 Ennis Independent School District

The Ennis Independent School District utilizes a combination of Fast Ethernet and Gigabit Ethernet technologies for provisioning access to curricular guides, multimedia instructional resources, Web services, and distance education programs.

4.26.15.2 Texas Education Service Center Region 10

A state-chartered organization in Richardson, the Texas Education Service Center Region 10 employs Ethernet and Fast Ethernet technologies for provisioning educational services to school districts in nine counties. Created as a replacement for a legacy network, this configuration supports cost-effective operations and enables applications such as videoconferencing, Web browsing, and teletutorials for improving student performance.

4.26.16 WASHINGTON

4.26.16.1 Washington State K–20 Education Network

The Washington State K–20 Education Network employs Fast Ethernet technology for interlinking schools, school districts, universities, community libraries, and state and local government agencies in a multiservice network configuration. In addition to facilitating access to Web resources and academic intranets at disparate locations statewide, this configuration expands curricular options, supports course offerings that were not previously available, fosters faculty teleresearch and telecollaboration, and delivers faculty in-service teletraining sessions.

4.26.17 PUERTO RICO

4.26.17.1 University of Puerto Rico Educational Network (UPRENET)

The University of Puerto Rico Educational Network (UPRENET) employs Ethernet and Fast Ethernet implementations at each campus location for enabling access to library resources, e-mail, and the Internet. Moreover, the UPRENET platform supports telecollaborative activities between students and faculty and enables administrators to monitor online course registrations. Teachers and staff at local K–12 schools access UPRENET via 56 Kbps links for sharing administrative information and provisioning Web applications in classrooms. UPRENET also enables participants in the Caribbean University Network (CUNET) to access Web resources.

4.27 INTERNATIONAL TELE-EDUCATION PROJECTS

4.27.1 AUSTRIA

4.27.1.1 Vienna University

Vienna University employs Fast Ethernet technology for providing access to the Austrian Academic Network and the Internet. Vienna University also provisions technical support for the Vienna Internet Exchange (VIX). VIX uses Ethernet and Fast Ethernet technologies to support bilateral peering arrangements and traffic exchange among academic networks that operate in Central and Eastern Europe and among Internet Service Providers (ISPs) such as GlobalOne, AboveNet, and AustriaONE.

4.27.2 CHINA

4.27.2.1 Chinese University of Hong Kong (CUHK)

The Chinese University of Hong Kong (CUHK) employs a backbone ATM network that works in concert with Ethernet LAN segments in academic and administrative departments across campus. The CUHK ResNet (Residential Network) for students in campus housing supports 10 Mbps Ethernet links to the campus ATM network and provisions reliable access to Web resources.

4.27.2.1.1   Hong Kong Internet Exchange

Developed by the Chinese University of Hong Kong, the Hong Kong Internet Exchange (HKIX) supports dependable transmission and reception of multimedia traffic throughout the metropolitan area via Fast Ethernet and ATM connections and provides access to the Hong Kong Academic and Research Network (HARNET).

4.27.2.2   City University of Hong Kong (CTNET)

The City University of Hong Kong (CTNET) employs a high-speed ATM backbone network that works in conjunction with Ethernet and Fast Ethernet segments to support Web browsing, tele-instruction, videoconferencing, and multimedia distribution at transmission rates reaching 155.52 Mbps (OC-3).

4.27.3 FINLAND

4.27.3.1 City of Tampere InfoCircle Metropolitan Network

Situated north of Helsinki, the City of Tampere employs an ATM core backbone network that works in concert with Ethernet, Fast Ethernet, and Gigabit Ethernet segments for interlinking municipal schools and libraries in a metropolitan network configuration called InfoCircle. In addition to serving as a municipal intranet for the City of Tampere, InfoCircle provides access to the Web, digital library resources, and online services for local residents. Moreover, InfoCircle supports telementoring, videoconferencing, and tele-education programs for students, teachers, and staff at Tampere public schools. The University of Tampere and the Technical University of Tampere also participate in this initiative.

4.27.3.2 North Kareleia

A sparsely populated province in Finland, North Kareleia employs an Ethernet, Fast Ethernet, ATM, and SDH (Synchronous Digital Hierarchy) configuration for provisioning access to distance education courses, teletraining programs, teletutorials, and teleseminars. This configuration also interworks with the Finnish University Research Network (FUNET).

4.27.4 JAPAN

4.27.4.1 Nagasaki Institute of Applied Research

The Nagasaki Institute of Applied Research employs an ATM backbone network that interoperates with Ethernet and Fast Ethernet segments in supporting tele-instruction, high-speed multimedia distribution to the desktop, and videoconferencing. An in-place FDDI network provisions backup services.

4.27.5 UNITED KINGDOM

4.27.5.1 Strathclyde University

A leading provider of postgraduate education in the United Kingdom, Strathclyde University uses a Gigabit Ethernet network as a replacement for a FDDI backbone to optimize internetwork connectivity. This Gigabit Ethernet solution provisions access to teleresearch and telecollaborative initiatives as well as connectivity to high-speed, high-performance networks such as SuperJANET4 (Super Joint Academic Network, Phase 4).

4.28 U.S. E-GOVERNMENT (ELECTRONIC GOVERNMENT) INITIATIVES

4.28.1 U.S. DEPARTMENT OF DEFENSE INFORMATION SYSTEMS AGENCY (DISA)

The U.S. Department of Defense Information Systems Agency (DISA) implements an Ethernet and ATM infrastructure for enabling videoconferencing and distance education as part of the DISA Global Fiber Network Initiative. This network configuration also supports on-demand video broadcasts featuring news reports, sports videos, and U.S. Department of Defense briefings.

4.28.2   CALIFORNIA

4.28.2.1   City of Santa Clara

The City of Santa Clara operates a Gigabit Ethernet MAN for interlinking municipal offices, the Santa Clara Adult Education Center, schools in the Santa Clara Unified District, and SOHO venues. Moreover, this Gigabit Ethernet MAN provisions access to police agency and fire department databases, utility services, municipal records and permits, and library resources, and supports data warehousing, tele-education programs, and Web browsing.

4.28.3   NEW YORK

4.28.3.1   Monroe County

Situated in upstate New York, Monroe County implements an ATM backbone network that operates in concert with Ethernet and Fast Ethernet LAN segments. The Monroe County network enables videoconferencing between local courts and prisons, and transmits satellite images to environmental officials monitoring the potential impact of natural disasters.

4.28.4   OKLAHOMA

4.28.4.1   City of Tulsa

The City of Tulsa employs an ATM backbone network with Ethernet segments for connecting municipal agencies and departments in a metropolitan network configuration. This metropolitan network supports administrative functions, law enforcement operations, utility services, and license and permit distribution. The municipal fire department employs the Tulsa MAN for delivery of teletraining videos on firefighting procedures. A network VLAN accessible via the Tulsa MAN enables residential users to reserve books at local public libraries.

4.29   U.S. TELEMEDICINE INITIATIVES

4.29.1   ILLINOIS

4.29.1.1   St. John's Hospital

Situated in Springfield, St. John's Hospital, a major teaching hospital affiliated with the Southern Illinois University School of Medicine, employs an FDDI backbone network that interworks with Ethernet and Fast Ethernet segments for supporting IP multicasts and access to patient records.

4.29.2   NEW YORK

4.29.2.1   University of Rochester Medical Center

The University of Rochester Medical Center employs an ATM backbone network that supports rates at 155.52 Mbps (OC-3) and interoperates with Ethernet, Fast Ethernet, and Gigabit Ethernet network segments. This network provisions access to patient records; enables transmission of x-rays, CAT and MRI scans, and angio-grams; and supports videoconferences and teleconsultations between hospital-based physicians and inmates at local jails.

4.30 INTERNATIONAL TELEMEDICINE PROJECTS

4.30.1 JAPAN

4.30.1.1 Saiseikai Kumamoto Hospital

The Saiseikai Kumamoto Hospital in Kumamoto City uses a Gigabit Ethernet solu-tion for enabling fast access to digitized patient records and online administrative services. In addition, this Gigabit Ethernet solution facilitates remote patient mon-itoring, teleconsultations, tele-education, and videoconferencing.

4.30.2 UNITED KINGDOM

4.30.2.1 Advanced Medical Imaging Network (AMInet)

Sponsored by St. Thomas and Guys Hospitals in London, the Advanced Medical Imaging Network (AMInet) initiative employs an ATM backbone that supports connections to Ethernet and Fast Ethernet network segments. AMInet provisions access to hospital intranets, videoconferences, and continuing education courses at rates reaching 2.488 Gbps (OC-48). In addition, AMInet supports links to the British Healthcare Association Network and the OMNI (Organizing Medical Networked Information) initiative sponsored by the National Library of Medicine.

4.31 ETHERNET PLANNING GUIDELINES

Planning a new Ethernet network or upgrading an already in-place Ethernet instal-lation involves assessment of performance requirements, analysis of current and projected applications, and examination of user needs to ensure the provision of practical and advanced network applications and services. The Ethernet implemen-tation process also involves accommodating the enterprisewide mission, goals, and objectives; establishing budget allocations; and selecting and implementing Ethernet components that can be integrated into the existing network configuration and support future expansion.

An effective Ethernet, Fast Ethernet, and/or Gigabit Ethernet network complies with Ethernet design guidelines that clarify maximum number of network nodes and maximum segment lengths that can be supported, and justifies the expense of net-work investment. Procedures for network administration, management, operations, maintenance, and security facilitate the transition to an upgraded or new Ethernet, Fast Ethernet, and/or Gigabit Ethernet configuration while also ensuring network availability and information integrity during the conversion process.

In-service training sessions and staff development courses promote skillful use of Ethernet technology and acceptance of network-enabled applications. Capabilities of an Ethernet, Fast Ethernet, and/or Gigabit Ethernet configuration should be carefully evaluated in pilot tests prior to full-scale implementation. The cutover process for network upgrades should minimize any required downtime to enable a smooth transition to the new infrastructure.

In the academic arena, Ethernet, Fast Ethernet, and/or Gigabit Ethernet implementations facilitate reliable and dependable access to an expanding array of Web and institutional resources. While the three Ethernets enable exciting options for information delivery to the desktop and in local and wider areas, utilization of Ethernet technology does not automatically guarantee its efficient and positive use in the educational domain. Successful deployment in the telelearning environment ultimately depends on the capabilities of Ethernet, Fast Ethernet, and/or Gigabit Ethernet to foster knowledge-building competencies and exploratory learning, quality education and focused research, and instructional innovation and creativity.

It is important to note that 10 Gigabit Ethernet is a relatively recent Ethernet innovation. As a consequence, capabilities and constraints associated with the use of 10 Gigabit Ethernet in the tele-education E-commerce, E-government, and corporate environments are not yet fully documented or understood.

4.32   SUMMARY

Ethernet, Fast Ethernet, Gigabit Ethernet, and 10 Gigabit Ethernet technologies enable the implementation of seamless integrated networks comprised of heterogeneous computing equipment and resources. The four Ethernets are distinguished by their support of reliable high-speed, volume-intensive transmissions in DANs, LANs, MANs, and WANs. This chapter explores capabilities of the four Ethernets in enabling reliable network performance in tele-education, telemedicine, and E-government applications and initiatives. Recent innovations in home networking are introduced. The role of the Ethernet suite of technologies in enabling home phoneline networks is described.

Ethernet is among the world's most pervasive, popular, and ubiquitous technologies. The four Ethernets provision frameworks for flexible network solutions that are scalable and interoperable with a diverse array of network technologies, protocols, and architectures, including ATM, SONET/SDH, Frame Relay, FDDI, Fibre Channel, WDM, and DWDM solutions. Ethernet networks employ CSMA/CD, utilize a specified frame format, enable dependable half-duplex and/or full-duplex transmission, provision network management procedures and security mechanisms, and foster interconnectivity to wireline and wireless networks through compliance with specifications endorsed by the IEEE.

Factors contributing to the accelerating numbers of Gigabit Ethernet implementations in telemedicine, E-government, and tele-education reflect Gigabit Ethernet's compatibility with previously established Ethernet and Fast Ethernet solutions and the need for faster networks to accommodate bandwidth demands of complex applications. Gigabit Ethernet is further distinguished by its self-healing architecture, support of VLANs for creating logical workgroups across the network, and capabilities in provisioning traffic prioritization for CoS assurances. Advanced protocols and switching technologies that support interoperability and interworking contribute to the convergence of Gigabit Ethernet and ATM networks and the emergence of

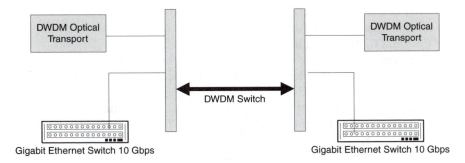

FIGURE 4.5 A link between a long-haul DWDM (Dense Wavelength Division Multiplexing) switch and a Gigabit Ethernet connection for enabling optical information transport via a 10 Gigabit Ethernet implementation.

10 Gigabit Ethernet. 10 Gigabit Ethernet optical solutions enable robust networking implementations that extend from the network core to the desktop and work in concert with WDM and DWDM technologies. (See Figure 4.5.)

Moreover, 10 Gigabit Ethernet optical configurations ensure consistent information throughput; support dependable voice, video, and data delivery; and provision bandwidth allocations to accommodate multiservice networking requirements. An understanding of distinctive features and functions of Ethernet, Fast Ethernet, Gigabit Ethernet, and 10 Gigabit Ethernet configurations is critical for enabling seamless and secure information transport and optimizing capabilities of tele-education, tele-medicine, and E-government solutions.

4.33   SELECTED WEB SITES

3Com. Small Business Case Studies.
Available:
http://www.3com.com/solutions/en_US/additional_details/smallbusiness_casestudy.html

10 Gigabit Ethernet Alliance (10 GEA) Home Page. Last modified on September 21, 2001.
Available: http://www.10gea.org/

10 Gigabit Ethernet Alliance. Gigabit Ethernet Alliance Archived White Papers. Last modified on September 17, 2001.
Available: http://www.10gea.org/Tech-whitepapers.htm

CEBus Industry Council. Home Plug and Play.
http://www.cebus.org/HomePnP.html

Ciena Corporation. Next Generation Metropolitan Ring Solution.
Available:
http://www.ciena.com/products/applications/metroring/index.asp

Cisco Systems. Technology Brief. Fast Ethernet-100 Mbps Solutions. Posted July 3, 2001.
Available:
http://www.cisco.com/warp/public/cc/so/neso/lnso/lnmnso/feth_tc.htm

Home Phoneline Networking Alliance. About HomePNA.
  Available: http://www.homepna.org/about/
IEEE 802.3 CSMA/CD. Last modified on August 29, 2001.
  Available: http://grouper.ieee.org/groups/802/3/index.html
IEEE 802.3ae 10 Gbps Ethernet Task Force. Last modified on September 20,
  2001.
  Available: http://grouper.ieee.org/groups/802/3/ae/index.html
Intel Networking and Communications Products. The Future of Ethernet is
  Everywhere.
  Available: http://www.andygrove.com/network/ethernet/
Nortel Networks. Optical Ethernet Solutions.
  Available:
  http://www.nortelnetworks.com/solutions/opt_ethernet/index.html
Open Services Gateway Initiative (OSGi). OSGi Home Page.
  Available: http://www.osgi.org/
Optical-Ethernet.com. Optical Ethernet Resources.
  Available: http://www.optical-ethernet.com/optical/index.htm
Spurgeon, Charles. Welcome to Charles Spurgeon's Ethernet Web Site. Last
  modified on June 25, 2001.
  Available: http://www.ots.utexas.edu/ethernet/ethernet.html
Technical Essence Webs. 10 Gigabit Ethernet Resource Site. Last modified on
  February 25, 2001.
  Available: http://www.10gigabit-ethernet.com/childweb/index.htm
University of New Hampshire Interoperability Lab. Gigabit Ethernet
  Consortium.
  Available: http://www.iol.unh.edu/consortiums/ge/index.html

5 Frame Relay (FR) and Fibre Channel (FC) Technologies

5.1 CHAPTER OVERVIEW

Chapter 5 presents an examination of the features, functions, and capabilities of Frame Relay (FR) and Fibre Channel (FC) technologies. Frame Relay and Fibre Channel platforms were developed in the 1980s for enabling fast transmission, diverse applications, and networking operations in local area and wider area environments. The chapter begins with an exploration of Frame Relay technical fundamentals, operations, standards, and representative initiatives. Following the FR examination, Fibre Channel configurations, applications, and implementations are described.

5.2 FRAME RELAY (FR) INTRODUCTION

Accelerating demand for dependable network access to current and next-generation Web services motivates continued interest in the utilization of Frame Relay (FR) networking solutions. Frame Relay is a standards-based, fast packet-switching telecommunications technology that enables dependable and reliable information delivery, IP (Internet Protocol) multicasts, seamless Web connectivity, and VPN (Virtual Private Network) deployment.

The following sections describe Frame Relay technical fundamentals, standards, merits, and constraints; the role of Frame Relay technology in enabling VPN (Virtual Private Network) deployment and the effectiveness of Frame Relay and ATM (Asynchronous Transfer Mode) solutions in supporting multimedia services. Also, representative Frame Relay initiatives are reviewed.

5.3 FRAME RELAY FOUNDATIONS

Frame Relay implementations support delay-sensitive voice and video transport and delay-insensitive data transmission. Developed in the 1980s, FR service was initially designed to support fast packet delivery and enable affordable Wide Area Network (WAN) implementations by enabling LAN-to-LAN connections. Because the FR platform was extendible, scalable, and flexible, FR technology was expected to replace leased line connections and X.25 implementations, and function in tandem with ISDN (Integrated Services Digital Network) in co-located networking environments.

5.3.1   Frame Relay and X.25 Technology

Frame Relay is a fast packet-switching service that handles higher traffic volumes than X.25 technology. Regarded as the forerunner to Frame Relay, X.25 is a packet-switching technology developed during the 1970s to support data transmission. As with Frame Relay implementations, X.25 solutions employ packet-switching technology and transports variable length frames or packets. Also known as a slow-packet technology, X.25 employs complex error correction and control mechanisms for information transport and supports data rates reaching 56 Kbps (Kilobits per second).

In contrast to X.25 technology, Frame Relay networks also support scalable transmission rates at speeds ranging from 56 Kbps, T-1 (1.544 Mbps) and E-1 (2.048 Mbps) to T-3 (44.736 Mbps) and E-3 (34.368 Mbps) for enabling voice, video, and data transmission. FR technology also eliminates complexities in the error correction and control process, reduces transmission errors, and compresses X.25 overhead. FR configurations support more effective bandwidth utilization and higher reliability in networking operations than X.25 networks.

X.25 networks employ the Physical Layer or Layer 1, the Data-Link Layer or Layer 2, and the Network Layer or Layer 3 of the Open Systems Interconnection Reference model for processing network transactions. The Frame Relay protocol supports an elegant two-layer architecture that enables networking operations at Layer 1 or the Physical Layer and Layer 2 or the Data-Link Layer of the OSI (Open Systems Interconnection) Reference Model for enabling higher speeds and faster throughput than X.25 solutions.

5.3.2   Frame Relay and ISDN (Integrated Service Digital Network)

In 1988, FR functions in enabling ISDN B (Bearer) Channel services for supporting bi-directional or full-duplex transmission of service data units (SDUs) through a network were clarified in the ITU-T (International Telecommunications Union-Telecommunications Standards Sector) I.222 and the ITU-T I.223 Recommendations. Initially, FR supported applications and operations in conjunction with ISDN in the same networking environment. As a consequence, the ITU-T I.223 Recommendation also defined FR procedures for interconnecting N-ISDN and B-ISDN LANs and approaches for enabling interoperability between Frame Relay and X.25 configurations. In 1990, the American National Standards Institute (ANSI) T1.606 specification for utilizing Frame Relay as a non-ISDN technology was endorsed.

5.4   FRAME RELAY FORUM

Organized in 1991, the Frame Relay Forum is a global consortium of carriers, vendors, users, and consultants. This consortium facilitates the development and implementation of Frame Relay services and configurations that operate in compliance with national and international FR standards. In addition, the Frame Relay Forum (FRF) develops IAs (Implementation Agreements) such as the FRF.13 IA, which establishes the framework for a service-level implementation agreement. The

FRF.13 IA also clarifies approaches for evaluating Frame Relay service in terms of networking performance, reliability, and response time.

IAs support dependable transmission of voice, video, and data via an FR infrastructure. In addition, IAs define approaches for interworking FR with broadband technologies including ATM. Participants in the Frame Relay Forum include equipment providers such as 3Com, Cabletron, and Cisco Systems and communications providers such as WorldCom, Ameritech, AT&T, INTELSAT, TeleDanmark, France Telecom, and Global One. In addition to the United States, chapters of the Frame Relay Forum are situated in member states of the European Union, Australia, New Zealand, and Japan.

5.5   FRAME RELAY TECHNICAL FUNDAMENTALS

Frame Relay (FR) is a low-cost mainstream telecommunications networking technology for dependably transporting a mix of voice, video, and data traffic. An FR infrastructure is flexible, scalable, and extendible and enables the easy addition or deletion of virtual connections in an FR network implementation.

Frame Relay (FR) networks transmit variable-length packets called frames. A frame consists of a payload that carries up to 4096 bytes and a header consisting of 6 bytes. The header contains overhead and addressing information. The header allocates bytes for the Data Link Connection Identifier), Forward Explicit Congestion Notification (FECN), Backward Explicit Congestion Notification (BECN), and the Discard Eligibility Indicator (DEI). The header also includes an extension field and a command/response field.

With FR, the error checking and control process is straightforward. Data recovery procedures are not employed. Any frame that is problematic is discarded. As a consequence, FR frames can be inadvertently lost or destroyed. Traffic delays in an FR network vary with frame size.

FR configurations transport bursty LAN traffic at relatively high speeds over long distances, support LAN-to-LAN interconnectivity, and facilitate traffic exchange between LANs and WANs. Frame Relay is an enabler of an array of applications, including e-mail, document imaging, mainframe-to-mainframe links, Electronic Data Interchange (EDI), bulk file transfer, voice telephony, facsimile (fax) transmission, data warehousing, and inventory management. Recent technical advances contribute to the implementation of video-over-Frame Relay service. This service facilitates applications that include videoconferencing, remote security and surveillance, IP (Internet Protocol) multicasts, and cable television programming distribution. (See Figure 5.1.)

5.6   FRAME RELAY OPERATIONS

Frame Relay service enables development of a scalable and flexible network architecture that effectively allocates bandwidth on an as-needed basis. FR networks work in conjunction with legacy, narrowband, and broadband technologies and architectures such as SNA (Systems Network Architecture), Ethernet, Fast Ethernet, Gigabit

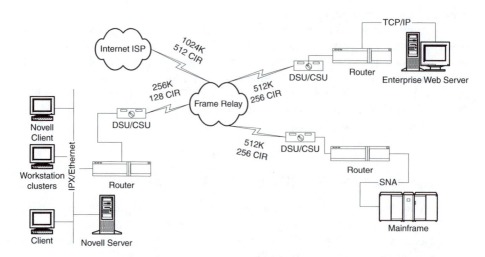

FIGURE 5.1 LANs that are interconnected by Frame Relay technology.

Ethernet, Fibre Channel, ISDN, DSL (Digital Subscriber Line), SMDS (Switched Multimegabit Data Service), ATM (Asynchronous Transfer Mode), and SONET/SDH (Synchronous Optical Network and Synchronous Digital Hierarchy). In addition, Frame Relay technology supports IPv4 (Internet Protocol version 4) and IPv6 (Internet Protocol version 6) operations.

Inasmuch as costs for transmitting voice, video, and data via a Frame Relay network are based on a flat rate, an enterprisewide FR network can cost-effectively support links to remote corporate or academic sites. FR technology allocates higher permanent bandwidth on specific circuits or connections. As a consequence, advanced FR data, video, and voice services can be supported by the in-place FR infrastructure. Additional network hardware and network upgrades generally are not required for service enhancements.

5.7 FRAME RELAY TECHNICAL FUNDAMENTALS

5.7.1 FRAME RELAY TRANSMISSION

Procedures for transmitting multimedia in Frame Relay and other packet-switching networks are defined by the ITU (International Telecommunications Union) H.323 Recommendation. This Recommendation clarifies approaches for encapsulating audio, video, and data in frames or packets that serve as envelopes for network transmission. In contrast to data and voice transmission, video-over-Frame Relay transport requires additional networking equipment such as Frame Relay codecs (coders and decoders) or conversion units. The connection-oriented Frame Relay packet interface protocol provisions a basic set of switching capabilities for transporting variable-sized frames via local and wider area networking configurations.

Frame Relay technology reduces the costs and complexity associated with designing and deploying multi-application multiprotocol networks by eliminating the need for redundant equipment and dependence on T-1 (1.544 Mbps) and E-1 (2.048 Mbps) and T-3 (44.736 Mbps) and E-3 (34.368 Mbps) leased lines for network services. By supporting an integrated network platform, FR technology also reduces the complexity of network management, administration, and maintenance functions.

5.7.2 VOICE-OVER-FRAME RELAY SERVICE

Originally described in the Frame Relay Forum (FRF) User-to-Network-Interface (UNI) and the FRF Network-to-Network Interface or Network-to-Node Interface (NNI) specifications, the Frame Relay protocol has been extended in recent years to support IP routing, LAN bridging, and SNA applications. In 1998, the Frame Relay Forum endorsed an Implementation Agreement (IA) for enabling FR voice transmissions.

This Implementation Agreement defines procedures for transmission of compressed voice within a Frame Relay frame payload and approaches for multiplexing voice and data payloads via a voice-over-Frame Relay PVC (Permanent Virtual Circuit) connection. In addition, the FR Forum determines methods to prioritize the transmission of voice and data frames entering the network and clarifies procedures to compensate for bandwidth delay limitations and network congestion. VoFR service eliminates international telephone toll charges for enterprises with sites in geographically distributed locations. (See Figure 5.2.)

5.7.3 PERMANENT VIRTUAL CIRCUITS (PVCs) AND SWITCHED VIRTUAL CIRCUITS (SVCs)

In FR networks, Permanent Virtual Circuits (PVCs) and Switched Virtual Circuits (SVCs) are logical channels or pathways that emulate actual physical channels or pathways over which voice, video, and data in FR-compliant frames are transported. FR frames employ Data Link Connections (DLCs) that contain user data or variable-length payloads, and Data Link Connection Identifiers (DLCIs) that perform multiplexing and addressing functions. The DLCI two-octet address field in the FR header indicates the logical PVC or SVC that will enable frame transmission to the destination address.

5.7.3.1 Permanent Virtual Circuits (PVCs)

With Frame Relay service, PVCs are permanently assigned for enabling data transmission from a point of origin to a specified endpoint. Network administrators and managers determine PVC service classes and endpoints based on application content and transmission requirements. Typically, multiple PVCs co-exist in a single User-to-Network Interface (UNI). Moreover, Frame Relay configurations support operations of multiple PVCs over a single optical fiber link for optimizing information delivery and facilitating multisite interconnectivity.

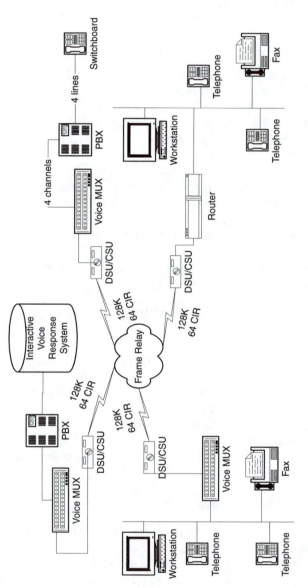

FIGURE 5.2 Voice services supported by Frame Relay technology.

5.7.3.2 Switched Virtual Circuits (SVCs)

Switched Virtual Circuits (SVCs) provide virtual channels or pathways on-demand. Developed to improve the effectiveness of network transport, SVCs typically support bursty video applications such as near-video-on-demand (NVOD). The Frame Relay Forum also developed the FRF.7 Implementation Agreement (IA) for supporting IP multicast services.

5.7.4 FRAME RELAY ENCAPSULATION

An FR network only works in concert with FR-compliant frames. As a consequence, FR interface devices encapsulate local network traffic into FR frames that then transit the FR network to the destination address. This process enables FR networks to interwork with diverse networking technologies. With PVCs, the encapsulation process is established prior to information transmission. With SVCs, the encapsulation process is initiated during call setup and call establishment. Multiprotocol encapsulation enables FR to interoperate with diverse technologies.

5.7.5 FRAME RELAY CONGESTION METHODS AND TECHNIQUES

In Frame Relay networks, congestion control methods and techniques eliminate service degradation and optimize traffic flow and network performance. Congestion control procedures such as FECN (Forward Explicit Congestion Notification) and BECN (Backward Explicit Congestion Notification) inform network nodes about frame corruption and network congestion. FECN alerts a network destination device that network congestion was experienced. By contrast, BECN informs a network source device that the network is experiencing bottlenecks.

5.7.6 COMMITTED INFORMATION RATE (CIR)

Frame Relay technology was initially designed for data transport. Improved compression techniques enable Frame Relay networks to effectively carry packetized voice and video traffic at or below the Committed Information Rate (CIR) as well. CIR is defined in terms of bits per second (bps) and establishes transmission rates and services supported for each Permanent Virtual Circuit (PVC). From the network perspective, the CIR references the amount of information that the FR configuration agrees to deliver at specified time intervals. From the user perspective, the CIR indicates bandwidth that will be required for accommodating networking applications.

Moreover, the CIR supports assessment of Quality-of-Service (QoS) guarantees and enables the provision of congestion recovery services by minimizing the occurrence of network gridlock and bottlenecks resulting from severe congestion. The CIR also fosters reliable transmission of delay-sensitive and delay-insensitive traffic.

5.7.6.1 Committed Information Rate (CIR) and Committed Burst Information Rate (CBIR)

The Committed Burst Information Rate (CBIR) supports random peaks in workflow generated by bursty applications. Frames transported in excess of the CIR and CBIR are tagged with a discard eligible (DE) bit. If congestion occurs, frames transmitted in excess of the CIR can be discarded. Charges for Frame Relay service are based on the CIR or guaranteed bandwidth instead of the distance data travel or the duration of the transmission.

5.7.7 FRAME RELAY DEVICES

Frame Relay service optimizes available bandwidth and enables dependable and reliable networking operations. Frame Relay devices that are highlighted in this section enable network topologies and architectures such as LANs, MANs, WANs, VPNs, intranets, and extranets, thereby reducing hardware-related expenditures. In the present-day high-performance, multiprotocol, multiservice networking environment, Frame Relay hardware also streamlines network operations and supports switched access for enabling remote users to connect to network resources.

5.7.7.1 FRADs (Frame Relay Access Devices)

FRADs are designed specifically to work with Frame Relay networks. FRADs are assemblers and disassemblers that link endpoints to the network and enable non-Frame Relay protocols to access Frame Relay services. In addition, FRADs provide the framing function by inserting the two-bit DLCI (Data Link Connection Identifier) into the Frame Relay frame header for network transport. FRADs multiplex data, video, and voice streams to circuits or access devices where frames are disassembled and transported via virtual circuits to the network node specified by DLCI header information.

FRADs facilitate network operations by supporting access and switching functions, congestion control techniques, multiprotocol communications, and network management. Furthermore, FRADs cost-effectively link customer premise equipment (CPE) such as multiprotocol routers to private, public, and mixed-mode Frame Relay networks.

5.7.7.1.1 Voice FRAD (VFRAD)

A voice FRAD (VFRAD) is a special type of FRAD that supports voice-over-Frame Relay (VoFR) transmission by employing compression algorithms to optimize bandwidth utilization. In addition, VFRADs enable encapsulation functions for VoFR payloads to facilitate dependable transmission.

5.7.7.2 FR Internetworking Devices

In addition to FRADs and VFRADs, routers, bridges, and switches are also popular Frame Relay internetworking devices that provision dependable and reliable FR transmission services. These devices work in concert with protocols that include the Routing Information Protocol (RIP) and the Open Shortest Path First (OSPF) Protocol in transporting FR packets across LANs, MANs, and WANs directly to destination addresses.

5.7.7.2.1 Frame Relay Switches

Frame Relay switches generally employ frame-switching or cell-switching technologies for transporting user information via an FR network. In addition to working in conjunction with FRADs, Frame Relay switches provision service for bursty LAN traffic, support bandwidth availability, and ensure dependable network performance by reducing network gridlock. Moreover, FR switches facilitate network integration and support reliable transport of delay-sensitive voice, video, and data over the Frame Relay network.

5.8   FRAME RELAY VIRTUAL PRIVATE NETWORKS (VPNS)

5.8.1   FRAME RELAY VIRTUAL PRIVATE NETWORK (VPN) OPERATIONS

The remarkable success of initiatives in the E-commerce (electronic commerce), distance education, E-government (electronic government), and telemedicine domains, and the popularity of applications involving telecollaboration and teleresearch drive migration from private network configurations that interlink fixed sites to Virtual Private Network (VPN) implementations that are accessible via a public network. In parallel with IP and ATM VPNs, an FR VPN employs a shared network such as the commodity or public Internet for enabling secure communications exchange among specified individuals and closed user groups.

FR VPNs (Virtual Private Networks) support transmission of private, time-sensitive, and time-insensitive voice, video, and data via PVCs (Permanent Virtual Connections) and SVCs (Switched Virtual Circuits) that emulate physical connections and securely extend FR services and applications to distant users regardless of their locations.

5.8.2   VPN SECURITY

Inasmuch as a Frame Relay VPN installation interfaces with public networks such as the Internet, security mechanisms and policies for safeguarding transmission integrity must be established prior to network implementation. Generally, FR VPNs employ combinations of security tools and techniques such as firewalls, encryption, passwords, biometric devices, and protocols to provide network security.

Firewalls isolate Frame Relay VPNs from Web intrusions and protect FR VPNs from unauthorized access via external networks such as the Internet. The encryption process involves encoding all data that are transmitted via Internet-to-FR VPN connections. Cryptosystems for VPN deployments are based on protocols such as DES (Data Encryption Standard), RSA (Rivest, Shamir, and Adleman), and Kerberos. Authorization and authentication mechanisms such as passwords and biometric identifiers ensure that only legitimate users access FR VPN resources.

5.8.2.1   Internet Engineering Task Force (ETF) Frame Relay Security Protocols

5.8.2.1.1   Layer 2 Tunneling Protocol (L2TP)

Endorsed by the IETF, the Layer 2 Tunneling Protocol (L2TP) works in concert with protocols that include the PPTP (Point-to-Point Tunneling Protocol), the VTP (Virtual Tunneling Protocol), the L2TP IP Differentiated Services Protocol, and the L2F (Layer 2 Forwarding Services Protocol) to ensure secure FR implementations.

5.8.2.1.2   MultiProtocol Label Switching (MPLS)

The IETF supports MPLS (MultiProtocol Label Switching) extensions for enabling LT2P (Layer 2 Tunneling Protocol) to interoperate with the IPSec (IP Security) Protocol in FR VPN implementations.

5.8.2.1.3   Internet Protocol Security (IPSec)

Defined by the Internet Engineering Task Force (IETF) IPSec Working Group, IPSec (Internet Protocol Security) supports utilization of mechanisms for protecting IP client protocols in VPN implementations. An Internet protocol for encryption and decryption, IPSec provisions cryptographic security services for supporting access control, user authentication, information integrity, and data confidentiality to safeguard networking operations at Layer 3 or the Network Layer of the Open Systems Interconnection (OSI) Reference Model.

5.8.3   FRAME RELAY TUNNELING OPERATIONS

Tunneling is designed to safeguard voice, video, and data transmissions in FR VPNs. The tunneling process involves encrypting Frame Relay frames that are then encapsulated into IP packets for transmission via a tunnel across a public network such as the Internet to destination addresses. A tunnel is a virtually dedicated point-to-point channel or specified pathway that enables secure FR VPN transmission.

Tunnel switches, gateways, routers, and concentrators available from vendors such as 3Com, Lucent Technologies, and Cisco Systems facilitate tunneling operations. At the destination site, FR packets are decrypted. Internet Engineering Task Force (IETF) tunneling specifications ensure the integrity of FR packets and support development and implementation of interoperable multivendor equipment.

5.8.4   FRAME RELAY VPN MERITS AND CONSTRAINTS

VPNs based on Frame Relay (FR) technology enable bandwidth-intensive applications and multicast services that are readily accessible via an array of narrowband and broadband communications solutions, thereby eliminating the need for expensive leased line connections. In addition to cost savings, FR VPNs also support straightforward network implementations and migration to new applications with fewer administrative and operational requirements than private networks. A Frame Relay VPN enables transmission rates at T-1 (1.544 Mbps) and T-3 (44.736 Mbps) in the United States and E-1 (2.048 Mbps) and E-3 (34.368 Mbps) in the European Union. VPNs are implemented via shared public networks. It is important to note that private network protocols and management policies also interwork with FR VPN implementations.

Frame Relay VPNs extend an enterprise network to telecommuters at SOHO (Small Office/Home Office) venues by working in conjunction with residential broadband technologies such as cable modem and DSL (Digital Subscriber Line). FR VPN solutions economically accommodate enterprisewide strategic and tactical requirements and consolidate network operations by eliminating the need for leased lines, multiple circuit connections, and redundant network equipment.

With a Frame Relay Virtual Private Networks, traffic shaping levels bursty traffic, thus, optimizing the performance of wider area networking connections. An enterprisewide network policy that guarantees Committed Information Rates (CIRs) for bandwidth allocations reflects enterprise priorities. Moreover, FR VPN implementations economically support expanded geographical coverage, increased network uptime, seamless networking operations and maintenance, and rapid addition and deletion of network users at geographically separated locations. FR VPNs are flexible and extendible and provision temporary, periodic, and permanent connectivity to the network core depending on enterprisewide requirements.

Despite the benefits, FR VPN deployment is also associated with problems and risks. With an FR VPN, network services operate on a single network that is shared by multiple users that can lead to security risks. Moreover, FR VPNs that are accessible via the Internet are also subject to Web-based cyberintrusions. Heavy network usage by multiple users contributes to unpredictable FR VPN performance and degradation in network services. Congestion on shared public networks such as the Internet can also lead to slowdowns in information transport and difficulties in ensuring network throughput; acceptable network response time; and voice, video, and data delivery guarantees in FR VPNs. Available from vendors such as ADC Kentrox and Cisco Systems, network monitoring and maintenance devices such as the DSU/CSU (Data Sensing Unit/Carrier Sensing Unit) generate measurements of bandwidth usage, overloaded circuits and switches, traffic delays, and FR service. These metrics enable resolution of network congestion and transmission delays and contribute to the provision and maintenance of reliable and dependable FR VPN services.

5.9 FRAME RELAY INTERWORKING IMPLEMENTATION AGREEMENTS (IAs)

To support increased implementation of FR technology, the Frame Relay Forum (FRF) develops Implementation Agreements (IAs) that ensure FR interoperability with diverse network technologies, protocols, architectures, and standards and establish a framework for implementing mixed-mode FR solutions. These IAs describe FR functions in enabling multiprotocol encapsulation, Physical Layer or Layer 1 interfaces, multicast services, and data compression.

In 1999, the Frame Relay Forum endorsed Implementation Agreements (IAs) for supporting Frame Relay as the dominant VPN platform. The FRF.15 IA defines end-to-end multilink aggregation and the FRF.16 IA describes the User-to-Network Interface (UNI) and the Network-to-Network or Network-to-Node Interface (NNI) for enabling multilink aggregation. Multilink aggregation enables scalable and symmetrical connectivity rates such as T-1 (1.544 Mbps) and T-3 (44.736 Mbps) and E-1 (2.048 Mbps) and E-3 (34.368 Mbps).

In addition to supporting multilink aggregation services in FR implementations, the Frame Relay Forum endorses the use of Frame Relay PVCs (Permanent Virtual Circuits) and SVCs (Switched Virtual Circuits). PVCs and SVCs provision more bandwidth than a single virtual circuit or physical connection and increase the total bandwidth available for time-sensitive bandwidth-intensive applications such as videoconferencing and on-demand video. In addition, FRF IAs also delineate FR congestion control strategies and methods for interworking FR with technologies that include IP, SONET/SDH, DSL, and ATM.

5.9.1   FRAME RELAY AND INTERNET PROTOCOL (IP)

An IP (Internet Protocol) network overlay enables a Frame Relay network to support IP multicasts. Importantly, IP multicasts also optimize network performance by significantly reducing the quantity of redundant network traffic. To enable IP multicasts, an FR network replicates and distributes single copies of software updates, news feeds, stock quotes, catalogs, management reports, newsletters, and content for kiosks, intranets, and extranets to specified reception sites.

FR solutions comply with networking protocols and specifications defined by standards organizations such as the ITU-T, the American National Standards Institute (ANSI), the Internet Engineering Task Force (IETF), and the Institute of Electrical and Electronics Engineers (IEEE).

5.9.2   FRAME RELAY AND SONET/SDH (SYNCHRONOUS OPTICAL NETWORK AND SYNCHRONOUS DIGITAL HIERARCHY)

The Frame Relay Forum endorses a Physical Layer Implementation Agreement, formally known as the FRF.14 IA, that describes guidelines for Frame Relay support of SONET/SDH (Synchronous Optical Network and Synchronous Digital Hierarchy) physical interfaces. FRF.14 enables FR-over-SONET/SDH transmission rates at 155.52 Mbps (OC-3) and 622.08 Mbps (OC-12) for optimizing the availability of FR service and the reliability of FR network performance. In addition to SONET/SDH, the FRF.14 IA describes Frame Relay interoperability with ISDN and ATM physical interfaces.

5.9.3   FRAME RELAY AND ASYNCHRONOUS TRANSFER MODE (ATM)

Frame Relay and ATM are connection-oriented technologies that support bandwidth efficiency, low latencies in transmissions, and development of extendible network configurations. Importantly, Frame Relay and ATM synergistically work together in enabling advanced network services, applications, implementations, and solutions.

Frame Relay-over-ATM service enables users to maintain their in-place FR networks and benefit from increased bandwidth. In addition, Cisco Systems, Newbridge Networks, and Hughes Network Systems support development of interoperable ATM and FR devices for enabling ATM network stations or nodes to seamlessly communicate with Frame Relay endpoint equipment.

5.9.3.1   The Frame Relay Forum and the ATM Forum

To ensure Frame Relay and ATM interoperability, the Frame Relay Forum works in concert with the ATM Forum in designing Implementation Agreements (IAs) that clarify approaches for achieving Frame Relay and ATM interconnections. In 1993, the Frame Relay Forum (FRF) and the ATM Forum jointly approved Frame Relay and ATM Interworking Implementation Agreements (IAs). The Frame Relay-to-ATM Network Interworking Implementation Agreement, also known as FRF.5 IA, describes solutions for mapping FR packets to ATM cells and approaches for transporting Frame Relay traffic over an ATM network.

In addition, IAs define procedures for enabling a Frame Relay-over-ATM network to support ATM and Frame Relay technologies and maintain FR virtual links including PVCs (Permanent Virtual Circuits) and SVCs (Switched Virtual Circuits). Moreover, these FRF IAs clarify capabilities of transmission services including CIR (Committed Information Rate), PIR (Peak Information Rate), VBR (Variable Bit Rate), and ABR (Available Bit Rate). FRF IAs also specify approaches for enabling the seamless transmission of multiprotocol LAN traffic from one Frame Relay network to another Frame Relay network via an ATM backbone network.

The Frame Relay Forum and ATM Forum also developed the Frame User-to-Network Interface (FUNI). FUNI defines an ATM interface that handles information in frame formats rather than in cell formats. FUNI endorsement contributes to development of new generations of ATM switches that support Frame Relay physical interfaces and increasing numbers of ATM implementations that interwork with Frame Relay installations.

5.9.3.2 Internet Engineering Task Force (IETF) FR Service Management Information Base (MIB) Working Group

The IETF (Internet Engineering Task Force) Frame Relay Service Management Information Base (MIB) Working Group works in concert with the Frame Relay Forum and the IETF ATM MIB Working Group in establishing MIB solutions that support interoperable FR and ATM services. In addition, the IETF FR Service MIB Working Group develops procedures for managing SVCs in a Frame Relay environment. Procedures that define a Management Information Base (MIB) module for supporting an FR User-to-Network Interface (UNI) and an FR Network-to-Network or Network-to-Node Interface (NNI) are also in development.

5.9.3.3 Advanced Networking Forum Australia (ANFA)

The Australia ATM Interest Group (AIG) and the Pacific Frame Relay Forum (PFRF) formed the Advanced Networking Forum Australia (ANFA). This Forum defines advanced networking services supported by Frame Relay and ATM, promotes FR and ATM interoperability, and clarifies techniques for voice, video, and data convergence on mixed-mode FR and ATM networks.

5.9.4 FRAME RELAY AND DIGITAL SUBSCRIBER LINE (DSL)

In 2000, the Frame Relay Forum endorsed a Service Level Management Implementation Agreement that provisions a framework for enabling DSL-to-FR links that are established via an ATM backbone network. Methods for scalability, installation, and minimally impacting the integrity of FR network applications and services are also in development. Moreover, procedures for maintaining FR service quality and functionality and monitoring and controlling FR network performance are under consideration.

5.10   WIRELESS FRAME RELAY SOLUTIONS

Wireless technologies and services extend Frame Relay capabilities to remote users on the move who are unable to access a landline network. For example, BellSouth Wireless Data supports wireline Frame Relay service that operates in conjunction with its Intelligent Wireless Network.

Research on interworking Frame Relay technology with WATM (Wireless ATM) architecture is sponsored by the ATM Forum and the Frame Relay Forum. Specific FR Implementation Agreements (IAs) for wireline and wireless FR networks are also in development.

In addition, the Frame Relay Forum investigates the viability of using landline FR technology in conjunction with geosynchronous or geostationary (GEO) satellite communications links. Moreover, Hughes Network Systems supports initiatives for interlinking wireline Frame Relay networks and satellite configurations employing VSAT (Very Small Aperture Terminal) solutions.

5.11   FRAME RELAY MARKETPLACE

Spiraling demand for global network services, worldwide communications coverage, and anytime and anywhere connectivity contributes to the development of advanced telecommunications solutions that provide access to FR network configurations. Communications providers active in this arena include Global One, Sprint, Verizon, and WorldCom.

Frame Relay is a viable solution for interlinking remote sites in an integrated platform and provisioning dependable access to VPN resources by authenticated users at distant locations.

Careful planning ensures that the FR installation runs smoothly and supports mission-critical applications. In addition to Global One, Sprint, Verizon, and WorldCom, communications carriers that provide an array of public and/or private services for Frame Relay networks include Ameritech, 3Com, Lucent Technologies, Newbridge Networks, BellSouth, GTE, and SBC.

5.11.1   AT&T

AT&T offers local Frame Relay service in Chicago, Detroit, and Indianapolis and long-distance Frame Relay service for enabling global interconnectivity at rates that include 1.544 Mbps (T-1) and T-3 and E-1 and E-3

5.11.2   CISCO SYSTEMS

Cisco Systems supports an FR VPN solution that supports applications across the enterprise and accommodates diverse protocols, architectures, and connectivity requirements. In FR VPN implementations, Cisco routing services reflect QoS parameters, thereby reducing network congestion and ensuring seamless throughput of mission-critical and delay-sensitive traffic. FR VPNs established by Cisco Systems are also scalable, robust, and flexible in enabling enterprisewide connectivity in local area and wider area environments.

5.11.3 E.SPIRE

e.spire provisions coast-to-coast Frame Relay and ATM applications in the United States via a multiservice platform that supports the consolidation of IP, Frame Relay, and ATM traffic onto a single network backbone. PictureVision employs e.spire Frame Relay service in combination with dial-up Internet access for its PhotoNet initiative. The PhotoNet initiative enables dependable transmission of digital photographs between PhotoNet photofinishers and designated customer Web sites nationwide.

5.11.4 GLOBAL ONE

A joint venture originally sponsored by France Telecom, Sprint, and Deutsche Telekom, Global One supports a range of narrowband and broadband network solutions including Global Frame Relay Service. In 2000, Sprint sold its share of Global One to France Telecom and Deutsche Telekom. Subsequently, France Telecom purchased Global One shares from Deutsche Telekom.

Currently managed by EQUANT, Global Frame Relay Service provided by GlobalOne facilitates local and wider area FR network implementations that support distance education, E-government (electronic government), and E-business (electronic business) applications. Global Frame Relay Service also supports LAN interconnectivity, LAN-to-WAN interworking, data warehousing, CAD/CAM (Computer Aided Design/Computer Aided Manufacturing) applications, and bulk file transfers.

In addition, Global Frame Relay Service fosters worldwide access to Frame Relay intranets, extranets, and VPNs. In 2001, Global Frame Relay service was available in approximately 128 countries. Global Frame Relay service supports transmission rates ranging between 9.6 Kbps (Kilobits) and 44.736 Mbps (T-3) rates.

Global Frame Relay solutions enable flexible prioritization of Frame Relay traffic and three classes of service (CoSs) for ensuring dependable delivery of applications with variable response times and QoS assurances and guarantees. In addition, Global Frame Relay Service supports Committed Information Rates (CIRs) for handling bursty traffic at rates ranging from 8 Kbps to 25 Mbps and facilitates the integration of Frame Relay networks with ATM installations.

FRADs enable dependable delivery of voice, data, fax and IP traffic over Global Frame Relay Service connections. Global Frame Relay Service also provides support for FR network installation, management, and maintenance.

5.11.5 GTS

GTS supports Frame Relay and ATM operations via the optical fiber GTS Trans-European Network. This network configuration employs an IP-over-DWDM infrastructure for provisioning access to broadband applications in member states throughout the European Union as well as in the Czech Republic, Slovakia, Hungary, the Ukraine, and Russia.

5.11.6 HyperStream

HyperStream operates a public Frame Relay networking service supported by nine national telephone companies in Canada.

5.11.7 Sprint

The Sprint Frame Relay Network supports voice and data services that include bursty LAN applications, VPN implementations, and LAN-to-LAN and LAN-to-WAN interconnectivity. As a consequence of its alliance with Deutsche Telekom and France Telecom, the Sprint Frame Relay Network provides Frame Relay service to customers worldwide.

Sprint also supports migration from Frame Relay networks to other platforms such as the Sprint ION (Integrated On-Demand Network) broadband solution to reduce network gridlock, congestion, and equipment costs; increase transmission rates and available bandwidth; and optimize resource utilization. Sprint ION enables fast access to multimedia services via a single connection and employs ATM-over-SONET and ATM-over-DWDM (Dense Wavelength Division Multiplexing) technologies.

5.11.8 Verizon

Verizon provisions national and international Frame Relay service for enabling transmissions at rates of 56 Kbps, 384 Kbps, 1.544 Mbps (T-1), and 44.736 Mbps (T-3). Verizon Frame Relay connections also provision backup service in case of natural or artificial disasters and foster access to multimedia applications, Web browsing, client/server computing, and peer-to-peer networks.

5.11.9 UUNET

UUNET, the first commercial Internet service provider (ISP) in Canada, sponsors Frame Relay service for promoting Internet connectivity at rates ranging from 56 Kbps to 512 Kbps for users with moderate traffic requirements. This service enables e-mail exchange, access to multimedia Web resources and applications, and VPN deployment. UUNET also supports implementation of an ATM backbone network that extends across Canada, and supports voice, video, and data transmission at rates reaching 44.736 Mbps (T-3).

5.11.10 WorldCom

WorldCom provides Frame Relay service to municipal governments, international corporations, and local businesses in more than 50 countries, including Canada, Belgium, Japan, Romania, France, and England. WorldCom Frame Relay solutions enable VoFR (voice-over-Frame Relay) services, backup support, and Frame Relay-to-ATM interworking connections. WorldCom also utilizes PVCs (Permanent Virtual

Circuits) for enabling high, medium, and low levels of service assurances in place of CIR (Committed Information Rate) and CBIR (Committed Burst Information Rate) solutions. In order to forestall VPN incursions, WorldCom monitors more than 1,000 Points of Presence (PoPs) on the Web to ensure that only authenticated users gain access to their VPNs.

5.12 FRAME RELAY TELE-EDUCATION INITIATIVES IN THE UNITED STATES

Telecommunications providers, research organizations, and academic institutions support Frame Relay trials and full-scale implementations. Representative FR initiatives in U.S. schools and post-secondary institutions are highlighted in the material that follows.

5.12.1 ARIZONA

5.12.1.1 Arizona State Public Information Network (ASPIN)

The Arizona State Public Information Network (ASPIN) provisions statewide networking services via telecommunications technologies that include Frame Relay. ASPIN also operates an ATM backbone network and provisions access to next-generation research initiatives such as vBNS+ (very high-performance Backbone Network Service Plus).

Participants in ASPIN include universities, rural community colleges, and K–12 schools, as well as the business community and local and state government agencies. For example, Arizona State University (ASU), the University of Arizona, Northern Arizona University (NAU), Central Arizona College, Cochise County Library District, Gila County Library, and the Catalina Foothills and the Sonoita School Districts take part in the ASPIN project. In addition, the Arizona Center for the Book, the Mojave Library Alliance, the Farm Institute at South Mountain, and the Phoenix Electronic Village Coalition also participate in ASPIN.

The ASPIN Network Information Center (NIC) at Arizona State University (ASU) provides instruction on Internet utilization to state residents in need of technical assistance. The ASPIN Network Operations Center (NOC) at ASU supports Frame Relay service and point-to-point leased line connections for enabling public organizations in La Paz, Maricopa, Pina, Yavapi, and Yuma Counties to access Web resources.

5.12.2 CALIFORNIA

5.12.2.1 California Institute of Technology (Cal Tech)

The California Institute of Technology (Cal Tech) employs Frame Relay technology for enabling connections to the Southern California Seismographic Network. This FR network supports real-time data delivery on seismic activities via VPN links to Cal Tech, public transportation and public utility agencies, and emergency response teams.

5.12.2.2 Monterey BayNet

Monterey BayNet provisions Frame Relay services to local public and private elementary and secondary schools, libraries, and media centers. Additionally, Monterey BayNet also supports Frame Relay applications at the Santa Cruz County Office of Education, Cabrillo and Crown Colleges, the Monterey Bay National Marine Sanctuary, and the Monterey Bay Aquarium Discovery Lab.

5.12.2.3 Pasadena Public Library System

The Pasadena Public Library System employs Frame Relay and ISDN technologies for facilitating public access to digital information resources.

5.12.2.4 San Bernardino, Inyo, and Monroe Counties

Installed in 1996, a Frame Relay network in northeastern California enables students and their instructors in public schools situated in San Bernardino, Inyo, and Monroe Counties to access curricular enrichment activities and Web resources.

5.12.2.5 San Diego County Office of Education

The San Diego County Office of Education sponsors a Frame Relay WAN that provisions access to a full range of instructional services and accommodates educational requirements of public school faculty, staff, and students within San Diego County. This network also meets the educational needs of community school students under the jurisdiction of the Juvenile Court. Public school personnel in San Diego County access resources on the San Diego County Office of Education FR network via the San Diego State University Frame Relay network.

5.12.2.6 San Diego Unified School District

The San Diego Unified School District uses a Frame Relay solution that supports network traffic consolidation, legacy applications, economical delivery of new teleservices for learner enrichment, and administrative functions. Applications enabled districtwide by this FR districtwide network include bulk file transfer, Web browsing, distance learning, student records management, and class scheduling. The San Diego Unified School District Frame Relay solution also provides a migration path to ATM.

5.12.2.7 San Mateo Community College and the Peninsula Library System

San Mateo Community College and the Peninsula Library System use Frame Relay service to interlink campus and library sites. This network provides access to data, voice, and video resources and connectivity to local public kiosk applications.

5.12.2.8 San Jose Education Network (SJEN)

The San Jose Education Network (SJEN) employs a mix of network technologies that include Frame Relay, ISDN, and Ethernet for interlinking high schools throughout the district. The SJEN supports multimedia tele-education applications and enables migration to high-speed, high-performance broadband networking services.

5.12.3 COLORADO

5.12.3.1 Cherry Creek School District Number 5

Cherry Creek School District Number 5 in Englewood utilizes a Frame Relay network for interlinking schools and administrative offices. This districtwide FR network provisions access to Web resources, multimedia applications, and a virtual repository of student records. Network personnel monitor circuit usage and match Committed Information Rates (CIRs) to actual bandwidth employed in order to control the cost of network operations.

5.12.4 IOWA

5.12.4.1 Cedar Rapids Community School District

The Cedar Rapids Community School District utilizes a Frame Relay WAN to enable access to the Internet, academic resources, and e-mail services.

5.12.5 KANSAS

5.12.5.1 *North Kansas City School District*

The North Kansas City School District employs Frame Relay services for interlinking local networks. In addition to Frame Relay, this district intends to implement ATM technology for providing additional bandwidth and capacity to support high-performance voice, video, and data applications.

5.12.6 MAINE

5.12.6.1 Maine School and Libraries Network

The Maine School and Libraries Network employs Frame Relay service for enabling students, faculty, staff, and administrators at schools and libraries statewide to access Web applications and distributed library resources. This configuration also supports student and faculty participation in telecollaborative distance learning projects.

5.12.7 MASSACHUSETTS

5.12.7.1 Springfield Public School System

The Springfield Public School System, the second largest school district in New England, employs a Frame Relay service that operates over a fiber optic infrastructure for enabling students and teachers to access the Internet. This configuration also provisions e-mail service and links to curricular enrichment activities. Local and state grants and E-rate (Education rate) funds contributed to the development of the Springfield Public School System FR network solution. This school system also intends to implement a wireless network to facilitate anytime access to Web resources. This wireless network will also provision links to systemwide educational services and applications.

5.12.7.2   Merimac Education Center Network (MECnet)

Situated in Chelmsford, the Merimac Education Center Network (MECnet) employs a Frame Relay network; an ISDN configuration provides backup services. The MECnet FR network enables students and their teachers to access K–12 educational enrichment tele-activities. In addition, this FR network fosters Web browsing, online research, teacher teletraining sessions, and e-mail exchange. Students and their parents or guardians must sign Acceptable Use Policies (AUPs) in order to access the Web via MECnet connections. Each MECnet school site has at least a single 56 Kbps connection that can be upgraded to 384 Kbps, depending on requirements.

5.12.8   NEW JERSEY

5.12.8.1   Atlantic County Library System (ACLS)

The Atlantic County Library System (ACLS) in Mays Landing provisions Frame Relay links to 17 school districts for supporting student, staff, and faculty access to digital library resources as part of the ACLS Project Connect initiative.

5.12.9   NEW YORK

5.12.9.1   New York State Colleges

New York State Colleges at Canton, Cortland, Morrisville, Oswego, Potsdam, and Cayuga utilize Frame Relay services for accessing regional and statewide networks such as SUNYNET (State University of New York Network).

5.12.10   NORTH DAKOTA

5.12.10.1   Bismarck Initiative

In Bismarck, a VSAT (Very Small Aperture Terminal) configuration that is integrated with a Frame Relay landline network enables students at local high schools to participate in Advanced Placement (AP) courses in Calculus, Spanish, and French with subject specialists at distant locations.

5.12.11   VIRGINIA

5.12.11.1   Henrico Public Schools

The Henrico Public School system employs Frame Relay as a backup solution for in-place Ethernet and Fast Ethernet installations.

5.12.11.2   Virginia Polytechnic Institute and State University (Virginia Tech)

The Virginia Polytechnic Institute and State University (Virginia Tech) uses Verizon Frame Relay-to-ATM service for interlinking campus sites. This Frame Relay-to-ATM solution enables universities and K–12 schools throughout the state to participate in

Net.Work.Virginia, a state-sponsored broadband network initiative. Prior to deployment of the Verizon solution, only institutions with ATM installations could participate in the Net.Work.Virginia initiative.

5.13 INTERNATIONAL TELE-EDUCATION INITIATIVES

5.13.1 CANADA

5.13.1.2 BCNET (British Columbia Network)

The British Columbia Network (BCNET) employs a mix of technologies, including Frame Relay and ATM, for supporting point-to-multipoint videoconferencing, access to Web courses, and teleresearch. Designed for businesses, government organizations, nonprofit agencies, and educational institutions, BCNET enables connectivity to networks that include the Internet and Ca*net II.

5.13.2 GREECE

5.13.2.1 National Technical University of Athens, University of Athens, and Aegean University

The National Technical University of Athens, the University of Athens, and the Aegean University explore Frame Relay and ATM functionality in facilitating teleworking, teletraining, telemedicine, and telemarketing applications.

5.13.3 JAPAN

5.13.3.1 SINET (Science Information Network)

Operated by the National Center for Science Information Systems in Japan, SINET (Science Information Network) provides general networking services for academic research. This configuration features ATM switching systems that work in conjunction with Frame Relay technology for enabling high-speed, high-performance multimedia transmission. SINET supports Web browsing, telecollaborative research, and dependable transmission of voice, data, and video with QoS (Quality of Service) guarantees. Participants in SINET include the Universities at Hokkaido and Okinawa Island.

5.13.4 MALAYSIA

5.13.4.1 Sabah.Net

Sabah.Net is a government and community Frame Relay and ATM network configuration that provides information services in the State of Sabah in Malaysia. Sabah.Net also supports Education.Net, a school network implementation, and enables VPN solutions, E-commerce activities, and E-government applications for the Telekom Malaysia Corporate Information Superhighway (COINS) initiative.

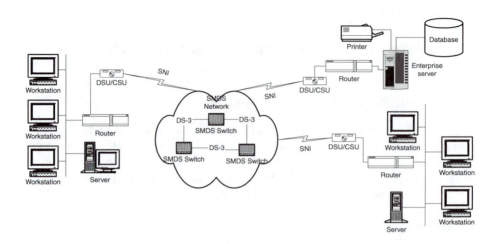

FIGURE 5.3 Several LANs that are interlinked by a Switched Multimegabit Data Service (SMDS) network. The Synchronous Network Interface connection requires implementation of a Data Sensing Unit/Carrier Sensing Unit (DSU/CSU) and a router for data transport.

5.13.5 SPAIN

5.13.5.1 RedIRIS
(National Research and Education Network or NREN of Spain)

Among the first major networking initiatives in Spain, the RedIRIS (National Research and Education Network in Spain) configuration supports teleresearch, distance education, and telecollaborative projects via an ATM backbone that works in concert with Frame Relay and SDH technologies. RedIRIS facilitates access to TEN-155 (Trans-European Network-155.52 Mbps) and enables connections to GÉANT, the next-generation European research and education network. RedIRIS also supports Web browsing and utilization of SMDS (Switched Multimegabit Data Service) to interconnect metropolitan networks developed by local research institutions. Academic participants in the RedIRIS initiative include the Universities of Barcelona, Vic, Girona, and Catalonia. (See Figure 5.3.)

5.14 U.S. GOVERNMENT FRAME RELAY INITIATIVES

5.14.1 GENERAL SERVICE ADMINISTRATION (GSA)

The U.S. Department of State General Service Administration (GSA) utilizes Sprint Frame Relay service for enabling voice, video, and data delivery to geographically distributed GSA sites.

5.15 U.S. ARMED FORCES FRAME RELAY INITIATIVE

5.15.1 U.S. Army Recruiting Command (USAREC)

The U.S. Army Recruiting Command (USAREC) deploys the Newbridge Network Frame Relay VPN solution for interlinking U.S. Army Recruiting Offices worldwide to USAREC headquarters situated at Fort Knox, Kentucky.

5.16 U.S. STATE AND LOCAL E-GOVERNMENT FRAME RELAY INITIATIVES

5.16.1 California

5.16.1.1 Solano County Network

Solano County employs Frame Relay, Fast Ethernet, and ISDN technologies for interconnecting LANs in Fairfield, Vallejo, and outlying locations. This countywide network supports voice, video, and data transmission and E-government applications. In addition, the Solano County Network enables local residents to access building and planning permits, business licenses, and county records, and also facilitates administrative functions such as data warehousing, document imaging services, and fund accounting.

5.16.2 Florida

5.16.2.1 Miami District Court in the Eleventh Judicial Circuit

The Miami District Court in the Eleventh Judicial Circuit employs a Frame Relay VPN solution that interoperates with Fast Ethernet technology. This VPN provides access to archival documents, court cases, legislative records, and caseloads.

5.16.3 Massachusetts

5.16.3.1 Commonwealth of Massachusetts

The Commonwealth of Massachusetts employs a Frame Relay network for interlinking 150 state agencies to dependable access to state-based information resources. This scalable, dependable, and reliable statewide configuration provisions sustained bandwidth and Committed Information Rates (CIRs) for bursty traffic; supports multimedia services; and enables authenticated users to access bandwidth-intensive database applications.

5.16.4 Nebraska

5.16.4.1 Nebraska State Government Initiative

Headquartered in Lincoln, the Nebraska State Government employs a Frame Relay network for supporting distributed networking applications and connectivity to information resources on the Web posted by city and county governments and state agencies. In addition, this extendible and flexible statewide network infrastructure provides universities and schools with direct access to digital resources maintained by the state library system, supports delivery of telemedicine services to rural hospitals, and facilitates dependable transmission of records to law enforcement personnel in rural Nebraska communities.

5.16.5 SOUTH DAKOTA

5.16.5.1 South Dakota WAN (Wide Area Network)

The South Dakota statewide WAN (Wide Area Network) enables state and local agencies and state residents to obtain information on government regulations, legislative activities, and state resources via a combination of technologies that include Frame Relay, POTS (Plain Old Telephone Service), ATM, ISDN, Ethernet, and Fast Ethernet. The South Dakota statewide WAN also facilitates dependable delivery of voice, video, and data traffic; LAN-to-LAN internetworking; and connections to the commodity Internet.

5.17 INTERNATIONAL E-GOVERNMENT FRAME RELAY INITIATIVES

5.17.1 CANADA

5.17.1.1 Department of Indian and Northern Affairs (INAC)

Headquartered in Ontario, the Department of Indian and Northern Affairs (INAC) employs a Frame Relay-over-ATM infrastructure for interlinking INAC sites across Canada. This Frame Relay-over-ATM configuration supports videoconferencing, voice-over-Frame Relay services, and registration of voters in the First Nations. In addition, INAC personnel use this network to track land claims and monitor the environment.

5.17.2 CHINA

5.17.2.1 Guangdong Post and Telecommunications Administration (GPTA)

The Guangdong Post and Telecommunications Administration (GPTA) utilizes a multiservice Frame Relay and ATM network configuration for provisioning access to data services and applications. This configuration interlinks corporate sites and government agencies in more than 100 cities and counties in Guangdong Province.

5.17.3 FRANCE

5.17.3.1 French National Employment Agency

The French National Employment Agency migrated from a public X.25 network to a Frame Relay VPN so as to effectively manage increased traffic loads, enhance network performance, and reduce transmission errors.

5.17.3.2 French National Social Security Agency

The French National Social Security Agency utilizes a Frame Relay VPN for processing claims and tracking expenditures for healthcare services at local branches.

5.17.4 SOUTH AFRICA

5.17.4.1 Voter Registration

In South Africa, a mixed-mode satellite and terrestrial Frame Relay network supports electronic transfer of voter registration data prior to elections.

5.17.5 WALES

5.17.5.1 Carmathensire County Government

In Wales, the Carmathensire County Government employs Frame Relay service for facilitating access to local government services. The Carmathensire County Government expects to replace the Frame Relay platform with an ATM platform for enabling sophisticated multimedia applications.

5.18 U.S. GOVERNMENT TELEMEDICINE INITIATIVES

5.18.1 VETERANS HEALTH ADMINISTRATION (VHA)

The Veterans Health Administration (VHA), an agency of the U.S. Department of Veteran Affairs (VA), supports utilization of a Frame Relay network that operates over fiber optic cabling for linking more than 600 VA facilities nationwide. The Frame Relay configuration enables these VA facilities to exchange data with medical centers, regional offices, and the VA Central Office.

In addition, this integrated communications network supports access to Web resources via a secured gateway. The VHA FR network also provisions links to electronic library resources, VA bulletin board postings, and clinical and administrative reports. An automated medical information exchange module enables the electronic transmission of information relating to veterans' benefits between the Veterans Benefits Administration Regional Offices and the VA medical centers.

This FR network also facilitates connectivity to the U.S. Department of Veterans Affairs home page, a Web site that provisions access to data relating to home loans, disability compensation, insurance, and education. Moreover, the VHA FR implementation supports links to the database maintained by TRICARE/CHAMPUS (Connectivity to Civilian Health and Medical Program of the Uniformed Services), the missing patient registry, and the patient inquiry and locator system. Medical professionals can access the Patient Data Exchange, an FR network module, for reviewing patient medical records. FR connections support videoconferencing and transmission of MRI (Magnetic Resonance Imaging) and CAT (Computerized Axial Tomography) scans between VA medical centers.

In addition to the FR configuration, the Veterans Health Administration also operates a satellite television network featuring two-way audio that enables VA medical centers and clinics to access healthcare programs. The Healthcare Informatics Telecommunications Network, the American Hospital Association, and the Joint Commission on Accreditation of Healthcare Organizations sponsor these healthcare televised programs.

5.19 INTERNATIONAL FRAME RELAY COMMUNICATIONS PROVIDERS

5.19.1 ARGENTINA

5.19.1.1 Telintar Norte

Telintar Norte, a unit of Telecom Argentina and Global One, provisions Frame Relay service over an ATM backbone for national and international businesses in Argentina. A low-cost solution, this Frame Relay configuration employs PVCs (Permanent Virtual Circuits) and CIRs (Committed Information Rates) for effectively handling bursty on-demand transmissions. Network access speeds ranging from 56 and 64 Kbps to 2 Mbps are supported.

5.19.2 BOLIVIA

5.19.2.1 Costa

A telecommunications provider in Santa Cruz, Costa implements a high-performance broadband backbone ATM network that works in concert with an in-place Frame Relay infrastructure. This scalable and extendible network configuration enables LAN interconnectivity, Internet access, and broadband applications, including full-motion videoconferencing. This infrastructure also supports corporate VPN implementations and differentiated classes of service (CoSs) for ensuring seamless transport of delay-sensitive and delay-insensitive network traffic.

5.19.3 BULGARIA

5.19.3.1 Bulgaria Telecommunications Company

The Bulgaria Telecommunications Company provisions a nationwide integrated data and voice network that utilizes a mix of technologies, including Frame Relay and ISDN, for providing communications services to residential customers, government agencies, and corporations. This network enables IP multicasts, Frame Relay VPN implementations, and LAN interconnectivity. Approaches for upgrading this infrastructure to ATM are in development.

5.19.4 HUNGARY

5.19.4.1 GTS Hungary

GTS Hungary provisions managed Frame Relay service supporting rates ranging from 8 to 128 Kbps for residential subscribers and rates ranging from 256 Kbps to

19 Mbps for business entities. VSAT systems and ISDN technology provide backup services. Subscribers purchase bandwidth as needed. This FR solution supports LAN-to-LAN interconnectivity, Web browsing, bulk file transfer, and links to international networks and works in concert with Packet-over-SDH (POS) implementations.

In addition, GTS Hungary operates an SDH GTSRing network in Budapest that provides integrated Frame Relay, X.25, and ATM services. This fiber optic ring supports VPN implementations, Ethernet and Fast Ethernet applications, voice-over-Frame Relay (VoFR) service, IP telephony, Web browsing, and peer-level information interchange via the Budapest Internet Exchange (BIX).

5.19.5 ISRAEL

5.19.5.1 Bezeq

Bezeq, the largest telecommunications provider in Israel, provisions Frame Relay services for corporate clients. ATM services are also available.

5.19.6 LATVIA

5.19.6.1 Lattelecom

Lattelecom, the Latvian national telecommunications provider, employs a hybrid X.25 and Frame Relay network configuration called the Lattelecom Digital Access Network (LDAN). Designed for residential and corporate users, the LDAN provisions connections to Web resources and the national electronic lottery.

5.19.7 MEXICO

5.19.7.1 Televisa

Televisa, a Mexican television corporation, employs Frame Relay service for interlinking corporate headquarters in Mexico City with offices in Chile, Argentina, Colombia, Peru, Puerto Rico, Ecuador, and the Untied States.

5.19.8 RUSSIA

5.19.8.1 Sovan Teleport

Sovan Teleport, a communications provider in Moscow, supports X.25, Frame Relay, and ATM services. The Sovan Teleport mixed-mode network configuration facilitates access to Web resources and VPN implementations and supports network services for entities located in the greater Moscow metropolitan area.

5.20 FRAME RELAY IMPLEMENTATION CONSIDERATIONS

Functions, applications, and services provided in a Frame Relay network implementation depend on institutional budget allocations, traffic profiles, the number of network users and their locations, and bandwidth and response time requirements.

The decision to implement a Frame Relay network also involves determining installation, management, service, and maintenance costs; usage fees; CIR (Committed Information Rate) and CBIR (Committed Burst Information Rate) charges; the number of PVCs and SVCs that require support; and the extent of FR network geographic coverage.

Additionally, costs for CPE (Customer Premise Equipment), software products, and FR devices such as FRADs, VFRADs, bridges, routers, and/or switches; the need for network infrastructure additions, modifications, and upgrades; and the availability of technical support and training sessions must be determined at the outset. Factors that also affect the decision to move forward with FR implementation include application requirements, user demand for FR services, and projected plans for broadband implementation. Academic institutions generally establish partnerships with vendors, ISPs (Internet Service Providers), and/or public communications providers such as Verizon, Sprint, and Global One to facilitate development and implementation of affordable FR network solutions.

5.21 FRAME RELAY SUMMARY

Frame Relay (FR) solutions facilitate reliable access to current and emergent initiatives in fields that include tele-education, teletraining, teleworking, electronic publishing, teleradiology, and electronic commerce. FR networks also support operations for small- and medium-sized businesses, the military, and federal, state, and local government agencies.

FR takes advantage of the low error rates in present-day optical fiber landline networks for supporting flexible networking configurations and feature-rich enhancements such as CIRs (Committed Information Rates), video-over-FR, IP multicasts, voice-over-FR (VoFR) services, and VPN implementations. Moreover, the Frame Relay platform facilitates network consolidation by enabling LAN-to-LAN connections, LAN integration with backbone networks, and LAN interconnectivity across WANs at T-1 (1.544 Mbps) and T-3 (44.736 Mbps) rates and at E-1 (2.048 Mbps) and E-3 (34.368 Mbps) rates.

Protocols such as ATM-DXI (ATM-Data Exchange Interface) and FUNI (Frame User-to-Network Interface) facilitate Frame Relay-over-ATM networking implementations for achieving real-time voice, video, and data transmission. The ATM-DXI and FUNI protocols also support integration of Frame Relay networks with SONET/SDH internetworking environments.

Frame Relay solutions are less expensive to implement than leased line services and enable bandwidth for VPNs, LAN-to-WAN integration, and volume-intensive applications.

Frame Relay networks facilitate dependable transmission of narrowband, broadband, and legacy network traffic and provide economical support of VPN applications that optimize user productivity in virtual environments. Frame Relay network platforms are scalable and flexible and work with multiple network technologies including X.25, SNA, IP, Ethernet, Fast Ethernet, ISDN, SMDS, SONET/SDH, and ATM.

Migration to a Frame Relay network involves utilization of FRADs and network management tools for monitoring network endpoints or nodes. Data compression techniques employed by Frame Relay technology support seamless transmission of bursty LAN traffic. Frame Relay-over-ATM solutions facilitate deployment of converged multiservice network configurations that support multimedia applications, VPN services, and dependable network performance. It is important to note that implementation and maintenance of Frame Relay-over-ATM solutions can be costly.

The Frame Relay Forum (FRF) develops Implementation Agreements (IAs) that enable standards-based Frame Relay installations. Frame Relay specifications also support the integration of heterogeneous equipment from multiple vendors into a single network implementation. However, a single vendor solution is the preferred method for FR network deployment.

Variations in transmission in a Frame Relay network are called jitter or slight network delays.

These delays contribute to difficulties in transmitting voice and video via an FR platform. Methods for eliminating jitter are not fully defined. Techniques for error-free support of video-over-FR and voice-over-FR and approaches for addressing problems associated with supporting Quality of Service (QoS) guarantees for Frame Relay VPNs are still in development.

5.22   FIBRE CHANNEL (FC) INTRODUCTION

A high-performance networking technology, Fibre Channel implementations foster dependable transmissions over long optical fiber cable lengths and feature flexible topologies. As an example, FC technology enables high-speed data exchange over a single-mode optical fiber link between network nodes that are separated by distances up to 10 kilometers. With FC extenders, a Fibre Channel network supports operations over a single-mode optical fiber connection at distances reaching 30 kilometers.

Fibre Channel technology interworks with protocols that include IP (Internet Protocol) and the Small Computer Systems Interface (SCSI) and supports massive network storage applications and services over diverse network platforms. Moreover, FC interconnects storage devices, servers, and network systems and subsystems for enabling volume-intensive applications at gigabit rates. Distinguished by its networking capabilities and robust performance, Fibre Channel is emerging as the premier interconnect technology for enterprises that require high-availability, improved connectivity, and the capability to rapidly transmit large volumes of data over local and extended distances. Accelerating demand for scalable storage solutions to protect the integrity of critical information resources contributes to FC installations in fields that include education, government, business, industry, and medicine.

5.23   FIBRE CHANNEL FOUNDATIONS

Originally, Fibre Channel was designed to work in conjunction with fiber optic cabling. After FC support of operations over twisted copper pair was accomplished, a Task Force of the International Standards Organization (ISO) renamed the technology by changing the English spelling of the word "Fiber" to the French spelling of the word "Fibre." This change maintains name recognition for Fibre Channel technology while limiting its direct association with fiber optic cabling.

Created by Hewlett-Packard, Sun Microsystems, and IBM, the Fibre Channel Systems Initiative (FCSI) was proposed in 1992. In addition, these companies also defined SCSI (Small Computer System Interface) and IP (Internet Protocol) interoperability within the context of the Fibre Channel specification.

5.24   FIBRE CHANNEL STANDARDS ORGANIZATIONS AND ACTIVITIES

5.24.1   AMERICAN NATIONAL STANDARDS INSTITUTE (ANSI)

Fibre Channel (FC) standards were ratified by the American National Standards Institute (ANSI) in 1994. Formally called ANSI X.3.230-1994, the FC specification provisions a framework for Fibre Channel SAN (Storage Area Network) implementation supported by a switched network fabric. In addition, the FC ANSI standard defines a Media Access Control (MAC) Layer for Fibre Channel and Gigabit Ethernet interoperability and approaches for interworking FC technology with SCSI and IP protocols.

5.24.1.1   ANSI Accredited Standards Committee

The American National Standards Institute appointed the Accredited Standards Committee as the primary committee responsible for further defining the Fibre Channel specifications. The ANSI Accredited Standards Committee works in conjunction with the National Committee for Information Technology Standards (NCITS) T11 Technical Committee in clarifying a base level of interoperability between FC SAN (Storage Area Network) switches, topologies, management services, and devices for enabling Fibre Channel standards-compliant implementations in multivendor environments.

5.24.2   FIBRE CHANNEL ASSOCIATION (FCA)

The Fibre Channel Association (FCA) is an industry group that promotes an understanding of Fibre Channel technology and fosters development and deployment of standardized Fibre Channel networks, services, and applications. The FCA also establishes standards for managing FC Storage Area Networks (SANs) and for monitoring Fibre Channel network components such as host bus adapters, switches, hubs, routers, and servers that reside on FC SANs. In addition, the Fibre Channel Association works with the IETF (Internet Engineering Task Force) in defining specifications for enabling Fibre Channel implementations to interwork seamlessly with IP technology.

5.24.2.1   FCA Fibre Channel Loop Community (FCLC) Working Group

A Fibre Channel Association (FCA) affiliate, the Fibre Channel Loop Community (FCLC) Working Group is an association of computer industry companies that provide marketing support for FC software developers, component manufacturers, communications companies, computer service providers, and systems integrators. The FCLC Working Group promotes utilization of Fibre Channel Arbitrated Loop (FC-AL) solutions, supports FC product interoperability, and actively participates in the FC standards development process.

5.24.3 FIBRE CHANNEL CONSORTIUM (FCC)

The Fibre Channel Consortium (FCC) conducts conformance tests at the Interoperability Laboratory at the University of New Hampshire and the Computational Science and Engineering Laboratory at the University of Minnesota for ensuring vendor compliance with FC specifications and FC Storage Area Network (SAN) requirements.

5.24.4 FIBRE CHANNEL INDUSTRY ASSOCIATION-EUROPE (FCIA-EUROPE)

The Fibre Channel Industry Association-Europe (FCIA-Europe) is an international alliance of FC professionals, systems developers, vendors, and manufacturers that supports production of a Fibre Channel infrastructure for enabling applications that foster video services, networking, storage, and SAN (Storage Area Network) administration and management. Moreover, the FCIA-Europe promotes implementation of a distributed FC infrastructure for enabling diverse IT (Information Technology) applications.

5.24.5 INTERNATIONAL STANDARDS ORGANIZATION (ISO)

In addition to ANSI endorsement, FC is also an approved ISO (International Standards Organization) specification.

5.25 FIBRE CHANNEL TECHNICAL FUNDAMENTALS

5.25.1 FIBRE CHANNEL TRANSMISSION

Fibre Channel (FC) is high-speed serial interface technology that interconnects equipment such as mainframes, workstations, supercomputers, peripherals, servers, and storage devices. The FC infrastructure enables information storage and bulk file transfer and supports dependable delivery of sequential blocks of data that consist of frames.

Fibre Channel frames vary in size from 36 to 2112 bytes. An FC frame consists of a header that includes addressing and error checking information and a payload for user data. An FC sequence refers to a set of related frames for the same operation. FC network transmissions support guaranteed voice, video, and data delivery.

FC hardware enables up to 65,536 frames to be concatenated into a single FC sequence, thereby enabling transport of very large blocks of data. FC networks feature high capacity, speed, reliability, flexibility, and dependability. These attributes contribute to the popularity of Fibre Channel solutions and FC utilization for long-haul communications services.

FC technology supports peer-to-peer connectivity between any pair of ports on an FC network. FC frames are transported between two ports in full-duplex mode.

The port initiating an information exchange is called the originator. The port answering the exchange is called the responder. A Fibre Channel infrastructure supports transmissions at rates that include 100 Mbps, 200 Mbps, and 2.12 Gbps. The Fibre Channel Association (FCA) also defines the FC 4.24 Gbps specification.

FC installations enable implementation of single-loop and dual-loop architectural configurations. Dual-loop architecture enables full-duplex rates between storage systems and servers and supports backup and disaster recovery services in the event of a loop failure or optical fiber cut. Fibre Channel networks enable operations at the Physical Layer or Layer 1, the Data-Link Layer or Layer 2, and the Network Layer or Layer 3 of the Open Systems Interconnection (OSI) Reference Model.

5.25.2 FIBRE CHANNEL SERVICES

Fibre Channel implementations interwork with high-performance LANs featuring ATM and Fast Ethernet technologies and interoperate with high-speed optical networks that employ SONET/SDH (Synchronous Optical Network and Synchronous Digital Hierarchy), WDM (Wavelength Division Multiplexing), and DWDM (Dense WDM) solutions. Fibre Channel also works in conjunction with IEEE 802.2 specifications for HIPPI (High-Performance Parallel Interface) hardware and protocols.

5.26 FIBRE CHANNEL ARBITRATED LOOP (FC-AL)

The Fibre Channel Arbitrated Loop (FC-AL) is a recent enhancement to the Fibre Channel specification. The FC-AL supports operations over twisted pair copper connections. An FC-AL topology includes at least one FC switch for interconnecting a number of end systems in a ring or arbitrated loop environment.

Endorsed by the Fibre Channel Association (FCA), FC-AL architecture supports up to 127 active devices per loop. These devices are either attached directly to switches or to hubs that in turn connect to switches. In a point-to-point configuration, FC-AL technology links hosts and devices directly via hubs and single-loop solutions.

Because each FC node functions as a repeater for every other node in a FC point-to-point implementation, one disconnected or malfunctioning network node can bring network activities to a halt. Typically, each node connected to a Fibre Channel Arbitrated Loop (FC-AL) has a port bypass circuit so that a failed device will not disrupt data transfer and corrupt data integrity. The FC switching fabric also enables FC-AL solutions to facilitate improved network performance and provision redundant services for large systems.

5.27 FIBRE CHANNEL OPERATIONS

Fibre Channel transmissions are carried by coaxial cable, shielded twisted pair (STP), hybrid fiber optic and coaxial cable (HFC) configurations, and single-mode and multimode optical fiber.

Fibre Channel employs dark fiber for linking remote nodes at full data rates. In the absence of dark fiber, a public access network bridge enables fast rates and transmission over long distances.

5.27.1 FIBRE CHANNEL NETWORK COMPONENTS

A FC network features one or more switching elements. Typically, FC switches are implemented in SANs (Storage Area Networks) for enabling reliable and dependable gigabit connections.

A collection of FC switching elements is called a switching fabric. FC topologies include point-to-point direct channel connections, shared media rings or arbitrated loops, and workgroup or campuswide switching fabrics. Basic network elements in an FC implementation include storage devices, switches, and supercomputers that are equipped with ports for enabling interconnectivity to diverse networking configurations and transmission media.

FC configurations support Virtual Interface Architecture (VIA) for enabling high-performance bandwidth-intensive transmission. VIA also works in concert with IP and SCSI on the same FC network infrastructure, thereby optimizing FC flexibility in enabling current and next-generation broadband applications. An FC configuration can be readily upgraded for enabling faster data rates, broader areas of coverage, and the inclusion of additional nodes by adding supplementary switches and ports to the switching fabric.

5.27.2 FIBRE CHANNEL CLASSES OF SERVICE (CoSs)

Fibre Channel switches route frames in accordance with their Class of Service (CoS) and destination addresses. Fibre Channel technology supports four CoSs (Classes of Service).

5.27.2.1 Fibre Channel Class-1

Fibre Channel Class-1 is a connection-oriented, circuit-switched service that provides dedicated links such as connections between two supercomputers for enabling transmission of time-critical, non-bursty data in a specified period of time. With FC Class-1, connections are terminated only after all FC frames have been transmitted in a sequential mode.

5.27.2.2 Fibre Channel Class-2 and Fibre Channel Class-3

Fibre Channel Class-2 features connectionless frame-switched transmission with guaranteed delivery and receipt confirmation. Fibre Channel Class-3 supports connectionless frame-switched service for point-to-multipoint multicast transmissions involving real-time broadcasts, packetized video, and video clips with synchronized audio. In contrast to Class-2, Class-3 does not provide delivery guarantees or confirmations of voice, video, and data transport.

With both FC Class-2 and FC Class-3, the network connection is busy only during frame transmission. When the link is idle, the transmission path is available for supporting frame transmission generated by other network nodes or end stations.

5.27.2.3 Fibre Channel Class-4

Still in development, Fibre Channel Class-4 is a connection service that transmits information via a defined physical path rather than a virtual path. FC Class-4 features guaranteed bandwidth for enabling applications such as high-definition television (HDTV) programming. Additionally, FC Class-4 supports isochronous services that carry real-time voice, video, and data traffic at a consistent rate with the information reaching its destination sequentially.

5.28   FIBRE CHANNEL AND SMALL COMPUTER SYSTEMS INTERFACE (SCSI)

The Small Computer Systems Interface (SCSI) is a set of ANSI standards for electronic interfaces that enable PCs (Personal Computers) to communicate with peripheral devices such as printers and scanners. Approved by ANSI in 1986, SCSI is the oldest peripheral interconnect technology that still is in widespread use.

In addition to printers and scanners, SCSI technology is integrated into disk drives, tape drives, removable disk drives, and optical disk drives that support CD-ROM (Compact Disk-Read Only Memory), DVD (Digital Versatile Disk), and WORM (Write Once, Read Many) applications. Disk array subsystems and plotters feature SCSI technology as well.

Small Computer Systems Interface (SCSI) technical solutions employ intelligent parallel I/O (Input/Output) buses for enabling communications between servers and storage devices. A recently approved specification, Ultra-2 SCSI supports multimedia transmission rates at 40 Mbps and up to eight devices on an 8-bit bus. Moreover, Ultra-2 SCSI implementations enable voice, video, and data transport rates at 80 Mbps and support a maximum of 16 devices on a 16-bit bus.

Typically, SCSI deployments foster information transmission over distances that extend to 6 meters and support point-to-point links between a storage device and a server.

Since its introduction in the mid-1980s, SCSI technology has served as the *de facto* physical and transport specification for storage arrays and disk drives. SCSI-compliant devices are generally used in conjunction with high-performance servers, PCs, notebook computers, and network workstations.

Fibre Channel and Small Computer Systems Interface (SCSI) implementations support similar functions and applications. FC operations are compatible with SCSI services. In addition, FC technology interworks with SCSI products. In a Fibre Channel installation, SCSI-compliant devices appear to a network server or workstation as if they were directly attached to the FC network. Moreover, FC implementations allow reuse of existing SCSI drivers with limited modification, and the Fibre Channel Protocol (FCP) for SCSI deployments enables network management and security operations and reduces bottlenecks and link congestion in hybrid Fibre Channel and SCSI environments.

In comparison to SCSI implementations, Fibre Channel solutions support faster transmission rates over greater distances. In addition, FC deployments provision sophisticated networking services for supporting the interface between servers and clustered storage devices and enabling interconnectivity with complex peripheral equipment.

5.29 FIBRE CHANNEL AND IP (INTERNET PROTOCOL)

Originally, the lack of clear-cut specifications for defining Fibre Channel operations in enabling IP encapsulation and address resolution contributed to the proliferation of proprietary solutions for using IP-over-FC implementations. As a consequence, the IETF (Internet Engineering Task Force) Network Working Group subsequently defined standards for enabling IP-over-Fibre Channel deployments.

In addition, approaches for mapping, routing, and encapsulating Fibre Channel frames into IPv4 and IPv6 packets, encapsulating ARP (Address Resolution Protocol) packets into FC frames, and procedures for interworking distributed islands of FC SANs over IP networks were also clarified. The IETF IP-over-Fibre Channel Working Group collaborates with the IETF Network Working Group in promoting development, acceptance, and implementation of IP-over-FC standards and specifications.

5.30 FIBRE CHANNEL (FC) NETWORK MANAGEMENT OPERATIONS

Fibre Channel systems employ Simple Network Management Protocol (SNMP) for facilitating effective network operations and supporting disaster recovery services and resource planning. By enabling storage and server system consolidation in network applications such as SANs (Storage Area Networks), FC technology streamlines network administrative services.

Additionally, Fibre Channel systems work in conjunction with Simple Name Service (SNS) for identifying each device that participates in the switching fabric. Moreover, FC systems also utilize Registered State Change Notification (RSCN) for issuing updates to FC network nodes on topology changes.

To reduce network downtime, FC systems employ latency metrics for fault isolation and error detection. At present, a commonly accepted method for monitoring SAN (Storage Area Network) elements and launching alerts in case of problems in a Fibre Channel WAN is not available. As a consequence, the Fibre Channel Association (FCA) plans to design and implement a universal method for performing SAN management functions. (See Figure 5.4.)

5.31 FIBRE CHANNEL (FC) STORAGE AREA NETWORKS (SANs)

Distinguished by its support of large numbers of devices, fast throughput, and network availability, Fibre Channel technology is the major enabler of SAN solutions. FC SAN solutions facilitate high-speed, bandwidth-intensive voice, video, and data transport over local and extended distances; provision a diverse array of connectivity options for enabling links between storage servers and storage devices; and support flexible, dependable, scalable, and redundant network operations.

5.31.1 FC SAN OPERATIONS

FC SANs are dedicated networks for resource storage that offload traffic from multipurpose multimedia networks; transport voice, video, and data at extremely

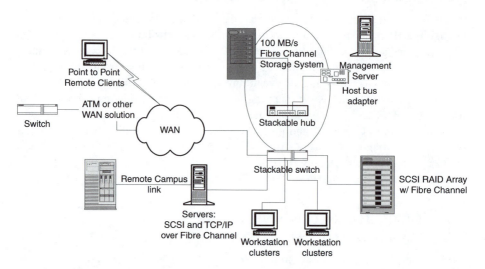

FIGURE 5.4 A Fibre Channel Storage Area Network (SAN).

high rates; and perform backup operations. FC SANs are reliable, extendible, and scalable in order to accommodate increasing numbers of storage devices and bandwidth-intensive voice, video, and data applications. Nodes in FC SAN (Storage Area Network) configurations are called storage devices. Each storage device must also be equipped with a Network Interface Card (NIC) in order to support connections to the switching fabric.

FC SANs support communications between hosts and storage devices within dedicated or shared network environments. FC SANs facilitate testbed functions and enable the evaluation of new technologies prior to their full-scale introduction into the workplace. In addition, FC SANs support disaster recovery operations and enable networks to operate continuously in the event of server crashes. Prior to the utilization of Fibre Channel technology in SAN configurations, SANs consisted of I/O (Input/Output) buses and I/O channels.

FC SANs are separate networks or subnetworks that consist of multiple servers situated at distributed sites. FC SANs also feature network components such as switches, adapters, routers, and hubs for fostering access to stored voice, video, and data applications.

FC SAN traffic is isolated on a separate section of the network from other network traffic. As a consequence, the speed and performance rates required for other essential network applications are not adversely affected by utilization of high-resolution media from a shared storage pool.

5.31.2   FC SAN APPLICATIONS

FC SANs support applications such as video editing, data warehousing, and online storage. FC SANs also enable disk mirroring, data backup services, bulk file transmission, and archival operations. Moreover, Fibre Channel SANs facilitate seamless access to audio and video clips and special effects for broadcast news programs; support desktop applications and film production; and provision ultrafast voice, video, and data transmission for enabling real-time telecollaboration in corporate environments. These flexible, efficient, and affordable special-purpose networks or subnetworks enable storage consolidation in fields that also include manufacturing, healthcare, banking, finance, and entertainment.

The accelerating popularity of volume-intensive services such as data warehousing, decision-support systems, and digital libraries drives implementation of FC SAN solutions. Increased utilization of multimedia objects, including bit-mapped images and data, voice, and video streams in interactive tele-education, teletraining, teleradiology, E-commerce, and E-government initiatives, also contribute to the demand for FC SAN implementations.

5.31.3 FC SAN VENDORS

5.31.3.1 Brocade Communications

Brocade Communications is a provider of FC SAN switching fabrics for large-scale SAN environments. Brocade Communications FC SAN solutions support dependable throughput and enable interoperability between FC switches manufactured by vendors supporting the Open System Fabric Initiative (OSFI).

5.31.3.1.1 Brocade Communications and Cisco Systems

Cisco Systems and Brocade Communications support FC SAN implementations over IP metropolitan and wider area networks. These implementations enable remote data backup, disaster recovery services, high-speed data replication, data mirroring, and ultra-fast digital content distribution.

In addition, these deployments enable SAN-to-SAN connectivity over DWDM (Dense Wavelength Division Multiplexing) metropolitan network configurations that are compliant with the National Committee for Information Technology Standards (NCITS) T11 Technical Committee recommendations.

5.31.3.2 Storage Area Networks, Limited

Available from Storage Area Networks, Limited, DiskLink Fibre Channel solutions promote the development of an enterprise storage infrastructure by removing geographic constraints associated with the use of SAN solutions. With DiskLink Fibre Channel installations, users conduct networking activities across ATM and/or Gigabit Ethernet networks. Host computers and servers are interlinked to storage devices such as RAIDs (Redundant Array of Interactive Disks), regardless of locations. Transmission rates at 622.08 Mbps (OC-12) are enabled. Also developed by Storage Area Networks Limited, the DiskLink intelligent SAN gateway supports local and wide area SAN interfaces for provisioning disaster recovery services, enabling consolidation of storage applications, and facilitating dependable access to digitized archival records.

5.31.4   Carnegie Mellon University (CMU) SAN Research

As part of the Active Networking for Storage Initiative sponsored by DARPA (U.S. Department of Defense Advanced Research Projects Agency), Carnegie Mellon University (CMU) benchmarks capabilities of active networks that interwork with network attached storage configurations. CMU also supports development of innovative storage network protocols and services such as network object caching and evaluates capabilities of FC SANs to enable secure voice, video, and data exchange in dynamic LAN and WAN environments.

5.31.5   Storage Network Industry Association (SNIA)

The Storage Network Industry Association (SNIA) defines SAN technical architectures and topologies, security and application specifications, and procedures for implementation of standards-compliant SAN solutions. In addition, the Storage Network Industry Association establishes specifications for interoperable SAN devices and supports development of SAN management procedures and automated SAN disaster recovery services. SNIA participants include 3Com, Hewlett-Packard, Hitachi Data Systems, IBM, Novell, Quantum, Seagate Technology, and Sun Microsystems.

5.32   FIBRE CHANNEL RAID (REDUNDANT ARRAY OF INTERACTIVE DISKS) SOLUTIONS

The integration of serial Fibre Channel technology with storage products supports deployment of robust information storage solutions. As an example, FC-supported RAID (Redundant Array of Interactive Disks) solutions facilitate dependable access to operational data maintained by data warehouses. Typically, RAID solutions are configured with redundancy to eliminate points of failure, ensure continuous data availability, and enable reliable, secure, and high-performance operations.

Fibre channel technology optimizes RAID performance and storage capacity, streamlines resource management activities, and conserves host computer resources. In addition, Fibre Channel technology fosters point-to-point connections between database and application servers and next-generation fault-tolerant RAID subsystems.

An FC-supported RAID implementation accommodates enormous video and information storage requirements while also enabling instantaneous access to high-resolution digitized images. As an example, professional baseball teams use FC-supported RAID technology for coaching, scouting, and game analysis.

5.32.1   RAID Advisory Committee

The RAID (Redundant Array of Interactive Disks) Advisory Committee includes vendors that are active in the FC arena as well. These vendors promote development of high-performance, cost-effective RAID solutions. For example, Infortrend Corporation manufactures and promotes product lines featuring FC connectivity for

RAID implementations. United Digital Limited provisions FC RAID solutions to accommodate consumer requirements, support transparent data sharing on FC-attached storage devices, and facilitate implementation of cross-platform FC SAN installations. The RAID Advisory Committee also supports FC-AL implementations.

5.33 COMPETITOR FIBRE CHANNEL SOLUTIONS

5.33.1 EtherStorage

Developed by Adaptec, EtherStorage enables affordable SAN construction with Ethernet technology, protocols, and interconnectivity devices such as Ethernet hubs, routers, and switches. EtherStorage works in concert with the Physical Layer or Layer 1, the Media Access Control (MAC) Layer or Layer 2, and the Network Layer or Layer 3 of the OSI Reference Model.

In addition, EtherStorage fosters encapsulation of SCSI commands and data through utilization of the SCSI Encapsulation Protocol (SEP) for enabling upper OSI Layer functions.

EtherStorage supports full-duplex transmissions at 200 Mbps over a Gigabit Ethernet link. The EtherStorage solution accommodates distributed storage requirements and also enables small- and medium-sized enterprises to leverage investments in on-site Ethernet infrastructures. A flexible local networking solution, EtherStorage accommodates networking requirements of large corporations with volume-intensive data applications as well.

5.33.2 High-Performance Parallel Interface (HIPPI) and Gigabit System Network (GSN)

5.33.2.1 HIPPI (High-Performance Parallel Interface) Technical Features and Functions

HIPPI (High-Performance Parallel Interface), also known as HIPPI-800, fosters information transport at gigabit rates and supports operations at supercomputer centers and in other data-intensive environments. HIPPI enables digital studio applications, special film effects, animations, and digital rendering in the post-production process. The ITU and ANSI endorse the HIPPI specification.

HIPPI employs a point-to-point local network technology that supports high-speed data storage and retrieval operations for short distances. HIPPI enables unidirectional rates reaching 800 Mbps over shielded twisted pair (STP) and works in conjunction with single-mode and multimode optical fiber. HIPPI transmissions extend to distances that reach 10 kilometers. HIPPI repeaters and extenders expand the reach of HIPPI transmissions to 20 kilometers via an optical fiber plant.

5.33.2.2 HIPPI Transmission Fundamentals

HIPPI supports transmission of streaming video, data, and audio; interoperability with such technologies as Fibre Channel and SONET/SDH; and high-bandwidth connections. HIPPI is an ANSI X3T9.3 standard. Serial HIPPI refers to the fiber-optic version of the HIPPI specification. HIPPI switches connect an array of computers and peripheral devices.

HIPPI solutions support IP datagram transmission, ARP (Address Resolution Protocol) for mapping an IP address to the appropriate Physical Layer address, and IP multicasts. The typical HIPPI local network is comprised of HIPPI switches and high-performance workstations. HIPPI utilizes a framing protocol for information transport.

5.33.2.3 Gigabit System Network (GSN) Technical Features and Functions

GSN (Gigabit System Network) is a next-generation HIPPI technology. Also called Super-HIPPI and HIPPI-6400, GSN employs a point-to-point topology for enabling unidirectional transmission of bandwidth-intensive files between network elements that can be separated by 10 kilometers. GSN fosters host-to-host communications in LAN environments that are significantly faster than Gigabit Ethernet, Fibre Channel, and HIPPI solutions.

5.33.2.4 GSN Applications

GSN (Gigabit System Network) supports fast transport of large amounts of volume-intensive information, including satellite images and real-time uncompressed HDTV (High Definition Television) programming. In addition, GSN enables applications such as data mining, data warehousing, transaction processing, seismic modeling, film production, video and film archiving, and network storage.

5.33.2.5 GSN Operations

GSN (Gigabit System Network) transmits voice, video, and data traffic in 32-byte micropackets and enables operations at the Physical Layer or Layer 1 of the Open Systems Interconnection Reference Model. GSN implementations achieve dependable and ultra-fast transfer rates with extremely low latencies by utilizing virtual channels that deliver voice, video, and data in accordance with throughput requirements. GSN also transports messages extending to 4 Gbps in length.

GSN enables full-duplex optical transport of voice, video, and data at rates reaching 6.4 Gbps. In full-duplex transmissions, GSN sets aside 3.6 Gbps overhead for encoding, addressing, signaling, and error control functions. GSN fosters operations over twisted copper connections at distances that extend to 50 meters and longer with the use of signal regenerators. GSN provides optimal transmission rates at 10 Gbps COC-192).

5.33.2.6 GSN Specification

The Gigabit System Network (GSN) specification was approved by ANSI in 1998. The National Committee for Information Technology Standards (NCITS) T11 Technical Committee develops and recommends approval of additional GSN specifications. Los Alamos National Laboratories, NASA, Disney, Ford Motor Company,

Boeing, TRW, and the European Organization for Nuclear Research (CERN) implement GSN and HIPPI solutions

5.33.2.7   HIPPI and GSN Installations

HIPPI and GSN installations employ switches, gateways, and interoperable Network Interface Cards (NICs). Initially designed for utilization in supercomputing environments, HIPPI and GSN support high-end workstation clustering, backbone connectivity, and large file transfers in sectors such as entertainment and manufacturing.

Moreover, HIPPI and GSN technologies work in concert with Ethernet, Fast Ethernet, Gigabit Ethernet, Frame Relay, Fibre Channel, IP, SONET/SDH, and ATM technical solutions and employ the ST (Scheduled Transfer) upper layer protocol. The ST protocol streamlines packet transmission by using a one-time connection setup to remember data transfers. ST technology supports the interconnection of GSN gateways to Fibre Channel SANs and FC RAID installations. ANSI also supports development of an SCSI-over-ST specification.

5.33.2.8   HPNF (High-Performance Networking Forum)

The High-Performance Parallel Interface (HIPPI) and the Gigabit System Network (GSN) are high-speed, bandwidth-intensive technologies that foster information transport at gigabit rates. These technologies support operations in high-volume networking environments and are endorsed by standards organizations such as the ITU and ANSI.

The High-Performance Networking Forum (HPNF) is an industry effort that advocates implementation of HIPPI and GSN solutions. Participants include Silicon Graphics, Sun Microsystems, Intel, Hewlett-Packard, IBM, Convex, Compaq, and Los Alamos National Laboratory.

5.33.3   INFINIBAND ARCHITECTURE

The InfiniBand Trade Association supports a specification for InfiniBand Architecture that interlinks servers with remote storage and network devices. Transmission rates reaching 2.488 Gbps (OC-48) are supported. This open channel data network architecture enables point-to-point connections over optical fiber and twisted copper pair wiring and supports multiple redundant paths between network nodes.

By using switched fabric architecture, InfiniBand technology eliminates the need to add and remove shared bus I/O cards. Because InfiniBand technology is modular and scalable, users purchase capacity as needed. In comparison to load-and-store communications approaches employed by shared bus I/O solutions, the high-performance, high-bandwidth InfiniBand solution enables low-latencies, secure operations, fast data sharing, QoS guarantees, and improved usability.

Participants in the InfiniBand Trade Association include Cisco, Hewlett-Packard, Dell, Compaq, Intel, IBM, Microsoft, and Sun Microsystems. In addition, 3Com, Adaptec, Hitachi, Lucent Systems, Nortel Networks, NEC, and Fujitsu-Siemens also take part in the InfiniBand Trade Association.

5.33.4 NETWORK ATTACHED STORAGE (NAS)

Network Attached Storage (NAS) solutions provision a shared pool of storage resources and services that are available to multiple clients and servers in heterogeneous operating system environments. With an NAS solution, problems associated with one server running out of storage capacity while other servers possess more storage capacity than actually needed are eliminated. NAS management activities are conducted from any point on the network or via the Internet. Moreover, NAS installations are optimized for economically provisioning dependable, bandwidth-intensive data delivery. In FC SANs, each network node must be equipped with software for enabling file sharing and exchange across heterogeneous networking environments. This requirement is eliminated with NAS implementations.

NAS solutions function in concert with IP-over-Token Ring, Ethernet, Fast Ethernet, FDDI (Fiber Data Distributed Interface), and ATM technologies. NAS implementations employ standards-compliant technologies, protocols, and architectures. In contrast to NAS solutions, FC-compliant devices such as switches are required for interlinking network nodes on FC SANs. As a consequence, FC SANs are typically more expensive to operate and maintain than NAS implementations.

Moreover, full-scale standards for FC SANs are still in development. As a consequence, FC SAN implementations are currently based on proprietary solutions and work best in homogeneous computing environments.

5.33.5 SERIAL STORAGE ARCHITECTURE (SSA)

Developed by IBM, Serial Storage Architecture (SSA) was promoted as a dependable and reliable high-performance storage solution in the early 1990s. As a consequence of a lack of industry support, SSA has remained a proprietary technology, with only IBM producing SSA devices.

5.33.6 SIO (SERIAL INPUT/OUTPUT) SOLUTIONS

Developed by a vendor consortium that includes Microsoft, IBM, Intel, and Hewlett-Packard, Serial/Input Output (SIO) solutions are based on next-generation I/O (Input/Output) architectures. These architectures foster transmissions at rates ranging from 500 Mbps to 6 Gbps. SIO solutions are easier to install and less costly to operate than comparable FC deployments.

5.34 FIBRE CHANNEL IMPLEMENTATION CONSIDERATIONS

Developed for enabling information technology (IT) applications, Fibre Channel networks support high bandwidth, low latencies, and quick response times for data-intensive applications and services. In serving as a campus backbone network, Fibre Channel technology provides gigabit speeds, guaranteed delivery, and scalable and dependable network performance.

A multiservice multifunctional solution, Fibre Channel networks interoperate with legacy LANs and enable high-speed, high-performance voice, video, and data transmissions. Moreover, Fibre Channel SANs interwork with next-generation networks such as Internet2 (I2).

Fibre Channel is regarded as the connectivity solution of choice for supporting applications involving high-speed storage, disk arrays, and multiple server clusters consisting of distributed storage systems attached to single servers. Moreover, Fibre Channel technology enables reliable SAN solutions and audio and video editing in the post-production process. In addition, FC is gaining recognition as a viable network backbone solution for facilitating gigabit LAN implementations.

Fibre Channel topologies vary, depending on information transport requirements of the sponsoring entity. A versatile networking solution, Fibre Channel works in concert with protocols and technologies that include FDDI, ATM, SCSI, and IP.

It is important to note that Fibre Channel specifications are not yet completely defined or implemented. FC solutions in provisioning of Quality of Service (QoS) guarantees are still in development. Moreover, FC capabilities in monitoring networked peripheral devices are not yet fully documented. In addition, FC equipment in multivendor environments is not always interoperable.

5.35 FIBRE CHANNEL SUMMARY

Designed as an all-purpose transport mechanism, Fibre Channel technology supports point-to-point connectivity and guaranteed multimedia delivery at rates that range from 100 Mbps and 200 Mbps to 4.24 Gbps. In the entertainment industry, Fibre Channel technology readily accommodates the special demands of video production and broadcast facilities. Clusters of hubs and switches employing Fibre Channel technology support advanced computing applications such as mathematical computations and simulations. Fibre Channel technology also excels in enabling high-speed, high-performance LAN implementations.

An ANSI (American National Standards Institute) and ISO (International Standards Organization) standard, Fibre Channel technology serves as the foundation for the Physical Layer or Layer 1 of Gigabit Ethernet implementations and LAN and MAN backbone networks. Designed to move large blocks of data, Fibre Channel installations employ point-to-point topologies for smaller deployments and a switched fabric for enterprisewide implementations.

Fibre Channel has a maximum transmission distance of 10 kilometers. With extenders, FC transmissions can be sent to a distance of 30 kilometers via single mode optical fiber. For longer distances, Fibre Channel works in conjunction with broadband technologies such as ATM or Gigabit Ethernet. Fibre Channel solutions also foster bandwidth-intensive operations in high-volume data environments.

FC SANs are dedicated networks for resource storage that offload traffic from multipurpose multimedia networks, transport data at extremely high rates, and perform backup and disaster recovery operations. Moreover, FC SANs are reliable, extendible, and scalable in order to accommodate increasing numbers of storage devices and bandwidth-intensive multimedia applications.

FC SANs are enablers of data warehousing, online storage, and video editing, and support massive video, voice, and data transfer for facilitating real-time telecollaboration in production-studio and broadcast environments. These flexible, efficient, and affordable special-purpose networks or subnetworks also enable storage consolidation. FC implementations enable sustained bandwidth for large file transfers and support gigabit links for mission-critical applications. Fibre Channel establishes a foundation and framework for enabling reliable and dependable SAN operations and services.

5.36   SELECTED WEB SITES

Alliance Datacom. Frame Relay Resource Center.
  Available: http://www.alliancedatacom.com/framerelay.asp
American National Standards Institute. ANSI online.
  Available: http://www.ansi.org/
AT&T. IP-Enabled Frame Relay Service.
  Available:
http://www.ipservices.att.com/products/productoverview.cfm?productid = ipframe
Equant Customer Care Center Products/Network Services/Data Services/Frame Relay.
  Available:
  http://www.equant.com/content/xml/prod_serv_frame_relay.xml
European Organization for Nuclear Research (CERN). CERN Fibre Channel homepage. Last modified in June 1998.
  Available: http://hsi.Web.cern.ch/HSI/fcs/
Fibre Channel Industry Association (FCIA) Europe. About the Fibre Channel Industry Association (FCIA) Europe.
  Available: http://data.fibrechannel-europe.com/about/index.html
Frame Relay Forum. Welcome to the Frame Relay Forum.
  Available: http://www.frforum.com/
High Performance Networking Forum (HNF) Home Page.
  Available: http://www.hnf.org/home.htm
High-Performance Parallel Interface (HIPPI). HIPPI Standards Activities.
  Available: http://www.hippi.org/
Lucent Technologies. Enhanced Frame Relay Solutions.
  Available: http://www.lucent.com/products/
Nortel Networks and EMC Corporation. Extending the Reach of Fibre Channel Storage Area Networks. Information at the Speed of Light. White Paper. August 25, 2001.
  Available: http://www.nortelnetworks.com/products/library/
nstor Technologies. Success Stories. Meeting Data Storage Needs of Seismic Proportions: A Fibre Channel RAID Solution.
  Available: http://www.nstor.com/stories.cfm
RedIRIS. About RedIRIS.
  Available: http://www.RedIRIS.es/RedIRIS/index.en.html
Storage Networking Industry Association (SNIA). SNIA Home Page.
  Available: http://www.snia.org/

Sun Microsystems. Fibre Channel vs. Alternative Storage Interfaces. An
  Overview. Technical Brief.
  Available: http://www.sun.com/storage/white-papers/fc_comp.html
University of New Hampshire Interoperability Lab. Fibre Channel Consortium.
  Available: http://www.iol.unh.edu/consortiums/fc/
WorldCom. Domestic Frame Relay Service.
  Available:
  http://www.worldcom.com/us/products/datanetworking/framerelay/domest
  ic/index.phtml

6 Digital Subscriber Line (DSL) and Powerline Networks

6.1 INTRODUCTION

The demand for inexpensive and dependable communications services that resolve local loop problems and enable access to bandwidth-intensive multimedia applications contributes to the development of residential broadband access networks. This chapter examines the capabilities of the DSL (Digital Subscriber Line) technology suite and powerline network solutions for enabling geographically separated subscribers to reliably access basic and sophisticated teleservices from dispersed residential and workplace venues.

DSL is also known as xDSL. The letter "x" stands for any one of the various DSL technologies that are grouped under the DSL umbrella. DSL technologies include ADSL (Asymmetric DSL), RADSL (Rate-Adaptive DSL), VDSL (Very High-Speed DSL), and ADSL.Lite (also known as Universal ADSL and G.Lite). In addition, HDSL1 (High Bit-Rate DSL, Phase 1), HDSL2 (High Bit-Rate DSL, Phase 2), SHDSL (Symmetric High-Bit Rate DSL or G.shdsl), SDSL (Symmetric or Single-Line DSL), IDSL (ISDN or Integrated Services Digital Network DSL), and CDSL (Consumer DSL) are also part of the DSL technology suite.

DSL technologies support links to a broad spectrum of Web portals and Internet services and applications such as e-mail exchange, IP telephony, and television programming. In addition, DSL is an enabler of tele-education, teleshopping, E-banking, teletraining, VPNs (Virtual Private Networks), video gaming, teleworking, telementoring, tele-entertainment, and telemedicine.

DSL technologies transform copper twisted pair lines into high-speed links for provisioning connections to basic and advanced communications networks via the local loop. By fostering fast information transmission over the PSTN (Public Switched Telephone Network), DSL eliminates the need for costly optical fiber installations and infrastructure upgrades.

Accelerating demand for quick and dependable access to the commodity or public Internet drives design and deployment of a wide range of powerline networking solutions as well. Powerline networks described in this chapter refer to networking configurations created by water, gas, and electric power companies with core businesses that do not fall within the traditional telecommunications domain. These broadband configurations utilize the ubiquitous powerline infrastructure to provision Internet connectivity, Web browsing, interactive videoconferencing video transport, and broadband access to multimedia applications and Web resources.

6.2   PURPOSE

This chapter provides an introduction to the family of DSL technologies and protocols. Trends in the standardization of this feature-rich technology suite are
explored. Technical attributes and capabilities of DSL technologies are presented.
DSL merits and operational constraints are indicated. Key DSL trials and initiatives
are described. Implementation considerations are reviewed. In addition, distinctive
characteristics of powerline networks are examined. Representative powerline initiatives that provision connectivity to broadband services are highlighted.

6.3   DSL FOUNDATIONS

Bandwidth bottlenecks and user frustrations drive the search for last-mile technologies that cost-effectively deliver voice, video, and data services. Data transmission
services supported by ITU-T-compliant voiceband (V.x) modems are limited in
provisioning fast transmission rates. For example, V.32 modems enable data rates
at 9.6 Kbps (Kilobits per second), V.34 modems support data rates ranging from
28.8 to 33.6 Kbps, and V.90 modems support data rates at 56 Kbps. Because the
aforementioned rates are provisioned only if the signal undergoes smooth analog-
to-digital conversion and the line quality is good, full rates are rarely attainable. The
inability of conventional voiceband modems to support reliable access to media-rich
applications contributed to the implementation of DSL solutions.

Developed by Bellcore in 1989, Digital Subscriber Line (DSL) technology
operates on telephone wires originally intended to provision voiceband communications. The term "line" refers to the PSTN link or local loop between a user location
and the local telephone exchange, also called the telephone company Central Office
(CO). Bell Atlantic and British Telecommunications launched initial DSL trials in
1993.

At the outset, DSL technology was regarded as an interim solution for transporting interactive television programming and video-on-demand (VOD) to single
family homes and high-rise apartment buildings. Fiber-to-the-home (FTTH) and
fiber-to-the-curb (FTTC) configurations were slated to replace DSL solutions. However, as a consequence of the expense and time involved in installing fiber optic
cabling directly to the subscriber location, FTTH and FTTC implementations were
not widely implemented.

In addition to FTTH and FTTC solutions, cable networks based on optical fiber
or a hybrid optical fiber and coaxial cable (HFC) infrastructure were also expected
to replace DSL as a residential broadband access solution. With the accelerating
popularity of the Web and the demand for fast access to broadband services from
the home and the workplace, cable modem technology is now a major DSL competitor solution.

In recent years, the tremendous growth in the utilization of computer networks
has contributed to a re-assessment of the distinctive features of the various DSL
technologies and their capabilities in cost-effectively delivering bandwidth-intensive

services to the customer premise. The endorsement of local loop unbundling in the United States and the European Union has also contributed to the evolvement of symmetric and asymmetric DSL solutions that vary in channel capacity, interactive capabilities, rates enabled, and maturity.

6.4   THE LOCAL LOOP

DSL technologies operate over the last mile or local loop. Local loop refers to that length of the copper phone line that interconnects the local telephone exchange and the customer premise. Loop length or the distance between a customer site and the local telephone exchange, line gauge or thickness, and line quality are among the variables that directly impact the availability and reliability of DSL service.

6.4.1   LOCAL LOOP UNBUNDLING

6.4.1.1   United States

In the United States, the Telecommunications Act of 1996 continues to accelerate DSL deployment through its support of deregulation of the communications industry and provision of regulatory and market strategies for local loop unbundling. Following the passage of this Act, Regional Bell Operating Companies (RBOCs) were required to unbundle the communications infrastructure extending from the customer premise to the local telephone exchange.

The unbundling process enabled multiple operators and service providers, in addition to the incumbent carrier, to access the twisted copper pair, the frequency spectrum of the local loop, and/or the digital bitstream for provisioning communications services and applications such as fast Web connections, interactive video-conferencing, and IP telephony. In the United States, local loop unbundling promotes competition in the broadband residential marketplace and eliminates monopolies established by the RBOCs.

6.4.1.2   European Union

The European Commission also mandated unbundled access to the local loop and elimination of monopolies maintained by incumbent operators in the European Union no later than 2000. In January 2001, Cable & Wireless Communications, First Telecom, Energis, WorldCom, and Telewest Communications initiated broadband residential trials in London, Edinburgh, Belfast, and Leeds. This process began with the submission of requests to install equipment at local telephone exchanges maintained by British Telecom, the incumbent operator.

In addition to accelerating the introduction of DSL solutions, local loop unbundling also encourages implementation of wireline and wireless cable networks. As with the DSL technology suite, cable modem technologies deliver synchronous and asynchronous multimedia traffic to disparate venues and support educational, medical, government, and business applications.

6.4.2 U.S. Federal Communications Commission (FCC)

In 1999, the U.S. Federal Communications Commission (FCC) adopted rules to promote competition for residential access networking services by directing local telephone companies to share their telephone lines with providers of high-speed Internet connections and other forms of voice, video, and data services. FCC rules enable competitive carriers to provision DSL services over the same telephony lines used concurrently by incumbent local exchange carriers to provision basic telephone service. The FCC also encourages implementation of advanced data services and enhancements to conventional POTS (Plain Old Telephone Service).

6.4.3 Deregulation and DSL Solutions

In the present-day deregulated telecommunications environment, the DSL technology suite promotes access to the commodity or public Internet; dependable connectivity to local, regional, statewide, national, and international networks; and delivery of voiceband services. DSL solutions also facilitate reliable network interconnectivity and delivery of voice, video, and data transmissions to residential sites and business locations and provision fast connections to broadband applications for an expanding client base that includes schools and universities, nonprofit agencies, and small- and medium-sized enterprises.

6.5 DSL TECHNOLOGIES

DSL technologies foster continuous connectivity to broadband applications such as Web training, virtual reality (VR) worlds, 3-D (three-dimensional) animations, interactive video-on-demand (IVOD), and dynamic learning environments. Global demand for instantaneous connectivity to Web resources and user frustration with accessing Web applications via conventional dial-up modems motivate implementation of DSL technologies for enabling high-volume information transmission via POTS links. DSL solutions are easily implemented and require only minor changes to the in-place PSTN (Public Switched Telephone Network) infrastructure. Moreover, ADSL.Lite can be implemented without any infrastructure alteration.

The PSTN employs frequencies between 300 Hz (Hertz) and 3.4 kHz (Kilohertz) of the 1.1 MHz (Megahertz) spectrum available for conventional telephony service. The DSL technology suite supports high-speed data transport simultaneously with voice calls by utilizing that portion of the bandwidth not generally used for conventional telephony service. With DSL solutions, digital data are sent in packets composed of digital bitstreams in the spectrum above the 3.4 kHz (Kilohertz) RF (Radio Frequency) band. Voice conversations occupy the spectrum between the 300 Hz and the 3.4 kHz RF frequencies.

DSL technologies are extendable, scalable, and compatible with technologies such as ISDN (Integrated Services Digital Network), IP (Internet Protocol), Frame Relay, ATM (Asynchronous Transfer Mode), SONET/SDH (Synchronous Optical Network and Synchronous Digital Hierarchy), Ethernet, Fast Ethernet, and Gigabit Ethernet. DSL installations support basic and sophisticated services and enable internetwork connectivity by leveraging the capabilities of the existing PSTN infrastructure. DSL popularity is reflected in the expanding numbers of DSL technologies and implementations.

It is interesting to note that DSL technologies differ in terms of features, functions, and operations supported. Although DSL technologies interwork with diverse narrowband and broadband solutions, these technologies are not always capable of interworking with one another.

6.5.1 DSL Transmission

DSL technologies support bus and star network topologies and virtual point-to-point links from a subscriber site such as a home, school, or business to the local telephone exchange over the existing copper wireline infrastructure. DSL deployment involves the use of a pair of compatible DSL modems or transceivers (receivers and transceivers) at the local telephone exchange and at the subscriber site for establishing a virtual private channel over the local loop.

6.5.2 Discrete Multitone Technology (DMT) Modulation

DSL modems optimize the amount of information that conventional telephone circuits can transport by using DMT (Discrete Multitone Technology). DMT modulation supports high throughput by adapting to changing line conditions and optimizing bandwidth utilization.

DMT divides available spectrum between the 0 kHz and 1.1 MHz (Megahertz) frequencies into 256 subchannels or tones for information transport. Each tone is modulated by QAM (Quadrature Amplitude Modulation), an advanced modulation scheme that creates four bits out of one baud. DMT employs QAM for optimizing signal strength against local loop impairments.

Every 4 kHz sub-channel in the RF (Radio Frequency) bands between the 0 kHz and 1.1 MHz frequencies carries up to 15 bits of data. Lower frequencies extending from 300 Hz to 3.4 kHz support POTS services. Middle frequencies are reserved for upstream and downstream full-duplex transmissions. Higher frequencies far above voice channel frequencies are set aside for bandwidth-intensive downstream transmissions.

DMT creates subchannels and divides the available spectrum on an ordinary copper wire between voice and data by using a digital technique known as Discrete Fast-Fourier Transform and C-OFDM (Coded-Orthogonal Frequency-Division Multiplexing). The European Telecommunications Standards Institute (ETSI) also endorses the use of C-OFDM for enabling delivery of DAB (Digital Audio Broadcast) with high-quality sound in mobile environments

In higher frequencies, signals are more attenuated than lower frequencies. As a consequence, DMT transmitters monitor subchannels to adjust transmissions to the characteristics of the phone line. DMT modulation supports service guarantees regardless of the distance signals travel over the local loop and provisions always-on connections. In addition, DMT modulation optimizes bandwidth utilization, enables services operating at Committed Information Rates (CIRs), and generates performance reports for network maintenance. In recognition of DMT capabilities, operational effectiveness, and robustness in noisy local loop environments, ANSI (American National Standards Institute) selected DMT as the basis of the ANSI T1.413 specification developed in 1993.

6.5.2.1 DMT and Discrete Wavelet Multitone Multicarrier (DWMT) Modulation

Based on DMT, DWMT (Discrete Wavelet Multitone Multicarrier Modulation) creates more isolation between subchannels than DMT for performance enhancement.

6.5.2.2 DMT Modulation and Carrierless Amplitude and Phase (CAP) Modulation

DMT (Discrete Multitone) modulation and CAP (Carrierless Amplitude and Phase) modulation are sophisticated digital signaling processing (DSP) techniques that enable point-to-point transmission over voice-grade telephone lines. DMT is the standard line code for DSL technologies. In addition to serving as the basis for the ANSI (American National Standards Institute) T1.413 specification, DMT modulation is also endorsed by the European Telecommunications Standards Institute (ETSI) and the International Telecommunications Union-Telecommunications Standards Sector (ITU-T). DMT manufacturers include Alcatel, Amati, Motorola, Texas Instruments, PairGain, Motorola, Orckit, ADI, and Aware.

CAP (Carrierless Amplitude and Phase) modulation is also a version of QAM (Quadrature Amplitude Modulation). In contrast to DMT, CAP is a proprietary technology. As a consequence, CAP chipsets are not always interoperable. Nonetheless, CAP modems for DSL operations are still in use by vendors such as Nokia.

In comparison to CAP, DMT transmissions are adaptable to changing bandwidth and line conditions. Moreover, DMT-supported DSL transmissions are resistant to crosstalk, universal thermal noise, impulse noise generated by electrical appliances, and RFI (Radio Frequency Interference) produced by AM radio band signals.

6.6 DSL STANDARDS ORGANIZATIONS AND ACTIVITIES

DSL technologies provide an unprecedented amount of affordable bandwidth over ordinary copper telephone lines. Standards groups in the DSL arena define specifications, procedures, methods, and approaches for supporting DSL equipment compatibility and interoperability, seamless communications over the local loop, internetworking operations, and economical and reliable multimedia delivery to the desktop.

6.6.1 COMMITTEE T1

Accredited by the American National Standards Institute (ANSI) and sponsored by the Alliance for Telecommunications Industry Solutions (ATIS), Committee T1 develops DSL specifications and clarifies approaches for DSL operations.

6.6.2 DSL FORUM

Established in 1994, the DSL Forum promotes worldwide implementation of residential broadband applications based on DSL technologies. Originally called the ADSL Forum, the DSL Forum contributes to technical advances in DSL technologies.

For example, in 2000, the DSL Forum endorsed a set of recommendations supported by the OpenDSL Consortium for enabling automatic configuration of customer premise equipment (CPE). These recommendations provision regular and dependable access to DSL services via a modem linked to an already-configured telephone line.

Study Groups and Working Groups sponsored by the DSL Forum explore DSL capabilities in interworking with ATM and IP and develop architectural specifications, interfaces, and procedures fostering DSL implementation and assessment. DSL Forum participants include Aware, Ascend, Cisco Systems, Diamond Lane, Ericsson, Nokia, Philips, Samsung, Sumitomo, 3Com, and Lucent.

6.6.2.1 Testing and Interoperability Working Group

Affiliated with the DSL Forum, the Testing and Interoperability Working Group clarifies the capabilities of DSL network configurations, interfaces, operations, and performance by developing tests for verifying DSL equipment interoperability and conformance to DSL standards and specifications.

6.6.2.2 Voice-over-DSL (VoDSL) Working Group

A DSL Forum affiliate, the Voice-over-DSL (VoDSL) Working Group sponsors development of voice-over-DSL (VoDSL) specifications. This Working Group conducts an extensive review of telephony requirements, including voice quality, reliability, and local loop issues, in order to identify VoDSL objectives, deliverables, applications, and service opportunities. Conventional DSL architecture provides one voice channel for telephony service and one large data pipe that supports a larger channel for downstream transmissions and a smaller channel for upstream data transport.

Circuit-switched VoDSL solutions convert voice signals into data-like packets that are then interleaved with other data packets and transmitted in a single DSL bitstream via the local loop. Bandwidth is dynamically allocated to various voice and data services on an as-needed basis with voice services receiving priority bandwidth allocations. Packet-switched VoDSL implementations utilize DSL bitstreams of variable length for transporting multiple voice and data packets in an integrated all-packet network environment.

The VoDSL Working Group defines parameters for Broadband Loop Emulation Services (BLES) and voice-over-MultiService Data Networks (VoMSDN). This Working Group also delineates network architecture requirements and recommendations for enabling interoperable telephony services via a DSL platform that also supports VoIP or IP (Internet Protocol) telephony applications. Based on the VoDSL Working Group's recommendations, the DSL Forum in 2000 endorsed IP (Internet

Protocol) and ATM specifications that work in concert with DSL for carrying VoDSL traffic.

6.6.3 OpenDSL Consortium and OpenDSL Initiative

Sponsored by the OpenDSL Consortium, the OpenDSL initiative streamlines the DSL implementation process and promotes widespread availability of DSL solutions in the marketplace. The OpenDSL initiative establishes a platform for easy installation of interoperable DSL equipment and auto-configuration of network elements, thereby fostering rapid service provisioning of plug-and-play DSL solutions.

The OpenDSL Consortium works in concert with the DSL Forum in developing DSL standards and specifications. Moreover, this Consortium also operates an independent certification program in conjunction with the OpenDSL Certification Laboratory for enabling vendors to verify equipment interoperability and standards conformance. Participants in the OpenDSL Consortium include DSL chipset and equipment manufacturers, service providers, and system integrators such as Cisco Systems, Intel, SBC, Efficient Networks, 3Com, Globespan, and Qwest.

6.6.4 International Telecommunications Union-Telecommunications Standards Sector (ITU-T)

The International Telecommunications Union-Telecommunications Standards Sector (ITU-T) endorses a series of Recommendations in the DSL domain. For example, the ITU-T G.992.1 Recommendation supports utilization of a filter for splitting data from voiceband signals. This process enables ADSL (Asynchronous Digital Subscriber Line) transmissions downstream or from the local telephone exchange to the customer premise at rates of 8 Mbps (Megabits per second). Moreover, the ITU-T G.992.1 Recommendation endorses the use of G.dmt (discrete multitone technology) and defines the interface between ADSL equipment and the local loop.

The ITU-T G.996 Recommendation defines methods and procedures for benchmarking performance and interoperability of DSL transceivers. The ITU-T G.997.1 Recommendation describes DSL functions in using SNMP (Simple Network Management Protocol) and DSL Physical Layer or Layer 1 operations. In addition, the ITU-T specifies approaches for enabling seamless DSL symmetric and asymmetric operations, methods for achieving higher DSL transmission rates, and procedures for supporting DSL operations over extended local loops.

6.7 ASYMMETRIC DSL (ADSL)

6.7.1 ADSL Foundations

ADSL (Asymmetric DSL) is the dominant DSL technology in the present-day DSL environment. Developed in 1994, ADSL technical capabilities are standardized by organizations that include the American National Standards Institute (ANSI), the European Telecommunications Standards Institute (ETSI), and the International

Telecommunications Union-Telecommunications Standards Sector (ITU-T). ADSL technology supports the consolidation of data, video, and voice traffic for transmission over the local loop and provisions QoS (Quality of Service) assurances.

ADSL leverages the in-place infrastructure to enable applications such as video security monitoring, interactive television programs, Web exploration, and videoconferencing, and concurrently supports the continuation of telephone conversations or fax transmissions. Officially known as full-rate ADSL, this technology is standardized in ITU-T Recommendation G.992.1. Endorsed by the ITU-T in 1998, this specification also supports utilization of a single terminal interface at the subscriber location.

6.7.2 ADSL OPERATIONS

ADSL employs a pair of modems or transceivers that are located on either end of the local loop, specifically at the local telephone exchange and at the subscriber site. ATU-R (ADSL Terminal Unit-Remote) refers to ADSL transceivers at subscriber venues. ATU-C (ADSL Terminal Unit-Central Office) refers to ADSL transmission equipment at the local telephone exchange. The ITU-T G.994 Recommendation establishes handshaking procedures for enabling dependable data exchange between ATU-R (ADSL Terminal Unit-Remote) and ATU-C (ADSL Terminal Unit-Central Office) devices.

At the subscriber premise, ATU-R (ADSL Terminal Unit-Remote) transceivers such as external and internal ADSL modems perform modulation and demodulation functions for optimizing transmission capabilities of copper wire telephone lines. In addition to ADSL modems, ADSL implementations at subscriber venues require utilization of personal computers (PCs) equipped with Ethernet NICs (Network Interface Cards) and associated wireline connections. ATU-R devices route voice, video, and data traffic to the local telephone exchange and support DSL Physical Layer or Layer 1 operations such as forward error correction and echo cancellation to minimize transmission disruptions in noisy PSTN (Public Switched Telephone Network) environments.

ATU-C (ADSL Terminal Unit-Central Office) devices include ADSL transceivers and DSLAMs (DSL Access Multiplexers) at the local telephone exchange. Virtual connections via a DSLAM interface enable connectivity between subscriber venues and the Internet or another network via the local telephone exchange. At the local telephone exchange, ATU-C devices such as hubs, bridges, and routers redirect voice calls to the PSTN (Public Switched Telephone Network) and transmit video, voice, and data to high-speed overlay IP (Internet Protocol) networks that route traffic to the public or commodity Internet or other high-speed backbone networks.

Backbone network technologies that work in concert with ADSL include ATM, SONET/SDH, IP, Fibre Channel, Frame Relay, and Gigabit Ethernet. These networks seamlessly transport ADSL frames or packets across local, municipal, and wider area networks to destination addresses. Additionally, telephone service remains available even if the ADSL transceiver at the subscriber premise is incapable of supporting data services.

6.7.3 DSL Access Multiplexers (DSLAMs)

6.7.3.1 DSLAM Features and Functions

ADSL implementation requires the installation of equipment that includes a pair of transceivers (or transmitters and receivers) at the subscriber site and the local telephone exchange. This installation serves as the foundation for virtual point-to-point DSL dedicated network connections that provision fast access to high-performance applications via the local loop.

At the local telephone exchange, DSL transmissions are redirected to high-speed backbone networking configurations or the public Internet by DSLAMs (DSL Access Multiplexers). DSLAMs also redirect voice calls to and from the PSTN, thereby eliminating the need for time-consuming dial-up operations required by conventional analog telephone modems.

In addition to traffic routing and distribution services, DSLAMs support transmission of streaming media from the Internet or other high-speed backbone network to designated DSL subscriber venues. DSLAMs enable framing operations for encapsulation of bit streams into DSL frames or packets for transmission and support basic network monitoring, administrative, and maintenance functions.

Advanced DSLAMs also support packet discards, traffic shaping, and traffic prioritization to enable reliable transmission of time-sensitive voice and video applications and data transport. These devices employ advanced protocols such as MPLS (MultiProtocol Label Switching) and work in concert with ATM SVCs (Switched Virtual Circuits) in hybrid ATM-over-ADSL configurations.

6.7.3.2 DSLAM Marketplace

Developed by Lucent Technologies and Nortel Networks, next-generation DSLAMs facilitate innovative service combinations such as DSL-over-Frame Relay (DSLoFR) and enable access to high-volume broadband applications such as IP multicasts, VOD (Video-on-Demand), and IVOD (Interactive VOD) with bandwidth assurances.

Cisco Systems provisions next-generation DSLAMs that enhance network scalability and incorporate network intelligence for enabling communications providers to support guaranteed delivery of high-quality ADSL services. Next-generation DSLAMs from Nokia enable ATM-VoDSL (ATM or Asynchronous Transfer Mode and voice-over ADSL) applications.

Available from Paradyne, ReachDSL solutions feature next-generation DSLAMs and long-loop reach technology that enables transmission between 256 and 768 Kbps at distances of 18,000 feet from the local telephone exchange.

6.7.4 POTS (Plain Old Telephone Service) Splitter

ADSL technology transforms present-day twisted pair telephone lines into virtual communications channels to support high-speed multimedia communications. ADSL installations eliminate the need to upgrade the in-place POTS (Plain Old Telephone Service) infrastructure and purchase signal regenerators. However, ADSL implementations require utilization of a POTS splitter that is integrated into the ADSL modem or transceiver at the subscriber location.

Also called a low-pass high-pass filter, a POTS splitter divides available bandwidth on a telephone line into two virtual channels for separating voice calls from data transmissions at the customer premise. In addition to eliminating signal interference, this process supports development of a virtual channel or circuit between the 300 Hz and 3.4 kHz frequencies for basic telephone service and a virtual channel of higher frequencies for multimedia transmission.

In addition, a channel separator divides the virtual multimedia channel into two virtual channels or circuits. The larger channel or circuit supports voice, video, and data transmission downstream as digital pulses in higher frequencies not used for telephone communications. The smaller circuit or channel enables full-duplex operations and information transport upstream.

6.7.5 ADSL Transmission Fundamentals

ADSL transmits delay-sensitive video, audio, and data traffic and delay-insensitive transmissions as sequences of frames containing variable-length packets. In the downstream path or from the local telephone exchange to the customer site, ADSL technology supports rates ranging from 1.544 Mbps (T-1) to 8 Mbps. For upstream transmission from the customer premise to the local telephone exchange, ADSL technology enables rates ranging from 16 to 640 Kbps.

In ADSL transmissions, frames are aggregated into blocks to which error correction codes are affixed. At the customer site, an ADSL modem corrects errors that occur during transmission based on limits previously defined by error correction codes. ADSL systems also employ advanced algorithms for enabling broadband speeds via twisted pair copper lines.

ADSL downstream speeds depend on the distance of the subscriber site from the local telephone exchange, wire gauge or thickness, and the condition of the in-place wireline plant. ADSL supports downstream rates reaching 8 Mbps (Megabits per second) at 9,000 feet, 6.312 Mbps at 12,000 feet, 2.048 (E-1) at 16,000 feet, and 1.544 Mbps (T-1) at 18,000 feet or 5.5 kilometers.

In parallel with other DSL technologies, ADSL is a high-speed always-on digital switching, routing, and signal processing technology that enables voice calls and fax transmission in RF bands between the 300 Hz and 3.4 kHz frequencies and multimedia transmission in the upper frequency range. Conventional voiceband modems compress voice, video, and data into a narrow range of frequencies for supporting information transmission via PSTN service. By contrast, ADSL employs Digital Signal Processing (DSP) for creating high-speed digital channels to optimize bandwidth capacity of copper telephone lines.

Discrete Multitone Technology (DMT) modulation enables asymmetrical information transport. It also employs a spectral mask that eliminates signal interference. When upstream and downstream frequency bands overlap in ADSL implementations, echo cancellation circuitry is incorporated into ADSL devices for diminishing the effects of signal mismatch and crosstalk.

ADSL service to a subscriber premise is provisioned via Unshielded Twisted Pair (UTP). Depending on the ADSL solutions available from the local telephone exchange, a POTS or ISDN channel can be used instead of an ADSL channel for information transport on the return path in the upstream direction.

6.7.6   ADSL and ATM

The explosion of network traffic and demand for high-speed access to broadband applications contribute to development of ADSL solutions that support interoperability with ATM configurations. ATM-over-ADSL services enable multimedia transport and applications such as interactive teleconsultations and real-time videoconferencing. In ATM-over-ADSL implementations, ATM functions as a Layer 2 or Data-Link Layer protocol over the ADSL access network and ATM cells are encapsulated in ADSL frames to facilitate transmission over the PSTN infrastructure. ADSL provisions Physical Layer or Layer 1 services. In addition, this hybrid network also supports CoS (Class of Service) with QoS (Quality of Service) assurances. The ANSI T1.415 specification standardizes ATM-over-ADSL operations.

6.7.7   ADSL Standards Organizations and Activities

6.7.7.1   ANSI (American National Standards Group) T1E1.4 Study Group

The ANSI (American National Standards Group) T1E1.4 Study Group develops DSL specifications for enabling nationwide deployment of interoperable ADSL solutions. In 1995, ANSI approved the T1.413 specification describing standardized ADSL operations. This specification also defined a single terminal interface, protocols for network operations, and parameters for network management to enable seamless ADSL implementations.

An expanded version of the original ANSI specification, the T1.413-1998 standard defines the ADSL interface with the customer premise. In addition, this standard describes enhancements to network performance, endorses ADSL transmission rates at 8 Mbps as opposed to the previously approved 6.1 Mbps rates, and supports ADSL utilization of spectral frequencies that are compatible with ATM operations. Moreover, the ANSI T1.413-1998 specification describes ATM-over-ADSL transmission services and delineates ADSL functions that are compatible with the ATM Adaptation Layer (AAL) of the ATM protocol stack and with ATM Unspecified Bit Rate (UBR) service.

The ANSI T1.413 standard is routinely reviewed, expanded, and updated to reflect technical advances. As an example, the European Telecommunications Standards Institute (ETSI) contributed an Annex to ANSI T1.413-1998 that describes requirements for ADSL implementations in the European Union. In addition, the ITU-T G.992.1 G.dmt (Discrete Multitone) Recommendation and the ITU-T G.992.2 ADSL.Lite Recommendation are based on the ANSI.T1.413-1998 ADSL specification and the work of ETSI Technical Committees.

6.7.7.2   International Telecommunications Union-Telecommunications Standardization Sector (ITU-T)

In 1998, the ITU-T (International Telecommunications Union-Telecommunications Standardization Sector) approved a series of Recommendations that describe ADSL capabilities in supporting high-speed broadband services via the local loop.

6.7.7.3   DSL Forum and the ATM Forum

Standards groups that contribute to the development of ADSL and ATM integrated services and applications include the ATM Forum and the DSL Forum. In addition, the ATM Forum defines an end-to-end ATM system that works in conjunction with ADSL technology for supporting information transport via the local loop. Originally called the ADSL Forum, the DSL Forum defines system configurations and interfaces for supporting transmission of ATM cells via ADSL networks.

6.7.7.3.1   Time-Division Multiplexing (TDM) Protocol

Endorsed by the ATM Forum and the DSL Forum, the TDM (Time-Division Multiplexing) protocol enables the encapsulation of ATM cells into ADSL packets and supports seamless information transport in mixed-mode ATM and ADSL configurations. In the downstream direction, the access node in an ATM-over-ADSL internetwork provides routing and demultiplexing capabilities. In the upstream direction, this access node performs multiplexing functions.

This approach supports secure delivery of basic and sophisticated services; access to applications such as IP multicasts, interactive games, and tele-courses; and multimedia transport with QoS (Quality of Service) guarantees. Transmission of ATM cells over an ADSL platform is consistent with the ITU ANSI T1.413 standard. At the local telephone exchange, the DSLAM functions as an ATM multiplexer to enable hybrid ATM-over-ADSL applications.

6.7.7.3.2   Point-to-Point Protocol (PPP)

The DSL Forum supports utilization of Point-to Point-Protocol (PPP) packets for encapsulating ATM cells in ADSL packets for transport via residential broadband access networks. Transmission of PPP frames in ATM-over-ADSL networks eliminates the need to run optical fiber to SOHO locations. PPP is a WAN (Wide Area Network) protocol that, in addition to ADSL and ATM, interworks with ISDN, SONET/SDH, Ethernet, Fast Ethernet, Gigabit Ethernet, and Frame Relay technologies.

6.7.8   U.S. ADSL Trials and Implementations

ADSL implementations require ADSL subscribers to be located within specified distances of the local telephone exchange. As a consequence, independent telephone companies, local exchange carriers, network vendors, and Network Service Providers (NSPs) evaluate ADSL network capabilities and performance in research trials and pilot implementations that are carried out in diverse locations among relatively small numbers of users. As an example, GTE sponsors ADSL trials at Purdue and Duke Universities; 3Com conducts ADSL trials at Princeton University; Verizon sponsors ADSL implementations at Georgetown University; and U.S. West evaluates ADSL capabilities in trials at the University of Wyoming.

6.7.8.1   Cisco Systems

Available from Cisco Systems, ATM25 ADSL modems support ATM-over-ADSL solutions that provision connections to Virtual Private Networks (VPNs) and facilitate

Quality of Service (QoS) levels for enabling telecommuting, distance learning, and video broadcast applications at rates of 25 Mbps.

6.7.8.2   GST Telecommunications

GST Telecommunications implements full-rate ADSL solutions in San Francisco that facilitate access to the GST Virtual Integrated Transport and Access (VITA) network. The VITA network supports data, voice, and video transmissions via a long-haul fiber optic ATM network that provisions communications services throughout the Western United States.

6.7.9   INTERNATIONAL ADSL TRIALS AND VENDOR IMPLEMENTATIONS

TeleDanmark, Helsinki Telephone Company, Deutsche Telekom, and France Telecom provision ADSL service throughout the European Union. In the Middle East, Bezeq, popularly known as Israel Telecom, offers ADSL service in Tel Aviv and Jerusalem. ADSL pilot trials are also conducted in Brazil, Argentina, Taiwan, Japan, Korea, Australia, and New Zealand.

6.7.9.1   Canada

6.7.9.1.1   New Brunswick Telephone (NBTEL)

In Canada, the New Brunswick Telephone (NBTEL) Company tests an ATM-over-ADSL solution for enabling access to the Video Active Network. Also known as VIBE, this high-performance network supports access to distance education teleprograms and facilitates interactive videoconferencing, E-commerce transactions, and home security services.

6.7.9.2   Finland

6.7.9.2.1   University of Tampere

In Finland, the University of Tampere conducts trials in conjunction with communications carriers and research institutions to assess the capabilities of ATM-over-ADSL configurations in enabling access to telemedicine and tele-education services and applications.

6.7.9.3   Greece

6.7.9.3.1   Hellenic Telecommunications Organization

In Greece, the Hellenic Telecommunications Organization, the National Technical University of Athens, the Aristotle University of Thessaloniki, and the Universities of Athens, Crete, and Patras evaluate ADSL capabilities in enabling access to Web services and delivery of multimedia applications such as video-on-demand (VOD).

6.7.9.4   Netherlands

6.7.9.4.1   Snelnet Project

In the Netherlands, the Snelnet Project explores the suitability of an ATM-over-ADSL platform for enabling residential users in Amsterdam and Utrecht to access music clips, concerts, telecourses, interactive movies, news items, weather-on-demand, film documentaries, cooking courses, simulations, and games. The Snelnet ATM-over-ADSL platform enables rates reaching 2.5 Mbps downstream and 384 Kbps upstream and supports links to SURFnet (National Research and Education Network or NREN of the Netherlands).

6.7.9.5   Singapore

6.7.9.5.1   SingTel (Singapore Telecommunications)

Sponsored by the Singapore government and a nationwide industry consortium, SingTel (Singapore Telecommunications) supports utilization of an ATM-over-ADSL platform that enables subscribers at SOHO venues to access Web resources, telebanking and E-commerce services, distance education courses, movies, and entertainment programs. SingTel also provisions links to Singapore ONE or 1-NET (One Network for Everyone), a public Web site featuring broadband applications.

6.7.10   ADSL IMPLEMENTATION CONSIDERATIONS

ADSL technology uses sophisticated modulation processes for transporting data, voice, and video traffic over unshielded twisted copper wire pair. This technology provides a constant connection and supports considerably faster transmission rates than a 56 Kbps analog modem. As an example, telecommuters employ ADSL solutions to access an office LAN at relatively the same rate as onsite employees using T-1 (1.544 Mbps) connections. In addition to enabling connectivity to broadband applications and transforming the way the Public Switched Telephone Network (PSTN) is employed, ADSL also makes lifeline services available via the basic telephone channel in case of emergencies.

Designed primarily for residential and SOHO subscribers, ADSL deployment typically involves expenditures for a technician to install a signal splitter at the subscriber site. In addition, ADSL implementation requires the installation of compatible ATU-R and ATU-C modems or transceivers at the customer premise and the local telephone exchange. DSLAMs at the local telephone exchange play a critical role in enabling fast connections and multimedia transmission. ADSL NSPs generally provision network management services and help-desk support.

In contrast to cable modem systems, ADSL networks establish virtual dedicated services that are not shared among multiple users. Factors that contribute to the decision to move forward with ADSL implementation include costs of the upgraded service and application requirements.

There are constraints associated with ADSL utilization. ADSL users must be within 18,000 feet of the local telephone exchange to obtain bandwidth benefits associated with this service. Users outside this radius experience degradation in bandwidth capacity and rates supported. ADSL capabilities also depend on the age and state of the local loop architecture and the quality of the telephone line. Moreover, load coils originally used to limit noise on telephone lines interfere with the quality of voice signals in upper ADSL frequencies. Power outages, snowstorms, and thunderstorms disrupt the integrity of ADSL transmissions as well.

6.8   ADSL.LITE

6.8.1   ADSL.LITE OVERVIEW

A streamlined version of ADSL, ADSL.Lite is an affordable solution for enabling fast network access, simultaneous data and voice transmissions, and always-on connections. Designed as a plug-and-play technology, ADSL.Lite eliminates the need for a splitter at the customer premise. As a consequence, ADSL.Lite is called splitterless ADSL as well as G.Lite and Universal ADSL.

6.8.2   ADSL.LITE TRANSMISSION FUNDAMENTALS

ADSL.Lite supports downstream transmission rates at 1.544 Mbps (T-1) and upstream transmission rates at 512 Kbps via unshielded twisted copper pair. ADSL.Lite enables faster transmission than analog voiceband modems operating at 56 Kbps or basic rate ISDN (BRI) supporting rates at 128 Kbps. As with full-rate ADSL, ADSL.Lite employs DMT modulation and works in concert with the in-place infrastructure at the subscriber site. ADSL.Lite is an open technology that provides an evolutionary path to full-rate ADSL deployment.

As with ADSL, ADSL.Lite is based on the ANSI T1.413 specification and supports transmission via the local loop to a distance of 18,000 feet. ADSL.Lite interoperates with full-rate ADSL as well. Full-rate ADSL and ADSL.Lite share the ability to automatically adapt transmission rates to the capability and quality of each line in order to expedite information transport. In contrast to ADSL, ADSL.Lite is less complex and features fewer overall installation requirements.

6.8.2.1   ADSL.Lite and DLC (Digital Loop Carrier) Solutions

A streamlined version of ADSL, ADSL.Lite is expected to work more effectively in DLC (Digital Loop Carrier) environments than ADSL. Also regarded as a competitor ADSL.Lite solution, DLC systems extend the length of the local loop to remote neighborhoods, thereby eliminating the need for modifications at the local telephone exchange. DLC solutions enable access to high-quality broadband services in the suburbs, new business complexes, and residential developments that are more than 18,000 feet from the local telephone exchange.

DLC installations consolidate twisted pair copper lines running between the subscriber premise and the local telephone exchange into a few transmission channels to support virtual point-to-point links. Remote Access Multiplexers (RAMs) increase DLC transmission rates. As with ADSL.Lite, DLC solutions eliminate the need for infrastructure upgrades and local telephone exchange installations such as DSLAMs.

6.8.3   ADSL.LITE STANDARDS ORGANIZATIONS AND ACTIVITIES

6.8.3.1   ITU-T (Internet Telecommunications Union-Telecommunications Standards Sector)

Endorsed by the ITU-T in 1999, the ITU-T G.992.2 Recommendation defines splitterless ADSL.Lite service and supports DMT modulation. Splitterless ADSL.Lite service indicates the absence of a separate filter at the entrance to the user site for splitting higher-frequency DSL signals from lower-frequency voiceband signals. Created as a consumer-oriented version of full-rate ADSL, ADSL.Lite technology can be implemented on a modem chipset or external modem that connects to a PC (Personal Computer) or information appliance through a Universal Serial Bus (USB) port.

6.8.3.2 Universal ADSL Working Group (UAWG)

An international consortium consisting of American, European, and Asian telecommunications carriers, the Universal ADSL Working Group (UAWG) was formed in 1998 by the DSL Forum to define parameters for ADSL.Lite operations. In 1999, the ADSL.Lite specification developed by UAWG was endorsed by the ITU-T as the G.922.2 Recommendation.

The ITU-T ADSL.Lite Recommendation supports features and functions that are compatible with the ANSI T1.413 standard. As a consequence, the UAWG endorsed the adoption of the ITU-T G.922.2 Recommendation by ANSI as an addition to the ANSI T1E.413 standard. Upon accomplishing its mandate, the Universal ADSL Working Group was disbanded and its activities were taken over by the DSL Forum.

6.8.4 U.S. ADSL.Lite Trials and Implementations
6.8.4.1 BellSouth

BellSouth evaluates ADSL.Lite functions at the Universities of Florida and Miami.

6.8.4.2 Covad Communications

Covad Communications supports ADSL.Lite services in New York City, Boston, Baltimore, Miami, Philadelphia, Pittsburgh, Minneapolis, Denver, Detroit, and Chicago.

6.8.5 International ADSL.Lite Trials and Implementations
6.8.5.1 Germany

In Germany, students at the University of Munster and Aachen Technical College participate in a Deutsche Telekom ADSL.Lite initiative. Rates at 1.5 Mbps downstream and 128 Kbps upstream are supported.

6.8.5.2 Singapore

Singapore is the first city in the Asia-Pacific region to conduct ADSL.Lite trials in consumer venues. Participants taking part in ADSL.Lite trials reside in apartment buildings situated between 7,200 and 14,100 feet from the local telephone exchange.

6.9 CONSUMER DSL (CDSL)

6.9.1 Consumer DSL (CDSL) Features and Functions

Rockwell Semiconductor Systems and Nortel Networks support development and implementation of CDSL (Consumer DSL) technology. As with ADSL.Lite, CDSL implementations support utilization of low-speed DSL modems that are bundled with traditional modem technology in all-purpose devices.

In locations where DSL services are available, CDSL technology supports downstream data rates of 1 Mbps and upstream rates of 128 Kbps at distances up to 18,000 feet. In the absence of CDSL services, CDSL modems provision access to conventional PSTN services. CDSL modems can be plugged directly into phone jacks to initiate network applications. A proposal for endorsing CDSL as a standardized low-cost splitterless DSL solution is under consideration by ITU-T Study Group 15.

6.10 RATE-ADAPTIVE ADSL (RADSL)

6.10.1 RADSL FEATURES AND FUNCTIONS

RADSL (Rate-Adaptive ADSL) technology supports symmetric and asymmetric information transmission. RADSL uses one wire pair for transmitting information at rates between 1.4 and 8 Mbps downstream and between 16 and 640 Kbps upstream. By employing DMT modulation, RADSL optimizes transport speeds and provides consistent service. In addition, RADSL adjusts information transport rates to changing line conditions and temperature fluctuations for enabling the maximum possible line speed at the time of the connection.

The ANSI T-1 Committee clarifies solutions for interfacing RADSL equipment to metallic loops using single-carrier modulation. Cisco Systems has developed an integrated router and RADSL DMT modem to support dependable data delivery, Web services, and PPP (Point-to-Point Protocol) connections.

6.11 HIGH-BIT RATE DSL, PHASE 1 (HDSL1) AND HDSL, PHASE 2 (HDSL2)

6.11.1 HIGH-BIT RATE DSL, PHASE 1 (HDSL1)

HDSL1 (High Bit-Rate DSL, Phase 1) provides an identical amount of bandwidth in downstream and upstream directions. This bi-directional symmetric transmission system is an extension of ADSL technology. However, HDSL1 and ADSL cannot function effectively in the same twisted pair bundle.

HDSL1 enables full-duplex symmetric transmission at rates reaching 384 Kbps over a single copper wire pair. By using two copper wire pairs, HDSL1 supports full-duplex rates reaching 1.544 Mbps (T-1) and 2.048 Mbps (E-1). An economical replacement for T-1 service, HDSL1 connects servers to the Internet and supports PBX (Private Branch Exchange) implementations. As with IDSL (ISDN DSL), HDSL1 also uses 2B1Q (Two Binary, One Quaternary) line-code modulation for data compression. HDSL1 transmissions extend between 12,000 and 15,000 feet without the use of repeaters for signal regeneration. The European Telecommunications Standards Institute (ETSI) supports specifications for enabling interoperable HDSL and SDH (Synchronous Digital Hierarchy) solutions.

6.11.2 HIGH-BIT RATE DSL, PHASE 2 (HDSL2)

A low-cost alternative to T-1 and E-1 solutions, HDSL2 (High-Bit Rate DSL, Phase 2) employs two pairs of twisted copper wiring for delivering full-duplex transmissions at 1.544 Mbps (T-1) and 2.048 Mbps (E-1) speeds. By expanding the usable bandwidth of a single copper pair, HDSL2 or two-wire HDSL2 fosters fast file transfers, multimedia delivery, and broadband communications services such as telecommuting, tele-education, and telemedicine. In addition, HDSL2 interoperates with other DSL deployments that employ HDSL1 and SDSL (Symmetrical or Single-Line DSL) and in-place telecommunications services. HDSL2 supports robust network performance by employing echo cancellation, burst correction, filters, and TCM (Trellis Code Modulation), a sophisticated modulation scheme that uses Forward Error Correction (FEC) to reduce local loop impairments.

6.11.2.1 HDSL2 Consortium

The HDSL2 Consortium encourages utilization of OPTIS (Overlapped PAM or Pulse Amplitude Modulation Transmission with Interlocking Spectra) modulation with HDSL2 systems. Based on 8-PAM (Pulse Amplitude Modulation) line code, OPTIS modulation maximizes network performance by reducing noise and crosstalk in wireline copper environments. This reduction is achieved by changing the HDSL1 line code from 2B1Q or 4-PAM (Pulse Amplitude Modulation) to PCM (Pulse Code Modulation) or 8-PAM. PAM converts analog signals into pulses and then transforms these pulses into 8-bit digital numbers. In 1999, the ANSI T1E1.413 Committee approved a draft standard, formally known as T1E1.4/99-006, based on OPTIS modulation.

Participants in the HDSL2 Consortium include ADC Telecommunications, Conexant Systems, Globespan, and Adtran. In addition, LevelOne Communications, PairGain Technologies, Teltrend, and Metalink participate in the HDSL2 Consortium. This Consortium conducts tests at the University of New Hampshire InterOperability Lab (IOL) for ensuring HDSL2 equipment interoperability and standards conformance.

6.12 SYMMETRIC HIGH-BIT RATE DSL (SHDSL)

6.12.1 SHDSL FEATURES AND FUNCTIONS

SHDSL (Symmetric High-Bit Rate DSL) supports fast Internet access and bandwidth-intensive applications and services such as corporate LAN connectivity, Web hosting, videoconferencing, and tele-entertainment. In parallel with other DSL technologies, SHDSL enables multimedia delivery to the customer premise, thereby bridging the gap between the optical fiber termination point at the FTTN (Fiber-to-the-Neighborhood) and the customer premise.

Also known as G.shdsl, SHDSL technology is based on HDSL2, specifications. As with HDSL2, SHDSL enables full-duplex rates at 1.544 Mbps (T-1) and 2.048

Mbps (E-1). In addition, SHDSL employs repeaters for enabling full-duplex rates at 192 Kbps over 40,000 feet. SHDSL also works in concert with ATM technology and supports the G.hs (handshake) protocol for call setup and call termination to ensure interoperable communications between compatible DSL devices.

6.12.2 SHDSL OPERATIONS

Vendors supporting SHDSL implementations include Newbridge Networks and Adtran. SBC Communications plans to integrate DSLAM equipment into remote terminals to further extend the availability of SHDSL service offerings.

6.12.3 SHDSL STANDARDS ACTIVITIES

The ITU-T Study Group 15 promotes development and acceptance of the SHDSL standard and works in collaboration with the ANSI T1E1.4 Study Committee and the ETSI TMC (Transmission and Multiplexing Committee) in facilitating SHDSL endorsement. The SHDSL specification provisions a framing mode that is compatible with multirate systems and supports utilization of TC-PAM (Trellis Coded-Pulse Amplitude Modulation) as the line code for extending the reach of SHDSL services. SHDSL is expected to provision services and applications available in HDSL1, HDSL2, SDSL (Symmetrical or Single Line DSL), and IDSL technical solutions, thereby eliminating the need for these technologies.

6.13 SYMMETRICAL OR SINGLE LINE DSL (SDSL)

6.13.1 SDSL CAPABILITIES

SDSL (Symmetrical or Single Line DSL) is a popular solution for enabling multimedia services that require identical downstream and upstream speeds. SDSL technology supports applications such as IP telephony, Web hosting, telebanking, teleworking, and videoconferencing, and works in concert with ATM and Frame Relay technologies. In parallel with ADSL, SDSL supports QoS guarantees.

SDSL consolidates data and voice traffic for effective information transmission and provisions increased bandwidth without compromising information integrity and network security. Because standards do not govern SDSL implementations, SDSL deployments are proprietary.

6.13.2 SDSL TRANSMISSION FUNDAMENTALS

As with HDSL2, SDSL (Symmetrical or Single Line DSL) enables full-duplex symmetrical transmission rates at 384 Kbps, 1.544 Mbps (T-1), 2.048 Mbps (E-1), and higher speeds in increments of 64 Kbps. In contrast to HDSL2, SDSL supports full-duplex transport via a single copper wire pair. Whereas ADSL supports services over distances that extend to 18,000 feet, SDSL transmissions are confined to distances between 10,000 and 12,000 feet. However, SDSL solutions can also support information transport over longer distances at lower rates.

6.14 VERY HIGH-SPEED DSL (VDSL)

6.14.1 VDSL TECHNICAL FUNDAMENTALS

VDSL (Very High-Speed DSL) is the fastest member of the DSL technology suite. VDSL works in concert with FTTN (Fiber-To-The-Neighborhood) and FTTC (Fiber-To-The-Curb) solutions, thereby bridging the gap between the copper wire infrastructure and the optical fiber plant. An FTTN or FTTC configuration features a mix of fiber optic cables that extend from the local telephone exchange to the neighborhood Optical Network Unit (ONU) or the last drop from a fiber optic junction point to the customer premise. VDSL connections interlink the ONU and the subscriber location.

6.14.2 VDSL APPLICATIONS

VDSL dependably supports symmetric and asymmetric transmission and advanced broadband applications such as multichannel television distribution, fast Internet connectivity, and telephony services. VDSL also enables video-on-demand (VOD), high-quality videoconferencing, HDTV (High-Definition Television) programming, and applications in telemedicine, E-commerce, and tele-education.

6.14.3 VDSL TRANSMISSIONS

VDSL supports operations in the spectrum between the 0.3 MHz (Megahertz) and the 30 MHz frequencies. Amateur radio operators and AM radio broadcasters use this RF (Radio Frequency) spectrum as well. As a result, VDSL transmissions are affected by near-end crosstalk and other forms of interference generated by adjacent radio broadcasts. To shield transmissions from interference generated by the PSTN infrastructure, VDSL technology utilizes fixed band allocations and equalizers.

VDSL is spectrally compatible with POTS, ISDN, and ADSL technologies. Asymmetric VDSL typically supports downstream transmissions at 12.96 Mbps at 4,500 feet, 26 Mbps at 3,000 feet, and 51.84 Mbps at 1,000 feet. In contrast to ADSL, VDSL supports higher data rates over shorter loop lengths. Upstream transmissions are supported at rates that include 1.6 Mbps at 4,500 feet, 2.3 Mbps at 3,000 feet, and 6.4 Mbps at 1,000 feet. In addition to the length of the local loop, VDSL transmission speeds are also dependent on the condition and gauge of the wireline infrastructure.

6.14.4 ASYMMETRIC VDSL AND SYMMETRIC VDSL

6.14.4.1 Asymmetric VDSL Capabilities

Asymmetric VDSL service is designed for utilization in multi-unit dwellings such as apartment buildings and condominiums and in multistory office buildings. In parallel with ADSL, Asymmetric VDSL technology supports asymmetrical transmission and enables delivery of high-speed, high-performance broadband applications. As with ADSL, asymmetric and symmetric VDSL solutions employ Frequency-Division

Duplexing (FDD). With FDD, signals transit distinct frequency bands upstream and downstream.

6.14.4.2 Symmetric VDSL Capabilities

In addition to asymmetric transmission capabilities, VDSL technology also supports symmetric or full-duplex transmission. VDSL technology is typically optimized for wire line lengths of less than 9,000 feet. With symmetrical or full-duplex VDSL service, transmissions at rates of 2 Mbps at 9,000 feet, 6.5 Mbps at 4,500 feet, 13 Mbps at 3,000 feet, and 26 Mbps at 1,000 feet are enabled.

6.14.5 VDSL Standards Organizations and Activities

6.14.5.1 DSL Forum

The DSL Forum supports an Emerging DSL Study Group for defining VDSL and SDSL architectures, capabilities, and services. The relationships of ADSL-to-VDSL and ADSL-to-SDSL technologies are also clarified. Approaches for extending the reach of VDSL transmissions via a hybrid optical fiber coaxial cable (HFC) infrastructure are also explored.

6.14.5.2 VDSL Alliance

The VDSL Alliance supports compliance of VDSL implementations with the ETSI VDSL Transceiver Specification (DTS 06003-2). In addition to ETSI, ANSI also endorses this DMT-compliant VDSL specification. VDSL Alliance participants include Alcatel, Newbridge Networks, Globespan, Toshiba, and Texas Instruments.

6.14.5.3 VDSL Coalition

The VDSL Coalition addresses VDSL technical issues and promotes implementation of dependable, reliable, and interoperable VDSL installations. Moreover, the VDSL Coalition identifies, defines, and recommends VDSL specifications to the ANSI T1E1.4 Working Group.

The VDSL Coalition also endorses the use of single carrier modulation (SCM) line code. SCM integrates CAP (Carrierless Amplitude and Phase) modulation and QAM (Quadrature Amplitude Modulation) for downstream VDSL transmissions. According to the VDSL Coalition, SCM solutions optimize network capabilities at high speeds over short transmission distances in the downstream direction, thereby enabling rapid reuse of the in-place local loop to support volume-intensive VDSL application requirements. ETSI endorses utilization of SCM and DMT modulation processes for VDSL implementations. The VDSL Coalition also promotes the utilization of low-cost and low-power VDSL transceivers and CAP and QAM solutions for VDSL system deployments.

The VDSL Coalition includes modem, semiconductor, and communications network equipment firms among its membership. VDSL Coalition participants include Globespan Technologies, Broadcom, Harris Semiconductor, Lucent Technologies, Metalink, Orckit Communications, and Rockwell International.

6.14.5.4   Full-Service Access Network (FSAN) Coalition

The FSAN (Full-Service Access Network) Coalition develops recommendations for standardizing VDSL operations for ANSI, ETSI, and ITU-T Study Group 15. Participants in the FSAN Coalition include Bell Canada, Malta Telecom, Bezeq, Telecom Italia, NTT (Nippon Telegraphy and Telephone Corporation), Alcatel, Fujitsu, SingTel, Swisscom, Chungwa Taiwan, and Deutsche Telekom.

6.14.5.4.1   FS-VDSL (Full Service-VDSL) Committee

Sponsored by the FSAN Coalition, the FS-VDSL (Full Service-VDSL) Committee supports FS-VDSL implementations over the local loop for enabling high-capacity, high-speed Internet connections, video entertainment, and voice telephony services, and promotes the international adaptation of an FS-VDSL standard. FS-VDSL technology builds on research accomplished by the FSAN (Full Service Access Network) Coalition.

   In addition, the FS-VDSL Committee also endorses utilization of an FTTN infrastructure that works in concert with the FS-VDSL deployment for enabling broadband transmissions.

   FS-VDSL Committee participants include Deutsche Telecom, France Telecom, Qwest, SBC, Cisco Systems, Motorola, Fujitsu, Lucent, and Nortel Networks.

6.14.6   VDSL TRIALS AND IMPLEMENTATIONS

6.14.6.1   Alcatel

Available from Alcatel, ASAM (Alcatel Subscriber Access Multiplexer) DSLAMs feature a common platform for ADSL and VDSL services for supporting ultra-high-speed Internet access and video-on-demand (VOD). Alcatel also supports ATM-over-VDSL transmission at rates reaching 60 Mbps.

6.14.6.2   Aware Communications

Aware Communications supports VDSL downstream rates at 26 Mbps and VDSL upstream rates at 3 Mbps over a single copper wire pair at distances up to 3,000 feet. Aware also provides VDSL solutions based on DWMT modulation that enable bi-directional rates of 10 Mbps over copper wire pair at distances up to 5,000 feet.

6.14.6.3   Broadcom

VDSL technologies work in concert with several modulation techniques. For example, Broadcom employs QAM (Quadrature Amplitude Modulation) for error correction and adaptive equalization to optimize VDSL performance. In addition, Broadcom supports a single chip VDSL transceiver that works in concert with FSAN (Full Service Access Network) requirements, features ATM interfaces, and fosters broadband transmission via the PSTN.

6.14.6.4   Telecom Portugal

Telecom Portugal supports ATM-over-VDSL implementations that support transmission rates at 51.84 Mbps over twisted copper pair. Students enrolled in the

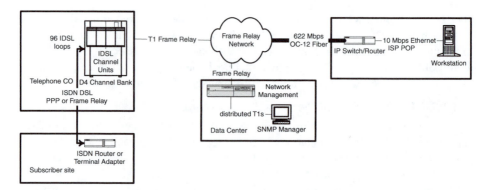

FIGURE 6.1 An IDSL (ISDN or Integrated Services Digital Network DSL), configuration. Communications carriers such as BellSouth and Verizon, and Network Service Providers (NSPs) such as America Online (AOL) and Microsoft Network (MSN), currently provision DSL services.

Department of Electronics at the University of Aveiro participate in this initiative. This VDSL configuration provisions fast connections to the Internet, library resources, and campus programs, and enables students and their teachers at public schools in Aveiro and Lisbon to access multimedia applications.

6.14.6.5 Texas Instruments

Texas Instruments offers a DSP (Digital Signal Processing) VDSL chipset based on SDMT (Synchronous Discrete Multitone) technology with TDD (Time-Division Duplexing) for enabling high-speed data, video, and voice delivery over ordinary UTP in noisy environments. With this technology, downstream speeds at 51.84 Mbps are enabled.

6.15 ISDN DSL (IDSL)

6.15.1 IDSL Capabilities

IDSL (ISDN or Integrated Services Digital Network DSL) supports information transport at 128 Kbps via a single copper wire pair. The maximum range of IDSL transmissions is 18,000 feet. Through the use of repeaters, IDSL service extends to 36,000 feet from the local telephone exchange to the customer premise. IDSL supports telecommuting, fast access to Web video applications, and connectivity to VPNs (Virtual Private Networks). (See Figure 6.1.)

6.15.2 IDSL and ISDN Parallels and Contrasts

As noted, IDSL (ISDN DSL) is an amalgam of ISDN (Integrated Services Digital Network) and DSL technologies. IDSL employs the 2B1Q line coding modulation also used in ISDN solutions. IDSL interworks with DLC and ISDN BRI (Basic Rate Interface) technologies and supports the ISDN BONDING (Bandwidth On-Demand Interoperability Group) process. As noted in Chapter 1, BONDING enables channel aggregation for achieving higher throughput via the PSTN.

In parallel with ISDN, ISDL supports autoSPID (Automatic Service Profile Identifier), thereby streamlining the service initiation process. In contrast to ISDN support of voice, video, and data services, however, IDSL supports data transport only. In order to move forward with IDSL utilization, a separate line must be installed at the customer premise.

ISDN is a switched service in which the origination and termination points support ISDN activities. By contrast, IDSL technologies enable point-to-point connections that are generally always available. IDSL charges are based on flat monthly fees rather than per minute usage.

IDSL is also known as ISDN-Basic Access (ISDN-BA) and dedicated ISDN (DISDN). ISDN and IDSL operate on telephone wires originally intended to support voiceband communications. Advances in DSP (Digital Signal Processing) technologies and innovations in line coding methods and algorithms fostered access to previously unused bandwidth capacity and enabled development of ISDN and IDSL solutions.

The ISDN component of IDSL enables services for legacy phone networks used by DLC (Digital Loop Carrier) customers. As a consequence, Lucent Systems, Adtran, and ADC support development of ISDL solutions that work in concert with DLC implementations.

6.16 ADVANCED DSL TRIALS AND VENDOR INITIATIVES

6.16.1 AWARE

Developed by Aware, wDSL (wavelet Digital Subscriber Line) provisions dynamic bandwidth on-demand and enables dependable throughput. Based on DWMT multicarrier modulation, this overlay network solution optimizes information transport in noisy environments. wDSL fosters symmetric transmission at rates of 1.544 Mbps (T-1) over a single UTP (Unshielded Twisted Pair) at distances up to 12,000 feet. With two unshielded twisted copper pairs, wDSL provisions rates at 4 Mbps at distances extending to 12,000 feet. With two unshielded twisted copper pairs, wDSL also supports transmissions at 1.544 Mbps (T-1) at distances reaching 18,000 feet.

6.16.2 NEWBRIDGE NETWORKS

Developed by Newbridge Networks, 3dSL supports voice services, VOD, IP multicasts, distribution of broadcast television and radio programming, and Internet applications with QoS (Quality of Service) guarantees via a platform consisting of ATM-over-ADSL technologies. 3dSL implementations in the United Kingdom enable tele-entertainment, teleshopping, video-on-demand, and digital multichannel broadcast programming.

6.16.3   NORTEL NETWORKS' 1-MEG MODEM SOLUTIONS

Nortel Networks provisions 1-Meg Modem solutions for SOHO venues and small- and medium-sized enterprises. An ADSL.Lite equivalent solution, 1-Meg Modem technology provisions high-speed splitterless service via digital modems, voice and data transmission via ordinary POTS lines, and always-on services. Transmission rates reaching 1.3 Mbps downstream and 320 Kbps upstream are supported over distances in excess of 18,000 feet.

The 1-Meg Modem interoperates with a maximum of two information appliances, such as two PCs or one PC and one printer. This plug-and-play solution employs 10BASE-T Ethernet protocols for enabling internetwork communications. With an inexpensive 10BASE-T Ethernet hub, two PCs readily share a single 1-Meg Modem device. For more than two devices, a router is necessary.

Because the 1-Meg modem uses spectrum in the lower frequencies to minimize signal loss, the need for a splitter and dedicated UTP (Unshielded Twisted Pair) cabling is eliminated. The U.S. Federal Communications Commission (FCC) certifies the safety of 1-Meg Modem equipment for in-home and office use.

6.16.3.1   Colorado State University (CSU)

At Colorado State University (CSU), capabilities of 1-Meg Modem solutions are evaluated in pilot tests. These tests benchmark performance of 1-Meg Modem implementations in provisioning always-on connectivity; dependable voice, video, and data services; and reliable access to Web resources via the local loop. CSU also utilizes this technology for enabling high-speed connections via ordinary twisted copper pair to clusters of adjacent buildings on campus. In the absence of FTTB (Fiber-to-the-Building) links, the 1-Meg Modem implementation at CSU is less costly to implement than T-1 leased lines and ISDN installations. In addition, capabilities of 1-Meg Modem solutions are also explored at Northern Illinois and Cornell Universities and the University of Michigan.

6.16.4   NORTEL NETWORKS AND ELASTIC NETWORKS ETHERLOOP SOLUTIONS

6.16.4.1   EtherLoop Technical Features

Developed by Nortel Networks and Elastic Networks, EtherLoop fosters transmission of Ethernet packets over standard twisted pair telephone lines at distances that extend to 21,000 feet. EtherLoop is an emerging residential broadband access solution that supports LAN extension and combines key features of Ethernet and DSL technologies. As with Ethernet, EtherLoop supports high-speed LAN connectivity and packet-switched services. In addition, EtherLoop employs Ethernet checksum for frame error checking and retransmission and readily interoperates with in-place Ethernet configurations.

As with DSL, EtherLoop is a modem-based technology that supports always-on connections and enables broadband operations via the PSTN. EtherLoop technology also interoperates with technologies such as ISDN, ADSL, ADSL.Lite, SDSL, and HDSL2. Like DSL, EtherLoop transmission speeds depend on loop length and conditions of the in-place wireline infrastructure. For example, EtherLoop technology enables information transport at rates reaching 8 Mbps at distances of 2,000 feet, 5 Mbps at distances of 8,000 feet, and 800 Kbps at distances of 21,000 feet. Also called EtherLoop DSL, EtherLoop employs spectrum between the 30 kHz and 3 MHz frequencies to facilitate information transport and utilizes adaptation algorithms along with QAM (Quadrature Amplitude Modulation) and QPSK (Quadrature Phase-Shift Keying) modulation to increase transmission rates and ensure transmission quality.

Typically, implemented in libraries, apartment buildings, multistory office buildings, and campus dormitories, EtherLoop solutions employ small portable modems developed by Elastic Networks for enabling interconnectivity with an Elastic Modem Multiplexer at the local telephone exchange. By separating voice and data calls, EtherLoop enables fast Internet access and facilitates VPN connections to corporate and academic intranets and extranets in the absence of additional software and equipment.

6.17   EUROPEAN COMMISSION TELEMATICS APPLICATIONS PROGRAM (EC-TAP) INITIATIVE

6.17.1   Cities Telecommunications and Integrated Services Project

Sponsored by the European Commission as part of the Telematics Applications Program (TAP), the CITIES initiative supported a telecommunications infrastructure featuring technologies that included ADSL, ATM, GSM (Global System for Mobile Communications), and ISDN. This multiservice infrastructure facilitated access to healthcare, tele-education, and E-government services. Local citizens, healthcare professionals, and emergency medical personnel in Rome, Marseilles, Madrid, and Brussels participated in this initiative. Multimedia kiosks, publicly available workstations, home-based PCs, and Web-adapted television sets provisioned access to CITIES resources.

6.18   EUROPEAN COMMISSION TELEWORK ONLINE PROGRAM

Sponsored by the European Commission, the European Telework Online Program supported DSL implementation for enabling telecommuters to access telecooperative team projects. In addition, this initiative provided access to tele-employment, telemarketing, and E-commerce services.

6.19   EUROPEAN COMMISSION ADVANCED COMMUNICATIONS TECHNOLOGIES AND SERVICE (EC-ACTS) PROGRAM

6.19.1   Advanced Multimedia Services for Residential Users (AMUSE)

The AMUSE initiative confirmed the capabilities of ADSL technology in facilitating access to ATM tele-entertainment, teleshopping, and tele-education services from SOHO venues. Capabilities of the ATM-over-ADSL platform in enabling transmissions at rates up to 8 Mbps downstream and 800 Kbps on the return path or upstream were verified. The Universities of Iceland, Bonn, and Stuttgart participated in this initiative.

6.19.2 BROADBAND URBAN RURAL BASED OPEN NETWORKS (BOURBON)

The BOURBON project demonstrated the viability of using an ATM-over-ADSL platform to support teleworking, multimedia content delivery, file transfer, and Computer-Supported Collaborative Work (CSCW) in small- and medium-sized enterprises.

6.19.3 TELESHOPPE

The TELESHOPPE initiative demonstrated the capabilities of ADSL, cable modem, and ISDN technologies in provisioning access to broadband teleshopping services. Project findings contributed to the development of electronic commerce applications for the European retail industry.

6.20 EUROPEAN COMMISSION INFORMATION SOCIETY TECHNOLOGIES (EC-IST) PROGRAM

6.20.1 E-PASTA

The E-PASTA initiative investigates the feasibility of implementing a secure and trusted home network platform (HNP) in smart homes via DSL, ISDN, and WAP (Wireless Application Protocol) platforms.

6.20.2 HAS VIDEO

The HAS VIDEO project explores capabilities of a tele-assistance Home Access Network (HAN) that supports delivery of real-time audio-over-IP and video-over-IP applications such as videotelephony via DSL and cable modem technologies.

6.21 DSL IMPLEMENTATION CONSIDERATIONS

DSL technologies foster dependable access to Web content and multimedia services by overcoming problems associated with Web gridlock and congestion. These technologies support teleresearch, telementoring, telemedicine, telecollaboration, teleworking, and teletraining applications and provision the bandwidth necessary for lifelong learners to participate in telecourses leading to undergraduate, advanced, and professional degrees via in-place twisted copper phone lines. ATM-over-ADSL implementations transform the PSTN into a multimedia broadband network.

In DSL transmissions, downstream rates depend on variables that include telephone wire condition, type, and thickness. Specifically, heavier 24-gauge wire is more effective in supporting information transport than thinner 26-gauge wire. UTP (Unshielded Twisted Pair) Category 5 copper wiring is currently recommended for DSL installations.

DSL transmissions via the local loop are also subject to attenuation, dispersion, and signal impairments. If the line exceeds the recommended length, the signal becomes distorted. Loading coils and telephone lines with mixed wire gauges cause reflections that interfere with the integrity of transmitted signals. Additional factors affecting information throughput include the electrical characteristics of the telephone equipment installed at the subscriber site, noise generated by wiring at the customer premise, and the presence of bridged taps in the local loop.

The process of planning, deploying, and maintaining DSL services via the PSTN is more complicated than provisioning simple dial-tone services. Any decision to implement DSL in the home, school, or workplace involves identification of the DSL technology that supports implementation goals and objectives, evaluation of the pricing structure, and assessment of DSL suitability in accommodating network performance, security, and service requirements.

Denial-of-service attacks on sites such as CNN, Ebay, and Yahoo in 2000 foster ongoing concern about the security of residential broadband access solutions supported by DSL and competitor residential broadband technologies. DSL utilization of dedicated virtual circuits contributes to the development of a more secure environment than cable modem solutions that foster point-to-multipoint shared network connectivity.

With DSL, private virtual circuits safeguard communications between the subscriber site and the Web. However, because DSL is an always-on technology, intrusion risks stem from the duration of the connection. To protect PCs from unwarranted attacks, installation of hardware firewall solutions and/or software firewall programs such as ZoneAlarm, BlackIce Defender, Norton Internet 2000, GuardDog, and VirusScan is a necessity.

6.22   COMPETING RESIDENTIAL ACCESS SOLUTIONS

Approximately 750 million phone lines are potentially capable of supporting DSL technology. However, it is important to keep in mind that DSL is not yet a universal local loop solution. DSL competes in the last-mile marketplace with competitor solutions that include Ethernet, ISDN, wireline and wireless cable networks, and Digital Loop Carrier (DLC) systems.

6.22.1   CABLE MODEM TECHNOLOGY

Currently, cable modem technology is the major DSL competitor as a last-mile broadband residential access solution. Wireline cable modem systems provide broadband Internet connectivity at high-speed rates over the in-place HFC (hybrid optical fiber/coaxial cable) infrastructure. As with DSL configurations, cable networks are always available and generally offer more raw speed than DSL technologies. However, bandwidth reductions resulting from the use of shared upstream and downstream channels by subscribers on the same network segment compromise this advantage. Actual rates supported by cable modem technology vary according to the number of users on the network segment at any particular point in time.

Safeguarding the integrity of information resources is always a critical concern in any type of network environment. As a consequence of supporting shared bandwidth and the use of a tree-and-branch topology, cable networks are not as secure as dedicated virtual point-to-point DSL systems. Shared bandwidth also raises significant concerns about privacy, information confidentiality, intellectual property protection, and the integrity of electronic commerce transactions.

6.22.2   WDSL (WIRELESS DSL)

Despite its name, WDSL (Wireless DSL) is a wireless cable MMDS (Multichannel Multipoint Distributed System) solution. WDSL supports point-to-point and point-to-multipoint connections for enabling high-speed, always-on, fast Internet access at symmetrical rates ranging from 128 Kbps to 1.544 Mbps (T-1). Designed for SOHO venues, WDSL also enables electronic commerce applications, VPN services, Web browsing, and Internet telephony.

6.22.2.1   WDSL Consortium

Organized in 2000, the Wireless DSL Consortium expects to leverage the capabilities of MMDS technology for enabling WDSL services over the last mile. Efforts are underway to develop an air interface based on the cable modem DOCSIS (Data over Cable Service Interface Specification). This specification includes an enhanced modulation scheme based on QAM (Quadrature Amplitude Modulation) and QPSK (Quadrature Phase-Shift Keying). Participants in the WDSL Forum include ADC Telecommunications, Conexant Systems, Gigabit Wireless, Intel, Vyyo, and Nortel Networks.

6.22.2.2   WDSL Vendor Activities

Wireless, Inc. supports implementation of WDSL solutions for business customers throughout West Virginia, Kentucky, and Ohio. MultiLink Wireless implements StarPort Network WDSL solutions that support rates at 512 Kbps in businesses in Fort Walton Beach and Pensacola, Florida. Vyyo, Inc. evaluates WDSL services in pilot tests in Mexico, and WorldCom sponsors WDSL trials in Boston.

6.23   DSL SUMMARY

The DSL technology suite is distinguished by its ability to support affordable and uninterrupted access to voice, content-rich data, and video applications over the PSTN. ADSL, VDSL, and SHDSL are among the most promising DSL technologies for provisioning integrated Internet, intranet, and extranet access and remote LAN connectivity.

DSL technologies transform ordinary telephone lines or twisted copper pairs into high-speed digital lines by using advanced digital line coding algorithms. DSL effectiveness in enabling digital transmission over twisted copper pairs stems from its use of DMT modulation, a process that enables one signal to modify the property of another signal.

Integration of DSL technologies with ATM core networks facilitates migration from narrowband services and analog video distribution to digital broadband services and HDTV (High-Definition Television) applications. DSL research trials and initiatives clarify practical requirements, risks, challenges, and opportunities associated with DSL deployment in actual environments.

The soaring popularity and profusion of Web multimedia services and applications contribute to Internet congestion and traffic gridlock. These bottlenecks undermine the capability of the underlying infrastructure to support dependable and reliable real-time voice, video, and data delivery to the home, school, and workplace and generate demand for DSL deployment. In addition, delays in laying fiber optic cabling over the last mile contribute to the popularity of DSL deployments.

An analog-to-digital transmission technology, DSL employs special modems or transceivers (transmitters and receivers) that are attached to twisted copper pair, thereby eliminating the need to rewire the line. DSL supports Internet access by rerouting traffic from voice to data networks so that phone service is not disrupted. Voice, video, and data in a DSL configuration travel to destination addresses via virtually dedicated point-to-point connections. As a consequence, DSL implementations provision higher levels of security dial-up than modem, cable modem, and powerline networking installations.

DSL solutions enable multimedia applications that include telebanking, teleshopping, video-on-demand (VOD), and remote LAN and WAN interconnectivity. As noted, there are distance limitations with DSL. For example, the maximum reach for ADSL service is 18,000 feet. Beyond this distance, information throughput diminishes significantly.

The DSL suite operates in a relatively noisy and hostile environment. As a consequence, specifications for DSL modems at the Physical Layer or Layer 1 delineate procedures for accommodating crosstalk and impulse noise, support latency techniques to minimize the delay in voice transport, and feature algorithms for forward error correction and line coding to achieve efficient transmission.

As noted, transmission via the existing copper plant is subject to an array of impairments that constrain operational effectiveness. Components in a POTS infrastructure that adversely affect DSL transmission include surge protectors, bridged taps, loading coils, and radio frequency interference (RFI) filters. Moreover, DSL service, transmission rates, and signal strength are impaired by impulse noise generated by powerlines, lightning strikes, broadcast transmitters, and fluorescent lighting. The state of the copper infrastructure and condition of the local loop affect the quality of transmissions as well. DSL solutions are designed to be extendible, interoperable, scalable, and easily implemented. However, installation of ADSL equipment such as a splitter at the customer premise requires the service of a technician.

Evaluation of DSL performance in pilot tests and research trials through the collection and analysis of traffic statistics is critical in supporting the ongoing evolvement of DSL technologies. Documentation of DSL capabilities contributes to further DSL enhancements, growing the in-place DSL infrastructure, and providing QoS (Quality of Service) guarantees for DSL feature-rich transmissions. Clearly, DSL faces competition in the networking arena. Competitor technologies include ISDN, satellite, wireline and wireless cable, and powerline solutions. Nonetheless, the popularity of DSL pilot tests and implementations demonstrate the accelerating popularity of DSL deployments and the viability of the DSL technology suite.

DSL solutions support dependable and affordable high-speed voice, video, and data services at the customer premise without costly infrastructure upgrades. However, research and experimentation in testbed environments remain necessary to standardize equipment and optimize DSL network performance. DSL standards continue to evolve in response to subscriber demand for access to current and emergent high-speed multimedia broadband services.

6.24   OVERVIEW OF POWERLINE NETWORKS

As noted, demand for ready access to current and emergent broadband networks and bandwidth-intensive interactive multimedia applications contributes to exploration and assessment of the DSL technologies, satellite services, wireline and wireless cable networks, and ISDN as last-mile solutions for overcoming limitations of the Public Switched Telephone Network (PSTN). Demand for residential access to broadband services generates development of powerline network solutions as well.

Powerline networks use the in-place wireline utility infrastructure for provisioning multimedia applications and network connectivity. Multiservice utility companies operating core businesses that do not fall within the traditional telecommunications domain are key supporters of these installations. The following sections examine, the expanding role of powerline networks sponsored by subsidiaries, partners, and affiliates of multiservice utility companies; introduce powerline networking fundamentals, capabilities, and constraints; and review the national and international powerline networking initiatives as well as the work of standards organizations in enabling interoperable powerline networks.

6.25   POWERLINE NETWORK FOUNDATIONS

Initially, powerline networks employed communications technologies to support implementation of automated meter reading systems for monitoring electricity consumption. Data were collected at the meter and then transmitted to the utility via low-voltage powerlines that also delivered electricity. In past decades, government-owned and government-supported public utility companies were distinguished by their monopoly status.

In the present-day environment, these monopolies are undergoing massive structural reorganizations in response to free-market forces. Diverse factors ranging from rate restructuring and downsizing to telecommunications deregulation and local loop unbundling promote utilization of in-place utility network infrastructures for provisioning access to telecommunications services and applications.

6.25.1   TELECOMMUNICATIONS ACT OF 1996

The Telecommunications Act of 1996 enables utility companies to participate in the telecommunications sector as long as telecommunications operations are not subsidized with funds from electric power operations. With the passage of this act, the FCC mandates the unbundling of different network elements, including the local loop, as well as the deregulation of utility companies. These entities are currently exploring solutions for provisioning basic and value-added communications services, from cable television to IP telephony. In addition, approaches for dependably delivering data, video, and voice transmissions at high speeds to SOHO and business venues via powerline networks are also examined.

6.26 POWERLINE NETWORK TECHNICAL FEATURES AND FUNCTIONS

A rapidly evolving local loop communications solution, powerline networks utilize in-place electrical wiring, adapters, and software products to support network operations and multimedia delivery to SOHO (Small Office/Home Office) venues. Powerline networks use in-place powerlines as communications channels through which electrical energy is transmitted for enabling voice, video, and data transport. At SOHO sites, external powerline modems are connected to PCs (Personal Computers). Links to electrical outlets and to the protocol translator that is attached to the home electricity meter are also established. The protocol translator sends electric signals over powerlines to a transformer at the public utility or powerline company substation. At the public utility substation, transformers change the values of electric current and voltage at which electrical energy is transmitted to voice, video, and data signals for transmission via the data concentrator. Subsequently, these signals travel over twisted copper pair or fiber optic cabling to destination addresses such as corporate intranets and extranets or the Internet.

Because powerline networks use already installed wiring for supporting transmission at the Physical Layer or Layer 1 of the OSI Reference Model and communications channels are in-place, the need for new wires, infrastructure upgrades, rights-of-way negotiations, and new construction is eliminated.

IBM demonstrated the viability of powerline solutions in the 1990s with the implementation of Arigo technology. This technology supported information transport over a 220-volt powerline for activating lights and security systems and enabling information exchange among intelligent home appliances that were plugged into electrical outlets.

6.27 POWERLINE HEALTH AND SAFETY ISSUES

6.27.1 FEDERAL COMMUNICATIONS COMMISSION (FCC)

In response to public health concerns about powerline utilization and the relationship between electromagnetic radiation fields that emanate from powerlines and cancer, the FCC develops specifications for maximum signal levels at specified frequencies for powerline operations. Powerline equipment must also be tested and certified acceptable by the FCC prior to deployment.

6.27.2 UNITED KINGDOM (U.K.) RADIOCOMMUNICATIONS AGENCY

According to the United Kingdom (U.K.) Radiocommunications Agency, radiation emission levels associated with powerline services are above recommended safety levels in the present-day environment. As a consequence, powerline technology vendors in the United Kingdom are developing new versions of powerline equipment with lower radiation emission levels. European legislation also prohibits the high-frequency modulation required for voice telephony services in present-day powerline installations.

6.28 POWERLINE STANDARDS ORGANIZATIONS AND ACTIVITIES

6.28.1 CEBus (CONSUMER ELECTRONIC BUS) SPECIFICATION

Developed by CIC, an international organization that promotes implementation of interoperable in-home powerline networks, the CEBus (Consumer Electronic Bus) specification for powerline carrier technology supports utilization of in-home, 120-volt, 60-cycle electrical wiring for transporting information between HAN (Home Area Network) devices. Formally known as EIA (Electronic Industries Alliance) IS-600, this specification is based on the utilization of spread spectrum technology for overcoming noise, signal attenuation, and signal distortion in HAN electrical powerline connections.

6.28.2 CONSUMER ELECTRONICS ASSOCIATION (CEA) HOME NETWORKING COMMITTEE (HNC)

The Consumer Electronics Association (CEA) Home Networking Committee (HNC), also called the R7 HNC, supports the development of standards, installation guidelines, and interfaces for an integrated home network that facilitates voice, video, and data transmission via powerlines. Standards for in-home powerline networks that support connectivity of PCs, peripherals, and consumer electronic devices; compliance with FCC guidelines; multimegabit transmission rates; and applications that include streaming audio and video and IP telephony are under consideration. Participants include Intel, Adaptive Networks, Sony, Cisco, and 3Com.

6.28.3 EUROPEAN TELECOMMUNICATIONS STANDARDS INSTITUTE (ETSI)

The European Telecommunications Standards Institute (ETSI) sponsors a powerline network project that facilitates development of specifications and standards for provisioning voice and data services via in-place private and public powerline networks. These standards support equipment interoperability, co-location of powerline networks and residential broadband access configurations based on technologies such as DSL and cable modem within the same in-home environment, and electromagnetic compatibility to avoid interference between powerline networks and other technologies operating in the same spectral frequencies.

6.28.4 HomePlug Powerline Alliance

A vendor consortium established in 2000, the HomePlug Powerline Alliance supports the design and implementation of in-home powerline networks. This Alliance also promotes utilization of standards-compliant HAN powerline products and equipment. Moreover, the HomePlug Powerline Alliance encourages development of scalable and extendible enabling technologies for home powerline configurations and adoption of HAN standards. Participants in the HomePlug Powerline Alliance include Panasonic, Texas Instruments, Cisco Systems, and Motorola.

6.28.4.1 HomePlug 1.0 Specification

In 2001, the HomePlug Powerline Alliance formally endorsed the HomePlug 1.0 Specification. This high-speed specification supports home powerline operations at 14 Mbps and development of interoperable devices such as PCs and peripherals that can connect to a home network via electrical outlets. Designed to accommodate the HomePlug Market Requirements Document (MRD), the HomePlug 1.0 Specification enables in-home applications such as streaming media, bulk file transfer, and IP telephony, and establishes requirements for interoperable HomePlug equipment.

6.28.5 International Powerline Communications Forum (IPCF)

The International Powerline Communications Forum (IPCF) is a worldwide forum that supports the global development and implementation of powerline network services and communications products.

6.28.6 PLC*forum*

Established in 2000, the PLC*forum* facilitates development of international powerline network solutions and specifications.

6.29 NATIONAL POWERLINE NETWORKING TRIALS AND VENDOR IMPLEMENTATIONS

Recent changes in regulatory and business environments enable utility companies to compete with established telecommunications providers. Utility companies active in the telecommunications domain typically outsource telecommunications functions to a third party or independently implement a full-service powerline telecommunications infrastructure. Powerline infrastructures generally feature high-speed fiber optic lines, underground conduits, and electricity poles, and employ spread spectrum technology to enable secure and robust network transmissions.

Vendors and communications carriers active in the powerline network arena include Ericsson and Siemens in the European Union and Korea Telecom in South Korea. Available from vendors and manufacturers such as VideoCom and Pulsar Technologies, powerline devices include chipsets that support transceiver functions, Data-Link Layer or Layer 2 application processors, modems, filter-suppressors, power transformers, interrupters for power shutdown, and electrical circuit analyzers that are specifically designed for powerline networking operations. Protective equipment such as AC (Alternating Current) powerline surge suppressors and filters safeguard network transmissions from AC spikes and eliminate noise generated by loose electrical connections, defective plugs and cords, and cellular phones. Representative powerline network implementations and vendor solutions are highlighted in this section.

6.29.1 ADAPTIVE NETWORKS

Powerline installations from Adaptive Networks foster data communications rates reaching 115 Kbps for enabling commercial, industrial, and consumer applications. These installations employ spread spectrum technology for enabling secure information transport. Chipsets support functions carried out by powerline transceivers. Adaptive Network solutions are also immune to powerline noise and attenuation, feature forward error correction and modulation and demodulation capabilities, and support up to 65,500 nodes on each powerline substation network segment.

6.29.2 AFN

A public utilities holding company, AFN leases telecommunications capacity to NSPs (Network Service Providers), competitive local exchange carriers (CLECs), and wireless communications companies. AFN powerline network solutions facilitate bandwidth-intensive voice, video, and data applications. These network solutions are implemented in small- and medium-sized underserved communities, including Morgantown, West Virginia, and Binghamton, New York.

6.29.3 ALCATEL

Alcatel provisions home powerline solutions for medium-voltage networks. In Germany, Alcatel supports LineRunner PDSL (PowerlineDSL) for enabling voice and data transmission on a 10,000- to 20,000-volt network. These medium-voltage networks support transmissions over distances that extend to 400 meters. The LineRunner PDSL solution enables voice, data, and video services via powerline networks that facilitate full-duplex or symmetrical rates reaching 2.32 Mbps. PDSL modules and equipment conform to safety regulations established by the German Technical Testing Agency for enabling operations on a 4,000- to a 20,000-volt powerline network.

6.29.4 ALTCOM

Altcom provisions a powerline LAN solution that features the same networking capabilities as conventional Ethernet LANs. With an Altcom solution, an in-home network is readily setup by plugging network-ready devices into electrical outlets. Applications such as remote file access and Web connectivity are concurrently supported. Altcom powerline networks operate over 220-volt powerlines and support data rates at 115 Kbps.

6.29.5 AMERICAN ELECTRIC POWER

One of the largest energy producers in the United States, American Electric Power replaced an aging microwave system with a high-speed fiber optic SONET backbone network. This optical fiber network supports broadband communications services in Charleston, West Virginia, and Columbus, Ohio.

6.29.6 C3 COMMUNICATIONS

Based in Austin, Texas, C3 Communications is a telecommunications subsidiary of Central and SouthWest Corporation (CSW), a Dallas-based public utility holding company. C3 Communications maintains a network division and a utility automation division. The utility automation division provides meter reading services for commercial and industrial clients; electric, gas, and water utility firms; and other energy service providers. The C3 Network Division enables communications carriers and NSPs in Dallas, Houston, San Antonio, Abilene, San Angelo, Laredo, and Corpus Christi to purchase excess optical fiber capacity on the C3 high-capacity fiber optic plant for providing in-home powerline network solutions.

6.29.7 CALIFORNIA PUBLIC UTILITIES COMMISSION

After the California Public Utilities Commission eliminated the monopoly status of utility companies, San Diego Gas and Electric initiated delivery of broadband communications services to customers in San Diego County and the southern portion of Orange County via the in-place optical fiber powerline plant. This network enables voice, video, and data transmission and facilitates access to broadband communication services and applications.

6.29.8 CITIZENS UTILITIES

Citizens Utilities provisions public utility services such as electric, gas, and water distribution and wastewater treatment. A subsidiary of Citizens Utilities, Electric Lightwave provides broadband powerline networking solutions in the Western United States. Proceeds from public utility divestitures contribute to the expansion of Electric Lightwave service offerings.

6.29.9 CITY OF VINELAND, NEW JERSEY, AND THE MUNICIPAL ELECTRIC COMPANY

The City of Vineland, New Jersey, the municipal electric company, the school district, and local government agencies joined together in creating a scalable and extendible ATM metropolitan network. This municipal network employs the in-place optical fiber powerline infrastructure for provisioning broadband voice, video, and data applications.

6.29.10 ENIKIA

Enikia supports an IAN (Information Access Network) powerline networking solution. This solution involves the use of Enikia powerline transceivers that work in concert with standard Ethernet controllers for enabling voice, video, and data transmission at 10 Mbps. An IAN installation readily integrates smart devices plugged into electrical outlets throughout the home into a powerline network configuration that supports E-commerce applications, home security teleservices, tele-entertainment, and utility resource management.

6.29.11 FLORIDA POWER AND LIGHT COMPANY (FP&L)

Florida Power and Light Company (FP&L) responded to deregulation in the utility and telecommunications domains by implementing a fiber optic network system that supports WDM (Wavelength Division Multiplexing), SONET, and ATM technologies. This infrastructure provides intrastate utility service. Excess capacity supports high-capacity broadband networking solutions.

6.29.12 INARI

A developer of powerline home networking solutions, Inari (formerly called Intelogis) provisions powerline networking solutions for transporting high-speed digital signals via existing AC electrical wiring in homes and high-rise condominiums. Inari supports development of powerline chipsets for communications providers and utilization of plug-in powerline networking solutions that transform every home electrical outlet into a network connection. Inari powerline network solutions support FDMA (Frequency-Division Multiple Access) and TDMA (Time-Division Multiple Access) modulation schemes.

6.29.12.1 Inari and Globespan

Inari and Globespan support a high-speed integrated powerline home network solution that fosters connectivity to DSL installations. By integrating powerline and DSL technologies, this solution enables multiple users of a home powerline network to connect simultaneously to the Web via one common Internet service account. Generally, a single provider provisions telephony service, network interconnectivity, and entertainment applications for Inari and Globespan residential networks.

6.29.13 INTELLON

The HomePlug 1.0 Specification is based on the Intellon powerline solution. Intellon powerline networks support operations at the Physical Layer or Layer 1 and the Data-Link Layer or Layer 2 of the OSI Reference Model. Moreover, Intellon powerline home networks support QoS assurances, voice-over-IP (VoIP) services, and streaming media, and comply with FCC emission standards. Typically, Intellon networks enable services in a 100 MHz block of spectrum between the 4.3 MHz and 20.9 MHz frequency bands. These networks employ CSMA/CA (Carrier Sense Multiple Access/Collision Avoidance) for enabling seamless transmissions and use DBPSK (Differential Binary Phase Shift Keying) and DQPSK (Differential Quadrature Phase Shift Keying) modulation.

6.29.14 LOWER COLORADO RIVER AUTHORITY

The Lower Colorado River Authority implements high-speed, high-performance metropolitan networks that use the in-place utility infrastructure for delivery of broadband applications to businesses, hospitals, medical clinics, chambers of commerce, government agencies, and K–12 schools and school districts.

6.29.15 NEWBRIDGE NETWORKS

Newbridge Network employs OFDM (Orthogonal Frequency-Division Multiplexing) technology for optimizing available bandwidth in powerline networking configurations.

6.29.16 NOR.WEB DPL (DIGITAL POWERLINE) SOLUTIONS

Digital PowerLine (DPL) solutions from Nor.Web DPL, a joint venture between United Utilities and Nortel Networks, supports information transport at 1 Mbps over powerline networks. This solution also requires the installation of base stations that are situated between the subscriber premise and the powerline company substation transformer. These base stations foster transmissions that extend to 300 meters. Nor.Web DPL supports information transport by translating radio frequency (RF) transmissions into data streams that transit powerlines directly adjacent to the electrical current.

The Nor.Web DPL solution features a network interface at the power company substation, a stand-alone powerline communications module, and a data unit at the customer premise. This configuration supports home automation, data backup and security services, E-banking, tele-entertainment, and high-speed Internet connectivity. Nor.Web DPL services are available in Singapore, Canada, the United Kingdom, Sweden, the Netherlands, and Germany.

6.29.17 SOUTH CAROLINA ELECTRIC & GAS (SCANA) COMMUNICATIONS

A South Carolina holding company with energy-based businesses, South Carolina Electric & Gas (SCANA) Communications uses an installed base of more than 4,500 miles of fiber optic cabling for broadband network services. SCANA also supports long distance telephone services and holds licenses for PCS (Personal Communications Services) in Memphis, Birmingham, Jacksonville, and Atlanta.

6.30 INTERNATIONAL POWERLINE IMPLEMENTATIONS

6.30.1 MAINE.NET COMMUNICATIONS PLC

An Israeli broadband communications service company, Main.net Communications PLC supports powerline networking solutions that operate over low-voltage powerline grids. Access to the Maine.net powerline network is achieved by plugging a specially adapted powerline modem into a home electrical outlet. This modem is also connected to a PC (Personal Computer) that is equipped with a USB (Universal Serial Bus) or an Ethernet NIC (Network Interface Card). Rates at 2.5 Mbps for supporting videoconferencing, IP telephony, and data transmission are supported. Main.net Communications PLC partners with local communications providers and utility companies to provision home powerline networking solutions in Germany, Poland, Sweden, Spain, France, Norway, Finland, and the Netherlands.

6.31 EUROPEAN COMMISSION INFORMATION SOCIETY TECHNOLOGIES (EC-IST) PROGRAM

6.31.1 PALAS (POWERLINE AS AN ALTERNATIVE LOCAL ACCESS)

The PALAS initiative supports development of a suite of service packages for the powerline communications market. This product suite provisions a technical foundation for the rapid deployment of powerline communications solutions.

6.32 ADDITIONAL OPTIONS: WATERWAY NETWORKS

In the European Union, inland waterways are distinguished by their continuity, lack of obstructions or competing infrastructures, and penetration into city centers. As a consequence, approaches for creating inland waterway telecommunications networks are in development.

As an example, in 1998, a consortium that includes Alcatel, LD COM (Louis Dreyfus Communication), VNF, and WorldCom installed a network of fiber-optic cabling along the bed of the River Seine in Paris and in the City of Paris for enabling tele-applications and teleservices. This consortium plans to extend the inland waterway network to adjacent regions in order to interlink major European cities, including Lille, Calais, Strasbourg, Lyon, Montpellier, Arles, and Geneva in a Regional Area Network (RAN) configuration.

6.33 POWERLINE NETWORK IMPLEMENTATION CONSIDERATIONS

Powerline networks function in harsh environments. Communications carried on AC (Alternating Current) powerlines are subject to spikes and surges that impair network performance. Atmospheric conditions such as hurricanes, tornadoes, thunderstorms, snowstorms, and lightning disrupt powerline network communications as well. Unpredictable noise generated by smart appliances contributes to multipath delay, signal attenuation, and low data rates. Power grid transformers also interfere with information transport by mixing up data signal sequences.

All users connected to the same powerline substation share bandwidth. As more users are added to the same network segment, the rate of transmissions for supporting bandwidth-intensive services such as media streaming declines. Powerline networks are not yet widely implemented. Although powerlines are in place, communications providers must still send out an individual to install a box near the electricity meter for initiating network services.

In the European Union, the power grid is based on 220-volt transformers. A single 220-volt transformer provisions teleservices to several hundred homes. By contrast, the power grid in the United States is based on 110-volt transformers. A single 110-volt transformer serves just ten homes. As a consequence, powerline network solutions in the United States are costly. Utility companies and communications providers are currently investigating the feasibility of implementing mixed-mode wireline and wireless powerline networking solutions.

6.34   POWERLINE NETWORK SUMMARY

Deregulation in the telecommunications sector contributes to the popularity of DSL and cable modem implementations as well as the utilization of powerline networks in HAN (Home Area Network) configurations for enabling high-speed, high-capacity broadband service delivery. Generally, utility companies deliver value-added telecommunications services via residential broadband access networks by selling excess capacity to communications carriers, establishing partnerships with companies in the communications sector, or creating subsidiaries and affiliates.

Regardless of location, homes are connected to electric utilities. In contrast to phone jacks, electrical outlets for powerline networks are situated throughout a customer premise. As a consequence, access to the powerline infrastructure for transporting high-speed, high-capacity multimedia services directly to HANs is readily available.

In this chapter, the expanding role of powerline networks sponsored by subsidiaries, partners, and affiliates of utility companies is examined. Powerline networking fundamentals, capabilities, and constraints are explored. Powerline networking attributes and initiatives are reviewed. The work of standards organizations in enabling interoperable powerline networks is noted. Recent advances in waterway networks are highlighted as well.

6.35   SELECTED WEB SITES

Adaptive Networks. Powerline Communications Networking.
   Available: http://www.adaptivenetworks.com/
DSL.com. Home Page.
   Available: http://www.dsl.com/
DSL Forum. Whitepapers.
   Available: http://www.adsl.com/
Consumer Electronics Association. Home Page.
   Available: http://www.ce.org/
Enikia. Solutions for Powerline Networking.
   Available: http://www.enikia.com/index.html
European Committee for Electrotechnical Standardization (CENELEC). About
   CENELEC.
   Available: www.cenelec.org/Info/about.htm

European Telecommunications Standards Institute (ETSI). PLT Summary. Last
modified on October 18, 2001.
Available: http://portal.etsi.org/plt/Summary.asp
European Telecommunications Standards Institute (ETSI). Telecom Standards.
Tutorial on xDSL Technologies.
Available: http://www.etsi.org/getastandard/home.htm
Everything DSL. Solutions for Everything in DSL.
Available: http://www.everythingdsl.com/
HomePlug Powerline Alliance. Home Page.
Available: http://www.homeplug.org/
Inari. Home Page.
Available: http://www.inari.com/
International Powerline Communications Forum. Powerline World: The PLC
Community.
Available: www.ipcf.org/
Net Access Corporation. DSL Reports.
Available: http://www.dslreports.com/
Orckit Communications, Ltd. DSL Knowledge Center.
Available: http://www.orckit.com/fr_newsa.html?/knowledge.html
Pacific Bell. DSL Internet Center.
Available: http://www.pacbell.com/DSL/
Paradyne Technology. DSL Technology.
Available: http://www.paradyne.com/technology/index.html
Powerline Communications Forum. Home Page.
Available: http://www.sigma-consultants.fr/plcforum/index.htm
Telechoice. DSL Background Information.
Available: http://www.xdsl.com/content/backgroundinfo/
Verizon. Residential and Business Applications.
Available: http://www.gte.com/dsl/industryapp-res.html

7 Cable Networks

7.1 INTRODUCTION

Consumer demand for ready access to Web resources, multimedia applications, and information on every topic imaginable undermines the capabilities of the PSTN (Public Switched Telephone Network) in supporting real-time and dependable voice, video, and data delivery via the local loop. At SOHO (Small Office/Home Office) venues, wireline transmission speed for accessing Web services and applications is ultimately dependent on the condition of the local loop. As noted, the local loop enables connections over the first mile between the subscriber premise and the local telephone exchange, also called the telephone company central office (CO).

At the local telephone exchange, servers and routers aggregate information for transport at multimegabit and multigigabit rates to and from high-performance networks. However, transmission rates between a SOHO site and the local telephone exchange default to transmission speeds enabled by local loop technology.

The inability to provision low-cost, high-speed network access over the local loop contributes to the implementation of DSL (Digital Subscriber Line), powerline, and cable network solutions. These broadband residential access networks overcome local loop constraints and enable interconnectivity over the first mile to high-speed, high-performance broadband networks.

7.2 PURPOSE

This chapter features an examination of the distinctive characteristics of wireline and wireless cable networks. Technical advances in cable networks are examined. Specifications, recommendations, and guidelines developed by standards organizations in the wireline and wireless cable networking domain are described. Distinctive attributes of wireline and wireless cable system configurations and representative examples of cable networking field trials and implementations are presented. Capabilities of wireline and wireless cable networks in enabling high-speed, high-performance teleservices are delineated. Representative wireline and wireless cable networking initiatives are highlighted.

7.3 FOUNDATIONS

Cable modem systems are increasingly popular technical solutions for accommodating trouble-free transmission requirements of delay-sensitive voice, video, and data traffic. Originally consisting of sealed coaxial cable lines, the cable infrastructure was installed to support CATV (Community Antenna Television) programming.

Initially, cable systems re-transmitted broadcast television signals via sealed coaxial cable links to single-family homes and apartment buildings that were unable to obtain quality television reception. As part of the implementation process, cable operators also installed large antennas on rooftops to improve cable broadcast reception.

In the 1940s, the National Television Systems Committee (NTSC) authorized the allocation of a 6 MHz (Megahertz) segment of the electromagnetic spectrum for each television channel. Cable television programming was introduced in Oregon and Pennsylvania in the 1950s. In the 1970s, cable operators expanded programming by televising satellite broadcasts and offered pay-per-view services to cable subscribers.

Beginning in the 1980s, the cable network platform was upgraded from coaxial cable to an HFC (Hybrid Optical Fiber Coaxial Cable) infrastructure to support delivery of additional cable television programs and enable improvements in signal quality by reducing line noise and signal attenuation. Auctions of spectral bands by the FCC (Federal Communications Commission) were initiated in 1996. At that time, the FCC also ruled that wireless cable providers could offer digitized services as long as operations of adjacent systems were not adversely affected. By the turn of the century, wireline and wireless cable networks routinely facilitated access to voice, video, and data services and interactive broadband applications in fields such as tele-education, telemedicine, E-business, and tele-entertainment.

7.4   TELECOMMUNICATIONS ACT OF 1996

With the passage of the Telecommunications Act of 1996, statutory and legal barriers preventing cable operators from entering the telephone industry and telephone operators from entering the cable industry were eliminated. This Act also paved the way for local loop unbundling, enhancements to the wireline cable infrastructure, and innovative applications such as IP (Internet Protocol) telephony, desktop videoconferencing, and WebTV.

Currently, traditional cable services are available from conventional telephone companies such as Ameritech, GTE, Southern New England Telephone (SNET), BellSouth, and SBC. Similarly, conventional cable operators such as Cox Communications support residential and commercial telephone service by provisioning long distance, local, and alternative access services in California, Arizona, Nebraska, Virginia, and New England.

Cablevision Systems Corporation delivers residential, regional, and long distance telephone services to households in Long Island, New York. Cable providers such as Jones Communications offer combined cable and telephony packages to subscribers. AT&T MediaOne, originally known as Continental Cablevision, provides cable modem service as well as digital telephone service in Los Angeles, Boston, Atlanta, and Jacksonville. Cable network operators generally sell cable modems to subscribers initiating cable system service.

7.5   CABLE NETWORK TECHNICAL FUNDAMENTALS

7.5.1   CABLE MODEM FEATURES AND FUNCTIONS

Cable modems take the form of external devices, internal devices, and interactive set-top boxes (STBs). Regardless of their structure, cable modems perform conventional signal modulation and demodulation operations at the subscriber premise for enabling voice, video, and data transport via a cable network.

In terms of operations, the cable modem is attached to the subscriber PC for enabling connectivity via coaxial cable or the upgraded HFC infrastructure to the CMTS (Cable Modem Termination System). The CMTS is located at the local cable operator facility. Adjacent to the CMTS, the headend or distribution hub at the local cable operator facility supports downstream transmissions over the local loop from the CMTS at the local cable operator facility to the subscriber premise over the local loop. (See Figure 7.1.)

7.5.2   CABLE MODEM OPERATIONS

Cable modems are constantly on, thereby eliminating the need for redialing to re-establish connectivity to cable provider services. As with cable modems, STBs enable e-mail, IP telephony, encryption, Web browsing, and connectivity to broadband services.

The terms "cable modem" and "set-top box" are used interchangeably. Cable-Labs supports an initiative to use the term "set-top box" as a replacement for the term "cable modem."

Between 500 and 2,000 subscriber households that are interlinked on a neighborhood cable network segment share bandwidth, upstream and downstream cable channels, and cable network services. To prevent cable signal interference, guard-bands or unused portions of the spectrum separate adjacent channels.

7.5.3   10 BASE-T ETHERNET AND ATM (ASYNCHRONOUS TRANSFER MODE) TECHNOLOGIES

In the United States, cable networks transmit IP packets in Ethernet frames via upstream and downstream channels. As a consequence, a 10BASE-T Ethernet NIC (Network Interface Card) is also installed in the subscriber PC (Personal Computer).

In the European Union, cable networks enable bi-directional transmission of ATM (Asynchronous Transfer Mode) fixed-sized cells. Therefore, ATM NICs (Network Interface Cards) are installed in customer premise PCs. As indicated in the examples that follow, cable vendors in the United States are also developing cable modem technology that interworks with ATM.

7.5.3.1   Com21 and Palo Alto Cable

Palo Alto Cable Corporation in San Francisco employs the Com21 Community Access System solution for delivery of ATM cable modem service to homes and

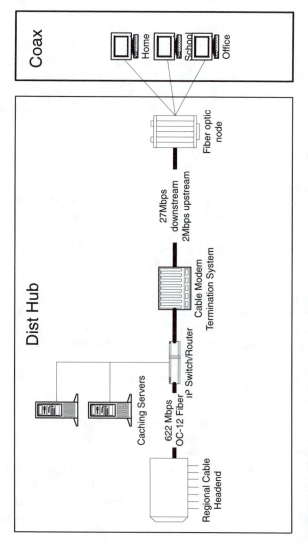

FIGURE 7.1 Cable network distribution architecture.

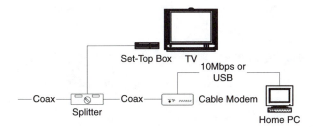

FIGURE 7.2 A typical home-based cable network configuration.

businesses in the Palo Alto metropolitan area. This system utilizes embedded ATM technology. Cable subscribers can order increased bandwidth and service levels from Palo Alto Cable Corporation to accommodate communications requirements.

7.5.3.2 Com21 and Terayon

Com21 partners with Terayon in fostering development of high-capacity cable modem solutions that interwork with ATM technology to provision QoS (Quality of Service) guarantees, high-speed information access and delivery, and improvements in network performance. In addition, this broadband ATM and cable modem solution enables bandwidth-intensive file transmission, teleworking, high-speed intranet and extranet connections, and interactive videoconferencing. (See Figure 7.2.)

7.5.4 CABLE MODEM TERMINATION SYSTEM (CMTS)

On the downstream path, a DOCSIS-certified CMTS (Cable Modem Termination System) amplifies, transforms, processes, and routes data via trunks and feeder cables from the NSP (Network Service Provider) to the subscriber PC based on the subscriber IP address. Transmission rates depend on such variables as the numbers of subscribers that are assigned to the same neighborhood cable network segment, the network architecture, and infrastructure robustness. On the upstream or return path, data are transported from the subscriber PC cable modem via feeder cables and trunks to the CMTS at the local cable operator facility. The CMTS then routes subscriber data to the NSP for transport to networks such as the Internet, extranets, and intranets, depending on the destination address. The CMTS performs an array of gateway, server, routing, management, and security functions to ensure trouble-free voice, video, and data transport. The CMTS also processes billing information and tracks network usage.

7.5.4.1 CMTS Marketplace

7.5.4.1.1 Nortel Networks
The CMTS available from Nortel Networks enables automatic cable modem registration and IP address assignment and fosters transmission rates at 36 Mbps over a single 6 MHz downstream channel or path with 64/256 QAM (Quadrature Amplitude Modulation). On the upstream channel, 10BASE-T Ethernet supports transmission

at 10 Mbps over eight separate upstream channels supporting QPSK or 16 QAM modulation.

7.5.5 HEADEND EQUIPMENT

Also called a distribution hub, headend equipment inserts voice, video, and data signals for transmission in the downstream direction or from the cable network to the subscriber premise. Headend equipment is situated next to the CMTS.

7.5.6 CABLE NETWORK OPERATIONS

Cable network configurations take the form of tree-and-branch topologies. Cable service for thousands of customers can be provided from a single CMTS (Cable Modem Termination System). With older coaxial cable systems, one-way amplifiers regenerate cable signals on the downstream path or from the cable provider to the subscriber household. With an HFC infrastructure, two-way amplifiers regenerate signals on the downstream and upstream paths.

7.6 WIRELINE CABLE NETWORK TRANSMISSION

7.6.1 CABLE NETWORK TRANSMISSION FUNDAMENTALS

Cable networks support multimedia transport via a coaxial cable plant or an upgraded HFC (Hybrid Optical Fiber Coaxial Cable) distribution infrastructure. Typically, downstream rates at 27 Mbps and upstream rates at 10 Mbps are supported by the HFC platform. The PSTN (Public Switched Telephone Network) or ISDN (Integrated Services Digital Network) BRI (Basic Rate Interface) service supports upstream transmission if the local infrastructure is not yet upgraded to an HFC plant.

In the United States, cable network systems enable access to broadband services by using separate channels for downstream and upstream transmissions. Each broadcast cable channel in the United States consists of a 6 MHz (Megahertz) slice of the RF (Radio Frequency) spectrum. Cable service initially supported asynchronous or one-way distribution of a common set of video signals downstream for enabling cablecasts.

Asymmetric cable transmission solutions remain popular in the present-day environment. With asymmetric transmissions, the downstream path receives a higher bandwidth allocation for transporting bandwidth-intensive streaming media from the cable operator to the subscriber premise than the upstream or return path from the subscriber premise to the cable operator facility.

Although a cable network can transmit multimegabits of voice, video, and data upstream from the subscriber premise to the cable operator facility, a PC is limited to enabling transmission rates at 10 Mbps in accordance with limits set by the 10BASE-T Ethernet NIC (Network Interface Card). Wireline cable transmissions are adversely affected by the amount of traffic on the neighborhood cable network segment, the quantity of shared bandwidth, the age and condition of the HFC plant or the coaxial cable wiring, faulty connections, and impulse noise generated by household appliances.

7.6.2 HYBRID OPTICAL FIBER COAXIAL CABLE (HFC) INFRASTRUCTURE

The flow of information in a conventional coaxial cable network is asymmetric. The flow of information on the upstream path is slower than the flow on the downstream path. To promote high-speed one-way or equivalent speeds in both directions, major cable companies replaced large segments of in-place coaxial cable with higher-capacity optical fiber in order to create an HFC infrastructure. This process was initiated during the 1980s when cable service providers obtained franchises or permits from local government agencies with permission to use local easements or rights-of-way for constructing HFC plants.

An HFC infrastructure supports high bandwidth transmission over long distances dependably and reliably. In comparison to coaxial cable, HFC installations reduce noise, attenuation, and the number of amplifiers needed for signal regeneration and thereby support improved signal quality and channel capacity.

Variations in transmission rates and network capacity depend on impulse noise, burst noise, and the number of concurrent users accessing a shared cable network segment. In contrast to DSL (Digital Subscriber Line) operations, cable service functions do not depend on user location on the HFC neighborhood network segment.

An HFC infrastructure supports voice, video, and data applications; videoconferencing; and streaming media services. An HFC infrastructure also enables video-on-demand (VOD), near-video-on-demand (NVOD), interactive-video-on-demand (IVOD), teleresearch, and interactive virtual classroom projects. Cable networks operating via an HFC infrastructure interwork with core infrastructure technologies such as SONET/SDH (Synchronous Optical Networks and Synchronous Digital Hierarchy). In addition, cable networks also interoperate with Ethernet, IP (Internet Protocol), and Asynchronous Transfer Mode (ATM) to optimize network performance and local loop operations. (See Figure 7.3.)

7.7 CABLE NETWORK PROTOCOLS

Cable networks take the form of branch-and-tree configurations with shared upstream and downstream channels. In cable networks, voice, video, and data signals are transported via a communal downstream channel to each subscriber household on a neighborhood cable network segment. For the upstream path, voice, video, and data signals from subscriber households on a neighborhood cable network segment are transported via a common upstream channel to CMTS headend equipment.

The rates at which voice, video, and data transit the cable network depend on the age and condition of the local loop and the number of users sharing the frequency. During periods of intense cable network usage by multiple subscribers, the rate of transmission slows down. The resulting congestion contributes to delays in network throughput and impedes network effectiveness in transporting time-sensitive voice and video traffic via the cable infrastructure to destination sites. The capabilities of protocols in supporting cable network operations are highlighted in the following sections.

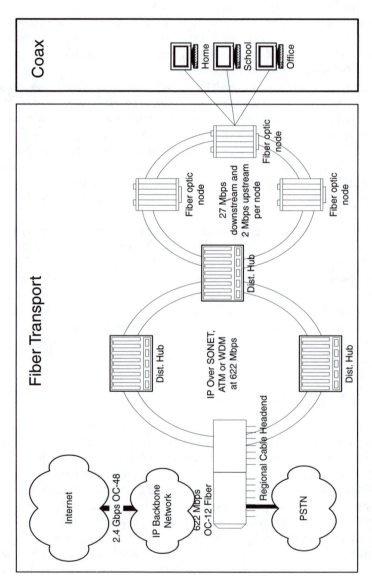

FIGURE 7.3 A HFC (Hybrid Optical Fiber Coaxial Cable) cable network.

7.7.1 TIME-DIVISION MULTIPLE ACCESS (TDMA) AND FREQUENCY-DIVISION MULTIPLE ACCESS (FDMA)

The Time-Division Multiple Access (TDMA) protocol promotes efficient bandwidth allocation by enabling multiple cable network nodes to share a single frequency. FA-TDMA (Frequency Agile-TDMA) modulation is endorsed in the DOCSIS 1.1 and the DOCSIS 2.0 specifications.

In contrast to TDMA operations, the Frequency-Division Multiple Access (FDMA) protocol dedicates a frequency to an individual network node. This approach is effective for handling bursty services such as voice conversations that transit the cable network infrastructure. With FDMA modulation, the virtually dedicated cable modem connection remains unaffected by network usage. However, the FDMA protocol fails to optimize utilization of available bandwidth for supporting volume-intensive multimedia services.

7.7.2 CODE-DIVISION MULTIPLE ACCESS (CDMA)

The Code-Division Multiple Access (CDMA) protocol employs spread spectrum technology for enabling voice, video, and data transport over cable networks. With spread spectrum technology, cable signals are spread over the entire width of each channel to support secure cable network transmissions.

CDMA comes in a variety of flavors, including Asynchronous-CDMA (A-CDMA) and Synchronous-Code-Division Multiple Access (S-CDMA). As with CDMA, A-CDMA and S-CDMA are distinguished by their immunity to noise impairments and provision of high-level security services.

7.7.2.1 Synchronous-Code-Division Multiple Access (S-CDMA)

Cable network transmissions transiting the HFC infrastructure in the upstream direction employ frequencies between the 5 and 42 MHz spectral bands. These transmissions are subject to impulse noise generated by microwave ovens, radios, televisions, refrigerators, PCs (Personal Computers), printers, and other electronic appliances at subscriber venues.

Developed by Terayon, the Synchronous-Code-Division Multiple Access (S-CDMA) protocol employs spread spectrum technology to ensure the integrity and reliability of upstream cable modem transmissions and eliminate or limit the adverse effect of impulse noise. The S-CDMA protocol supports secure upstream transmission at rates of 10 Mbps (Megabits per second) on each 6 MHz cable channel. Each payload is transported in multiple 64 Kbps (Kilobits per second) data streams. Moreover, S-CDMA technology facilitates seamless operations in channels that are adjacent to the shared downstream and upstream channels by using Quadrature Phase Shift Key (QPSK) modulation. DOCSIS 2.0 specifications endorse S-CDMA operations.

7.7.3 QUADRATURE PHASE SHIFT KEY (QPSK) AND QUADRATURE AMPLITUDE MODULATION

The Quadrature Amplitude Modulation (QAM) and Quadrature Phase Shift Key (QPSK) modulation schemes support transmission of fixed-sized ATM cells and variable-sized IP and Ethernet packets via the cable network. These protocols enable increased channel capacity in cable networks that support transmissions over an HFC platform. The QAM and QPSK modulation schemes support signal viability so that voice, video, and data are transported intact to destination addresses. Endorsed by the ITU-T (International Telecommunications Union-Telecommunications Standards Sector), the QAM and the QPSK modulation schemes also facilitate transmissions in DSL (Digital Subscriber Line) configurations and satellite networks.

The QAM (Quadrature Amplitude Modulation) modulation scheme enables dependable downstream transmissions of voice, video, and data in cells or packets, and facilitates the process of encoding digital data into analog signals by using a single signal to represent a 4-bit string. Cable networks that employ 64 Quadrature Amplitude Modulation (QAM) enable transmission at 27 Mbps and higher rates in the downstream direction and support operations in spectrum between the 50 and 750 MHz frequencies. With 256 QAM, a standard 6 MHz cable channel can carry up to 40 Mbps of information.

For upstream transmissions, 6 MHz cable channels are developed out of unused spectrum between the 5 and 42 MHz RF (Radio Frequency) bands. These channels are adjacent to channels carrying cable television signals. With QPSK modulation, a standard 6 MHz cable channel can carry up to 10 Mbps of information on the upstream or return channel in noisy environments.

7.7.4 Orthogonal Frequency-Division Multiple Access

Introduced in 1970, the OFDM (Orthogonal Frequency-Division Multiplexing) protocol is a robust spectrally efficient multicarrier RF (Radio Frequency) modulation technique that divides a single channel into several channels for data transport. Immune to propagation impairments and impulse interference, OFDM technology enables effective wireless and wireline cable network services over the last mile or local loop. OFDM fosters HDTV (High-Definition Television) transmission in Japan and the European Union, supports Digital Video Broadcasting (DVB) programs, and facilitates ADSL (Asynchronous Digital Subscriber Line) operations.

The European Telecommunications Standards Institute (ETSI) develops DVB specifications for digital terrestrial television broadcasts based on the OFDM protocol. In addition, the OFDM protocol supports wireless LAN (WLAN) implementations and third-generation (3G) wireless network operations.

7.7.5 Vector-Orthogonal Frequency-Division Multiplexing (V-OFDM)

A Radio Frequency (RF) technology, V-OFDM is a fixed broadband wireless cable network protocol that supports multivendor cable service and robust signal transmission in point-to-point and point-to-multipoint networks in urban and suburban environments. Developed and promoted by the Mobile Data Association (MDA) Consortium, the V-OFDM protocol reorders MDS (Multichannel Distribution System), MMDS (Multichannel Multipoint Distribution System), and LMDS (Local Multipoint Distribution System) signals so that these signals arrive in constant streams at destination addresses.

The V-OFDM protocol employs spatial and frequency diversity to eliminate multipath signal fading and attenuation resulting from signal interference and obstruction. V-OFDM modulation supports broadband wireless applications such as high-speed Internet connectivity, IP telephony, and VPN (Virtual Private Network) implementations. In addition, V-OFDM enables symmetric transmission of video, voice, and data to residences, small business, small- and medium-sized enterprises, and schools.

The Broadband Wireless Internet Forum (BWIF) supports utilization of V-OFDM (Vector-Orthogonal Frequency-Division Multiplexing) and DOCSIS (Data-over-Cable Service Interface Specification) standards to enable operations in broadband fixed wireless access (FWA) LMDS and MMDS networks. Cisco Systems, Motorola, KPMG Consulting, Texas Instruments, Toshiba Corporation, Bechtel Communications, Samsung Electronics, and Broadcom Corporation also participate in an industry coalition that promotes utilization of standardized V-OFDM operations. In addition, electronics suppliers, system integration providers, and chip manufacturers support V-OFDM implementation.

7.8 CABLE NETWORK ARCHITECTURE

In terms of architectural conformance with the seven layers of the OSI (Open Systems Interconnection) Reference Model, a cable network configuration supports information transport at Layer 1 of the OSI Reference Model or the Physical Layer. For Layer 2 or the Data-Link Layer, cable networks employ the Medium Access Control (MAC) protocol to facilitate collision detection and retransmission, error detection and error recovery, and timing and synchronization functions.

At the Network Layer or Layer 3, wireline cable modem configurations enable transmission of IP (Internet Protocol) packets via the HFC infrastructure. By contrast, wireless cable modem configurations support transmission of IP packets as electromagnetic signals through free space.

At the Transport Layer or Layer 4, cable networks operate in conjunction with the User Datagram Protocol (UDP) and the Transmission Control Protocol (TCP). At the upper layer of the cable modem protocol stack, cable networks employ SNMPv3 (Simple Network Management Protocol version 3) for managing and administering network operations.

7.9 CABLELABS (CABLE TELEVISION LABORATORIES)

7.9.1 BACKGROUND

7.9.1.1 Multimedia Cable Network System (MCNS)

A vendor alliance, the Multimedia Cable Network System (MCNS) defined specifications for implementing interoperable cable networks and clarified approaches for enabling telephony services, fax transmission, network security, and device management. The MCNS vendor coalition also developed the Data-over-Cable System Interface Specification (DOCSIS) and endorsed the use of variable length packets to enable interoperability of Ethernet and IP technologies with the cable network infrastructure. MCNS participants included Cox Communications, Tele-Communications, Inc. (TCI), Comcast Cable, Time Warner, MediaOne, and Rogers Cable System. CableLabs superseded the Multimedia Cable Network System.

7.9.1.2   IEEE 802.14 Working Group

The IEEE 802.14 Working Group currently develops standards, protocols, and architectures for accomplishing data transport over HFC networks and defines the Media Access Control (MAC) or Layer 2 and Physical Layer or Layer 1 protocols. The IEEE 802.14 Working Group also supports operations that are compliant with DOCSIS specifications and endorses utilization of the Quadrature Phase Shift Key (QPSK) and the Quadrature Amplitude Modulation (QAM) protocols.

Originally, the IEEE 802.14 Working Group selected ATM technology for facilitating transmission of voice, video, and data as fixed-sized cells between the Cable Modem Termination System (CMTS) at the local cable operator facility and the cable modem installation at the customer premise. This Working Group also supported the use of optional Physical Layer modulation techniques, including FA-TDMA (Frequency Agile-Time-Division Multiplexing Technology) promoted by Broadcom Corp and S-CDMA (Synchronous-Code-Division Multiplexing) developed by Terayon.

As a consequence of concerns about the slow progress of the IEEE 802.14 Working Group in developing cable specifications and inclusion of optional modulation methods in the Working Group proposal, the DOCSIS Certification Board moved forward with DOCSIS specifications without consulting the IEEE 802.14 Working Group. It is interesting to note that this Board subsequently included the FA-TDMA operations as part of DOCSIS 1.1 specifications and FA-TDMA and S-CDMA modulation as part of DOCSIS 2.0 specifications. The inability of the IEEE 802.14 Working Group to move forward with cable network specifications also contributed to the organization of the MCNS and subsequent formation of CableLabs.

7.9.2   CABLE TELEVISION LABORATORIES (CABLELABS) OVERVIEW

CableLabs (Cable Television Laboratories) is a development and research consortium that represents South American and North American cable operators employing HFC infrastructures for multimedia distribution. Participants include TCI, Comcast, AOL Time Warner, and AT&T MediaOne. This consortium clarifies approaches for enabling cable network interoperability, enhanced signal security, multiple user access to cable networks, and dynamic assignment of bandwidth on-demand.

Moreover, CableLabs defines specifications for transmission of data over cable networks, cable network functions, and interoperable high-speed cable modems. CableLabs sponsors development and implementation of the Data-over-Cable Service Interface Specification (DOCSIS), as well as the OpenCable initiative to facilitate development of advanced interoperable digital set-top boxes (STBs) and the PacketCable project to clarify approaches for enabling IP (Internet Protocol) services over the HFC (Hybrid Optical Fiber Coaxial Cable) infrastructure.

7.9.3 CableLabs Data-Over-Cable Service Interface Specification (DOCSIS)

Originally called the Multimedia Cable Network System (MCNS) specification, DOCSIS specifications were regarded as the *de facto* cable standard in the United States prior to ITU-T (International Telecommunications Union-Telecommunications Standards Sector) endorsement. A second-generation cable standard approved by the ITU-T, DOCSIS 1.1 specifications foster cable system implementations that support integrated voice, data, and video services in educational institutions, businesses, hospitals, government agencies, and SOHO (Small Office/Home Office) venues.

7.9.3.1 DOCSIS 1.1 Features and Functions

7.9.3.1.1 Basic DOCSIS Services

An extension of original DOCSIS 1.0 standards, DOCSIS 1.1 specifications serve as the foundation for the ITU-T J.112 Recommendation on cable networks. Interoperable cable networks based on DOCSIS 1.1 specifications support diverse services including high-speed Web access, streaming audio and video, videoconferencing, LAN-to-LAN interconnectivity, video-on-demand (VOD), cablecasts, and Virtual Private Networks (VPNs) applications. In addition, DOCSIS 1.1 specifications enable seamless service initialization and dynamic QoS (Quality of Service) assurances.

7.9.3.1.2 Security Services

DOCSIS 1.1 specifications also support utilization of the Baseline Privacy Key Management (BPKM) protocol based on DES (Data Encryption Standard) for cable modem authentication and baseline privacy cryptographic measures to safeguard voice, video, and data transmissions between the CMTS (Cable Modem Termination System) and the cable modem. Baseline privacy measures also safeguard cable network transmissions against service theft.

DOCSIS 1.1 specifications support the utilization of a 48-bit IEEE MAC (Medium Access Control) Layer identifier for each cable modem. This 48-bit unique identifier ensures that cable subscribers only obtain authorized services. However, this identifier does not authenticate the identity of the cable modem or the subscriber. Therefore, cyberintruders can employ bogus cable modems that masquerade as authenticated cable modems to obtain cable services.

In addition, DOCSIS 1.1 specifications define a set of interfaces for the network side of the Cable Modem Termination System (CMTS), the operations support system, and the cable modem-to-CPE (Customer Premise Equipment). Moreover, DOCSIS 1.1 specifications clarify the RF (Radio Frequency) interface between the cable network and the cable modem at the customer premise as well as the interface between the CMTS headend and the cable network.

DOCSIS 1.1 specifications support the use of SNMPv3 (Simple Network Management Protocol version 3) for enabling administrative services, network management, and support for essential cable network operations. DOCSIS 1.1 specifications also enable packet filtering functions for enhanced security, support utilization of the Internet Security (IPSec) protocol, and enable compliance with the Internet Group Management Protocol (IGMP) to facilitate distribution of IP broadcasts and multicasts to designated IP addresses.

7.9.3.1.3   DOCSIS 1.1 Implementations

DOCSIS 1.1 specifications promote implementation of compact interoperable cable modems that incorporate Universal Serial Bus (USB) ports and interwork with standards-compliant Cable Modem Termination Systems (CMTSs) to support scalable network solutions featuring tiered or multiple service classes with minimum and maximum data rates. DOCSIS 1.1 implementations enable transmission rates between 27 Mbps and 36 Mbps downstream and between 320 Kbps and 10 Mbps upstream. Inasmuch as cable service is shared, transmission rates decline as more users gain access to the same neighborhood cable network segment.

7.9.3.1.4   DOCSIS 1.1 Cable Modems

DOCSIS 1.1 specifications standardize communications protocols and formats for cable modems and related devices, thereby ensuring an open market for cable subscriber equipment. Non-vendor-specific or open regulations such as DOCSIS 1.0 and DOCSIS 1.1 promote the production of affordable cable modems. DOCSIS specifications also ensure multivendor interoperability.

7.9.3.15   VoIP (Voice-over-IP)

Procedures for provisioning VoIP (Voice-over-IP) or Internet telephony were omitted in DOCSIS 1.0 specifications. However, DOCSIS 1.1 specifications rectify this omission by clarifying approaches for implementing standardized VoIP teleservices and describing methods for digitization, compression, packet transmission, and decompression. By using the in-place HFC infrastructure for supporting packet telephone services, cable networks eliminate the need for stand-alone HFC telephony devices. In addition to encouraging implementation of real-time VoIP services, CableLabs promotes the utilization of support services for enhancing cable packet telephony operations and development of IP specifications in the cable modem domain.

CableLabs also supports deployment of integrated networks such as CABLESPAN developed by Tellabs. CABLESPAN is a multiservice communications network for enabling transmission of IP telephony and IP data via the HFC infrastructure at rates of 10 Mbps, thereby eliminating the need to implement a separate overlay IP network.

A telephone call is a time-sensitive network application that can be adversely affected by millisecond transmission delays. As a consequence, CableLabs also promotes utilization of QoS (Quality of Service) assurances, compression techniques, and prioritization schemes to eliminate distortions and ensure on-time delivery of voice packets.

7.9.3.2   DOCSIS 2.0 Specifications

A third-generation (3G) set of cable standards currently in development, DOCSIS 2.0 specifications support S-CDMA (Synchronous-Code-Division Multiple Access) and A-TDMA (Advanced Frequency Agile Time-Division Multiple Access), also called FA-TDMA (Frequency Agile-Time-Division Multiple Access), protocol operations. DOCSIS 2.0 specifications are compatible with DOCSIS 1.0 and DOCSIS 1.1 specifications for cable modems and CMTSs (Cable Modem Termination Systems). The inclusion of FA-TDMA and S-CDMA technologies in DOCSIS 2.0 specifications facilitate implementation of advanced Physical Layer modulation methods for enabling seamless transmission of bandwidth-intensive voice, video, and data over the local loop from the CMTS to the subscriber premise.

In terms of upstream transport, DOCSIS 1.0 specifications enable 5 Mbps of throughput capacity on the upstream path for each 6 MHz cable channel that operates over HFC networks. DOCSIS 1.1 specifications facilitate upstream transmissions at 10 Mbps for each 6 MHz channel. DOCSIS 2.0 specifications enable upstream transmissions at 30 Mbps for each 6 MHz channel.

7.9.4 CableLabs PacketCable Initiative

Sponsored by CableLabs, the PacketCable initiative supports the use of IP technology to facilitate access to Web resources and provision dependable delivery of voice, video, and data transmission over the HFC plant. Additionally, the PacketCable initiative enables cable subscribers to access Web servers that store or cache frequently accessed Internet content and employ popular Web browsers such as Netscape Navigator and Microsoft Internet Explorer, audio-streaming tools such as RealAudio, and video streaming tools such as QuickTimeLive. In 2000, Lucent began trials of a multifunctional PacketCable platform to support delivery of converged IP voice, video, and data applications over the local loop.

7.9.5 CableLabs Go2Broadband Service

A service of CableLabs, the Go2Broadband service facilitates implementation of a portfolio of business-to-business electronic commerce (E-commerce) applications to enable sales of cable equipment and services by cable NSPs (Network Service Providers) and cable manufacturers. Go2Broadband also supports implementation of standards-compliant DOCSIS cable modems.

7.9.6 CableLabs CableHome Initiative

Sponsored by CableLabs, the CableHome initiative supports the establishment of a home network infrastructure that facilitates delivery of cable-based multimedia services in the home environment. The CableHome network platform provides consistent levels of cable service across local loop transport technologies that include 10BASE-T Ethernet, ATM, Local Multipoint Distribution System (LMDS), and Multichannel Multipoint Distribution Service (MMDS). Moreover, CableHome supports the integration of non-CableHome devices to the CableHome network, and resolves issues associated with cable architecture, QoS (Quality of Service) guarantees, network address management, cable network operations, and cable network security. CableHome architecture operates independently of the underlying physical technologies.

In 2001, vendors participating in the CableHome initiative developed the Cable-Home QoS (Quality of Service) specification for extending capabilities of DOCSIS and PacketCable capabilities into the home environment. This specification defines a standardized approach for establishing and maintaining service sessions that feature differentiated QoS levels across the home network platform and the cable network infrastructure.

7.9.7 CERTIFICATION OF CABLE PRODUCTS

DOCSIS and CableLabs certified products such as cable modems and CMTSs (Cable Modem Termination Systems) support essentially identical functions. DOCSIS is the name of the CableLabs specification. CableLabs certified indicates the conformance of cable modems and headend equipment to the DOCSIS specification. Cable modems that feature CableLabs certified seals conform to CableLabs requirements and completed interoperability tests successfully. Standardization activities are rather recent in the cable network marketplace. As a consequence, cable providers such as AtHome (@Home) also put into place proprietary standards to make sure that cable modems and CMTSs (Cable Modem Termination Systems) interoperate with the AtHome proprietary operating system. AtHome uses a three-stage approval process for assessing cable modem capabilities. Cable modems at Level 3 are evaluated in trials; cable modems at Level 2 are tested in the field; and cable modems at Level 1 are approved for commercial deployments.

7.9.8 OPENCABLE SPECIFICATION

Developed by Cable Labs, the OpenCable specification defines the architecture for building next-generation digital set-top boxes (STBs) to enable wireline cable service in multivendor environments. This specification also describes functional requirements for enabling STBs to transport MPEG-2 (Moving Picture Experts Group-2) television signals and IP packets via bidirectional cable networks.

OpenCable set-top boxes (STBs) incorporate high-performance microprocessors with real-time operating systems to provision interactive services and transmission and delivery of streaming media at the subscriber premise. These STBs also work with plug-ins and open Internet specifications such as HTML (HyperText Markup Language) and VRML (Virtual Reality Modeling Language). Moreover, an interface based on the IEEE 1394 specification enables connections between standards-compliant STBs and other digital devices such as DVD (Digital Versatile Disc) players and high-definition television (HDTV) sets.

7.9.8.1 IEEE 1394 High-Performance Serial Bus Specification and Universal Serial Bus (USB) Technology

The IEEE 1394 High-Performance Serial Bus specification and USB (Universal Serial Bus) technology support plug-and-play installation of peripheral devices. USB technology and the High-Performance Serial Bus specification enable the cable industry to readily implement high-speed broadband services and applications to support IP telephony, videoconferencing, and CableHome implementations.

7.9.8.1.1   IEEE 1394 High-Performance Serial Bus Specification

The IEEE 1394 specification clarifies parameters for interlinking multimedia peripherals such as digital video camcorders, digital cameras, printers, and massive storage
devices via a high-performance serial bus to a single computer for enabling support
of high-performance data, video, and audio in-home applications. This specification
also establishes a flexible bus management system to facilitate the interworking of
such devices as digital videoconferencing systems and set-top boxes (STBs). Chip
manufacturers and electronic firms that include Adaptec, Texas Instruments, and
Sony develop standards-compliant IEEE 1394 devices. The IEEE 1394 specification
is based on FireWire technology developed by Apple for Power Macintosh computers.

7.9.8.1.2   Universal Serial Bus (USB) Technology

The IEEE 1394 High-Performance Serial Bus Specification and USB (Universal
Serial Bus) technology are complementary solutions for supporting interconnectivity
of multiple peripherals to a single computer. Universal Serial Bus (USB) technology
enables transmission rates reaching 12 Mbps and interconnects up to 127 devices
to a PC. By contrast, the IEEE 1394 serial bus specification enables the interconnectivity of a maximum of 62 devices and fosters transmission rates reaching 200
Mbps. USB technology is used primarily with lower-powered devices such as mice,
keyboards, modems, and joysticks. IEEE 1394 connections are used with higher-
speed, higher-powered multimedia peripherals.

*7.9.8.1.3   IETF (Internet Engineering Task Force) IP-over-IEEE (International
Electrical and Electronics Engineers) 1394 Working Group*

The IETF IP-over-IEEE 1394 Working Group defines communications specifications
for enabling delivery of voice, video, and data to hosts across the global Internet in
real-time by non-traditional networking devices such as televisions, cameras, and
videocassette recorders (VCRs). The IETF IP-over-IEEE 1394 Working Group also
clarifies procedures to support IP multicasts, the Address Resolution Protocol (ARP),
and the Multicast Channel Allocation Protocol (MCAP).

7.9.8.2   OpenCable Marketplace

Originally passive devices, STBs enable a diverse array of interactive multimedia
applications and broadband services. As an example, PowerTV supports a standards-
compliant STB that is used by AOL Time Warner to support access to digital cable
programming, near video-on-demand (NVOD), and broadband networks. AT&T
MediaOne, Rogers, Comcast, Videotron, and Adelphia employ Explorer 2000, an
OpenCable-compliant STB, for provisioning multimedia services and IP multicast
distribution.

Broadband Access System produces an OpenCable-compliant STB that aggre-
gates vast volumes of cable modem traffic for transmission via the HFC infrastructure
to IP backbone networks and the public or commodity Internet. This STB also
features WAN (Wireless Access Network) and LAN (Local Area Network) interfaces
and works in concert with DOCSIS-certified CMTS (Cable Modem Termination
System) equipment.

7.10   U.S. DIGITAL TELEVISION (DTV) OPERATIONS

7.10.1   DIGITAL TELEVISION (DTV) FUNDAMENTALS

DTV is an advanced television (ATV) solution that defines HDTV (High-Definition Digital Television) operations and HDTV broadcasts with high-resolution data, video, and sound. DTV transmissions are accessible via a PC or a HDTV-compliant television set. Subscribers using DTV installations can access programs with captions in multiple languages and retrieve newspapers, stock market updates, and interactive educational material on-demand.

7.10.2   U.S. FEDERAL COMMUNICATIONS COMMISSION (FCC)

The U.S. Federal Communications Commission (FCC) supports a DTV (Digital Television) specification that standardizes transmission of digital television broadcasts. DTV is slated to replace standard analog television broadcasts by 2002. The FCC also requires broadcasters to support digital television programs on-air by 2002. Inasmuch as educational institutions operate public television stations and broadcast tele-education programs, developments in digital television are also expected to impact the content and format of distance education programming.

Because DTV uses fewer channels to broadcast programs, broadcasters are required to return a portion of the spectrum presently in use for analog teleprograms to the FCC. The FCC intends to distribute this spectrum to designated police agencies and firefighter brigades to provision public safety broadcasts.

In 2000, the FCC issued guidelines on RF emission safety and the effect of antenna towers on community public health. These guidelines also specify procedures for ensuring that local antenna facilities are in compliance with FCC limits for human exposure to RF electromagnetic fields.

7.11   EUROPEAN CABLE TELEVISION LABORATORIES (EURO CABLE LABS)

7.11.1   EUROCABLE LABS OVERVIEW

Sponsored by the European Commission, EuroCableLabs is an international research and development consortium consisting of television system operators, cable manufacturers, cable service providers, and cable suppliers. EuroCableLabs develops advanced equipment for cable system implementations, verifies cable equipment interoperability, benchmarks cable operations, and provides data relating to best practices in cable networking utilization.

As with CableLabs, EuroCableLabs promotes wireline cable network implementation and encourages deployment of broadband voice, video, and data services over the HFC infrastructure. Moreover, CableLabs and EuroCableLabs address operational, legislative, judicial, and fiscal problems affecting cable network implementations and disparities between standards-compliant and proprietary cable networking solutions. As with CableLabs, EuroCableLabs also encourages universal adoption of its specifications as worldwide standards.

7.11.2 EuroCableLabs Initiatives

EuroCableLabs supports development of cable modem products based on DVB (Digital Video Broadcasting) specifications. The European Cable Communications Association (ECCA) encourages cable operators across the European Union to employ the EuroModem with ATM (Asynchronous Transfer Mode) NICs (Network Interface Cards) and the EuroBox platforms.

EuroModem and EuroBox enable flexible delivery of broadcast television content, bi-directional communications services, streaming media delivery, voice services such as VoIP or IP telephony, and VPN implementations. The EuroModem and EuroBox platforms are readily integrated into DVB environments via an Independent Network Adapter (INA) that controls terminal functions.

7.11.3 EuroDOCSIS

Developed by EuroCableLabs, EuroDOCSIS specifications enable cable operators and cable subscribers in the European Union to utilize interoperable cable system products. CableLabs does not certify or endorse EuroDOCSIS products.

7.11.4 EuroModem Specification

Based on the ETSI (European Telecommunications Standards Institute) cable modem standard, the EuroModem specification is endorsed by the ITU-T, the EuroModem Consortium, the DVB/DAVIC (Digital Video Broadcasting and Digital AudioVisual Council) Consortium, and EuroCableLabs. EuroModems are available in two formats. Class A EuroModems enable PC interconnections for supporting fast access to the Internet and time-critical applications such as videoconferencing. Class B EuroModems interoperate with telephony devices to provide voice telephony services with QoS assurances.

Approved in 1999, the EuroModem specification describes approaches for enabling IP multicasts and interactive multimedia applications and facilitates on-demand access to television broadcasts, music, and radio programs. EuroModems interoperate with the installed base of cable networks in the European Union and EuroBox devices or STBs. In addition, the EuroModem specification establishes guidelines for cable network implementations, operations, administration, and maintenance. Participants in the EuroModem Consortium include Deutsche Telecom, Cablecom, France Telecom, Telenor Avidi, Cablelink, Helsinki Media, and TeleDanmark.

7.11.4.1 International Electrotechnical Commission (IEC) and the European Committee for Electrotechnical Standardization (CENELEC)

The International Electrotechnical Commission (IEC) and the European Committee for Electrotechnical Standardization (CENELEC) work jointly in establishing EuroModem specifications. Participants in this joint initiative include the Czech National Standards Institute, the Hellenic Organization for Standardization, the Belgium Electrotechnical Committee, and the Finnish Electrotechnical Standards Association.

7.11.5   EuroBox Specification

Endorsed by the EuroBox Consortium, the EuroBox 4.0 specification complements the EuroModem specification. The EuroBox 4.0 specification defines the features and functions of a standard interactive set-top box (STB) and a standard EuroModem or integrated receiver and decoder (IRD) for facilitating on-demand access to multimedia services and radio, music, and broadcast television programs. Easy-to-use and affordable, EuroBox 4.0 devices also support telephone operations and cable network connections, and work in concert with the Open TV API (Applications Programming Interface) and standardized DVB (Digital Video Broadcasting) Multimedia Home Platforms (MHPs). Participants in the EuroBox Consortium include TeleDanmark, Cablecom, Deutsche Telekom, Helsinki Media, and Telia.

7.11.6   Digital AudioVisual Council (DAVIC)

Endorsed by EuroCableLabs, the Digital AudioVisual Council (DAVIC) established specifications for interoperable and interactive audiovisual broadcasts between 1994 and 1999. DAVIC supported the use of IP multicasts for distributing video and audio and making content easily accessible within and between DAVIC-compliant networks. Further, DAVIC developed procedures for authentication, copyright protection, privacy, and network monitoring to ensure secure online sessions and transactions.

DAVIC specifications serve as the present-day framework for interactive video broadcasts and video-on-demand (VOD) solutions. DAVIC also clarified approaches for multimedia retrieval and delivery via multiservice IP networks; procedures for accessing Web resources, television programs, and interactive video broadcasts; and methods for QoS (Quality of Service) management of interactive audio and video on multiservice IP networks.

In addition, DAVIC published the DAVIC TV anytime and anywhere specification for integrating DTV programming and Internet content. The International Standards Organization (ISO) and the International Electrotechnical Commission (IEC) endorsed DAVIC TV as an international standard.

7.11.7   Digital Video Broadcasting and Digital Audio-Visual Council (DVB/DAVIC) Consortium

The DVB/DAVIC (Digital Video Broadcasting and Digital AudioVisual Council) Consortium continues the work initiated by DAVIC. This Consortium supports development of interoperable cable products and services and implementation of DVB specifications in the cable network arena. The DVB/DAVIC Consortium also endorses the use of the HFC infrastructure for downstream traffic and the Local Multipoint Distribution System (LMDS) for upstream traffic.

7.11.8   Digital Video Broadcasting (DVB) Project

Endorsed by EuroCableLabs and sponsored by the European Telecommunications Standards Institute (ETSI), the Digital Video Broadcasting (DVB) project is an international initiative that supports the development of standards and guidelines enabling transmission of digital television signals via wireline and wireless cable networks, satellite systems, and microwave configurations. DVB also describes procedures for accessing interactive broadband services and delivery of MPEG-2 compliant broadcasts. DVB standards also clarify the role of Cable Modem Distribution Systems (CMTSs) in supporting distribution of television programming and multimedia signals. Cable companies in the European Union, South America, Australia, and Asia endorse DVB specifications.

7.11.8.1   DVB-C (Digital Video Broadcasting-Cable Only)

DVB-C (DVB-Cable Only) clarifies procedures for transporting audio, video, and data services over wireline cable networks and wireless cable networks based on the Local Multipoint Distribution System (LMDS) and the Multichannel Multipoint Distribution System (MMDS). Moreover, DVB-C specifications define channel coding, packet formats, and cable network operations and support EuroBox and Euro-Modem interoperability.

7.11.8.2   DVB-S (Digital Video Broadcasting-Satellite) and DVB-T (Digital Video Broadcasting-Terrestrial)

In addition to DVB-C, the DVB initiative defines common interfaces and international specifications for satellite (DVB-S) and terrestrial (DVB-T) services. DVB-S describes channel coding, frame format, and modulation functions for tele-applications provided by satellites that operate in the 11 GHz and 12 GHz spectral bands. DVB-T indicates approaches for enabling digital terrestrial broadcasts.

7.11.8.3   DVB-CI (Digital Video Broadcasting-Common Interface)

Based on specifications that include DVB-T (DVB-Terrestrial) and DVB-S (DVB-Satellite), DVB-CI (DVB-Common Interface) describes interfaces for CATV (Cable Television) and SMATV (Satellite Master Antenna Television) headend equipment and approaches for enabling SMATV installations in apartment complexes and local neighborhoods.

7.11.8.4   DVB-D (Digital Video Broadcasting-Data)

DVB-D (DVB-Data) facilitates utilization of interoperable MVDSs (Multipoint Video Distribution Systems) for enabling data broadcasts.

7.11.8.5   DVB-RCC (Return Channel for Cable Service)

The DVB-RCC (Return Channel for Cable Service) presents a framework for enabling bi-directional communications via cable networks in a specification endorsed by the ITU-T as Annex A to the ITU-T J.112 Recommendation. This Annex presents guidelines for establishing MAC (Medium Access Control) and Physical Layer interfaces, QoS guarantees, and cable network security.

7.11.9 DVB Multimedia Home Platform (MHP) Group

To enable interoperability of computer, broadcasting, and consumer electronics devices such as the EuroModem, the EuroBox, television sets, PCs, and laptops in the home environment, the DVB (Digital Video Broadcast) Multimedia Home Platform Group developed the Multimedia Home Platform (MHP). This standard facilitates implementation of interactive digital television services and DVB-MHP-compliant digital STBs (Set-Top Boxes).

7.11.10 EuroCableLabs and CableLabs: Parallels and Contrasts

CableLabs and EuroCableLabs sponsor research initiatives to determine the effectiveness of IP-over-cable solutions in delivering voice, video, and data transmission in real-time and evaluate capabilities of wireline and wireless cable deployments. Both groups endorse utilization of the HFC infrastructure for supporting interactive broadband services, support MPEG-2 compliance, and implement the QAM (Quadrature Amplitude Modulation) and the QPSK (Quadrature Phase Shift Key) protocols to facilitate reliable cable network transmission.

Historical differences in analog television standards adopted by the Europe Union and the United States contributed to the development of DOCSIS (Data Over Cable Service Interface Specification) and EuroDOCSIS specifications. These differences also are reflected in the U.S. cable modem and the EuroModem, and in the U.S. STB and the EuroBox in the European Union. Cable products in development are designed to be compatible with ATM and IP technologies, thereby enabling Quality of Service (QoS) guarantees for rapid transfer of broadband services and time-sensitive material. In the United States, cable products are compliant with 10BASE-T Ethernet specifications. In addition, the advantages and limitations of using an ATM platform with a wireline cable network solution are under consideration by CableLabs. The viability of a hybrid fiber radio (HFR) deployment as a last-mile enabler for cable network service is also explored. A universal standard for cable network deployment over the last mile or local loop based on a consolidated CableLabs and EuroCableLabs solution is expected in the long term.

7.12 STANDARDS ORGANIZATIONS AND ACTIVITIES

7.12.1 Cable Broadband Forum

The Cable Broadband Forum is a nonprofit alliance that promotes utilization of cable broadband networks and services for enabling high-speed access to the Internet, telecommuting, videoconferencing, and IP telephony. The Cable Broadband Forum also endorses the efforts of CableLabs, the National Cable Television Association (NCTA), the Society of Cable Telecommunications Engineers (SCTE), and the Internet Engineering Task Force (IETF). Cable Broadband Forum participants include AT&T MediaOne, Microsoft, Cisco Systems, Intel, and AOL Time Warner.

7.12.2 EUROPEAN TELECOMMUNICATIONS STANDARDS INSTITUTE (ETSI) HOME NETWORKS SPECIFICATIONS

The European Telecommunications Standards Institute adopted the TS 101.224 HAN (Home Area Network) specification for supporting Multimedia Home Platform (MHP) operations based on the work of the Multimedia Home Platform Group. In addition, this specification clarifies approaches for establishing connections between HANs (Home Area Networks). HANs that are MPEG-2-compliant transport video, data, and audio; support IP-over-ATM services; and interwork with cable networks and DSL (Digital Subscriber Line) implementations.

7.12.2.1 ETSI HLN (Home Local Network)

In addition, the ETSI HAN specification establishes a framework for a scalable and extendible home local network (HLN) based on the IEEE 1394 standard. An HLN links information appliances within rooms and between rooms in clusters of subnetworks and interconnects these clusters of subnetworks into an integrated home area network (HAN). ATM technology supports HLN connections to external networks such as the Internet via the local loop. Transmission rates at 25 and at 51.84 Mbps, depending on user requirements, are supported.

7.12.3 INTERNATIONAL TELECOMMUNICATIONS UNION-TELECOMMUNICATIONS STANDARDS SECTOR (ITU-T)

In the cable arena, the International Telecommunications Union-Telecommunication Standards Sector (ITU-T) develops specifications for transmission of television signals via analog and digital circuits, interoperable digital television applications, and the telephone-interface for upstream transmissions. The ITU-T Study Groups define techniques for utilization of electronic program guides, evaluate capabilities of MPEG-2 (Moving Picture Experts Group-2) toolkits for webcasting, and develop technical solutions such as cable networks and DSL to safeguard transmissions distributed to the home over the local loop. In addition, the ITU-T Study Groups define specifications for eliminating transmission disruptions resulting from delay, noise, jitter, echo, and packet loss, and establish guidelines for supporting IP telephony service. Moreover, the ITU-T Study Groups develop Recommendations for interoperable set-top boxes (STBs) and clarify approaches for implementation of cable network applications such as video banking.

7.12.3.1 ITU-T Video Quality Experts Working Group

The ITU-T Video Quality Experts Working Group develops algorithms that represent QoS (Quality of Service) guarantees for cable network applications and defines capabilities of asymmetric cable networks that support on-demand distribution of cable television programming.

7.12.3.2   ITU-T Study Group 9

The ITU-T Study Group 9 develops cable television specifications for endorsement by the ITU-T. As an example, this Study Group defined the home digital networking interface for the ITU-T J.117 Recommendation that was approved in 1999.

7.12.3.3   ITU-T H.323 Recommendation

Approved by the ITU-T in 1996, the ITU-T H.323 Recommendation supports video, audio, and data transmission across IP networks; streaming audio and video services; multimedia applications; and bandwidth-on-demand. In addition, this specification defines requirements for video and audio communications in LANs (Local Area Networks) that do not provision QoS (Quality of Service) guarantees and clarifies procedures for implementing the Real-Time Protocol (RTP), the Real-Time Control Protocol (RTCP), and the Resource Reservation Protocol (RSVP).

The ITU-T H.323 Recommendation is not linked to a specific network infrastructure or hardware product. As an example, cable television set-top boxes (STBs), IP telephone handsets, and PCs (Personal Computers) feature ITU-T 323-compliant platforms and cable modem, DSL, and Frame Relay networks support ITU-T H.323-compliant services such as video-over-IP and voice-over-IP (VoIP).

The ITU-H.323 Recommendation facilitates the use of VoIP technologies, applications and services defined by the Voice-over-IP (VoIP) Forum. Also called IP telephony, VoIP technologies enable real-time analog voice transmissions via IP networks. The transmission process begins with the use of compression algorithms for creating small digital data streams at the point of call initiation. These streams are then formatted and compressed into digital data packets for network transmission. At the destination address, decompression algorithms reverse the process and decompress the packets into data streams that are converted back to analog voice signals at the destination address. Approved in 1998, the ITU-T H.323v2 (ITU-T H.323, Version 2) Recommendation is an extension to the ITU-T H.323 Recommendation.

7.12.3.4   ITU-T J.83 Recommendation

The ITU-T J.83 Recommendation defines channel coding, framing structure, and digital signal modulation for television signals distributed by cable networks. In addition, this Recommendation clarifies functions of MPEG-2 (Moving Picture Experts Group-2) transmission, forward error correction mechanisms, and QAM (Quadrature Amplitude Modulation) services.

7.12.3.5   ITU-T J.117 Recommendation

Based on the IEEE 1394 standard, the ITU-T J.117 Recommendation establishes a framework for a home network that interlinks a maximum of 63 devices such as VCRs (videocassette recorders), television sets, set-top boxes (STBs), and PCs (Personal Computers) via a four- or a six-wire connection.

7.12.3.6 ITU-T G.902 Recommendation

The ITU-T G.92 Recommendation describes generic guidelines that support network management operations and maintenance services for broadband residential access networks such as cable modem configurations.

7.12.4 INTERNET ENGINEERING TASK FORCE (IETF)

An international standards organization, the IETF (Internet Engineering Task Force) supports implementation of open Internet standards to enable development of a global information infrastructure. IETF participants include vendors, researchers, operators, and network designers.

7.12.4.1 IETF IPCDN (IP-over-Cable Data Network) Working Group

The IETF IP-over-Cable Data Network (IPCDN) Working Group develops standards for implementation of IP-over-cable networks. This Working Group also standardizes SNMP (Simple Network Management Protocol) MIBs (Management Information Bases) to support cable network administration and management services and provision IP multicasts with QoS assurances. Moreover, the IP-over-Cable Data Network Working Group fosters implementation of standards-compliant cable network equipment, a telephone-return interface for upstream transmissions, and symmetric and asymmetric cable network operations.

7.12.4.2 IETF Uniform Resource Locator (URL) Registration Working Group

The IETF URL (Uniform Resource Locator) Registration Working Group defines approaches for defining URLs in a television context and steps for URL registration. URLs (Uniform Resource Locators) enable the recording and playback of television programs and refer to audio, video, and data streams as applications or events.

7.12.5 MOVING PICTURE EXPERTS GROUP (MPEG)

Established in 1988 as a joint International Stands Organization and International Electrotechnical Commission (ISO/IEC) Working Group, MPEG (Moving Picture Experts Group) describes a suite of technical specifications that govern video and audio compression. MPEG specifications also define video and audio coding formats for enabling representation of video and audio sequences in the form of compact coded data. Selected MPEG specifications are highlighted in the subsections that follow.

7.12.5.1 Selected MPEG Specifications

7.12.5.1.1 MPEG-1 and MPEG-2

The MPEG-1 specification describes syntax, compression, and synchronization functions for coded representation of audio and video packets that apply to video-coded compact discs and CD-ROM (Compact Disc-Read Only Memory) formats featuring progressive video sequences that are not interlaced. The MPEG-2 specification describes syntax, compression, and synchronization functions and the use of time-stamps for coded representation of voice, video, and data streams. Designed for interlaced or progressive video sequences, MPEG-2 establishes Quality of Service (QoS) requirements for enabling sequenced data, audio, and video delivery. MPEG-2 specifications apply to television programming, moving pictures, radio broadcasts, and DVDs (Digital Versatile Discs).

7.12.5.1.2   MPEG-4, MPEG-7, and MPEG-21

MPEG-4 serves as the *de facto* standard for delivering Web-based multimedia content. MPEG-7 defines the Multimedia Content Description Interface (MCDI). MPEG-21 promotes utilization of advanced multimedia resources across heterogeneous network environments and establishes a multimedia framework for enabling dependable access to and interactivity with multimedia objects. MPEG-21 also clarifies approaches for content creation, distribution, and production and procedures for intellectual property management.

7.12.6   NATIONAL CABLE TELEVISION ASSOCIATION (NCTA)

The National Cable Television Association (NCTA) monitors cable network developments and infrastructure improvements. This organization has joined with the broadcast industry to implement the TV Parental Guidelines rating system. The American Academy of Pediatrics, the National Association of Elementary School Principals, the National Education Association, the American Medical Association, and the National PTA (Parents Teachers Association) also endorse this rating system.

7.12.6.1   Cable in the Classroom Initiative

Sponsored by the National Cable Telecommunications Association (NCTA), the Cable in the Classroom initiative fosters distribution of high-quality, commercial-free educational programming and online resources to approximately 81,000 schools. In addition, the Cable in the Classroom High-Speed Education Connection program provides free broadband access to the Web for teachers and students in K–12 schools and libraries where cable modem service is available.

7.12.7   WIRELESS COMMUNICATIONS ASSOCIATION (WCA)

Originally known as the Wireless Cable Association International (WCAI), the Wireless Communications Association (WCA) supports innovations in the implementation and delivery of broadband video and bi-directional voice and data services. LMDS (Local Multipoint Distribution System) and MMDS (Multichannel Multipoint Distribution System) operators in France, Australia, Mexico, Russia, and Brazil participate in this association.

7.12.8 WORLDWIDE WEB CONSORTIUM (W3C)

The Worldwide Web Consortium (W3C) designs Web specifications such as the Broadcast Markup Language (BML) for describing television content on the Web. The W3C Television and Web Interest Group coordinates BML initiatives with the Advanced Television Systems Committee (ATSC), the Association of Radio Industries and Businesses (ARIB), the IETF URL (Uniform Resource Locator) Registration Working Group, the Advanced Television Enhancement Forum (ATEF), and the European Broadcast Union (EBU). Approaches for standardizing cable modem equipment, determining the number of users that can be effectively supported by a neighborhood cable network segment, and forecasting the point at which Quality of Service (QoS) is negatively affected in cable networks are under consideration. Metrics for evaluating audio, video, and data throughput and network response time and procedures for reducing packet loss, latency, and jitter on cable networks are in development as well.

7.13 CABLE NETWORK MARKETPLACE

Vendors supporting DOCSIS 1.0 and DOCSIS 1.1 standards-compliant cable modems include 3Com, General Instrument, Hewlett-Packard, Hughes, Intel, IBM, Bay Networks, AT&T MediaOne, Adelphia, Cabletron Systems, and Motorola. In addition, DOCSIS-compliant cable modems are also available from Newbridge Networks, Cisco, NextLevel Systems, Samsung, Toshiba, Nortel Networks, and Thomson Consumer Electronics. Terayon tests cable modem initiatives in Japan and Belgium. Com21 participates in cable modem field trials in Switzerland.

Cable network configurations enable Web browsing, utility monitoring, E-commerce transactions, interactive tele-education programs, telephone services, video-on-demand (VOD), and cablecasts. (See Figure 7.4.)

7.13.1 ATHOME (@HOME) NETWORK

AtHome Network (@Home Network) is a cable Network Service Provider (NSP) currently owned by cable operators including Comcast Corporation, Rogers Cablesystems, AT&T, and Shaw Communications. Based in Redwood City, California, AtHome Network delivers services to approximately 5 million broadband subscribers via a DWDM (Dense Wavelength Division Multiplexing) network that supports transmission rates at 2.488 Gbps (OC-48) in an area of coverage that extends to 15,000 miles. AtHome Network partners with Real Networks in producing high-quality voice, video, data, and imaging applications, and with Segasoft and Liquid Audio in developing Web multimedia content.

7.13.2 AT&T MEDIAONE

AT&T MediaOne brings high-speed Internet connectivity to all schools in its service areas and provisions access to Cable in the Classroom. AT&T MediaOne also

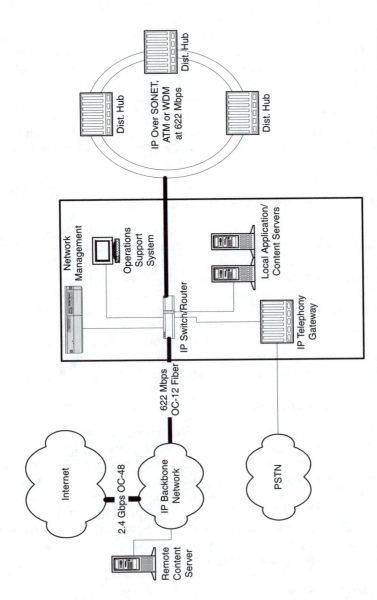

FIGURE 7.4 Configuration that supports IP overlays and works in conjunction with SONET, ATM, and WDM technologies.

provides basic cable service; commercial-free academic programming and instructional support materials, and works in concert with A&E, Discovery, Weather, and ESPN Networks in developing course content and Web resources. The AT&T Media-One COOL classroom initiative informs teachers and students about Web-based educational applications and cable television family viewing programs.

In addition, AT&T MediaOne provisions links to the Real Education initiative for enabling access to undergraduate courses and certificate programs on the Web. Supporters of the Real Education program include the Universities of Pennsylvania, Colorado, and Drexel, and San Francisco, Eastern Michigan, and Connecticut State Universities.

7.13.3 Telewest Communications, Cable & Wireless Communications, and NTL

In the United Kingdom, Telewest Communications, Cable & Wireless Communications, and NTL provision wireline cable services via an HFC infrastructure. These services support access to a television mall that features entertainment, music, news, home shopping, television programs, interactive games, video-on-demand (VOD), travel assistance, educational applications, and E-commerce services. British Airways, Littlewoods Home Shopping Group, and Barclays Bank provision content for this mall. MPEG-2 video compression enables cinema-quality viewing.

7.14 WIRELINE CABLE COMPETITOR SOLUTIONS

Wireline cable systems compete with technologies such as satellite, ISDN, and ADSL (Asynchronous Digital Subscriber Line) networks in supporting access to broadband networks from SOHO venues. As noted in Chapter 6, ADSL is a high-speed broadband residential access technology that supports information transport via the wireline infrastructure already in place for the Public Switched Telephone Network (PSTN). ADSL implementation involves the use of filters to split the existing phone lines into three frequency channels. These channels or circuits support traditional telephone service and enable upstream and downstream transmissions as long as the subscriber is no more than 18,000 feet from the local telephone exchange. If the distance from the local telephone exchange exceeds 18,000 feet, transmission rates decline. An ADSL modem failure only affects the virtual connection over the local loop between the customer premise and the local cable facility.

With cable network implementation, the first user on a neighborhood cable network segment generally receives excellent service. However, each additional subscriber adds traffic to the network segment. If subscribers overload the neighborhood cable network segment with traffic, network services are adversely affected, resulting in transmission slowdowns, bottlenecks, and a noticeable decrease in system reliability and dependability. Because a subscriber shares upstream and downstream cable network channels with other users, a cable system outage at one subscriber premise can cause cable outages at every subscriber household on the same neighborhood cable network segment. Inasmuch as cable channels are shared, security problems occur with greater frequency on cable networks than with ADSL installations.

7.15 WIRELINE CABLE NETWORK IMPLEMENTATION CONSIDERATIONS

Wireline cable configurations enable robust Internet, intranet, and extranet connectivity and real-time access to broadband services. Although early cable adopters seem satisfied particularly with downstream data transport, individuals evaluating cable systems for institutional implementations must be cautious. Cable configurations that enable multimedia deployment were deployed in the late 1990s. Standards are not yet fully developed or universally accepted for interconnecting cable equipment from different suppliers. As a consequence, distributed cable network equipment may not be interoperable. Information transmission from one cable operator's system to another may not be feasible. Suitable options for facilitating reliable and dependable voice, video, and data transport via cable networks are in development.

Because a cable network employs a shared communications platform, information transmissions are subject to degradation as more users are added to the network segment. Cable network operations can also be compromised by outages due to natural disasters such as earthquakes, snowstorms, and hurricanes, and computer and communications problems on the neighborhood cable network segment.

Despite the expanded network bandwidth and capacity associated with HFC installations, technical problems in supporting end-to-end connectivity and ensuring the availability of return channel bandwidth for upstream transport can compromise network performance. Impulse noise, inadvertent fiber cuts, and the condition of the in-place HFC plant can also contribute to signal corruption, attenuation, and degradation. To facilitate troublefree transmission in noisy environments, cable NSPs and MSOs (MultiService Operators) use digital compression technology to increase transmission efficiency and improve network response time.

Wireline cable networks are major contenders for bringing broadband access to diverse populations of users such as homebound learners and telecommuters in residential environments. Currently, 10BASE-T Ethernet is the most popular cable modem interface specification for cable modem installations at SOHO venues in the United States. As a consequence, the speed of the cable connection is automatically limited to 10 Mbps despite the capabilities of cable networks in supporting transmissions at substantially faster rates. Cable service is not universally available. Cable modem subscribers are limited to using cable operators that provision cable service in their neighborhoods and purchasing cable modems that are compliant with the in-place cable configurations. These cable modems may not be compliant with DOCSIS specifications.

To counteract cyberinvasions, the DOCSIS and the EuroDOCSIS specifications define baseline privacy specifications to sustain information integrity and data privacy across the shared cable medium. Nonetheless, cable networks are still susceptible to cyberinvasions by cyberhackers who can gain access to network files and directories maintained by all users sharing the same neighborhood cable network segment.

A nationwide cable network infrastructure is not yet available in the United States. As a consequence, a cable modem solution for a school, a school district, or a university is currently confined to a service area administered by a single cable service provider or cable operators participating in joint partnerships. Therefore, prior to full-scale implementation, pilot tests for evaluating cable network capabilities in enabling network interconnections, interoperable services, and broadband voice, video, and data delivery must also be conducted.

7.16 WIRELESS CABLE NETWORKS

7.16.1 FEATURES AND FUNCTIONS

Advances in technology and demand for fast access to broadband networks drive development of wireless cable networking solutions. Wireless cable service eliminates the need to rebuild, repair, replace, and/or upgrade the in-place coaxial cable or HFC infrastructure. As with landline cable operations, wireless cable deployment involves allocation of channel capacity for delivery of voice, video, and data signals and support of high-speed access to the Internet.

Wireless cable networks complement services supported by wireline cable implementations. As an example, wireless cable systems support interconnectivity to HFC backbone networks for transporting multimedia signals over the local loop, enabling interactive television programming, and supporting IP telephony.

7.16.2 INSTALLATIONS

In broadband fixed wireless cable network transmissions, the reception points or endpoints are stationary. As a consequence, broadband fixed wireless cable transmissions enable users to access network connections at anytime and from anyplace via mobile terminal devices.

Wireless cable network solutions are flexible, scalable, extendible, and affordable. These broadband fixed wireless access (FWA) solutions dependably provision connectivity to high-speed networks in a metropolitan area or at an isolated location, thereby eliminating costs and delays associated with securing easements and rights-of-way in order to modify or upgrade the in-place wireline infrastructure.

In addition, wireless cable operations extend wireline cable service to geographic areas where installing a wireline infrastructure is not permitted or economically feasible. For instance, wireless cable network solutions are used in Eastern Europe where fixed wireline infrastructure services are generally not available and in historic cities such as Florence, Jerusalem, and Venice where wireline cable installations are not permitted.

7.16.3 OPERATIONS

Wireless cable networks transmit voice, video, and data as electromagnetic signals through the air in the super-high frequencies of the electromagnetic spectrum. A basic wireless cable system consists of a transmitter site, the signal path, and the

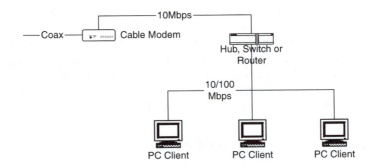

FIGURE 7.5 An office/school cable network configuration.

reception site. The transmitter site modulates digital signals onto microwave channels for broadcast to reception sites. Reception sites range from office buildings and hotels to condominiums and single-family homes. These sites are equipped with special rooftop or window antennas that are linked by coaxial cable to down-converters.

The feeder network extending from the antenna to the down-converter transmits microwave signals via the cable television band to the cable modem inside the home, school, or workplace. In the United States, the cable modem demodulates cable signals and transports these signals to a television set or to a PC via a 10BASE-T Ethernet link via a standard 10BASE-T Ethernet connection and Category 5 copper wiring. Actual throughput is limited in wireless and wireline cable networks to the rates supported by the 10BASE-T Ethernet link. Throughput is also affected by the amount of network traffic, the capabilities of the PC operating system, and the software configuration that is locally employed. (See Figure 7.5.)

7.16.4 WIRELESS CABLE SERVICES

With asymmetric wireless cable networks, POTS, ISDN, or DSL technology enables information transport on the return path. Upgrading asymmetric wireless cable networks to enable two-way transmission involves installation of a transverter at the customer premise to ensure data, voice, and video transmission on the return path in the upstream direction.

In addition, two-way symmetric wireless cable solutions supporting broadband applications and multimedia delivery are available from wireless cable operators that include Wireless One and Integrity Communications, CAI Wireless, DirectNet, Look Communications, Cellular Vision, General Instrument, Metro.Net, NextLevel Systems, and New Media Communications. As with wireline cable networks, wireless cable networks feature a variety of configurations and services.

7.17 MULTICHANNEL DISTRIBUTION SYSTEM (MDS)

Initially designed as a video program service, Multichannel Distribution System (MDS) supports high-speed multimedia transmission and cable network operations in the spectrum between the 2.150 GHz and 2.162 GHz frequencies. MDS signals are not affected by atmospheric conditions. As a consequence, cable customers in rural, underserved, suburban, and urban locations use this service. MDS implementations require a direct line-of-sight between transmitters and receivers.

MDS solutions enable downstream rates that range between 750 Kbps and 11 Mbps. Typically, a PSTN link supports upstream transmission.

Wireless cable operators generally aggregate the available MDS spectrum for providing up to 200 MHz of bandwidth or the equivalent of approximately 34 analog television channels for enabling advanced networking applications and services. The Instructional Television Fixed Service (ITFS), the Multichannel Multipoint Distribution System (MMDS), and the Local Multipoint Distribution System (LMDS) networks are based on MDS technology.

7.17.1 MDS in Action

7.17.1.1 Antenna Hungaria

In Budapest, Antenna Hungaria supports MDS implementations to provision Internet access and high-speed broadband services. The Antenna Hungaria system provides DVB and MPEG-2 (Moving Picture Experts Group-2) services and asymmetric transmissions. POTS (Plain Old Telephone System) links foster transmission on the return path in the upstream direction.

7.17.1.2 DirectNET

In Fort Lauderdale, Florida, DirectNET delivers high-speed wireless broadband Internet services in the MDS spectrum to business establishments in the downstream direction. Subscribers employ POTS or ISDN BRI (Basic Rate Interface) connections for enabling transmissions via the return channel.

7.18 INSTRUCTIONAL TELEVISION FIXED SERVICE (ITFS)

Instructional Television Fixed Service (ITFS) networks support bi-directional or full-duplex services in the spectrum between the 2.500 and 2.596 GHz frequencies via 6 MHz channels. ITFS broadcasts consist of multidirectional signals that are transmitted via direct-line-of-sight technology over large geographic areas of coverage from broadcast or microwave towers to reception sites that are equipped with special television antennas and converters for receiving ITFS programming.

The FCC regulates utilization of ITFS frequencies and grants licenses for ITFS operations. Based on FCC rulings approved in 1998, MDS licensees can provide high-speed, high-capacity symmetric networking services and broadband applications such as videoconferencing and continuing tele-education courses in the ITFS spectrum. Moreover, wireless cable operators also use channel capacity originally reserved for Instructional Television Fixed Service (ITFS) to broadcast educational programs.

7.19   MULTICHANNEL MULTIPOINT DISTRIBUTION SYSTEM (MMDS) AND LOCAL MULTIPOINT DISTRIBUTION SYSTEM (LMDS)

7.19.1   MMDS AND LMDS OVERVIEW

MMDS (Multichannel Multipoint Distribution System) and LMDS (Local Multipoint Distribution System) technologies enable fixed wireless broadband transmissions and employ protocols that include TDMA, FDMA, and OFDM. In MMDS and LMDS configurations, satellite and cable programming is distributed to headend equipment at the local cable operator facility. MMDS and LMDS networks transport video, voice, and data signals within multiple contiguous or overlapping cells.

MMDS and LMDS implementations are easily deployed, bypass local loop congestion, and eliminate costs associated with optical fiber installation. Moreover, MMDS and LMDS solutions support on-demand bandwidth to accommodate subscriber requirements.

Obstructions such as dense tree cover, hills, tall buildings, vegetation, and foliage hinder MMDS and LMDS reception. As a consequence, multiple transceivers are used in locations where line-of-site reception is blocked. In addition, atmospheric gases, rainstorms, and blizzards adversely impact MMDS and LMDS operations. Multipath signal distortion and signal interference from adjacent and overlapping cells also negatively impact the reliability and dependability of MMDS and LMDS network solutions. (See Figure 7.6.)

7.20   MULTICHANNEL MULTIPOINT DISTRIBUTION SYSTEM (MMDS)

7.20.1   MMDS SERVICES

Also called Multipoint Microwave Distribution Service and Multichannel Multipoint Distribution Service, Multichannel Multipoint Distribution System (MMDS) initially supported analog television signal transmission in the downstream direction. Currently, MMDS solutions provide broadband fixed wireless access cable service in areas of low-density population where installation of a conventional coaxial cable plant or an HFC infrastructure is disruptive and costly. MMDS solutions employ channels that are 6 MHz wide and support licensed and licensed-exempt operations in the ultra-high frequency (UHF) spectrum. MMDS licenses are available at FCC auctions for every Basic Trading Area (BTA) in the United States.

7.20.2   MMDS OPERATIONS

MMDS implementations that operate in licensed spectrum between the 2.596 and 2.644 GHz frequencies and between the 2.686 and 2.689 GHz frequencies are deployed in countries that include Ireland, Mexico, and the United States. As with MDS and LMDS (Local Multipoint Distribution System) implementations, MMDS

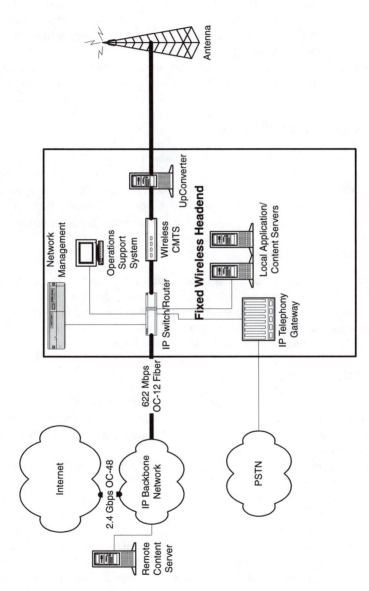

FIGURE 7.6 A broadband residential fixed wireless access (FWA) cable network solution.

configurations require installation of a relatively large number of repeaters and transmitters for transporting microwave signals via a direct-line-of-sight pathway to small antennas mounted on rooftops at subscriber venues.

An MMDS implementation consists of destination devices such as receivers or transceivers and antennas at the customer premise to enable access to a mix of voice, video, and data services, and cable, terrestrial, and satellite programs. Headend equipment includes satellite signal reception devices and a radio transmitter equipped with an omni-directional antenna that is installed at the highest point in the designated area of coverage. The area of coverage can extend to a radius of 100 kilometers if the terrain is flat. Generally, however, the MMDS coverage area extends to a radius of 50 kilometers. (See Figure 7.6.)

7.20.3   MMDS and ITFS Services

Because the MMDS and ITFS spectral allocations overlap, MMDS licensees can access ITFS channels through lease agreements. MMDS licensees can also acquire licenses for as many as eight unused ITFS frequencies in a BTA as long as eight frequencies remain available for ITFS service. A portion of each MMDS 6 MHz channel available in the ITFS spectrum is reserved for educational programming.

7.20.4   MMDS Applications

Also called wireless cable, MMDS networks support interactive services and applications such as electronic banking, online shopping, Web access, interactive games, video-on-demand (VOD), near-video-on-demand (NVOD), and delivery of tele-education courses and teletests consisting of multiple-choice questions. Generally, MMDS television programming is based on satellite feeds.

7.20.5   MMDS Vendor Initiatives

With the evolvement of video technology into a digital format, Sprint transformed MMDS analog video channels into 99 digital channels. Each channel transports streaming voice, video, and data at 10 Gbps. In addition, Sprint and WorldCom offer MMDS service in Phoenix, Arizona.

Heartland Wireless provisions MMDS service in Sherman, Texas, and CAI Wireless offers MMDS service in New York City and the greater Washington, D.C. metropolitan area. Nucentrix Broadband Networks conducts MMDS field trials in Austin, Texas.

Cisco Systems supports licensed MMDS deployments with V-OFDM modulation in spectrum between the 2.596 and 2.644 GHz frequencies and license-exempt MMDS implementations in the 5.7 GHz Unlicensed-National Information Infrastructure (U-NNI) frequencies. MMDS solutions based on the use of cable modems with 64 QAM technology support downstream transmission rates at 27 Mbps via licensed-exempt bands and transmission rates reaching 1 Gbps via licensed RF (Radio Frequency) bands.

7.21 LOCAL MULTIPOINT DISTRIBUTION SYSTEM (LMDS)

7.21.1 LMDS Features and Functions

Also called Local Microwave Distribution Service and Local Multipoint Distribution Service, Local Multipoint Distribution System (LMDS) supports fixed wireless access point-to-multipoint networking solutions. LMDS networks deliver a full range of broadband services to businesses, single-family homes, and multistory apartment buildings and condominiums.

A high-capacity, scalable, and flexible technology, LMDS technology accommodates residential, healthcare, library, school, and business networking requirements. LMDS installations support telemedicine, tele-education, and teleworking applications, and enable videoconferencing, video monitoring, and Video on-Demand (VOD).

LMDS broadband fixed wireless access solutions are flexible, reliable, inexpensive, and dependable, and overcome local loop barriers in provisioning access to high-speed, bandwidth-intensive voice, video, and data applications. LMDS network solutions also support direct broadcast of satellite programs and full-duplex transmissions.

7.21.2 LMDS Operations

LMDS employs a point-to-multipoint architecture for optimizing spectral efficiency and utilizes multiple adjacent or overlapping cells for information distribution within a radius of three to five miles. LMDS technology supports operations in the high-frequency millimeter waveband between the 27.5 and 29.5 GHz frequency block. In this spectrum, wavelengths vary in size from one to ten millimeters.

Situated in the center of a small cluster of LMDS cells, the hub or base station consists of transceivers affixed to towers spaced several kilometers apart for enabling transport of wireless traffic to and from the customer premise. Each hub provisions broadband fixed wireless access services to several thousand SOHO venues. Hubs are interlinked via optical fiber terrestrial connections to support mixed-mode ATM and wireless cable applications.

Each LMDS customer premise is equipped with a terminal station consisting of a small directional antenna mounted on the rooftop. A basic network interface unit (NIU) supports signal modulation, demodulation, and in-building wireline interface applications. A radio link with direct line-of-sight requirements interconnects the network terminal to the hub.

In the business sector, LMDS networks support downstream rates ranging between 51.84 Mbps and 155.52 Mbps. LMDS solutions work in concert with backbone network technologies such as Ethernet, Fast Ethernet, Frame Relay, ATM, and Packet-over-SONET/SDH (POS). As with MMDS transmissions, LMDS signals are transported over a longer range and at higher power in licensed spectrum than in license-exempt spectrum.

7.21.3 LMDS Implementation Considerations

Lack of standards, high costs, and direct line-of-sight requirements hamper LMDS implementation. In addition, system manufacturers differ on whether to employ TDD

(Time-Division Duplexing) or FDD (Frequency-Division Duplexing) technology. TDD supports a single communications channel with shared bandwidth. By contrast, FDD requires the utilization of two separate channels for upstream and downstream transmission and virtual point-to-point connections via the local loop. LMDS providers such as Teligent, XO Communications, and Winstar provision LMDS services in the 28 GHz spectral band.

7.21.4   U.S. FEDERAL COMMUNICATIONS COMMISSION (FCC) LMDS AUCTIONS

The FCC auctions spectrum in 1.3 GHz increments in spectrum between the 27.5 and 28.35 GHz frequencies, between the 29.1 and 29.25 GHz frequencies, and between the 31.075 and 31.25 GHz frequencies for LMDS operations. As a consequence of these auctions, LMDS spectrum is licensed by more than 490 Basic Trading Areas (BTAs).

LMDS license holders develop two-way or symmetric broadband fixed wireless access networks for supporting a combination of services and applications. LMDS auction winners include WNP Communications, Cortelyou Communications, and Eclipse Communications.

7.21.5   LMDS VENDOR INITIATIVES

In contrast to the expensive and time-consuming process of installing a hybrid optical fiber coaxial cable (HFC) infrastructure for conventional cable networks, LMDS NSPs (Network Service Providers) readily implement turnkey LMDS solutions. Representative LMDS operators supporting service in delimited areas in the United States include BellSouth, Gateway Telecom, South Central Telephone Cooperative, NextLink, Tri Corners Telecommunications, and Hybrid Networks. Cisco Systems, Bosch Telecom, and SpectraPoint Wireless provision LMDS solutions for LMDS license holders in Australia. Formus Communications provisions LMDS services in Budapest, Hungary, and Strasbourg, France.

7.21.5.1   Alcatel

Alcatel LMDS solutions support scalable networking services for small- and medium-sized businesses, SOHO venues, and apartment buildings in suburban and urban locations. Alcatel LMDS solutions enable fast access to the Internet, intranet, and extranet; Quality of Service (QoS) guarantees; and LAN-to-LAN interconnectivity. In addition, Alcatel LMDS installations facilitate voice, video, and data transport from high-speed ATM and Frame Relay backbone networks to and from the customer premise. Alcatel LMDS implementations also feature data encryption and channel coding capabilities for enabling secure multimedia delivery.

7.21.5.2   Eagle Wireless International

Eagle Wireless International has developed a wireless set-top box (STB) that transmits broadband voice, video, and data at rates reaching 11 Mbps via wireless cable networks. This solution supports LMDS and MMDS operations in spectrum between the 2 and 2.5 GHz frequencies.

7.21.5.3   Korea Telecom

Korea Telecom evaluates LMDS capabilities in a point-to-multipoint wireless networking testbed that supports voice, video, and data distribution in major Korean cities. Spectrum between the 24 and 31 GHz frequencies support testbed services. Vans equipped with mobile testing equipment monitor radio frequency (RF) propagation and collect data on LMDS network throughput. The ability of LMDS to deliver high-quality, high-speed services in conjunction with ATM technology is also examined.

7.21.5.4   Netro Corporation

Netro Corporation supports LMDS applications in the 28 GHz spectral band in Idaho, Hawaii, and Oregon. Based on packet-switching technology, this LMDS solution provisions fast Internet access and dependable delivery of data, video, and voice services.

7.21.5.5   WavTrace

WavTrace implements point-to-multipoint LMDS networks that employ TDD (Time-Division Duplexing) technology for supporting video, data, and voice transport via a single channel. TDD technology allocates bandwidth dynamically on-demand for handling bursty traffic in the upstream and downstream directions. In contrast to TDD, FDD (Frequency-Division Duplexing) employs two separate and dedicated channels for enabling full-duplex communications. With FDD, one channel is designed for reception and the other for transmission. Channel capacity is established at the time of implementation by predefining limits on information throughput. Artificial guardbands separate transmission and reception frequencies. WavTrace also partners with the Virginia Polytechnic Institute and State University (Virginia Tech) in provisioning LMDS services.

7.22   MULTIPOINT COMMUNICATIONS SYSTEMS (MCS)

7.22.1   MCS Features and Functions

Sponsored by the Canadian Network for the Advancement of Research, Industry, and Education (CANARIE), the MCS (Multipoint Communications System) Alliance promotes deployment of MCS configurations that support delivery of multimedia services to rural and urban locations across Canada. An MCS network is the Canadian equivalent of an MDS implementation in the United States. As with MDS deployments, MCS networks foster access to E-commerce applications and instructional and interactive analog and/or digital television broadcasts; connections to the Internet, intranets, and extranets; and fast and dependable voice, video, and data transmission to businesses, schools, and libraries.

MCS service enables operations in spectrum between the 2.500 and 2.686 GHz frequencies. Industry Canada and the FCC support the mutual use of MCS digital and analog systems that operate in spectrum between the 2.500 and 2.686 GHz frequencies within 80 kilometers of the border between the United States and Canada.

7.23   LOCAL MULTIPOINT COMMUNICATIONS SYSTEM (LMCS)

7.23.1   LMCS Services

Local Multipoint Communications System (LMCS) implementations are the Canadian equivalent of LMDS solutions in the United States. LMCS technology supports E-commerce transactions, cable television programming, and tele-instruction. LMCS signals are distributed from a central station via intercellular connections or wireless broadband radio relays to and from SOHO and business venues situated within a radius between four and five kilometers. The LMCS infrastructure features a mesh topology consisting of overlapping cells.

In response to increased demand for high-speed local access connections, Industry Canada also makes spectrum in the 24 and 38 GHz frequency bands available for LMCS implementations. LMCS license holders include MaxLink Communications and Call-Net Enterprises. LMCS field trials are conducted in cities in Canada, Brazil, and the United States. An Institute of Industry Canada, the Communications Research Center evaluates capabilities of LMCS operations.

7.23.2   LMCS in Action

7.23.2.1   Videotron

Videotron supports access to high-speed Web services in Montreal and in Eastern Quebec townships with LMCS equipment supplied by Motorola. Norsat International supports development of low-cost network architecture for enabling affordable LMCS deployments across Canada.

7.23.2.2   WI-LAN

WI-LAN of Calgary, Alberta, holds patents on LMCS wireless broadband technologies that include MC-DSSS (Multicode-Direct-Sequence Spread Spectrum) and W-OFDM (Wideband-Orthogonal Frequency-Division Multiplexing). MC-DSS solutions optimize utilization of spread spectrum technology for enabling secure, high-speed transmissions. W-OFDM deployments enable seamless service between transceivers that work in conjunction with Fast Ethernet and ATM technologies. W-OFDM multiplexing services are based on OFDM (Orthogonal Frequency-Division Multiplexing) technology.

7.24   MULTIPOINT VIDEO DISTRIBUTION SYSTEMS (MVDS)

7.24.1   MVDS Fundamentals

ETSI designated frequency bands above 10 GHz for MVDS (Multipoint Video Distribution System) installations in the European Union. Currently, MVDS implementations enable operations in spectrum between the 40.5 and 42.5 GHz frequencies. MVDS technology also works in concert with MPEG-2 technology.

MVDS implementations provision access to interactive broadband services, teleshopping applications, pay-per-view broadcasts, and multichannel cable programs. Available from Swisscom, MVDS service facilitates delivery of video programming to remote villages in Switzerland. In the United States, the FCC auctions MVDS spectrum. MVDS technology extends the reach of satellite systems and cable configurations and works in concert with DVB/DAVIC standards and specifications.

As with MDS, LMDS, and MMDS, MVDS technology employs a cellular point-to-multipoint radio system that transports multimedia services from a hub or central transmitter to local neighborhoods. Large-sized cells support MMDS and MVDS operations and enable information services in an area of coverage that extends to 5 kilometers.

7.24.2 European Conference of Postal and Telecommunications Administration (CEPT) and Multimedia Wireless Systems (MWS)

In 1999, the European Conference of Postal and Telecommunications Administration (CEPT) endorsed the recommendation by the European Radio Communications Committee for using spectrum between the 40.5 and 43.5 GHz frequencies to support Multimedia Wireless System (MWS) implementations based on technologies that include Multipoint Video Distribution System (MVDS). MWS (Multimedia Wireless Systems) are cellular point-to-multipoint radio systems that deliver multimedia applications such as videoconferencing and video-on-demand (VOD) to SOHO venues. CEPT encourages deployment of MWS services by member states in the European Union and, as indicated, allocates frequency bands for MWS initiatives.

7.25 WIRELESS AND WIRELINE CABLE NETWORK SYSTEMS INITIATIVES

Continuing innovations in cable system technologies, cable modems, and distributed cable networks facilitate an interesting range of initiatives and activities. In the field of education, cable technology serves as a platform for supporting interactive distance learning solutions involving groupware communications, Web browsing, educational television programming, video services, and multimedia distribution. In the business sector, cable modem technology fosters teleworking and telecommuting services. The initiatives that follow highlight the capabilities of wireline and wireless cable system solutions in addressing information communications requirements. It is interesting to note that wireline and wireless cable service operations are often widely disbursed and available locally only in selected neighborhoods.

7.25.1 Arizona

7.25.1.1 Mojave CC (Community College) Connectivity Initiative

The Mojave CC (Community College) Connectivity Initiative fosters development of a cable network based on an HFC infrastructure to provision access to tele-education, enrichment classes, and Internet access by students and their teachers at K–12 schools. The Mojave CC Connectivity project also supports implementation of a wireless cable network that interlinks four K–12 schools in rural communities in Northwestern Arizona.

7.25.2 CALIFORNIA

7.25.2.1 California Institute of Technology (Cal Tech)

California Institute of Technology (Cal Tech) employs a VWLAN (Virtual Wireless Local Area Network) for enabling off-campus faculty working at home to readily access imaging applications, printers, databases, and other educational resources via wireline cable network connections. Charter Communications, the cable operator for Cal Tech, furnishes cable modem service that supports shared bandwidth on the upstream and downstream channels.

7.25.2.2 University of California at Berkeley (UC Berkeley)

The University of California at Berkeley (UC Berkeley) operates an asymmetric high-bandwidth wireless cable configuration that fosters data delivery at rates reaching 30 Mbps on the downstream path via a broadcast data channel superimposed on a standard 6 MHz television channel. Each reception site requires a directional antenna and a cable modem that is interfaced with a PC. A POTS link enables upstream transmissions on the return path.

7.25.3 FLORIDA

7.25.3.1 Duval County School System

AT&T MediaOne provides high-speed cable network services in the Duval County School System in exchange for 1200 square feet of land in ten locations. This acreage is used for HFC infrastructure installations. The Duval County School System cable network configuration features a VPN that interlinks 152 schools and facilities.

7.25.4 GEORGIA

7.25.4.1 Georgia Public Broadcasting

Georgia Public Broadcasting and Clayton College and State University offer an Associate of Arts degree program featuring telecourses that are delivered via an HFC cable network system.

7.25.4.2 University System of Georgia (USG)

At the University System of Georgia (USG), cable networks enable staff development activities and tele-education course delivery to SOHO venues. A USG participant, Dalton College uses local educational access cable channel services to produce and deliver college telecourses for credit to lifelong learners.

7.25.5 HAWAII

7.25.5.1 Hawaiian Institutional Network (I-Net)

As part of their cable franchise contracts, commercial cable companies in Hawaii provide public access channels for educational programming. These companies are also required to contribute programs to support the Hawaiian I-Net (Institutional Network). I-Net interlinks public-sector entities such as schools, universities, research centers, and government agencies in a statewide network configuration.

7.25.5.2 University of Hawaii

The University of Hawaii Information Technology Services Division offers courses for credit and educational cablecasts via the University of Hawaii public access cable network. Tele-education programs are presented in real-time. The public access cable network features return channels with audio capabilities so that students can interact with instructors by telephone. The University of Hawaii public access cable system also supports delivery of an Associate of Arts Degree Program, personal enrichment classes, and job teletraining services.

7.25.6 LOUISIANA

7.25.6.1 Tulane University Cable Network (TUCAN)

The Tulane University Cable Network (TUCAN) is a university-operated and a university-owned wireline cable system that supports links to educational programs, interactive teleconferences on special topics, a campus video bulletin board, and Web resources.

7.25.7 MASSACHUSETTS

7.25.7.1 Massachusetts Institute of Technology (MIT)

MIT Cable Television features pre-recorded and live broadcasts of current events, help sessions, and class lectures. Classrooms, offices, and dormitories on campus are among facilities connected to the MIT cable configuration.

7.25.7.2 University of Massachusetts at Lowell

The University of Massachusetts at Lowell utilizes an interactive cable network for delivery of curricular enrichment and professional training teleprograms to four campuses and fifteen local public school districts. In addition to the cable network infrastructure, microwave and satellite technologies support student and faculty interactivity at geographically separated sites.

7.25.8 MICHIGAN

7.25.8.1 AT&T MediaOne Connections Program

AT&T MediaOne provisions high-speed cable network services at no charge to public schools in Detroit through the Connections Program. The Connections Program supports unlimited access to the AT&T MediaOne HFC infrastructure and transmission rates at 1.5 Mbps downstream and 300 Kbps upstream.

7.25.8.2 Mona Shores School District

The Mona Shores School District uses an ATM backbone network that works in concert with individual Ethernet school LANs for enabling access to bandwidth-intensive multimedia applications and services that accommodate the diverse learning styles of district students. Information transport rates at 155.52 Mbps are supported. This configuration also supports delivery of teleclasses and telecoursework via cable channels to homebound students with disabilities.

7.25.8.3 Pace Telecommunications Consortium

Situated in Northern Michigan, the Pace Telecommunications Consortium includes the Littlefield, Mackinaw City, Pellston, Central Lake, and Boyne City public school districts among its membership. This Consortium operates an ITFS network that provisions data, video, and voice services and interactive instruction.

7.25.9 MISSOURI

7.25.9.1 Big Horn Project

Sponsored by FOCUS (Fiber Optic Consortium United Schools), the Big Horn project provisions delivery of preschool teleprograms and tele-education courses to a tribal college and public schools in rural and geographically isolated communities. The Big Horn ITV (Instructional Television) configuration operates in conjunction with an ATM infrastructure for enabling access to enrichment courses in mathematics, science, fine arts, and foreign language; adult education courses in technology, business, and entrepreneurship; and teleseminars on ranching, agriculture, and economic development.

7.25.9.2 High Plains Education Consortium (HPEC)

The High Plains Education Consortium (HPEC) includes rural school districts in central Montana among its membership. The HPEC network infrastructure employs cable networks for provisioning access to tele-education programs. HPEC schools share teachers and educational resources for enabling students to take required courses that otherwise would not be available as a consequence of budgetary and enrollment constraints. Plans for linking HPEC to college and university networks in Eastern and Western Missouri are under consideration.

7.25.9.3 Monnett Public School System

Sponsored by the Missouri Department of Elementary and Secondary Education, the Monnett Public School System provisions cable network service in a metropolitan network configuration for enabling teachers and staff to share educational resources and, thereby meet state accreditation requirements. This network also enables high school students to participate in college programs.

7.25.9.4 *Scobey Public School System*

In a FOCUS initiative in Northeastern Montana, the Scobey Public School System utilizes cable ITV programs to train volunteer firefighters and emergency medical personnel and provide access to continuing education teleprograms for accountants, farmers, and nurses. In addition, the Scobey Public School System sponsors cable ITV telecourses in advanced placement (AP) calculus, foreign languages, and agriculture for high school students.

7.25.9.5 Southwest Missouri Cable TV

In connection with the FCC mandate that cable operators provide public benefits to local communities in their service areas, Southwest Missouri Cable TV supports links to tele-education applications and services in regional schools.

7.25.10 MONTANA

7.25.10.1 Salish Kootenai College (SIC)

The Media Center at Salish Kootenai College (SKC) enables the Reservation population to access SKC TV Public Television Programs featuring documentaries and distance education courses. In addition, SKC also produces teleprograms for the American Indian Higher Education Consortium (AHEC) Satellite Network. The Department of Nursing at SKC also sponsors access to nursing education for Native Americans and rural populations. An RN (Registered Nursing) to BSN (Bachelor of Science in Nursing) program in a distance education format enables Registered Nurses to remain employed while attending virtual cable teleclasses leading to the BSN degree.

7.25.11 NEW JERSEY

7.25.11.1 MercerNet

A Cable in the Classroom School of the Future project, MercerNet interlinks community college campuses, a science center, schools, school districts, and libraries in Mercer County, New Jersey. Sponsored by the MercerNet Consortium, this cable network implementation provisions access to the Web and supports high-speed voice, video, and data transmission via a fiber optic backbone that operates at 10 Mbps. MercerNet also provisions links to interactive distance learning classrooms situated in high schools throughout the county. Mercer County Community College provides technical support services for MercerNet programs.

7.25.12 SOUTH CAROLINA

7.25.12.1 South Carolina Educational Television (SCETV)

SCETV (South Carolina Educational Television) originally utilized four ITFS channels for delivery of educational programs to K–12 schools, community colleges, universities, state agencies, and businesses. This installation was replaced by a 32 digital satellite channel system to handle increased system demand. Currently, the four ITFS channels support transmission of tele-education programs over delimited distances to local schools and school districts.

7.25.13 TENNESSEE

7.25.13.1 Anderson County Schools

Anderson County Schools use a cable configuration to interlink homes, apartment buildings, schools, and businesses in a low-cost metropolitan initiative called the Jericho Project. This initiative facilitates teacher and parent teleconferences and fast Internet access for approximately 6,700 students in the Anderson County school district.

7.25.14 VIRGINIA

7.25.14.1 Henrico Public Schools

At Henrico Public Schools, one of the largest public school systems in Virginia, a hybrid fiber optic coaxial cable network, already in place for cable television transmission, supports delivery of televised distance learning courses to students at local high schools.

7.25.14.2 Virginia Polytechnic Institute and State University (Virginia Tech)

Following participation in a FCC LMDS auction in 1998, Virginia Tech (Virginia Polytechnic Institute and State University) became the first university in the United States to own four LMDS licenses. These LMDS licenses cover 16,507 square miles of territory in Virginia as well as portions of Tennessee and North Carolina.

7.25.14.2.1 Virginia Tech and WavTrace

Virginia Tech partners with WavTrace in offering LMDS applications. In an initial LMDS trial involving utilization of WavTrace equipment, Virginia Tech enabled delivery of full-duplex, high-speed voice, data, and video traffic from a hub site on campus to three off-campus office buildings. This trial demonstrated the viability of using a simplified LMDS architecture that is affordable and easy to install, as well as the flexibility of TDD services in enabling applications in spectral bands considered too small for FDD operations.

Virginia Tech and WavTrace also offer LMDS service to rural homes and businesses in Blacksburg, Virginia. This LMDS configuration supports interactive full-motion video, IP telephony, and voice, video, and data transmission at full-duplex rates reaching 4.5 Mbps.

7.25.14.2.2 Virginia Tech and the Center for Wireless Telecommunications

Virginia Tech and the Center for Wireless Telecommunications formed the LMDS Research Consortium for LMDS licensees, manufacturers, service providers, and vendors. This Consortium facilitates deployment of low-cost and easy-to-use LMDS services at SOHO venues and schools in rural communities, and evaluates LMDS capabilities in provisioning broadband telemedicine, distance education, and video-conferencing applications in regional LMDS testbed implementations.

7.26 INTERNATIONAL CABLE NETWORK TELE-EDUCATION INITIATIVES

Wireless and wireline cable network configurations are implemented worldwide in countries that include Brazil, Singapore, France, Australia, Slovenia, and Ecuador. Representative international cable network initiatives in the tele-education arena are highlighted in this section.

7.26.1 CANADA

7.26.1.1 Canadian Online Exploration and Collaborative Environment for Education (COECEE)

The COECEE initiative supports the integration of cable television, cable telephony, and the Internet into an interoperable telelearning system for enabling Canadians, regardless of location, to equitably access teletraining programs and skills development services. The Open Learning Agency in British Columbia and Simon Fraser University participate in this initiative.

7.26.1.2 Ontario Colleges of Applied Arts and Technology, Brock University, and TVOntario (TVO)

The Ontario Colleges of Applied Arts and Technology, Brock University, and TVOntario (TVO) provision a certificate and an undergraduate degree program in adult education for working adults. Faculty at the Ontario Colleges of Applied Arts and Technology and Brock University develop and design courses. TVO (TV Ontario) then produces videotapes of these courses featuring faculty, subject experts, and occasionally members of Toronto's Second City acting troupe. These videotapes are subsequently broadcast over cable television channels to groups of learners at designated community college locations.

7.26.1.3 Province Learning Network (PLnet)

British Columbia uses Delta Cable and Coast Cable Communications for provisioning access to the PLnet (Province Learning Network) initiative. PLnet supports broadband connections to cultural, scientific, and educational programs and organizations throughout the province.

7.26.2 GERMANY

7.26.2.1 University of Hanover

At the University of Hanover, dormitories are equipped with bi-directional cable networks that are interlinked to the university's broadband backbone network via

10BASE-T Ethernet connections. The cable network infrastructure supports tele-commuting applications, fast Web connectivity, and access to telelectures distributed on the German Broadband Science Network. Downstream rates at 1 Mbps and upstream rates at 500 Kbps are supported. Each student pays a nominal fee for dedicated bandwidth and cable services.

7.26.3 HUNGARY

7.26.3.1 Antenna Hungaria

Antenna Hungaria operates an MMDS network in Budapest for provisioning high-speed Internet connections and access to intranet services and educational resources. Rates downstream reach 52 Mbps; standard telephone service supports upstream transmissions. The MMDS network is compliant with DVB and MPEG-2 specifi-cations and also supports satellite program distribution.

7.26.4 SPAIN

7.26.4.1 Open University of Catalan

The Open University of Catalan in Barcelona employs a cable network configuration to deliver distance learning telecourses to off-campus students.

7.27 U.S. CABLE NETWORK TELEMEDICINE INITIATIVES

7.27.1 MONTANA

7.27.1.1 Eastern Montana Telemedicine Network (EMTN)

The Eastern Montana Telemedicine Network (EMTN) employs an interactive HFC cable network infrastructure for enabling healthcare providers in rural communities to access Web resources. This network also provisions access to advanced college preparation courses.

7.27.2 TEXAS

7.27.2.1 University of Texas Medical Branch (UTMB)

The University of Texas Medical Branch (UTMB) provides cable television service to patient rooms, student dormitories, classrooms, and auditoriums. Education chan-nels distribute information on medical procedures to patients and their families. UTMB also downlinks satellite videoconferences and continuing education pro-grams for real-time viewing or viewing by tape delay, depending on faculty, staff, and student requirements. The satellite configuration features a 4.5-meter movable C-band and Ku-band antenna. Programs are distributed directly to conference centers or to remote sites via the UTMB videoconferencing network.

7.28 EUROPEAN COMMISSION TELEMATICS APPLICATIONS PROGRAM (EC-TAP)

7.28.1 DOMESTIC INTERACTIVE TELEMATIC EDUCATION AND LEARNING (DOMITEL)

The DOMITEL initiative illustrated the capabilities of a wireline cable network that operated over an HFC infrastructure in enabling cultural minorities and adult learners to access telelectures, teletraining programs, teletutorials, videoconferences, and tele-education courses. The DOMITEL project also demonstrated the effectiveness of an HFC wireline cable network in distributing professional development courses to unemployed graduate engineers in Lahti, Finland, and teletraining courses to women returning to work in Dublin, Ireland.

7.28.2 ELECTRONIC LEARNING ENVIRONMENT FOR CONTINUAL TRAINING AND RESEARCH (ELECTRA)

The ELECTRA initiative provided researchers, tutors, students, home learners, and medical practitioners situated in the European region of Meuse-Rhine with access to advanced multimedia training materials and educational programs via wireline cable networks. Findings from the ELECTRA initiative contribute to the development of a virtual European university.

7.29 EUROPEAN COMMISSION ADVANCED COMMUNICATIONS TECHNOLOGIES AND SERVICES (EC-ACTS) PROGRAM

7.29.1 ADVANCED RESOURCE MANAGEMENT IN SERVICE INTEGRATED AND MULTI-LAYERED HFC ACCESS NETWORKS (AROMA)

The AROMA project confirmed the capabilities of wireline cable networks in provisioning access to an ATM-over-SDH core network and supporting delivery of broadband services over the last mile between the local cable operator facility and the subscriber premise.

7.29.2 ATM APPLICATIONS OVER HYBRID OPTICAL FIBER COAX (ATHOC)

The ATHOC project verified the flexibility, extendibility, and dependability of cable networks in enabling interoperations with ATM, IP, and SDH (Synchronous Digital Hierarchy) technologies. This initiative also validated the performance of the ATHOC HFC infrastructure in effectively supporting full-duplex transmission rates at 34 Mbps, facilitating access to Web resources, and delivering voice, video, and data applications to SOHO venues.

7.29.3 INTEGRATED BROADBAND COMMUNICATIONS ON BROADCAST NETWORKS (IBCoBN)

The IBCoBN initiative demonstrated the feasibility of using broadband cable networks for enabling video telephony and videoconferencing applications and dependable access to Web resources. In addition, this project validated capabilities of residential cable networks in fostering links to teleshopping, teletraining, and teleworking applications. Senior citizens and individuals with disabilities participated in IBCoBN trials in Belgium, Germany, France, Spain, Portugal, Russia, and the United Kingdom.

7.30 EUROPEAN COMMISSION INFORMATION SOCIETY TECHNOLOGIES (EC-IST) PROGRAM

7.30.1 MTV

The MTV television initiative supports implementation of network architecture for enabling personalized digital television services. This initiative examines capabilities of an interoperable platform that works in compliance with open standards including the DVB (Digital Video Broadcast) MHP (Multimedia Home Platform) in supporting multimedia services.

7.30.2 VIDEOGATEWAY

The VIDEOGATEWAY project enables the design and development of an advanced video gateway prototype that operates over cable network, Gigabit Ethernet, DSL, and ATM platforms. This prototype supports the exchange of narrowband analog streaming video traffic on the public Internet with broadband digital streaming video traffic on the next-generation Internet. In addition, the VIDEOGATEWAY initiative facilitates MPEG and DVD operations and provisions on-demand access to real-time video and audio applications.

7.31 SUMMARY

The first mile refers to the local loop or connection between a residence, school, or business and the local telephone exchange where a communications link supports connectivity to backbone networks that provision access to broadband applications and Web resources at fast rates. Barriers and constraints associated with the existing PSTN infrastructure in accommodating user demand for access to bandwidth-intensive distance learning and teletraining applications over the first mile contribute to accelerating deployment of wireline and wireless broadband cable networks. In addition to supporting access to multimedia services and dependable delivery of voice, video, and data to SOHO and workplace venues, cable networks reliably enable E-business, telemedicine, tele-instruction, and staff teletraining activities.

Despite technical advances in cable network technologies, problems persist with wireline and wireless cable network implementations that hinder the universal adoption of cable solutions. Protocols and standards in the cable network arena are still in development. Interoperability remains an issue. A cut in the HFC infrastructure or the loss of above-ground cable in a storm brings cable service to every subscriber on a neighborhood cable network segment to a halt.

Cable subscribers on a neighborhood cable network segment share bandwidth and network resources. As a consequence, the first subscriber on an HFC neighborhood network segment generally receives excellent service. However, as additional subscribers are added to the shared network segment, the capacity of the HFC link drops and network performance, security, and Quality of Services (QoS) assurances are negatively affected. Nonetheless, cable networks affordably facilitate links to broadband applications and remain viable solutions for overcoming local loop impediments.

Although limited in terms of geographic scope and user participation, wireline and wireless cable network trials and full-scale implementations demonstrate the effectiveness of broadband cable networks. Cable network configurations foster reliable connectivity to extranet and intranet configurations as well as the Internet; enable fast access to Web resources; and distribute a broad range of telecourses and instructional teleprograms for curricular enrichment. In addition, wireline and wireless cable network installations enable access to required high school courses so that high school seniors can graduate with a general education diploma and college preparation courses in subjects such as foreign languages, mathematics, and science.

Wireline and wireless cable networks support delivery of professional, technical, and vocational teletraining initiatives and provide access to lifelong learning, special-interest, and healthcare programs. By provisioning readily available, inexpensive, and easy-to-use interactive voice, video, and data services, wireline and wireless cable networks are popular local loop solutions for enabling access to broadband services and applications.

7.32 SELECTED WEB SITES

Cable in the Classroom (CIC). All About CIC.
  Available: http://www.ciconline.com/section.cfm/2
CableLabs. Current Projects. Go2Broadband.
  Available: http://www.cablelabs.com/go2/
CableLabs. Current Projects. CableHome.
  Available: http://www.cablelabs.com/cablehome/
CableLabs. Current Projects: Cable Modem.
  Available: http://www.cablemodem.com/
CableLabs. Current Projects. PacketCable Specification.
  Available: http://www.packetcable.com/specifications.html
CableLabs. OpenCable. Documents.
  Available: http://www.opencable.com/documents.html
CableLabs. Project Primer. DOCSIS.
  Available: http://www.cablemodem.com/docsisprimer.html
CableLabs. Revolutionizing Cable Technology.
  Available: http://www.cablenet.org/welcome.html
Digital Audio Visual Council. DAVIC Workplan and Specifications.
  Available: http://www.davic.org/speci.htm

DigiTerra Broadband. Broadband Compass.
    Available: http://www.digiterrabroadband.com/products/compass.htm
Digital Video Broadcasting (DVB). DVB Technology.
    Available: http://www.dvb.org/dvb_technology/index.html
EuroCableLabs. Home Page.
    Available: http://www.eurocablelabs.com/EuroCableLabs.html
European Cable Communications Association (ECCA). Home Page.
    Available: http://www.ecca.be/fs.htm
National Cable and Telecommunications Association (NCTA). NCTA Home
    Page.
    Available: http://www.ncta.com/
TComLabs. The Euro-DOCSIS Certification Lab. Home Page. Last modified
    on August 28, 2001.
    Available: http://www.tcomlabs.com/
Wireless Communications Association International. LMDS Overview.
    Available: http://www.wcai.com/lmds.htm
Wireless Communications Association, International. MMDS Overview.
    Available: http://www.wcai.com/mmds.htm

8 Cellular Technologies and Networks

8.1 INTRODUCTION

Accelerating demand for cellular telephony services and instantaneous access to network resources regardless of time constraints or subscriber location or mobility contributes to the wide array of cellular technologies, architectures, protocols, and solutions. Cellular networks enable increasingly affordable and customizable voice, data, and video transport via cellular phones. Cellular phones are also called mobile, wireless, portable, and/or compact handheld communicators, appliances, and/or devices. This chapter uses the aforementioned terms interchangeably.

In mobile communications networks, voice, video, and data are transported from source to destination through the air as electromagnetic signals, thus eliminating the need for wireline connections. Cellular solutions enable a diverse array of overlapping and complementary services and applications. As an example, D-AMPS (Digital-Advanced Mobile Phone Services) and PCS (Personal Communications Service) facilitate bi-directional or full-duplex short message exchange, voicemail, paging services, and access to wireless and/or wireline network connections via compact cellular communicators. These solutions also support frequency reuse, thereby enabling multiple subscribers within the same cell coverage area to use cellular telephony services concurrently. In addition, cellular communications implementations ideally facilitate global roaming, thereby enabling cellular subscribers to make and receive cellular phone calls at anytime and from everywhere.

8.2 PURPOSE

This chapter examines the distinguishing attributes and capabilities of cellular access systems, technologies, protocols, and architectures. The chapter also highlights standards groups and activities in the cellular communications domain; describes first-generation (1G), second-generation (2G), and third-generation (3G) cellular system initiatives; explores the capabilities of advanced mobile communications solutions that provision links to a diverse range of bandwidth-intensive wireless applications; and highlights research projects and practical initiatives enabling broadband cellular applications in fields that include education, government, medicine, and business.

Wireless technologies and networks such as WPANs (Wireless Personal Area Networks), WLANs (Wireless LANs), WMANs (Wireless MANs), and WWANs (Wireless WANs) initially developed apart from cellular technologies and solutions. In the present-day deregulated telecommunications environment, wireless networks and cellular communications solutions employ Mobile IP (Internet Protocol) platforms based on IPv6 (IP version 6). This chapter describes the capabilities of cellular solutions in supporting Mobile IP and IPv6, overlapping services and applications, convergent broadband solutions, Web browsing, virtually unlimited numbers of uniquely identified IP addresses, and seamless voice, video, and data delivery to subscribers on the move.

As noted, this chapter focuses on cellular communications technologies, capabilities, applications, and initiatives. Chapter 9 explores WPAN (Wireless Personal Area Networks, WHNs (Wireless Home Networks), WLAN (Wireless Local Area Networks), WMAN (Wireless Metropolitan Area Networks), and WWAN (Wireless Wide Area Networks) technologies, protocols, architectures, and implementations for enabling access to wireline and/or wireless resources. Chapter 10 examines satellite technologies, networks, applications, services, and solutions.

8.3 FOUNDATIONS

In 1843, Alexander Bain, a Scottish educator, philosopher, clockmaker, and psychologist, confirmed the feasibility of fax transmission. In 1876, the year marking the first centennial celebration of the United States, Alexander Graham Bell invented the telephone. In 1881, Bell demonstrated the use of twisted pair copper lines for carrying electrical phone signals. These discoveries contributed to the emergence of the Public Switched Telephone Network (PSTN) and subsequent innovations in cellular, personal, and mobile communications technologies.

In 1888, Heinrich Hertz established procedures for transporting energy by air. Subsequently in 1894, Guglielmo Marconi confirmed the capabilities of radio communications in maintaining contact with ships at sea.

Mobile telephony service in the United States was introduced by AT&T in the 1940s. In 1979, Nordic Mobile Telephone (NMT) launched first-generation (1G) analog cellular telephone services in Sweden and Norway. In 1983, Bell Labs introduced AMPS (Advanced Mobile Phone Service) in the United States. A 1G cellular analog solution, AMPS established the present-day foundation for cellular implementations.

Ever since Alexander Bain demonstrated the feasibility of fax solutions and Alexander Graham Bell invented the telephone, the cellular domain has been characterized by change. This change is reflected in the emergence of feature-rich mobile communications devices and services, and is characterized by innovations in cellular technologies, architectures, and protocols for enabling individuals on the move to access voice, video, and/or data resources via compact multifunctional cellular communicators at any time and from every place.

8.4 CELLULAR OVERVIEW

8.4.1 CELLULAR TECHNICAL FUNDAMENTALS

A basic cellular network supports signal transmission from a transmitter or cellular phone to a receiver or base station within a specified area of coverage or a cell. Each cell has one base station. The transmission process can be initiated by a voice call on the cellular phone. The cellular phone converts the voice call into electronic signals that are transmitted via a direct line-of-sight path to the base station antenna affixed to a tall building or cellular tower. At the base station, the electronic signals are switched to wireless and/or wireline networks, depending on their destination addresses.

Each subscriber is assigned separate forward and reverse channels that employ two frequencies for the duration of the phone call. A forward channel supports voice transmissions from the base station to the cellular phone. A reverse or return channel enables transmissions from the cellular phone to the base station.

Cellular networks support a common core of network functions that foster voice, imaging, video, and data services, and provide communications coverage via a honeycomb of cells in specified geographical locations. Cellular communications coverage depends on population density, topology of the terrain, regulatory authorities, and subscriber communications requirements.

Cellular structures include picocells, small cells that support operations inside homes and offices. Macrocells provide communications services in department stores, office buildings, multistory hotels and condominiums, and corporate complexes in densely populated cities in a 2-kilometer coverage area. In the open countryside, a macrocell facilitates communications coverage in an area that extends to 40 kilometers.

Signal strength and robustness between the cellular phone and the base station can be adversely affected by atmospheric conditions; construction of new or remodeled facilities; geographic barriers such as hills, ravines, deserts, trees, and canyons; and antenna height, location, and design. Depending on the RF (Radio Frequency) range in which cellular services operate, cellular transmissions can also be seriously impacted by co-channel interference that occurs when multiple subscribers use the same frequency band. Moreover, signal spikes and surges, adjacent channel interference, and spillover that is caused by intensive user activity in contiguous frequency bands also negatively impact the quality of voice calls. Additional sources of signal impedance include path loss or a reduction in the power of the transmitted signal, multipath fading resulting from reflections of electromagnetic signals off objects, and shadow fading caused by an impenetrable object that blocks a direct line-of-sight transmission.

8.4.2 CELLULAR HANDOFFS

Cellular or mobile communications involves the transmission of data, video, or voice of a specified quality over a radio communications link. Inasmuch as the geographical area covered by a cellular network is divided into cells, handoffs are required as cellular subscribers move from one cell to another. Cellular handoffs or handovers enable voice, video or data service continuation whenever the mobile device used by an individual on the move changes cell location. In some instances, handoffs occur during times of concentrated activity when subscribers and cellular communicators are stationary.

Cellular handoffs involve switching an ongoing call or transmission to a different communications channel. Cellular technologies and protocols, including Time-Division Multiple Access (TDMA), Code-Division Multiple Access (CDMA), and Frequency-Division Multiple Access (FDMA), foster handoff applications. Frequent handoffs contribute to short-term disturbances that result in signal errors, low throughput, disconnects, and retransmissions. The Enhanced Throughput Cellular (ETC) protocol and the Microcom Network Protocol, Class 10, Enhanced Cellular (MNP 10 EC) specification minimize the occurrence of disconnects and signal impairments resulting from recurrent handover operations.

8.4.3 MOBILE IP (INTERNET PROTOCOL)

Developed by the IETF (Internet Engineering Task Force), Mobile IP (Internet Protocol) enables a mobile network node to change a point of attachment to the Web while maintaining the same IP address. Based on TCP/IP (Transmission Control Protocol/Internet Protocol), Mobile IP can provision seamless subscriber services across public and/or private wireline and/or wireless network configurations. In Mobile IP operations, packets are initially routed to the home network address of a mobile node and then transmitted to the destination address if the mobile node is away from the home network.

By using Mobile IP, cellular subscribers maintain links to the Web regardless of location or mobility. Mobile IP enables operations that support IPv4 Internet Protocol version 4) and IPv6 (Internet Protocol version 6). As with IPv6, Mobile IPv6 facilitates sophisticated autoconfiguration and routing procedures, and expands the available addressing space, thereby making vast numbers of uniquely identified IP addresses available.

8.4.4 CELLULAR DEVICE SECURITY

Fraudulent use of cellular communicators contributes to the development of technological solutions for authenticating the identity of subscribers such as Personal Identification Numbers (PINs), roamer verification procedures, RF (Radio Frequency) fingerprinting, and SIM (Subscriber Identity Module) cards based on smart-card technology.

Fraudulent alterations of cellular devices led to FCC (Federal Communications Commission) regulations that require a unique Electronic Serial Number (ESN) for every cellular phone and resulted in a federal law that was passed in 1994. According to this statute, individuals altering telecommunications instruments and equipment are subject to fines up to $50,000 and 15 years imprisonment.

8.5 RADIO FREQUENCY (RF) SPECTRUM

8.5.1 RF SPECTRUM BASICS

The RF (Radio Frequency) spectrum consists of electromagnetic waves generated by the sun and the Earth. Electromagnetic waves are grouped in terms of frequencies and wavelengths in a continuum ranging from the very low frequency bands between the 9 and 30 kHz (Kilohertz) spectral block to the super high frequency bands between the 30 and the 300 GHz (Gigahertz) spectral block.

Mobile and personal communications technologies employ RF waves that make up the electromagnetic spectrum. Utilization of the RF spectrum is regulated by national and international government agencies to optimize the effectiveness of communications operations and eliminate communications problems associated with signal interference.

8.5.2 SPECTRAL ALLOCATIONS

The same RF bands can support diverse applications and services, depending on legislative mandates and government requirements. For example, in the United States, the 1.900 GHz frequency band supports terrestrial microwave and PCS (Personal Communications Service) activities. In the European Union, the 1.900 GHz band enables Digital Enhanced Cordless Telecommunications (DECT) applications.

Moreover, in the European Union, spectral bands between the 2.000 and 2.170 GHz frequencies facilitate the provision of Universal Mobile Telecommunications Systems (UMTS) 3G (third-generation) services. In the United States, spectral bands between the 2.000 and 2.170 GHz frequencies enable mobile satellite communications operations for facilitating 2G (second-generation) and 3G (third-generation) cellular services.

To unify cellular operations, the U.S. Federal Communications Commission (FCC), the European Telecommunications Standards Institute (ETSI), and the International Telecommunications Union-Telecommunications Standards Sector (ITU-T) participate in the World Radiocommunications Conference (WRC). The WRC coordinates spectral allocations for enabling universal global roaming and ubiquitous access to cellular communications services.

8.5.3 U.S. FEDERAL COMMUNICATIONS COMMISSION

8.5.3.1 FCC Spectral Allocations

In the United States, the Federal Communications Commission (FCC) designates unlicensed or license-exempt and licensed RF spectral bands for wireless applications and services. License-exempt ISM (Industry, Scientific, and Medical) bands support PCS (Personal Communications Service) solutions in spectrum between the 902 and 928 MHz frequencies, between the 2.400 and 2.483 GHz frequencies, and between the 5.725 and 5.875 GHz frequencies.

Licensing agreements from regulation authorities such as the FCC are not necessary for conducting wireless activities in ISM (Industry, Scientific, and Medical) bands. However, devices operating in ISM bands must conform to FCC regulations and requirements. ISM bands generally enable license-exempt communications services and applications between short-range devices that are in close proximity. In addition to license-exempt bands, the FCC designates licensed bands for mobile and personal communications. Utilization of these bands requires FCC approval. Typically, licensed bands in the extremely high and super-high frequencies support spread spectrum, microwave, and satellite operations and services.

FCC spectral allocations facilitate implementation of numerous applications by federal government agencies such as the National Aeronautics and Space Administration (NASA) and the U.S. Department of the Interior. These agencies use available spectrum for mandated services that include satellite tracking, oceanic exploration, transportation, resource management, and public safety. In addition, the U.S. Department of Defense utilizes spectral allocations for enabling information transport between airfields and military bases and facilitating aircraft command and control applications.

8.5.3.2   Public Safety Communications Systems

The FCC sets aside spectrum in the 800 MHz RF (Radio Frequency) spectral bands for public safety communications systems. To ensure seamless operations, the FCC develops procedures for eliminating signal interference resulting from the operations of nearby Commercial Mobile Radio Systems (CMRSs), Personal Communications Service (PCS) solutions, and wireless networks. The Federal Emergency Management Association (FEMA) employs emergency radio networks that operate in the 800 MHz RF spectrum for enabling communications to disaster sites.

8.5.3.3   SuperNet and the National Information Infrastructure (NII) RF Bands

In 1997, the FCC set aside the SuperNet and National Information Infrastructure (NII) spectrum between the 5.15 and 5.35 GHz frequencies and between the 5.725 and 5.875 GHz frequencies. These ISM bands support license-exempt, high-speed communications services in short-haul network configurations.

8.5.3.4   The FCC and the PCS Spectrum

In 1994, the FCC initiated auctions of spectrum to support PCS (Personal Communications Services) such as interactive television programs, advanced paging services, and mobile data applications. PCS bands are licensed in accordance with frequency range and region. Initially, the FCC designated the bands between the 1.850 and 1.900 GHz frequencies for PCS applications. As a consequence of the Telecommunications Act of 1996 and subsequent telecommunications deregulation, the FCC supported additional auctions of spectrum in the 900 MHz and 1.800 GHz frequency blocks for enabling PCS deployments.

8.6   CELLULAR COMMUNICATIONS STANDARDS ORGANIZATIONS AND ACTIVITIES

8.6.1   ALLIANCE FOR TELECOMMUNICATIONS INDUSTRY SOLUTIONS (ATIS)

The Alliance for Telecommunications Industry Solutions (ATIS) defines operating procedures and standards for the telecommunications industry. ATIS Committees monitor FCC (Federal Communications Commission) rules and regulations. ATIS participants include telecommunications service providers, software developers, and manufacturers.

8.6.2 AMERICAN NATIONAL STANDARDS INSTITUTE (ANSI)

The American National Standards Institute (ANSI) promotes development of standards and specifications in the telecommunications arena by building consensus among diverse public and private agencies and organizations.

8.6.2.1 ANSI T1 Standards Committee

Accredited by ANSI, the T1 Standards Committee defines technical guidelines and specifications for enabling interoperable telecommunications systems and devices. In addition, the T1 Standards Committee develops standards for wireless and/or mobile communications systems and services, and works in concert with other North American and international standards organizations such as the ATIS.

8.6.2.1.1 Global Standards Collaboration (GSC) Group

The ANSI T1 Standards Committee is a founding member of the Global Standards Collaboration (GSC) Group. Also accredited by ANSI, the GSC Group coordinates development of regional specifications for current and next-generation cellular communications services.

8.6.3 CELLULAR TELECOMMUNICATIONS AND INTERNET ASSOCIATION (CTIA)

The Cellular Telecommunications and Internet Association (CTIA) sponsors anti-fraud initiatives, monitors FCC actions, addresses cellular health and safety concerns, certifies standards-compliant cellular devices, authenticates roaming management agreements, and provides cellular inter-carrier billing services. The CTIA sponsors E-commerce applications, Personal Communications Services (PCS) implementations, and Commercial Mobile Radio Services (CMRS) deployment. CTIA participants include network operators, software developers, and communications carriers. In 2000, the Wireless Data Forum (WDF) merged with the CTIA.

The CTIA Wireless Foundation supports public-sector initiatives such as the Communities on Phone Patrol (COPP) for dispersing cellular phones to volunteer groups patrolling neighborhoods. Also a CTIA Wireless Foundation initiative, ClassLink distributes cellular phones to classroom teachers for reporting medical and security emergencies.

8.6.4 EUROPEAN TELECOMMUNICATIONS STANDARDS INSTITUTE (ETSI)

8.6.4.1 Terrestrial Trunked Radio (TETRA) Specification

Developed in 1995 by the European Telecommunications Standards Institute (ETSI), the Terrestrial Trunked Radio (TETRA) specification supports Private Mobile Radio (PMR) networks and Public Access Mobile Radio (PAMR) systems for enabling point-to-point and point-to-multipoint full-duplex transmissions. TETRA implementations work in concert with TDMA (Time-Division Multiple Access) and IP; support concurrent voice and data transmissions; and foster group, private, emergency, and/or individual calls, short message services (SMS), and alphanumeric paging operations. TETRA communications services also support police, ambulance, and fire-fighting operations.

The TETRA infrastructure fosters deployment of security, utility, transportation, and military applications and enables fast call setup, closed group communications, packet-switched and circuit-switched data and voice transmissions, and radio-to-radio communications. TETRA applications are implemented in spectral bands between the 380 and 400 MHz frequencies, between the 410 and 430 MHz frequencies, between the 450 and 470 MHz frequencies, and between the 870 and 920 MHz frequencies.

The TETRA Project Team at the University of Twente in the Netherlands conducts TETRA and TETRA-2 (TETRA-Phase 2) research projects. TETRA-2 implementations support voice calls, data transport, encryption, and global roaming services in a 200-kilometer coverage area. The TETRA-2 air interface interworks with second-generation (2G) GSM (Global System for Mobile Communications) solutions and advanced 2G, or 2.5G deployments based on GPRS (General Packet Radio Service) technologies. TETRA-2 solutions are also compatible with third-generation UMTS (Universal Mobile Telecommunications Systems) and 3GSM (Third-Generation GSM) technologies; support high-speed data rates; facilitate closed user group communications; and employ low bit rate voice codec technology for enhancing the quality of voice services.

8.6.4.1.1   TETRA in Action

In Iceland, a TETRA network provisions communications coverage in the City of Reykjavik. In Sweden, TETRA services provide extended communications coverage in rural communities and geographically isolated locations on the Island of Stikla. Moreover, a TETRA-compliant mobile radio network provides cellular voice and data services for local residents on Gotland, an island located 90 kilometers from the Swedish mainland. In Croatia, a TETRA network supports paging, telephony services, and transmission of mission-critical data to public safety, emergency, security, and border protection personnel.

In the transportation sector, TETRA solutions facilitate automatic vehicle location and fleet management operations. The Mini Metro Underground Network in Copenhagen, the Athens International Airport, and the London Underground utilize TETRA implementations for updating transportation schedules, monitoring traffic, and enabling passengers, employees, and security personnel to access integrated cellular communications network services. These systems also support passenger capacity planning, crowd control, incident management, and emergency services.

8.6.5   MOBILITY FOR EMERGENCY AND SAFETY APPLICATIONS (MESA)

Originally called the Public Safety Partnership Project (PSPP), the MESA (Mobility for Emergency and Safety Applications) initiative supports implementation of emergency medical services and civil defense and law enforcement applications via the TETRA network. Moreover, MESA sponsors mobile robotics projects for enabling the disarmament of bombs or explosives and containment of chemical spills.

8.7 ADVANCED MOBILE PHONE SERVICE (AMPS) FEATURES AND FUNCTIONS

Introduced in 1983, Advanced Mobile Phone Service (AMPS) is a first-generation (1G) circuit-switching mobile cellular solution that uses RF (Radio Frequency) spectrum for transmission of voice communications. AMPS technology is based on the analog cellular standard developed by Bell Labs, the predecessor of AT&T.

Also called N-AMPS (Narrowband-AMPS), AMPS defines the process for initiating and completing analog cellular calls. In a traditional AMPS solution, the mobile communicator transmits shortwave radio signals to the base station in spectrum between the 824 and 849 MHz frequencies. For the return path, the base station transmits short-wave radio signals to the mobile communicator in spectrum between the 869 and 894 MHz frequencies.

The base station also serves as the interface to the mobile switching center. This center switches calls to the PSTN (Public Switched Telephone Network), supports cellular system operations, and provides handoff services so that a mobile subscriber can seamlessly move from cell to cell throughout the service area and maintain cellular connections. Generally, AMPS handoffs take less than 30 seconds.

AMPS also supports call forwarding and call waiting services and provisions nationwide communications coverage. AMPS installations employ FDMA (Frequency-Division Multiple Access) for enabling analog transmission. AMPS services are relatively inexpensive. However, as a consequence of poor signal quality, AMPS cellular phone calls lack the clarity of D-AMPS cellular implementations.

8.8 D-AMPS (DIGITAL-ADVANCED MOBILE PHONE SERVICE)

The digital version of AMPS is called the Digital-Advanced Mobile Phone Service (D-AMPS). As with GSM (Global System for Mobile Communications), D-AMPS is a second-generation (2G) technology that supports cellular communications services via dual-mode digital cellular phones that support two types of technologies. For example, in the United States, D-AMPS supports operations over the cellular band in the 800 MHz frequencies and the PCS (Personal Communications Service) band in the 1900 MHz or 1.900 GHz frequencies.

8.8.1 TIME-DIVISION MULTIPLE ACCESS (TDMA) OPERATIONS

In D-AMPS configurations, TDMA (Time-Division Multiple Access) technology defines a digital air interface for enabling multiple users to share a single RF channel in the 800 MHz spectral block, divides a D-AMPS channel into six timeslots, and assigns each voice conversation or session to a separate timeslot. Formally known as ANSI-136, TDMA also employs digitization and compression for transporting bursty data and voice traffic in frames across the D-AMPS network. TDMA solutions optimize bandwidth utilization in cellular networks such as D-AMPS by supporting dynamic timeslot allocations. With this approach, a timeslot not in use is reassigned to an active session or voice call. Moreover, TDMA implementations enable data retrieval and telephony services both inside and outside a facility. In addition to

D-AMPS, TDMA is a key enabler of Personal Digital Cellular (PDC) and GSM solutions.

TDMA operations are specified in the TIA (Telecommunications Industry Association) IS-54 (Interim Standard-54) specification and endorsed by ANSI. In comparison to analog-only AMPS operations based on FDMA (Frequency-Division Multiple Access) technology, D-AMPS implementations employ the TDMA protocol to enable high-quality voice calls, faster transmission rates, and additional bandwidth, services, and capacity.

8.9 CELLULAR DIGITAL PACKET DATA (CDPD)

8.9.1 CDPD FEATURES AND FUNCTIONS

Cellular Digital Packet Data (CDPD) technology supports an open architecture that enables location-independent data delivery to mobile cellular subscribers. D-AMPS cellular networks employ CDPD technology to carry data when voice transmissions are idle and provision Mobile IP services. CDPD also works in concert with AMPS, GSM, TDMA, and CDMA technologies.

CDPD technology employs a frequency-hopping modulation scheme for transmitting data in small packets or bursts, and supports full-duplex transmissions. CDPD solutions provide connectivity to the Internet with performance similar to that supported by a dial-up Point-to-Point-Protocol (PPP) connection. CS (Circuit-Switched) CDPD implementations provide dedicated links for high-volume data transport. CDPD installations require MDBS (Mobile Data Base Stations) and MDGs (Mobile Data Gateways) for enabling transmissions via the in-place cellular infrastructure. Ideally, CDPD solutions enable transmission rates at 19.2 Kbps. However, as a consequence of overhead allotted for forward error control (FEC) operations, actual rates of 10 Kbps are achieved.

CDPD solutions facilitate e-mail exchange, E-commerce applications such as credit card verification, and IP multicasts, and provision access to stock quotes, industry announcements, and bank rates. Law enforcement personnel use CDPD implementations to access departmental records, rules, and regulations and query databases maintained by the National Criminal Information Center (NCIC). These resources provision information on outstanding warrants, missing persons, wanted persons, and potential suspects. Moreover, CDPD networks enable law enforcement personnel to verify property ownership, drivers' licenses, and license plate numbers.

8.9.2 CELLULAR DIGITAL PACKET DATA (CDPD) FORUM

Participants in the CDPD Forum include AT&T, GTE, Verizon, Novatel Wireless, Cingular, Padcom, Cerulean Technologies, and Sierra Wireless. This Forum develops specifications and standards for CDPD implementations.

8.9.3 CDPD MARKETPLACE

Developed by cellular communications carriers in the 1990s, CDPD implementations were originally designed to compete with emergent PCS solutions. It is interesting to note that CDPD installations currently compete with GPRS (General Packet Radio Service) deployments in the cellular marketplace. CDPD services are implemented worldwide in countries that include the United States, New Zealand, Venezuela, Indonesia, Peru, Israel, and China.

8.9.3.1 AT&T

In 1999, AT&T announced the development of an integrated voice and data cellular IP phone called MobileAccess T250. Designed for subscriber groups that include financial analysts and transportation providers, this portable CDPD device employs TDMA technology; supports data transmission and real-time interactive access to the Internet, intranets, and extranets; and enables two-way messaging services, voice calls, and voicemail notification.

8.9.3.2 Verizon

Verizon CDPD devices support access to calendars, address books, news bulletins, schedules, and e-mail. These CDPD devices also support SMS (Short Messaging Service), access to specified mobile Web services, and high-quality telephone calls.

8.10   CODE-DIVISION MULTIPLE ACCESS (CDMA)

8.10.1   CDMA FEATURES AND FUNCTIONS

Developed by the CDMA Development Group (CDG), CDMA (Code-Division Multiple Access) digital cellular spread spectrum technology employs different pseudo-random digital sequence codes for distinguishing multiple calls or data packets sharing the same frequencies. These sequence codes spread the digital bits that comprise the call or data packets across the available spectrum and enable each call or data packet to occupy a single frequency. The base station and the mobile unit share a pseudo-random digital sequence code. At call termination, this code is discarded.

Moreover, because CDMA technology effectively blocks interference from conflicting signals, CDMA transmissions are generally robust, reliable, and secure. As with D-AMPS installations, CDMA solutions support high-quality voice calls and dependable and robust data transmission. CDMA implementations enable secure cellular telephony service, on-demand bandwidth, WLL (Wireless Local Loop) applications, single cell frequency reuse, and wide area communications coverage. Also called N-CDMA (Narrowband-CDMA) and TIA IS-95 (Telecommunications Industry Association-Interim Standard 95), first-generation CDMA implementations employ small-sized cells.

8.10.1.1 Telecommunications Industry Association (TIA)

The Telecommunications Industry Association (TIA) approved the CDMA (Code-Division Multiple Access) digital multiple access specification, formally called TIA IS-95, in 1993. In terms of operations, TIA IS-95 divides the radio spectrum into a series of carrier frequencies or channels. Each channel is 1.25 MHz wide. In CDMA implementations, a full-duplex channel consists of two 1.25 MHz RF bands. One channel supports transmissions from the base station to the mobile communicator on the downlink path, and the other channel enables transmissions from the mobile communicator to the base station on the return or uplink path.

8.10.2 CDMA Marketplace

Communications carriers and vendors that participate in the CDMA marketplace include GTE, Sprint, Qualcomm, Lucent, Ameritech, Motorola, and Cingular.

8.10.2.1 Bell Mobility

Bell Mobility employs CDMA technology for enabling digital PCS applications. Codes are assigned to every voice call and every packet to safeguard transmissions from interception.

8.10.2.2 Kyocera

In 2001, Kyocera Corp. introduced a dual-mode cellular device that supports voice calls and digital data transmission via CDMA networks.

8.10.2.3 Nortel

Developed by Nortel Networks, CDMA Metro Cell installations enable communications in suburban and urban environments in the 800 MHz and 1.900 GHz frequencies. Nortel also supports CDMA Rural Cell solutions for providing extended communications coverage in isolated and sparsely populated rural areas. In addition, Nortel operates CDMA Micro Cell networks that provision communications services in shopping malls, airport terminals, office buildings, and sports arenas.

8.11   cdmaONE AND cdma2000

8.11.1   cdmaOne

Developed by the CDMA Development Group (CDG), cdmaOne and cdma2000 build on the capabilities of traditional CDMA technology. As with CDMA, cdmaOne operations are defined by the TIA IS-95 standard, support transmission rates that range from 64 to 144 Kbps, and facilitate transmission via 1.25 MHz-sized channels.

8.11.2   cdma2000

A third-generation cellular communications solution, cdma2000 technology provisions control mechanisms that balance QoS (Quality of Service) requirements for enabling dependable delivery of voice calls and data transmissions. cdma2000 implementations also interoperate with cdmaOne solutions. The 1xRTT (ITU Radio Transmission Technologies) specification serves as the basis for cdma2000 operations. This specification defines procedures for enabling circuit-mode and packet-mode transmissions at rates of 2.4 Mbps, with an average throughput at 880 Kbps in broadband fixed wireless access (FWA) networking environments and 14.4 Kbps in mobile environments.

cdma2000 enhancements enable channel aggregation, simultaneous transmission of data and voice traffic, and the ability to mix and match voice, data, and video signals within a single cell. cdma2000 deployments also establish a core air interface for providing third-generation cellular services and work in concert with 3G UMTS (Universal Mobile Telecommunications Systems) solutions.

8.11.3 cdma2000 Marketplace

Nortel Networks sponsors a cdma2000 initiative for enabling fast voice and data services, e-mail exchange, and E-commerce applications. In 2001, Novatel Wireless introduced a third-generation PC (PCMCIA or Personal Computer Memory Card International Association) card based on the lxRTT specification for implementation on the Sprint PCS nationwide cellular communications network. Qualcomm, Lucent, and Sprint also promote development and implementation of 3G cdma2000 systems and solutions.

8.12 WIDEBAND-CODE-DIVISION MULTIPLE ACCESS (W-CDMA)

8.12.1 W-CDMA Operations and Services

W-CDMA (Wideband-Code-Division Multiple Access) technology employs a standardized air interface that works in concert with ANSI-41 specifications. W-CDMA implementations support both circuit-switched and packet-switched operations. W-CDMA solutions also provision Quality of Service (QoS) assurances for multimedia applications and enable transmission rates between 8 Kbps and 2 Mbps.

Links between a W-CDMA access network and a core GSM (Global System for Mobile Communications) network are established through the utilization of ATM (Asynchronous Transfer Mode) mini-cell transmission protocols. W-CDMA deployments do not directly interoperate with air and network interfaces supported by cdmaOne or cdma2000 networks.

W-CDMA technology enables migration to third-generation cellular communications solutions such as 3GSM (Third-Generation Global System for Mobile Communications) and UMTS (Universal Mobile Telecommunications Systems) and features an advanced intelligent network for enabling global roaming. Moreover, W-CDMA implementations optimize use of the RF spectrum and support symmetric and asymmetric transmissions for enabling fast connections to Web resources, multimedia transport, and dependable access to VPNs (Virtual Private Networks).

W-CDMA solutions facilitate videoconferencing, short messaging services (SMS), interactive paging, and voice telephony applications. In addition, W-CDMA solutions enable videotelephony services and access to interactive entertainment. W-CDMA third-generation communicators support E-commerce transactions, teleshopping, electronic publishing, and E-banking (electronic-banking) services.

8.12.2   W-CDMA STANDARDS ORGANIZATIONS AND ACTIVITIES

The International Telecommunications Union-Telecommunications Standards Sector (ITU-T), the European Telecommunications Standards Institute (ETSI), and the Association of Radio Industries and Business (ARIB) in Japan develop W-CDMA specifications. In addition, communications carriers in Asia, the European Union, and New Zealand support the use of W-CDMA technology for enabling third-generation (3G) mobile and personal communications systems solutions.

8.12.3   W-CDMA MARKETPLACE

8.12.3.1   NEC Corporation

NEC Corporation supports implementation of compact W-CDMA mobile video-phones. Each third-generation device features a mobile phone handset with a viewer, screen, microphone, and video camera for enabling high-quality video and audio transmission, access to bulk data files, and video streaming services. The NEC mobile videophone employs Bluetooth technology for supporting the wireless link between the telephone handset and the viewer screen. By using omnidirectional radio signals, Bluetooth technology eliminates the need for directional infrared connections and connection cables.

8.12.3.2   Nokia

Nokia supports utilization of multiservice, multifunctional IP radio access networks (IP-RANs) that enable IPv6 implementation, differentiated QoS (Quality of Service) levels, and secure transactions in W-CDMA implementations. In addition, Nokia IP-RAN installations work in concert with GSM and EDGE (Enhanced Data Rates for Global Evolution) technologies and wireless LAN (WLAN) solutions.

8.12.3.3   NTT DoCoMo

NTT DoCoMo supports a next-generation network platform based on W-CDMA technology that features the Advanced-Intelligent Network (A-IN) for enabling global roaming, satellite communications services, and interoperability with ATM configurations. NTT DoCoMo and MBNS Multimedia Technologies conduct trials benchmarking W-CDMA effectiveness in the Malaysia Multimedia Super Corridor District near Kuala Lumpur. Test outcomes contribute to the development of stan-dards-compliant IMT-2000 (International Mobile Telecommunications-Year 2000) implementations.

8.13   WIRELESS APPLICATION PROTOCOL (WAP)

8.13.1   WAP CAPABILITIES, OPERATIONS, AND SERVICES

The Wireless Application Protocol (WAP) is a *de facto* global standard that supports Web microbrowsing, Web content delivery, and cellular telephony services via cel-lular handheld digital devices such as pagers, cellular communicators, and PDAs

| Wireless Application Environment (WAE) |
| Wireless Session Protocol (WSP) |
| Wireless Transaction Protocol (WTP) |
| Wireless Transport Layer Security (WTLS) |
| Wireless Datagram Protocol (WDP) |
| Bearers (e.g., Data, SMS, USSD) |

FIGURE 8.1 The WAP protocol stack.

(Personal Digital Assistants). In addition, WAP promotes utilization of the Wireless Markup Language (WML), Mobile IP, and E-commerce services.

WAP applications operate in concert with a broad range of second- and third-generation cellular technologies, including Global System for Mobile Communications (GSM), General Packet Radio Service (GPRS), CDMA, PHS (Personal Handyphone Systems), Time-Division Multiple Access (TDMA), and Digital Enhanced Cordless Telecommunications (DECT). As with cellular technologies such as GPRS (General Packet Radio Service), HSCSD (High-Speed Circuit-Switched Data), and EDGE (Enhanced Data Rates for Global Evolution), WAP technology facilitates data transmission at 19.2 Kbps and higher rates. (See Figure 8.1.) As with the OSI Reference Model and the TCP/IP Protocol Suite, the WAP protocol stack defines a series of layers to enable sophisticated services and basic (BEARER) applications such as SMS, data transmission, and USSD (Unstructured Supplementary Services Data). USSD enables transmission of text messages that contain up to 182 characters in length and works in concert with GSM solutions.

8.13.2 WAP FORUM AND THE WAP 2.0 SPECIFICATION

Developed and endorsed by the WAP Forum, the WAP 2.0 specification consists of a set of standards that define WAP architecture, functions, operations, and services. WAP implementations use binary transmission for achieving effective data compression and dependable throughput.

The WAP Forum also promotes universal implementation of multifunctional WAP services and interoperable WAP devices that accommodate the communications requirements of multiple user communities. The WAP Forum works with the World Wide Web Consortium (W3C) and the Internet Engineering Task Force (IETF) to ensure that future versions of standard protocols such as IP, HTTP (HyperText Transfer Protocol), HTML (HyperText Markup Language), and TCP (Transmission Control Protocol) are compatible with WAP standards. WAP Forum participants include communications carriers, infrastructure providers, software developers, and vendors such as Nokia, Motorola, Ericsson, and Phone.com.

8.13.3 WIRELESS MARKUP LANGUAGE (WML)

WAP uses the Wireless Markup Language (WML) for defining content displayed on WAP communicators. WML displays from two lines to a full screen of text; supports connections to information services such as weather forecasts, tele-entertainment, and intranets; and facilitates easy navigation in the absence of a full-size keyboard.

WML also supports XML (Extensible Markup Language). XML is a meta-language developed by the W3C that features a set of rules for defining specific applications. In addition, the W3C also supports XHTML (Extensible HyperText Markup Language) for enabling content to be shared across network devices including PCs (Personal Computers), laptops, TVs, PDAs, and mobile phones.

8.13.4 MOBILE STATION APPLICATION EXECUTION ENVIRONMENT (MExE)

WAP supports development of the Mobile Station Application Execution Environment (MExE) for standardizing mobile telephony operations. As with WAP, MExE works with second-generation (2G) GSM services and third-generation (3G) UMTS applications. MExE also supports utilization of smart phones that feature intelligent customer menus and Customized Applications for Mobile Networks Enhanced Logic (CAMEL) services such as voicemail and prepaid roaming applications. Sponsors of the MExE initiative include Nortel, Motorola, Lucent, and Nokia.

8.13.5 PERSONAL DIGITAL CELLULAR (PDC) SOLUTIONS

WAP technology is an enabler of PDC (Personal Digital Cellular) service in Japan. Moreover, NTT supports development of a WAP-compliant PDC (Personal Digital Cellular) Packet Data Communications System (PDC-P) that enables multiple subscribers to use a single radio channel concurrently. Rates at 28.8 Kbps are supported.

8.13.6 WAP MARKETPLACE

Verizon, Sprint PCS, AT&T, Nextel, and Ericsson support development of standards-compliant WAP multiservice, multifunctional handsets that feature an on-screen keyboard or stylus for accessing a notepad, e-mail, address book functions, and a calendar. These communicators provision wireless connections to Web resources, sports scores, and headline news; support roaming service in approximately 120 countries; and work in conjunction with software products such as Lotus Notes and Microsoft Outlook.

Nokia and VISA enable subscribers to make secure Internet payments via WAP cellular communicators. In addition, Nokia and MobiNil support a GPRS (General Packet Radio Service) core network in Egypt that interoperates with WAP-compliant cellular phones in provisioning subscriber access to Mobile IP resources.

For subscribers with WAP mobile phones, NextBus supports retrieval of data on bus and train arrivals and departures, road conditions, and traffic problems. Sonera provisions links to data on local accommodations, restaurants, landmarks, and festivals. Available from Airtuit, BlueMoon middleware enables mobile subscribers to access corporate databases, Web sites, and intranets via WAP-compliant cellular communicators.

8.14   PERSONAL COMMUNICATIONS SERVICES (PCS)

8.14.1   PCS Features, Functions, and Solutions

Personal Communications Service (PCS) solutions enable individuals to readily access a personalized set of communications services via an array of cellular communicators. As subscribers move through a cell or area of coverage, an antenna picks up signals and forwards them to a base station with links to a wireless and/or wireline network. PCS deployment involves the utilization of digital cellular technologies and versatile cellular communicators that enable voice services, Web browsing, and Mobile IP applications.

In the United States, PCS describes a set of completely digital cellular technologies operating in spectrum between the 1.850 and 1.990 GHz frequencies. This spectral block includes three 30 MHz allocations, three 10 MHz allocations, and three 20 MHz allocations for license-exempt applications. In the United States, the higher RF block is called the PCS band and the lower frequency block is called the cellular band. This distinction is arbitrary. PCS applications are enabled in each spectral band that provisions digital cellular services.

In 2000, the European Union initiated 3G PCS operations in the International Mobile Telecommunications (IMT) band. Created to accommodate a range of user requirements at an affordable price, PCS implementations in the European Union are customized solutions that build upon present-day and emergent cellular topologies, standards, and architectures. In addition to personal and device mobility, PCS solutions support service profile management to provision selected services at the mobile phone or portable device and location specified by the subscriber.

PCS implementations employ overlapping clusters of microcells and picocells to accommodate demand for extended cellular services. Transmission rates at 36 Kbps and higher are enabled in densely populated urban areas. The Global System for Mobile Communications (GSM) specification and Cellular Digital Packet Data (CDPD) technology support PCS operations, applications, and services.

Lower costs for PCS deployments in comparison to analog cellular services such as AMPS (Advanced Mobile Phone Services) contribute to widespread usage of PCS technology. Narrowband PCS applications employ digital cellular technology for advanced messaging services and two-way paging operations. Broadband PCS applications enable basic and sophisticated WLAN services. By providing anytime and anywhere access to network services, D-AMPS, GSM, and UMTS (Universal Mobile Telecommunications Systems) solutions support a variety of PCS-related implementations.

8.14.2   PCS Marketplace

8.14.2.1   Motorola and AT&T

Motorola and AT&T support PCS phones equipped with cellular voice and non-voice capabilities. These phones enable connectivity to specially designed wireless Web sites, foster fax transmission and reception; enable access to stock quotes, horoscopes, winning lottery numbers, and news headlines; and support Web browsing, voicemail, caller ID, and call waiting services in cities throughout the United States.

8.14.2.2   Nokia

PCS communicators are available in a variety of shapes and configurations and support an array of services and applications. For example, Nokia communicators combine telephony capabilities with the functionality of small handheld personal computers (PCs) or miniature laptops. These multiservice devices enable fax transport, voicemail, paging operations, short messaging services (SMS), and Web browsing.

8.14.2.3   Omnipoint

Omnipoint supports a wireless PCS system that works in concert with GSM networks and competes with wireline ISDN configurations for market share.

8.14.2.4   Sprint

Sprint subscribers use WAP-compliant Internet-ready PCS communicators equipped with microbrowsers for accessing wireless Web services and applications.

8.15   GLOBAL SYSTEMS FOR MOBILE COMMUNICATIONS (GSM)

8.15.1   GSM TECHNICAL FUNDAMENTALS

A second-generation cellular solution, the Global System for Mobile Communications (GSM) employs digital technologies and protocols for enabling users on the move to access telecommunications services. GSM was initially called the European 900 MHz digital cellular system or GSM 900. In the present-day environment, GSM technology provisions widespread cellular coverage and supports operations in spectrum between the 1.805 and 1.880 GHz frequency bands. GSM service is also known as GSM 1800 and DCS (Digital Communications Service) 1800 in the European Union.

Second-generation GSM dual-band cellular phones support digital services in the 900 MHz frequencies and the 1.800 GHz frequencies in the European Union. In addition, GSM tri-band cellular phones enable operations in the 900 MHz and the 1.800 GHz frequencies in the European Union and in the 1.900 GHz PCS frequencies in the United States. In the United States, GSM service is also known as PCS 1900. (See Figure 8.2.)

GSM communicators enable access to e-mail, fax, the Web, extranets, and intranets; provision a migration path to 3G networks; and support advanced second-generation or 2.5 generation cellular services such as GPRS (General Packet Radio Service) and MExE (Mobile Station Execution Environment). As with UMTS, GSM deployments utilize Generic Radio Access Networks (GRANs) for provisioning cellular links to wireless and wireline narrowband and broadband applications.

8.15.2   BASIC GSM CONFIGURATION

A basic GSM configuration includes a mobile station that is carried by the subscriber and a base station or transceiver that monitors the radio link with the mobile station. The network infrastructure switches calls between mobile users and wireline or

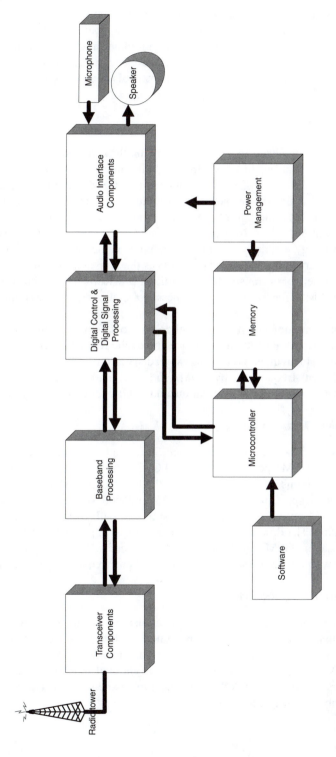

FIGURE 8.2 Data transfer between major components within a cellular device such as a cellular phone or advanced PDA.

mobile networks and supports management functions. Network performance, operations, and setups are monitored and administered by an operations and maintenance center. The mobile station and the base station employ a radio link or air interface for enabling transmissions. GSM uses FDMA (Frequency-Division Multiple Access) and TDMA (Time-Division Multiple Access) protocols and technologies for assigning carrier frequencies to base stations. (See Figure 8.3.)

8.15.3   GSM TRANSMISSION

Personal mobility is realized in GSM deployments through the use of smart cards featuring SIMs (Subscriber Identity Modules) that uniquely identify each cellular subscriber. Each SIM works in conjunction with a GSM-compliant handset and supports universal roaming. Because any individual can access the radio medium, security is a critically important element in a mobile communications network. In a GSM configuration, SIMs also support user authentication and data encryption.

GSM technology enables users to plug cellular phones into mobile computers and access the Internet, intranets, and extranets as they travel in countries with GSM coverage. GSM networks transport data traffic at 9.6 Kbps and enable short messaging service (SMS) for transmission of messages that contain a maximum of 160 characters.

GSM implementations employ channel coding to safeguard signals from artificial and natural electromagnetic interference and cyclic redundancy code (CRC) for error detection. FH (Frequency-Hopping) spread spectrum technology supports utilization of different frequencies by TDMA frames, guards against signal fading, conserves power usage at the mobile station, and minimizes co-channel interference.

8.15.4   GSM OPERATIONS

GSM developed independently in the United States and in the European Union. In the United States, GSM operates in the 1.900 GHz band. Outside the United States, GSM fosters operations in spectrum between the 890 and 915 MHz frequencies for the uplink and in spectrum between the 935 and 960 MHz frequencies for the downlink. A GSM uplink describes the transmission path from the mobile station to the base station. The downlink refers to the transmission path from the base station to the mobile station. GSM signals are carried on channels. A channel includes a pair of frequencies with one frequency for reception and one frequency for transmission.

8.15.5   SATELLITE TECHNOLOGIES

Satellite services foster access to GSM applications outside areas of cellular coverage. For example, ACeS, a provider of integrated satellite-GSM networks, employs a geostationary (GEO) satellite configuration for enabling subscribers to access terrestrial GSM voice, fax, and data applications via dual-mode satellite and GSM communicators within Asia. ACeS communicators also employ standardized GSM SIMs (Subscriber Identity Modules).

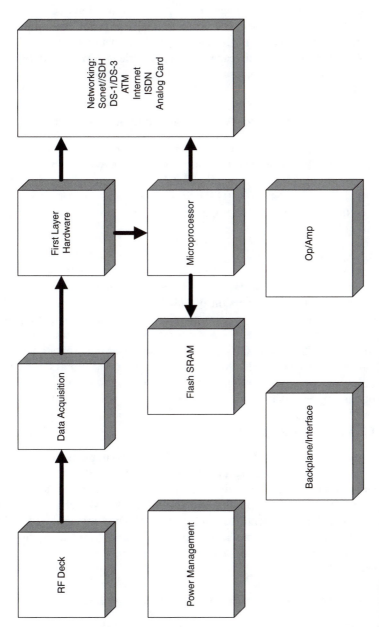

FIGURE 8.3 Cellular base station components and functions.

8.15.6   GSM Configurations

8.15.6.1   GSM 900 Networks

GSM 900 networks support equipment interoperability and service connectivity through the use of the Mobile Application Part (MAP) protocol. Communications providers such as Global Services enable national and international roaming on GSM 900 networks. This internetworking capability is available to AMPS subscribers that use communications providers holding memberships in the North American Cellular Network (NACN).

8.15.6.2   GSM and NACN (North American Cellular Network)

In addition to supporting connectivity to GSM 900 networks, the North American Cellular Network (NACN) provides cellular backbone network services that are compatible with X.25 protocols in approximately 7,500 cities worldwide. NACN uses the Global Signaling Network (GSN) for provisioning direct call delivery to subscribers on the move and enabling operations that are compliant with the ANSI-41 specification.

8.15.6.3   GSM 1.900 GHz

Utilized primarily in the United States, GSM 1.900 GHz technology fosters inter-connectivity with the European version of GSM technology. This connectivity is accomplished by upbanding or converting the European standard frequency of 900 MHz to 1900 MHz or 1.900 GHz. To take advantage of this conversion capability, Motorola developed GSM 1.900 GHz portable devices that interoperate with GSM 900 communicators. As with GSM 900 communicators, GSM 1.900 GHz portable devices enable an array of applications that include short messaging services (SMS) consisting of alphanumeric characters, data encryption, and high-quality voice services.

8.15.7   GSM Applications and Services

The GSM platform supports basic, value-added, and supplementary services. Moreover, the GSM infrastructure enables transmission and reception of text, graphics, slow-moving diagrams, voice, and short video sequences, and supports videotelephony, fax, and SMS applications. In addition, GSM technology enables emergency calls, caller ID (identification), multiparty conferencing, call waiting, call forwarding, call hold, and message notification. GSM implementations also provision seamless mobility inside and outside and operate in conjunction with satellite networks.

8.15.7.1   Short Messaging Service (SMS) and Cellular Broadcasts

The GSM platform supports short messaging service (SMS) and cell broadcasts. Also called text messaging, voicemail plus, PCS messaging, and cellular message teleservice (CMT), SMS facilitates transmission of a brief alphanumeric message to and from a cellular communicator. SMS operates in concert with the ANSI-41 signaling guidelines and NACN configurations. An extension to ASNI-41, ANSI-41D defines high-priority teleservices such as user authentication and point-to-point basic or bearer services.

Cellular broadcast messages range from one to fifteen pages in length, with each page containing a maximum of 93 alphanumeric characters. Because GSM networks are optimized for supporting circuit-switched voice calls, this technology is limited in effectively transmitting extremely short data bursts.

8.15.7.2 Unstructured Supplementary Services Data (USSD)

In parallel with SMS, Unstructured Supplementary Services Data (USSD) solutions support transmission of alphanumeric messages via a GSM network configuration and employ GSM channels for data delivery. Each USSD message contains a maximum of 182 characters. In contrast to store-and-forward SMS operations, USSD is a session-oriented service. After a session is set up, the radio link remains open until released by a user or an application. GSM standards clarify USSD capabilities and services.

8.15.8 GSM STANDARDS ORGANIZATIONS AND ACTIVITIES

8.15.8.1 Conference of European Posts and Telegraphs Administration (CEPT)

In the early 1980s, a Study Group appointed by the Conference of European Posts and Telegraphs Administration (CEPT) initiated development of a pan-European Public Land Mobile Telecommunications System (PLMTS), subsequently called Global System for Mobile Telecommunications (GSM).

8.15.8.2 European Telecommunications Standards Institute (ETSI)

The European Telecommunications Standards Institute (ETSI) supports GSM and UMTS standardization efforts and coordinates standards activities in the cellular communications domain with the International Telecommunications Union-Telecommunications Standards Sector (ITU-T). ETSI also sponsors projects such as the TIPHON (Telecommunications and Internet Protocol Harmonization Over Networks) initiative for enabling data exchange between individuals using wireline IP networks and GSM subscribers using circuit-switched networks.

ETSI defines GSM digital mobile specifications and works with the GSM Association in the standardization process. The Global Mobile Suppliers Association (GSA) also contributes to the development of GSM standards.

8.15.8.3 GSM Association

The GSM Association promotes widespread deployment of dual-band and tri-band GSM handsets and cellular phones. The GSM Association also facilitates implementation of GSM-related technologies, including GPRS, EDGE, and 3GSM.

8.15.8.4   GSM Roaming Forum

The GSM Roaming Forum promotes the development and utilization of interoperable network platforms and mobile handsets that enable seamless roaming across national boundaries and frequency bands. This Forum also supports GSM interoperability with CDMA, TDMA, and iDEN (integrated Digital Enhanced Network) technologies. Participants in the GSM Roaming Forum include the GSM Alliance, the CDMA Development Group (CDG), and the Universal Wireless Communications Consortium (UWCC).

8.15.8.5   International Telecommunications Union-Telecommunications Development Sector (ITU-D)

The International Telecommunications Union-Telecommunications Development Sector (ITU-D) supports utilization of GSM portable communicators for provisioning access to voice and data services. Teleworking, telemedicine, and distance learning organizations in the European Union, the United States, Japan, and Australia participate in ITU-D initiatives.

8.15.8.6   International Telecommunications Union-Telecommunications Standards Sector (ITU-T)

The International Telecommunications Union-Telecommunications Standards Sector (ITU-T) endorses Recommendations for enabling universal implementation of 2G and 3G cellular communications technologies that include GSM, UMTS, and IMT-2000 (International Mobile Telecommunications-Year 2000).

8.15.8.7   Mobile Data Association (MDA)

Sponsored by the CTIA, the Mobile Data Association (MDA) developed the V-OFDM (Vector-Orthogonal Frequency-Division Multiplexing) protocol. MDA also promotes implementation of standards-compliant GSM voice and data services and applications. MDA participants include Intel, Ericsson, Nokia, Compaq, IBM, and Microsoft. European GSM equipment manufacturers and network operators such as CellNet, Mobilfunk, and T-Mobil participate in the MDA as well.

8.15.8.8   Mobile Data Initiative Next-Generation (MDI-NG) Consortium

The Mobile Data Initiative Next-Generation (MDI-NG) Consortium supports development of affordable and interoperable cellular communicators that utilize packet-switched network technologies for provisioning seamless connections to Web services. MDI-NG Consortium participants include the GSM Association, Intel, Fujitsu, Siemens, Hewlett-Packard, and Motorola.

8.15.8.9   North American GSM Alliance

The North American GSM Alliance endorses the use of multiple cellular technologies and protocols, including W-CDMA in conjunction with third-generation implementations. Participants in this Alliance include SBC, Verizon, and Cingular. From 1999 to 2000, the North American GSM Alliance conducted trials of 3G networks based on W-CDMA technology in downtown Montreal. These trials verified W-CDMA capabilities in enabling soft handovers and voice and data applications in densely populated metropolitan environments.

8.15.8.10 Telecommunications Industry Association (TIA)

Accredited by the American National Standards Institute (ANSI), the Telecommunications Industry Association (TIA) promotes the development of voluntary industry specifications for telecommunications products. TIA takes part in international standards activities sponsored by the ITU, the Inter-American Telecommunication Commission (CITEL), and the International Electrotechnical Commission (IEC). The TIA encourages implementation of 3G standardized cellular products and works with the ITU-T in developing specifications for 3G implementations.

8.15.8.10.1 TIA Wireless Internet Protocol Partnership (WIPP)

Sponsored by TIA, the Wireless Internet Protocol Partnership (WIPP) sponsors development of the Wireless Internet Protocol (WIP) specification. WIP interworks with GSM, TDMA, CDMA, and satellite protocols and technologies and enables access to Web services via laptops, desktops, personal computers (PCs), and handheld cellular communicators. Ericsson, Lucent, Motorola, Siemens, Nortel Networks, and 3Com participate in WIPP initiatives.

8.15.9 GSM MARKETPLACE

8.15.9.1 MediaOne International

MediaOne International provisions GSM service in joint partnerships with communications carriers in Hungary, Russia, the Czech Republic, Malaysia, Poland, and India.

8.15.9.2 Nortel

In 2000, Nortel Networks installed a GSM 900 network throughout Xinjiang Province in China and introduced a GSM 1800 network in the provincial capital of Shijiazhaung. Nortel also expanded capabilities of GSM cellular networks in Taiwan and Hong Kong.

8.15.9.3 Telenor Satellite Tracking System

The Telenor Satellite Tracking System employs CellPoint for enabling GSM cellular networks to transport voice and data and determine positions of handheld mobile phones. Telenor Satellite Tracking System services are available worldwide in the Nordic Countries, South Africa, Central and Eastern Europe, Namibia, and the United States. The Telenor Satellite Tracking System also supports environmental monitoring and public safety services via LEO (Low-Earth Orbit) systems such as Argos and Orbcomm that interoperate with GSM solutions.

8.15.9.4 Total Access Corporation (TAC) and United Communication Industry Public Company Limited (UCOM)

In Thailand, TAC (Total Access Corporation) and UCOM (United Communication Industry Public Company Limited) operate a nationwide GSM network for more than one million mobile subscribers.

8.15.10 GSM IMPLEMENTATION CONSIDERATIONS

GSM subscribers access GSM services provisioned by multiple communications providers with equipment that theoretically can be purchased from multiple vendors that manufacture GSM standards-compliant devices. GSM is a popular communications solution throughout the European Union, South America, Asia, Australia, New Zealand, Africa, and the Middle East.

GSM supports telephony services, access to Web resources, and utilization of advanced 2G or 2.5G solutions such as MExE and GPRS (General Packet Radio Service). Moreover, GSM accommodates mobility requirements of users on the move and supports transition to third-generation cellular networks.

Problems preventing full-scale GSM implementation include potential health risks, identity theft, and security incursions. Regarded as the mobile communications equivalent of ISDN (Integrated Services Digital Network) technology, GSM is predominantly a narrowband technology. As a consequence, GSM services are restricted in terms of applications provided and rates supported. GSM cellular equipment from multiple vendors is not always interoperable. GSM service is not yet universally available and GSM standards are still in development.

Additional roadblocks to worldwide GSM deployment include insufficient support for global roaming, poor connections, substandard infrastructure services in local and metropolitan environments, and configuration problems in SOHO venues. Specifications developed by GSM standards organizations and vendor enhancements to GSM technology are designed to resolve these problems. In addition, in 2000, the GSM Association endorsed the Security Accreditation Scheme (SAS) to standardize GSM security procedures and ensure the integrity of GSM operations.

8.16 GENERAL PACKET RADIO SERVICE (GPRS)

8.16.1 GPRS FEATURES AND FUNCTIONS

Challenges associated with providing packet radio access services via GSM devices contributed to the development of General Packet Radio Service (GPRS). An advanced second-generation solution also known as 2.5G, GPRS is a packet-switched network overlay that operates on top of a GPRS circuit-switched network for enabling access to diverse non-voice applications by subscribers on the move. GPRS subscribers employ GPRS mobile communicators and IP addresses for accessing GPRS applications.

The number of GPRS subscribers sharing radio resources in a single cell depends on the robustness of the air interface and types of applications supported. An always-on technology, GPRS (General Packet Radio Service) potentially enables data retrieval and data transmission across mobile telephony networks at rates reaching 171.2 Kbps by using eight timeslots. However, GPS packets travel in different directions to reach the same destination, thereby contributing to delays in GPRS transmissions.

General Packet Radio Service (GPRS) implementations employ a GPRS Support Node for monitoring the location of mobile stations, verifying the identity of mobile subscribers, and provisioning support services. GPRS technology utilizes GMSK (Gaussian Minimum-Shift Keying) modulation and works in concert with the TDMA (Time-Division Multiple Access) protocol, the ANSI IS-136 specification, and EDGE (Enhanced Data Rates for Global Evolution) solutions in provisioning high-speed wireless data services. Moreover, Ericsson supports development of GPRS installations that work in concert with IPv6 (Internet Protocol version 6).

8.16.2 GPRS Applications

GPRS technology fosters connections to public and private networks that employ protocols such as X.25 and TCP/IP, and supports Web browsing, group chat sessions, links to multimedia applications, and transmission of bursty and intermittent data traffic. GPRS technology also enables remote monitoring and remote control of home appliances and facilitates e-mail services, connections to remote LANs, and customized delivery of sports scores, flight information, news headlines, horoscopes, lottery results, and jokes. In addition, GPRS deployments support transmission of photographs and still images and enable document sharing and collaborative tele-working applications.

In the Netherlands, GPRS technology provides links to GigaPort, a high-speed research and educational network. Dutch universities, research centers, and corporations participate in the GigaPort initiative.

8.16.3 GPRS M-Services

In 2001, the GSM Association endorsed the implementation of the Mobile Services (M-Services) initiative for enabling GPRS subscribers to access Mobile Internet services and digital content via always-on and always-available GPRS communicators. M-Services work in conjunction with open wireless standards such as WAP and provision connections via integrated cellular and handheld devices to wireless Internet applications and corporate intranets and extranets.

In addition, M-Services support enhanced Web browsing via cellular communicators with uniform keypads, consistent screen displays, and buttons with standardized functions. M-Services also enable streaming video, short messaging service (SMS), e-mail, and multimedia messaging service (MMS), and provision access to downloadable graphics, games, music, ring tones, wallpaper, and screen savers posted on GSM cellular networks.

8.16.4　GPRS and High-Speed Circuit-Switched Data (HSCSD)

GPRS deployments interwork with IP networks, enable IPsec (Internet Protocol Security), and support VPN (Virtual Private Network) implementations. GPRS solutions also provision network operations, administration, and management (OA&M) functions.

GPRS networks enable transmission of packets with the same destination address in different directions. This best-effort delivery service results in lost packets and delays in transmission of time-sensitive information. As a consequence, High-Speed Circuit-Switched Data (HSCSD) technology is employed for enabling GPRS cellular devices to access broadcast-quality video and telephone-quality voice applications.

HSCSD service is an enhanced GSM solution that enables circuit-switched services, real-time videoconferencing, and access to voice, video, and data applications by establishing a permanent virtual link between two sites for information transport. Mobile GSM handsets equipped with GPRS and HSCSD capabilities support higher transmission rates than generic GSM handsets by enabling utilization of more than one timeslot per TDMA frame. For example, with HSCSD, as many as four GSM timeslots or channels operating at 14.4 Kbps apiece can be bonded together for achieving total transmission rates of 57.6 Kbps. Moreover, HSCSD also interoperates with other circuit-switched networks and technologies including the Public Switched Telephone Network (PSTN) and the Integrated Services Digital Network (ISDN).

8.16.5　GPRS Standards Organizations and Activities

8.16.5.1　European Telecommunications Standards Institute (ETSI)

Standardized by ETSI, GPRS technology optimizes utilization of radio resources by enabling fixed virtual connections. With GPRS implementations, radio resources are employed only when subscribers transmit or receive data. As a consequence, multiple GPRS subscribers can share bandwidth and obtain communications services within the same cell.

8.16.5.2　GPRS (General Packet Radio Service) Alliance

The GPRS Alliance promotes utilization of GPRS services in 3G mobile environments. This Alliance also supports development of an evolved GPRS core network that works in conjunction with advanced video, data, and voice-over-IP applications. Participants in the GPRS Alliance include IBM, Ericsson, Oracle, Lotus, Symbian, and Palm Computing.

8.16.6　GPRS Marketplace

Manufacturers and communications providers that support GPRS-based M-Services include Nokia, Ericsson, France Telecom, Motorola, Alcatel, and Samsung. GPRS vendor-supported initiatives are implemented in countries that include China, Germany, France, Austria, Turkey, and Hungary.

8.16.6.1   Beijing Mobile Communications Corporation and Motorola

Motorola and Beijing Mobile Communication Corporation (BMCC) support development of a GRPS infrastructure that works in conjunction with WAP (Wireless Application Protocol)-enabled smart phones.

8.16.6.2   Nokia

The Nokia Artus product family supports implementation of Multimedia Messaging Service (MMS) for PCS enrichment that works in concert with WAP and GPRS backbone networks. MMS builds on the SMS framework by including the addition of video and voice clips to conventional messaging services. With MMS, songs, electronic greetings, electronic postcards, and memoranda can be sent from device-to-device with instantaneous delivery. MMS applications are also supported by UMTS implementations. In addition, Nokia supports utilization of Internet operating system router software to support advanced Mobile IPv6 functions in GPRS networks and 3G IMT-2000 (Internet Mobile Telecommunications-Year 2000) cellular deployments.

8.16.6.3   Swiss Mobilcom

Swiss Mobilcom implements a nationwide GPRS backbone network that supports services in conjunction with an in-place GSM infrastructure, fosters connectivity to the Internet at rates reaching 115 Kbps, and enables migration to third-generation cellular communications networks.

8.17   DIGITAL ENHANCED CORDLESS TELECOMMUNICATIONS (DECT)

8.17.1   DECT Technical Fundamentals

A flexible digital radio access technology, Digital Enhanced Cordless Telecommunications (DECT) enables cordless communications in corporate and residential venues. DECT systems operate in spectrum between the 1.880 and 1.920 GHz frequencies. The DECT Multimedia Access Profile (DMAP) facilitates development of affordable in-home DECT communicators. DECT implementations interwork with the Public Switched Telephone Network (PSTN) and ISDN (Integrated Services Digital Network) and Ethernet technologies in provisioning links to Web resources via the local loop or the first mile from the subscriber premise to the local telephone exchange. DECT also supports data encryption for safeguarding the integrity of communications services.

8.17.2   DECT and GSM

The Digital Enhanced Cordless Telecommunications (DECT) standard was adopted by the European Telecommunications Standards Institute (ETSI) in 1998, a year prior to the approval of the GSM specification. Interestingly, the popularity of GSM mobile telephony service sparked interest in DECT technology. ETSI also supports a Generic Access Profile (GAP) that serves as the basic DECT air interface.

Dual-mode GSM and DECT communicators enable cordless access to voice, video, and data applications and the Web. As with GSM, DECT provides communications services in a specific location, such as a manufacturing plant, by dividing the coverage area into a cellular mesh configuration that enables a GSM mobile subscriber to initiate or respond to cellular phone calls anywhere within a specified area of coverage. In parallel with GSM, DECT also employs TDMA (Time-Division Multiple Access) modulation for enabling signal transport.

8.17.3   ETSI DECT Project

The ETSI DECT Project facilitates deployment of DECT IP applications such as IPv6 and voice-over-IP. In addition, the DECT Project supports DECT-based mobile communications services and utilization of DECT Packet Radio Service (DPRS) to enable multimedia transmission at 2 Mbps. Next-generation DPRS technology for transporting voice, video, and data wirelessly over the local loop at rates reaching 10 Mbps is also in development. ETSI-sponsored interworking documents verify DECT interoperability with ISDN and GSM, and support the establishment of DECT as the cordless technical component for UMTS technology and IMT-2000 implementations.

8.17.4   DECT Applications and Services

DECT (Digital Enhanced Cordless Telecommunications) configurations are interoperable, extendible, and scalable, and provision wireless services in isolated rural locations and in densely populated local environments. DECT capacity and coverage areas expand to accommodate communications requirements generated by additional subscribers. DECT also supports multimedia transmission via WLAN (Wireless Local Area Network) implementations.

DECT (Digital Enhanced Cordless Telecommunications) systems are deployed worldwide in hotels, hospitals, retail stores, multistory office buildings, apartment complexes, airports, conference centers, and exhibition halls. DECT solutions provision voice services, WLL (Wireless Local Loop) and broadband FWA (Fixed Wireless Access) solutions, and packet data transmission; provision automatic security alerts in case of flood or fire; and enable remote control of household applications.

Automobile manufacturing facilities in Sweden, hospitals in the Netherlands, supermarket chains in the United Kingdom, hotels in San Francisco and Germany, and grocery warehouses in Australia deploy DECT systems. DECT implementations in Hungary, France, Indonesia, Japan, Italy, New Zealand, China, Egypt, South Africa, and India enable access to narrowband and broadband FWA (Fixed Wireless Access) configurations.

8.17.5   DECT and Personal Handyphone System (PHS) Solutions

As with DECT, Personal Handyphone System (PHS) solutions employ multipoint microwave two-way or full-duplex digital radio access technology that provides connections to ISDN and the PSTN. In Hong Kong, dual-mode DECT and PHS communicators enable Web browsing and e-mail exchange at rates reaching 64 Kbps. Like DECT, PHS technology supports encryption, efficient use of the radio spectrum, and dependable delivery of voice and data digital transmissions.

8.17.5.1   NTT DoCoMo and the National University of Singapore

NTT DoCoMo and the National University of Singapore evaluate capabilities of dual-mode PHS and DECT communicators in supporting service enhancements such as call hold, call transfer, voicemail, and conference calls; delivery of message alerts; and provision of broadband FWA solutions.

8.17.6   DECT Forum

The DECT (Digital Enhanced Cordless Telecommunications Forum) was organized in 1997 by merging the DECT Operators Group with the Global DECT Forum. Participants in this Forum include DECT equipment suppliers, manufacturers, and telecommunications operators and communications carriers worldwide. DECT products include digital cordless telephones, fully featured systems that include handsets with menu-driven user interfaces, cordless terminal adapters, and answering machines.

8.18   SOFTWARE DEFINED RADIO (SDR)

Software Defined Radio (SDR) refers to collections of software and hardware technologies such as baseband modems, modular software, signal processing algorithms, and antennas that facilitate development of SDR solutions. These solutions support voice calls, data delivery, and dependable access to personalized services and Web applications. In addition, SDR implementations employ a generic architecture for enabling a direct or one-to-one relationship between the access network and the portable communicator. Tri-band radios that foster GSM 900, GSM 1800, and PCS 1900 services or, optionally, tri-band radios that enable GSM 900, GSM 1800, and W-CDMA services are examples of SDRs in action.

8.18.1   SDR Implementations and Applications

SDR (Software Definable Radio) platforms provision secure and ubiquitous communications support, universal coverage, and seamless roaming services for subscribers using 2G, 2.5G, and 3G cellular communications networks. SDRs work in concert with protocols and technologies that include CDMA, TDMA, W-CDMA, GSM, UMTS, and Bluetooth. Moreover, SDRs can be reprogrammed for receiving and transmitting voice, video, and data signals on a wide range of frequencies so that, for example, subscribers can switch from GSM 900 and GSM 1800 cellular services to PCS and D-AMPS cellular networks.

8.18.2   SDR Operations

In SDR implementations, software performs functions previously carried out in hardware such as signal generation and detection. An SDR is flexible, programmable, precise, and insensitive to environmental impacts.

In addition to maximizing utilization of spectral frequencies allocated for wireless transmission, SDRs also support expanded access to high-capacity broadband applications. Moreover, SDRs foster implementation of basic and advanced wireless services and eliminate the need for expensive components that are employed in hardwired radios. In SDR installations, service architectures can be scaled to support higher data rates and larger user populations. SDR popularity drives development of new protocols for supporting radio resource management and universal mobile functions.

8.18.3   SDR Forum

Created in 1996, the Software Defined Radio (SDR) Forum develops SDR elements, technical specifications, and open architectures for advanced wireless systems. Vendors in the SDR Forum include Verizon, Motorola, ITT Industries, Exigent, Raytheon, and QuickSilver Technologies.

8.18.4   Enhanced Software Radio Forum (ESRF)

The Enhanced Software Radio Forum (ESRF) encourages collaborative design and deployment of Intelligent Software Definable Radio (ISDR) solutions for third-generation mobile and personal communications systems such as UMTS. Sponsored by ETSI and the IEEE (International Electrical and Electronics Association), the ESRF develops procedures for utilizing broadband SDR technology to support seamless access to cellular services and delivery of voice, video, and data transmissions with QoS (Quality of Service) guarantees.

8.19   IMT-2000 (INTERNATIONAL MOBILE TELECOMMUNICATIONS-YEAR 2000)

8.19.1   IMT-2000 Fundamentals

A multifunctional, multiservice 3G cellular communications specification, IMT-2000 (International Mobile Telecommunications-Year 2000) facilitates seamless information delivery and global roaming via cellular communications networks. Endorsed by the ITU-T (International Telecommunications Union-Telecommunications Standards Sector), IMT-2000 enables voice telephony, instant messaging, short messaging services (SMS), multimedia messaging services (MMS), wide area paging, Web browsing, E-commerce applications, e-mail, videoconferencing, and ubiquitous access to information resources. IMT-2000 also facilitates migration from 2G and 2.5G systems to 3G systems that are designed to support full-scale global services.

8.19.2   IMT-2000 Infrastructure

IMT-2000 (International Mobile Telecommunications-Year 2000) implementations use a mesh cellular architecture that includes microcells with low power requirements for teleservice delivery in heavily populated metropolitan areas, picocells for indoor use, and macrocells for conventional cellular networking applications. IMT-2000 technology enables stationary subscribers and subscribers on the move to access narrowband and broadband applications.

Envisioned as a global standard for 3G mobile systems everywhere, IMT-2000 technology interoperates with 2G and 2.5G solutions such as GSM, D-AMPS, and GPRS, and interworks with IP and ANSI-41 WIN (Wireless Information Network) configurations. IMT-2000 promotes the establishment of a global mobile and personal communications platform capable of supporting multimedia services and radio interfaces via a universal infrastructure.

8.19.3 DISTINCTIVE IMT-2000 ATTRIBUTES

8.19.3.1 Universal Personal Telecommunications (UPT) Numbers and Services

Each IMT-2000 subscriber is assigned a UPT (Universal Personal Telecommunications) number that supports personal mobility. Based on a personalized service profile, UPT numbers enable subscribers on the move to receive and/or access personal telecommunication services on any 3G terminal and in any network as long as the communications carrier is a designated UPT service provider.

8.19.3.2 IMT-2000 Communicators

IMT-2000 compact pocket communicators combine features of a computer, television, newspaper, library, address book, credit card, personal diary, and telephone. In addition, IMT-2000 devices enable ubiquitous links to telecommunications services, Web resources, intranets and extranets, and voice, video, and data applications.

8.19.3.2.1 Samsung Electronics
Samsung Electronics supports a 2-inch reflective display for use on IMT-2000 cellular communicators. This color display provisions access to high-resolution video, still images, video-on-demand (VOD), and television broadcasts.

8.19.4 IMT-2000 CONFIGURATIONS

IMT-2000 fosters development of an integrated global framework consisting of interconnected terrestrial and satellite networks for enabling wireless access to teleservices at any time and from any place, and supporting seamless communications services in urban, suburban, rural, maritime, and aeronautical environments. IMT-2000 local area and wider area wireless networking solutions consist of personal cellular communicators or wireless terminals that enable access to network configurations such as ad hoc peer-to-peer networks, intranets, extranets, and the Internet.

IMT-2000 configurations consist of terrestrial and satellite components that employ Radio Transmission Technologies (RTTs). IMT-2000 RTTs include CDMA, cdma2000, DECT, UTRA (Universal Terrestrial Radio Access), and W-CDMA. Additional RTTs that enable 3G IMT-2000 applications include TD/CDMA, a hybrid system based on TDMA and CDMA technologies, and UWC-136 (Universal Wireless

Communications-136), a 3G TDMA interface developed by the Universal Wireless Communications Consortium (UWCC).

As with the GSM satellite component, the IMT-2000 satellite component defines a set of satellite interfaces for optimizing performance of mobile and personal communications systems in a wide range of geographical environments. S-UMTS (Satellite-UMTS) is regarded as a core satellite technology for 3G IMT-2000 deployments. S-UMTS technical solutions are examined in Chapter 10.

8.19.5   IMT-2000 STANDARDS ORGANIZATIONS AND ACTIVITIES

IMT-2000 provides a global framework for implementation of standardized 3G cellular communications standards. Major vendor forums, focus groups, and standards organizations in China, the European Union, Japan, Korea, and the United States take part in the IMT-2000 standardization process.

8.19.5.1   Third Generation Partnership Project (3GPP)

8.19.5.1.1   3GPP Fundamentals
The Third Generation Partnership Project (3GPP) develops 3G standards in partnership with internationally recognized standards organizations. 3GPP participants include ETSI, the ITU, ANSI, TIA, the Telecommunications Technology Association (TTA) in Korea, the Association of Radio Industries and Business (ARIB) in Japan, and the China Wireless Development Group (CWDG). The Third Generation Partnership Project (3GPP) supports seamless access to 3G wireless services via advanced multimode cellular communicators and coordinates 3G spectral allocations.

In addition, the 3GPP develops approaches for interworking 2G and advanced 2.5G communicators with 3G communicators and supports migration from 2G and advanced 2G or 2.5G to 3G implementations, global roaming, universal access to multimedia applications, and ubiquitous telephony services. Moreover, the 3GPP endorses implementation of IP technology and utilization of an IP infrastructure that interworks with 3G technologies; seamless delivery of voice, video, and data; and development of UTRAN enhancements.

8.19.5.1.2   3GPP2 (Third-Generation Partnership Project-Part 2)
The Third Generation Partnership Project-Part 2 (3GPP2) supports utilization of 3G technologies in conjunction with wireless radio communications systems and continues the work of the 3GPP. In 2000, 3GPP2 completed work on the cmda2000 specification as part of the IMT-2000 initiative. Additional contributors to 3GPP and 3GPP2 specifications include the CDMA Development Group (CDG), the GSM Association, the GSM Forum, and the UMTS Forum.

8.19.5.2   International Telecommunications Union-Telecommunications Standards Sector (ITU-T) and IMT-2000 Recommendations

The International Telecommunications Union-Telecommunications Standards Sector (ITU-T) endorses a series of terrestrial and satellite specifications for enabling IMT-2000 implementations to support wireless data, video, and voice transport at 2 Mbps and higher rates. The ITU-T Recommendations for IMT-2000 also support security services based on USIM (Universal Subscriber Identity Module), UPT numbers, and Unique Mobile Equipment Identifiers (UMEI).

Moreover, the ITU-T clarifies procedures for data encryption and decryption to maintain the privacy and integrity of voice and data transmissions. ITU-T also develops guidelines to support point-to-point short messaging service (SMS), over-the-air teleservice activation and HSCSD (High-Speed Circuit-Switched Data) applications, and endorses utilization of radio air interfaces and radio link protocols that provide affordable access to IMT-2000 services.

In addition, the ITU-T develops specifications for call forwarding, call holding, and call waiting services; mobile number portability; and CAMEL (Customized Applications for Mobile Network Enhancement Logic) applications. IMT-2000 also delineates procedures for forming closed user groups to enable restricted communications. Furthermore, IMT-2000 Recommendations by the ITU-T describe technical guidelines for UMTS network architectures and requirements for supporting the evolvement of GSM platforms to UMTS configurations.

In 1999, the ITU-T adopted a series of IMT-2000 Recommendations for provisioning 3G mobile multimedia and cellular networking services. As an example, the ITU-T Q.1521 Recommendation describes network requirements and signaling protocols for enabling Universal Personal Telecommunications (UPT). The ITU-T Q.1531 Recommendation specifies UPT security requirements; and the ITU-T Q.1721 Recommendation describes information flow procedures for enabling end-to-end intersystem IMT-2000 services. Moreover, the ITU-T Q.1731 Recommendation clarifies requirements for radio independent technologies and their support of the IMT-2000 Layer 2 radio air interface. The ITU-T Q.1751 Recommendation delineates IMT-2000 signaling requirements for the network-to-network interface (NNI) protocol. In addition, ITU-T Recommendations for IMT-2000 endorse the use of EDGE and DECT protocols and technologies.

8.19.5.3 International Telecommunications Union-Radio Communications Sector (ITU-R)

The International Telecommunications Union-Radio Communications Sector (ITU-R) establishes network architectures and radio interface requirements for IMT-2000 installations, specifications for high-definition television (HDTV) programming, and Recommendations for broadband fixed wireless access (FWA) networking solutions. The ITU-R also identified spectral bands between the 1.885 and 2.025 GHz frequencies and between the 2.110 and 2.200 GHz frequencies for IMT-2000 installations.

Major features of IMT-2000 are described in the ITU-R M.687 Recommendation and the ITU-R M.816 Recommendations. These Recommendations clarify functions of a single, universal, small pocket cellular device; support expansion of mobile services; and establish a modular and scalable architecture for enabling mobile subscribers to access wireless and wireline networks.

In 1999, the ITU-R approved a series of terrestrial and satellite radio interface specifications for enabling migration from second-generation and 2.5G mobile and personal communications systems such as GSM and GPRS to 3G IMT-2000 networks.

In addition, ITU-R Recommendations defined procedures for adapting IMT-2000 solutions to accommodate communications requirements of subscribers in developing countries. The ITU-R Recommendations also described security procedures, radio interface functions, radio subsystem attributes, transmission rates, and voice-band requirements for IMTS-2000 implementations.

8.19.5.4   World Radiocommunications Conference (WRC)

In 1992, the World Radio Communications Conference (WRC) identified spectrum between the 1.885 and 2.025 GHz frequencies and between the 2.110 and 2.200 GHz frequencies for IMT-2000 applications. The WRC also clarifies services of spectrum between the 1.980 and 2.010 GHz frequencies and between the 2.170 and 2.200 GHz frequencies for third-generation mobile satellite services that will be available worldwide by 2005. In addition, the WRC promotes development of a global seamless radio infrastructure capable of supporting IMT-2000 services provisioned by wireless configurations ranging from fixed wireless access (FWA) networks to multi-application ad hoc mobile systems.

8.19.5.4.1   WRC-2000

At the World Radiocommunications Conference (WRC) in 2000 (WRC-2000), spectral allocations above the 71 GHz spectral block were allocated for scientific initiatives pertaining to radio astronomy, satellite probing of the earth's natural land and water resources, and space research. In addition, WRC-2000 established spectral allocations for the satellite portion of IMT-2000 (International Mobile Telecommunications-2000) in spectrum between the 1.525 and 1.559 GHz RF (Radio Frequency) bands; between the 1.610 and 1.660 GHz RF bands; and between the 2.4835 and 2.500 GHz RF bands. Moreover, spectral allocations between the 2.500 and 2.520 GHz frequencies and between the 2.670 and 2.690 GHz frequencies were identified for the satellite component of IMT-2000 as well.

In 2001, based on WRC-2000 recommendations, the ITU-T designated spectrum between the 805 and 960 MHz frequencies, between the 1.710 and 1.885 GHz frequencies, and between the 2.500 and 2.690 GHz frequencies for the terrestrial component of IMT-2000.

In accordance with WRC-2000 recommendations, the ITU-T endorsed utilization of spectrum between the 1.850 and 2.025 GHz frequencies and between the 2.110 and 2.200 GHz frequencies for IMT-2000 UMTS implementations. As a consequence of spectral shortages, these spectral allocations currently share spectrum with in-place radio services and cellular communications systems.

As with WRC, the ITU-T also endorses the use of spectrum between the 2.700 and 2.900 GHz frequencies for IMT-2000 3G (third-generation) mobile and personal communications systems. However, this RF allocation currently enables airport surveillance radar operations in the United States, Germany, Brazil, Norway, Jamaica, and Australia. In addition, this RF frequency block also supports weather forecasting and life safety applications provisioned by NEXRAD (Next-Generation Radar) systems in the United States. As a consequence of spectral overlaps, the United States opposes utilization of spectrum between the 2.700 and 2.900 GHz frequencies for IMT-2000 implementations.

In the United States, additional ITU-T spectral allocations designated for IMT-2000 currently support diverse cellular communications services and applications. As an example, IMT-2000 allocations in frequencies between the 800 and 900 MHz enable U.S. cellular communications service. Moreover, IMT-2000 allocations in the 1.700 GHz frequency block foster national security and public safety service operations sponsored by the U.S. Department of Defense. Additionally, IMT-2000 allocations in the 2.500 GHz spectral band facilitate wireless data operations and delivery of instructional television programming throughout the United States. As a consequence, the FCC supports development of complementary approaches for spectral sharing and reallocation and promotes utilization of alternative spectral bands for supporting 3G cellular network deployments and global roaming solutions.

8.19.5.4.2 Operator Harmonization Group (OHG)

Communications operators that hold membership in the Operator Harmonization Group (OHG) monitor spectral allocations for IMT-2000 applications to prevent IMT-2000 allocations from disrupting in-place operations. The OHG also promotes development of affordable 3G telephony devices. Verizon, China Mobile Telecom, Japan Telecom, NTT DoCoMo, and SingTel participate in OHG activities.

8.20 UNIVERSAL MOBILE TELECOMMUNICATIONS SYSTEM (UMTS)

8.20.1 UMTS FOUNDATIONS

Developments in GSM technical capabilities, Personal Communications Services (PCS), and GPRS contribute to the creation of the Universal Mobile Telecommunications System (UMTS). UMTS is a third-generation digital cellular communications system that is standardized in the European Union by the ITU-T and ETSI. In addition to GSM, PCS, and GPRS, protocols and technologies that enable UMTS deployment include TDMA, CDMA, W-CDMA, DECT, SDR (Software Definable Radio), and Global Positioning Systems (GPS). Designated as the core technology in the IMT-2000 3G telecommunications suite, UMTS uses an evolved GSM platform, formally known as 3GSM, as the framework for provisioning 3G services at rates between 384 Kbps and 2.4 Mbps.

Moreover, UMTS and GSM technologies enable interoperable applications, support access to GSM and UMTS services via dual-mode UMTS and GSM communicators, and provide flexible bandwidth for voice, video, and/or data applications. As noted, GSM serves as the mobile component of wireless N-ISDN (Narrowband-Integrated Services Digital Network). UMTS functions as the mobile component of wireless B-ISDN (Broadband-ISDN).

8.20.2 UMTS CAPABILITIES

UMTS technology enables next-generation cellular communications, services, and implementations that support seamless communications coverage, on-demand

bandwidth, ubiquitous access to multimedia configurations, high-quality voice services, and customizable capabilities for accommodating unique subscriber requirements. UMTS solutions feature a multiservice platform that enables development
of ad hoc or stand-alone wireless networks. The UMTS platform interworks with
narrowband and broadband WLANs (Wireless LANs), WMANs (Wireless MANs),
and WWANs (Wireless WANs). UMTS implementations also enable voice recognition and paging services, video telephony, voicemail, SMS, multimedia messaging
(MM), and fax delivery and provision access to videoconferencing, videotelephony,
and teleshopping applications.

Additionally, UMTS universal multimode multiband communicators support
telecommuting and operations in fields that include teletourism, E-commerce, telemedicine, and E-government. In 1999, Finland became the first country to grant
UMTS licenses to communications operators. UMTS services are available in Finland and in Japan. (See Figure 8.4.)

8.20.3   UMTS COMMUNICATORS

UMTS communicators range from low-cost pocket devices to sophisticated cellular
phones and PDAs (Personal Digital Assistants) that support access to a rich combination of voice, video, and data services. UMTS communicators function as transceivers, work in concert with cordless base stations in the home or workplace, and
establish links to cellular and/or satellite communications networks when subscribers
move outside of SOHO environments.

UMTS communicators also provision alphanumeric addressing and private numbering schemes as well as phone number portability. Only one UMTS communicator
and one universal phone number are required to access multimedia services, regardless of subscriber location and mobility. Because a single UMTS communicator
supports multiple functions concurrently, the need for additional cellular devices
and fixed wireline phones is eliminated. UMTS solutions enable development of
virtual home and virtual workplace networking environments featuring personalized
services for UMTS subscribers whenever a UMTS portable device is utilized or a
SIM module is removed from one UMTS communicator and inserted into another
UMTS communicator.

8.20.4   UNIVERSAL TERRESTRIAL RADIO ACCESS (UTRA) TECHNOLOGY

UMTS networks employ the Universal Terrestrial Radio Access (UTRA) interface
for enabling seamless 3G operations and services. UTRA supports FDD (Frequency-
Division Duplex) and TDD (Time-Division Duplex) radio operations and operates
in concert with W-CDMA solutions.

UTRA is an evolved GSM third-generation radio access specification defined
by the ITU-R M 1457 Recommendation. UTRA FDD refers to IMT-2000 CDMA
Direct Sequence Spread Spectrum (DSSS) technology. UTRA TDD refers to IMT-
2000 CDMA TDD.

The Third-Generation Partnership Project (3GPP) develops UTRA architectures,
protocols, and topologies and supports the design and implementation of a global

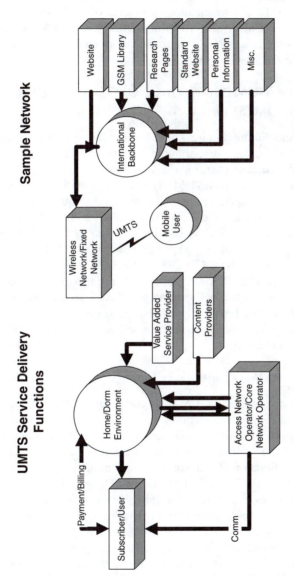

FIGURE 8.4 The sample network shows connections via an international backbone to a GSM Library, personal information, and Web sites.

UTRAN (UTRA Network). 3GPP also delineates procedures for mapping ANSI-41 services on top of UTRAN configurations and methods for enabling transparent operations between cdma2000 and UTRAN implementations.

In contrast to GSM systems that employ circuit-switching technology for low-bit rate data transmission, UMTS supports packet-switched and circuit-switched network services. UMTS communications providers offer flat rate and pay-per-bit payment options. The Universal Terrestrial Radio Access (UTRA) interface for UMTS standards-compliant portable communicators enables subscriber access to diverse voice, video, and data services.

8.20.5   UMTS OR T-UMTS (TERRESTRIAL-UMTS) AND S-UMTS (SATELLITE-UMTS) OPERATIONS

Originally called the Future Public Land Mobile Telecommunications Systems (FPLMTS), UMTS technology supports implementation of ubiquitous mobile multimedia solutions in a range of frequencies. As an example, the terrestrial component of UMTS or T-UMTS (Terrestrial-UMTS) enables operations in spectral bands between the 1.710 GHz (Gigahertz) and the 1.885 GHz frequencies, the 2.010 GHz and the 2.025 GHz frequencies, the 2.110 GHz and the 2.170 GHz frequencies, and the 2.500 to 2.690 frequencies. The satellite component of UMTS or S-UMTS (Satellite-UMTS) fosters operations in spectral bands that include the 1.980 GHz to the 2.010 GHz frequencies, the 2.170 GHz to the 2.200 GHz frequencies, and the 2.670 to the 2.690 frequencies. S-UMTS operations are also examined in Chapter 10.

UMTS trials are conducted in European and Scandinavian countries, including Monaco, Germany, Austria, Finland, and Italy. These pilot implementations benchmark network performance, verify protocol functions, and validate UMTS support of wireless broadband services.

8.20.6   UMTS AND IMT-2000

A 3G mobile system specification, UMTS is compliant with IMTS-2000 specifications. As noted, UMTS implementations support Web browsing, global roaming, and access to IP (Internet Protocol) applications. Moreover, UMTS initiatives enable stationary and mobile subscribers to access customized voice, video, and data applications via multifunctional, multiservice cellular communicators.

In accordance with IMT-2000 objectives, UMTS provides secure and reliable communications services and features the capability to accommodate virtually any communications requirement. Moreover, in parallel with IMT-2000, UMTS provisions high-quality voice telephony operations and supports dependable access to wireless and/or wireline network configurations. To promote UMTS acceptance in markets outside the European Union, the European Telecommunications Standards Institute (ETSI) and member states in the European Union have endorsed a European Union Parliament directive stating that UMTS is the core technology for IMT-2000.

8.20.7 UMTS Standards Organizations and Activities

Subscriber demand for increased access to mobile applications contributes to standardization of 3G UMTS systems. As with IMT-2000, UMTS technical solutions are developed by the European Telecommunications Standards Institute (ETSI) and the World Radiocommunications Conference (WRC) and approved by the ITU-T.

8.20.7.1 International Telecommunications Union-Radio Communications Sector (ITU-R)

The International Telecommunication Union-Radio Communications Sector (ITU-R) promotes the equitable and economical utilization of the RF (Radio Frequency) spectrum and development of a global information infrastructure for coordinating operations of current and next-generation wireless systems in space and terrestrial environments. The ITU-R supports the technological convergence of telecommunications and information technologies, the widespread utilization of PCS solutions, and the development of innovative television and sound broadcasting systems and services. Moreover, the ITU-R enables deployment of standardized mobile satellite services and implementation of standards-compliant second-generation and third-generation cellular communicators that work in concert with GSM and UMTS technologies.

8.20.7.2 Mobile Wireless Internet Forum (MWIF)

The Mobile Wireless Internet Forum (MWIF) supports implementation of standardized 3G (third-generation) wireless IP networks. 3G IP mobile networks optimize spectrum usage and enable subscribers to wirelessly access video, data, and voice services via standards-compliant UMTS communicators.

8.20.7.3 UMTS Forum

The UMTS Forum fosters development of interoperable wireless networks that are universally available and accessible. In addition, the UMTS Forum supports utilization of a distinctive numbering system that provides each UMTS subscriber with one universal telephone number and one handset for enabling total mobility. This unique number also enables UMTS subscribers to be contacted instantaneously regardless of location. The UMTS Forum monitors licensing requirements for 3G systems.

Specifications for UMTS developed by the UMTS Forum build on GSM standards and support next-generation IMT-2000 (International Mobile Telecommunications-Year 2000) goals and objectives. As a consequence, a graceful migration from 2G systems such as GSM and DECT and 2.5G solutions such as GPRS to 3GSM and 3G UMTS services is supported.

Manufacturers, network operators, regulators, and value-added service providers participate in UMTS Forum activities. In 1999, the UMTS Forum and the IPv6 Forum agreed to promote utilization of IPv6 for third-generation mobile services and work with the IETF in making IPv6 services universally available. As noted,

IPv6 (Internet Protocol Version 6) supports the availability of vast numbers of IP addresses to overcome IPv4 addressing constraints. Communications operators and service providers that participate in the IPv6 initiative include Ericsson, Nokia, NTT, Cisco, AT&T, Deutsche Telekom, and WorldCom.

8.20.7.4   UMTS and the World Radiocommunications Conference (WRC)

Sponsored by the ITU-T, the World Radiocommunications Conference (WRC), originally known as WARC (World Administration Radiocommunications Conference), initiated work on UMTS specifications in 1992. WRC also established the foundation for global roaming and ubiquitous mobile communications services and applications. As noted, the WRC currently identifies frequency spectrum for development and deployment of satellite and terrestrial components for enabling T-UMTS and S-UMTS solutions.

8.21   TIME-DIVISION MULTIPLE ACCESS-ENHANCED DATA RATES FOR GLOBAL EVOLUTION (TDMA-EDGE) TECHNOLOGY

8.21.1   TDMA-EDGE APPLICATIONS AND SERVICES

Approved by the ITU-T in 2000, TDMA-EDGE (Time-Division Multiple Access-Enhanced Data Rates for Global Evolution), popularly known as EDGE, is a 3G cellular solution defined in the UWC-136 (Universal Wireless Communications-136) specification for radio telecommunications technology (RTT). Developed by TDMA operators, the Universal Wireless Communications Consortium (UWCC), and the GSM Alliance, EDGE technology is an enhanced version of TDMA technology. In addition to the ITU-T, the American National Standards Institute (ANSI), the TIA, and the European Telecommunications Standards Institute endorse the EDGE specification. Capabilities of 3G TDMA-EDGE deployments are evaluated by Nortel Networks and AT&T in trials conducted in cities across the United States.

TDMA-EDGE configurations deliver circuit-switched voice services and packet-switched data services in real-time via channels that support dual-band analog and digital operations. As with TDMA, EDGE promotes convergence between GSM and D-AMPS technologies. TDMA-EDGE solutions also facilitate migration to UMTS implementations. TDMA-EDGE services are deployed in spectrum between the 1.800 and 1.900 GHz higher frequency bands and between the 800 and 900 MHz lower frequency bands, depending on the subscriber's geographic location. As noted, in the United States, lower-frequency allocations constitute the cellular band and the higher-frequency allocations constitute the PCS band. In the European Union, the United States PCS band is called DCS (Digital Communications Service) 1800 or, optionally, GSM 1800.

TDMA-EDGE implementations enable transmission rates up to 384 Kbps in pedestrian microcellular environments and in low-speed vehicular environments. In higher-speed vehicular environments, the TDMA-EDGE platform supports transmission rates at 14.4 Kbps.

8.21.2 UNIVERSAL WIRELESS COMMUNICATIONS CONSORTIUM (UWCC)

The Universal Wireless Communications Consortium (UWCC) promotes TDMA-EDGE implementation in 3G wireless networks that employ UMTS technology. The UWCC also supports cellular communications projects that facilitate conversion of AMPS and D-AMPS systems into converged TDMA-EDGE and WIN (Wireless Information Network) solutions. Endorsed by the UWCC, ANSI, and the Global WIN Forum (GWF), WIN provisions enhanced TDMA subscriber services in wireless environments.

8.22   3GSM (THIRD-GENERATION GSM)

Third-generation cellular systems such as 3GSM and UMTS support communications services and sophisticated multimedia applications such as video-on-demand (VOD), music, movies, high-speed multimedia transport, mobile Web applications, and cellular postcards, and feature a next-generation radio–air interface. Currently in development, 3G handsets feature a next-generation radio–air interface and convenient viewing screens, and range in size from a basic cellular phone to a handheld computer.

The GSM Association supports migration from second-generation GSM and advanced GSM or 2.5G services such as GPRS to 3GSM solutions and endorses the inclusion of 3GSM as an integral component of the ITU IMT-2000 technology suite. In addition, the GSM Association participates in the Third Generation Partnership Program (3GPP) and works with the UMTS Forum in developing 3G systems.

8.23   EUROPEAN COMMISSION ADVANCED COMMUNICATIONS TECHNOLOGIES AND SERVICES (EC-ACTS) PROGRAM

The European Commission Advanced Communications Technologies and Services (EC-ACTS) program sponsored an array of initiatives benchmarking capabilities of advanced GSM and UMTS services in real-world environments. Findings contributed to the development of mobile multiservice platforms for voice, still images, text, and full-motion video applications.

8.23.1 MOBILE MULTIMEDIA SYSTEMS (MoMuSys)

The MoMuSys (Mobile Multimedia Systems) initiative supported implementation of mobile broadband applications and interactive video and audio services via cellular networks in the educational, medical, and financial service sectors. Moreover, this initiative benchmarked Quality of Service (QoS) guarantees, evaluated capabilities of wireless ATM (WATM) solutions, and assessed satellite support of real-time multimedia delivery via portable multimedia communicators.

8.23.2 SECURITY FOR PERSONAL COMMUNICATIONS TECHNOLOGY (ASPECT)

ASPECT trials contributed to an understanding of UMTS capabilities in supporting universal roaming and promoted development of a common billing format. In addition, the ASPECT project fostered integration of 2G security mechanisms into 3G UMTS implementations. The ASPECT initiative also sponsored research trials for evaluating capabilities of smart cards and biometric techniques in provisioning subscriber identification and authentication services.

8.24 EUROPEAN COMMISSION INFORMATION SOCIETY TECHNOLOGIES (EC-IST) PROGRAM

As with the European Commission ACTS (EC-ACTS) Program, the EC-IST (Information Society Technologies) Program sponsors an array of initiatives that enhance applications provisioned by second-generation and third-generation cellular network implementations. The initiatives that follow provide an introduction to enhancements in second-generation cellular communications capabilities and highlight development efforts leading to innovations in third-generation UMTS networks.

8.24.1 BRAIN

The BRAIN initiative facilitates development of broadband radio access technology for enabling seamless connections to current and emergent IP multimedia applications. Designed as a wireless extension to GSM, GPRS, EDGE, and UMTS implementations, BRAIN is designed to provision links to voice, video, and/or data services supported by IP networks at rates reaching 2 Mbps in hard-to-reach locations such as airports, exhibition halls, conference centers, and train stations.

8.24.2 CAUTION

The CAUTION initiative supports development of reliable and advanced capacity management and administrative services in present-day and next-generation cellular networks. Approaches for eliminating traffic congestion and gridlock are examined. In addition, procedures for enabling seamless network operations and dependable network performance in cellular networks based on technologies including WAP, GPRS, GSM, and UMTS are explored.

8.24.3 IPv6 WIRELESS INTERNET INITIATIVE (6WINIT)

The 6WINIT (IPv6 Wireless Internet Initiative) project verifies capabilities of a pan-European IPv6-compliant wireless Internet. The wireless Internet platform employs a mix of technologies, including GPRS and UMTS, for enabling access to mobile E-commerce (M-commerce) and multimedia applications.

8.24.4 TIGRA

The TIGRA project utilizes spectrum between the 800 and 900 MHz frequencies for supporting next-generation TETRA-based Public Mobile Radio (PMR) and public safety services. TETRA systems currently operate in the overcrowded spectral block between the 380 and 430 MHz frequencies within the European Union.

8.25 ON THE HORIZON:
4G BROADBAND CELLULAR SOLUTIONS

In 2000, AT&T and Nortel Networks initiated development of 4G (fourth-generation) cellular networking solutions. These next-generation cellular networks were designed to work in concert with wideband transceivers, multibeam antennas, sophisticated modems, Software Definable Radios (SDRs), and advanced power amplifiers for enabling dependable delivery of streaming video and audio applications.

In addition, AT&T supports an asymmetric networking configuration called 4G Access. Expected to be available by 2005, 4G Access employs W-OFDM (Wideband-Orthogonal Frequency-Division Multiplexing) for the downlink channel and EDGE technology for the uplink or the return channel. Transmission rates at 10 Mbps for stationary systems and 384 Kbps for mobile systems will be supported. Nortel Networks supports development of a 4G wireless network that enables voice, video, and transport at rates of 20 Mbps.

Ericsson expects to implement 4G cellular systems that provision access to Web resources at rates reaching 100 Mbps. NTT DoCoMo also supports development of 4G systems that facilitate access to data, voice, and full-motion video applications at transmission rates between 2 Mbps and 155.52 Mbps.

8.26 CELLULAR HEALTH AND SAFETY CONSIDERATIONS

8.26.1 HEALTH AND SAFETY ISSUES

Health and safety issues related to the use of cellular communicators are increasingly the focus of media attention. Reports on the potential role of cellular phone use in causing short-term memory problems, dizziness, nausea, headaches, biochemical stress, vision impairments, high blood pressure, and cancerous brain and jaw tumors appear regularly in the popular and scientific press. Some findings suggest that exposure to electromagnetic fields (EMFs) generated by cellular devices may result in headache, birth defects, miscarriages, behavioral changes, Alzheimer's disease, and Parkinson's disease.

According to the United Kingdom National Radiological Protection Board, cellular phone use is associated with significant absorption of microwave energy in the eyes. The brain, nose, tongue and surrounding muscles may also be affected. The Swedish National Institute of Working Life identified a link between the time and the number of calls per day and the prevalence of fatigue, facial tingling sensation, headache, and warming sensations on or around the ear.

The Canadian Wireless Telecommunications Association (CWTA) addresses health-related issues associated with handheld devices, antenna installations, and utilization of mobile phones for emergency 911 calls. According to CWTA investigations, sufficient evidence to link adverse health effects with exposure to radiowaves emitted by cellular telephony systems has not yet been established.

Sponsored by the World Health Organization (WHO), the Electromagnetic Frequency (EMF) Project supports the documentation of research relating to health effects of exposure to electromagnetic frequencies (EMFs). Results from WHO research are mixed. The data are difficult to interpret. Experiments that work in one research center are difficult to replicate in other laboratories.

8.26.2 U.S. FCC (FEDERAL COMMUNICATIONS COMMISSION) HEALTH AND SAFETY GUIDELINES AND STANDARDS

The U.S. FCC (Federal Communications Commission) defines health and safety standards for cellular phone use and establishes maximum allowable exposure limits for cellular devices operating in spectrum between the 300 kHz and 100 GHz frequencies. The FCC also establishes specific absorption rate (SAR) limits for cellular devices that operate near the body.

According to the FCC, cellular communicators must conform to guidelines for human exposure to RF electromagnetic fields established by the IEEE, ANSI, and the National Council on Radiation Protection and Measurements.

Moreover, the FCC can require the routine environmental evaluation of cellular devices and services to ensure compliance with RF radiation exposure limits. As an example, concerns about the safety of handheld cellular devices and the compliance of these devices with FCC safety guidelines have resulted in new procedures for testing and evaluating handheld devices by the FCC and the IEEE in 1999. Moreover, the FCC currently requires the routine environmental evaluation of satellite, radio, and cellular services and assessment of antenna placement to ensure compliance with FCC limits for radiation limits.

8.27 SUMMARY

In this chapter, distinguishing characteristics of 1G, 2G, 2.5G, and 3G cellular communications solutions are examined. Capabilities, merits, and operational constraints of cellular communications technologies, protocols, and standards are delineated. Health and safety concerns associated with cellular communications implementations are reviewed. The role of national and international standards organizations, such as the IEEE, the ITU-T, ANSI, and ETSI, in developing cellular specifications that facilitate multimedia delivery and bandwidth on-demand are explored.

Cellular communicators provide freedom of location to the subscriber, flexibility in day-to-day activities, and rapid and dependable access to voice, video, and/or data services. Demand for universal connectivity and seamless mobility contribute to the development of second-generation (2G) and 2.5G cellular solutions based on technologies such as D-AMPS, GMS, and GPRS. Capabilities of 2G and 2.5G implementations in enabling cellular subscribers on the move to access voice calls, stock quotes, weather forecasts, news headlines, Web resources, and e-mail are examined. Third-generation (3G) IMT-2000 cellular solutions based on technologies such as UMTS that provision ubiquitous access to mobile services via sophisticated next-generation, multifunctional, small and lightweight pocket communicators are reviewed. Major services provisioned by these compact communicators facilitate global roaming, voice, data, and video transmission; and dependable access to integrated and personalized services and networking configurations.

8.28 SELECTED WEB SITES

Association of Radio Industries and Business (ARIB). Outline of Activities.
Available: http://www.arib.or.jp/arib/english/index.html
CDMA Development Group (CDG). 3G. Detailed Information.
Available: http://www.cdg.org/3gpavilion/info.asp
CDMA Development Group (CDG). CDMA Technology.
Available: http://www.cdg.org/tech/tech.asp
Cellular Telecommunications and Internet Association (CTIA). Wireless E-
Commerce Home Page.
Available: http://www.wirelessdata.org/ecommerce/index.asp
Digital Enhanced Cordless Telecommunications (DECT) Forum. DECT
Overview.
Available: http://www.dectweb.com/dectforum/aboutdect/aboutdect.htm
European Commission. IST Program Projects.
Available: http://www.cordis.lu/ist/projects.htm
European Commission. The ACTS Information Window.
Available: www.de.infowin.org
European Telecommunications Standards Institute (ETSI). DECT. Project
Summary. Last modified on September 25, 2001.
Available: http://portal.etsi.org/dect/Summary.asp
European Telecommunications Standards Institute (ETSI). Project Mesa.
Aims.
Available: http://www.projectmesa.org/IE/gen_info/aims.htm
European Telecommunications Standards Institute (ETSI). TETRA Summary.
Available: http://portal.etsi.org/tetra/Summary.asp
Global System for Mobile Communications (GSM) Association. An Overview
of GSM Technology. GSM World. Last modified on September 25, 2001.
Available: http://www.gsmworld.com/technology/index.html
International Telecommunications Union (ITU). What is IMT-2000? Last
modified on October 11, 2000.
Available: http://www.itu.int/imt/what_is/imt/index.html
Mobile Data Initiative, Next Generation (MDI-ng). About MDI-ng.
Available: http://www.mdi-ng.org/about.html
Software Definable Radio (SDR). Forum. Welcome to the SDR Forum
Website.
Available: http://www.mmitsforum.org/
Third Generation Partnership Project (3GPPP). 3GPP-Third Generation
Partnership Project. Last modified on August 30, 2001.
Available: http://www.3gpp.org/
Universal Mobile Telecommunications System (UMTS). What is UMTS?
Available: http://www.umts-forum.org/what_is_umts.html
Wireless Application Protocol (WAP) Forum. WAP 2.01 Technical Paper.
Posted in August 2001.
Available: http://www.wapforum.org/what/WAPWhite_Paper1.pdf

9 Wireless Technologies and Networks

9.1 INTRODUCTION

WPANs (Wireless Personal Area Networks), WLANs (Wireless Local Area Networks), WMANs (Wireless Metropolitan Area Networks), and WWANs (Wireless Wide Area Networks) operate at various rates and levels of complexity; feature a multiplicity of architectures, protocols, and topologies; and vary in size and capacity. As with cellular networks, these configurations provision links to Web resources and narrowband and broadband network applications such as e-mail, E-banking transactions, distance education, and telehealthcare treatment via RF (Radio Frequency) waves in free space, thus eliminating requirements for fixed wireline connections.

The profusion of wireless devices reflects the popularity of cellular and wireless teleservices in addressing communications requirements in home, school, government, business, and hospital environments. Accelerating demand for instantaneous access to the Web and enterprisewide intranets and extranets, regardless of the technology employed, time constraints, or user and/or terminal location, and mobility motivates ubiquitous implementation of integrated cellular and wireless networks that interoperate with landline networks.

As noted in Chapter 8, wireless networks developed apart from cellular networks. In the present-day environment, wireless and cellular networks support overlapping services and applications, employ identical or closely related portable devices, and facilitate development of unified network implementations. Currently in development, these amalgamated networking solutions are expected to enable pervasive connections to networking resources and employ Mobile IP (Internet Protocol) platforms based on IPv6 for provisioning vast numbers of Internet addresses to accommodate the proliferation of wireless network nodes. Technologies such as Bluetooth, 3GSM (Third-Generation Global System for Mobile Communications), and UMTS (Universal Mobile Telecommunications Systems) are expected to support the realization of a global network infrastructure that provisions persistent connectivity to networking resources at any time and from every place.

9.2 PURPOSE

In this chapter, wireless network technologies, protocols, standards, and operations are examined. Capabilities of WPANs, WLANs, WMANs, and WWANs are explored. Representative wireless networking initiatives in education, business, government, healthcare, and SOHO (Small Office/Home Office) environments are described.

Innovative wireless networking solutions such as ad hoc or freestanding networks and broadband fixed wireless access (FWA) configurations are delineated. Projects and programs in the research arena that enhance wireless network performance and system functions are highlighted. Challenges associated with enabling effective wireless network implementations in dynamic environments and approaches for supporting wireless network deployments in academic venues are explored.

9.3   WIRELESS NETWORK TECHNOLOGIES

Demand for remote data access, Web services, pervasive communications links, and ubiquitous computing capabilities contributes to the implementation of a broad spectrum of wireless network solutions. The effectiveness of wireless network solutions in accommodating current and expected user requirements depends on such factors as budget allocations, geographical coverage, information flow, service availability, and wireless network operations and performance.

Wireless network configurations employ infrared, laser, spread spectrum, microwave, and satellite technologies for supporting voice, video, and data transmission in local area and wider area environments. Wireless networks operate in dynamic and flexible environments for accommodating personal and terminal mobility. These networks promote access to traffic tips, weather reports, stock quotes, bank accounts, e-mail, Web services, and enterprisewide networks. As with wireline networks, wireless networks facilitate diverse applications such as videoconferencing, telecollaborative research, tele-instruction, E-government (electronic government) activities, and E-commerce (electronic commerce) services. Key technologies in the wireless networking domain are now examined.

9.4   INFRARED TECHNOLOGY

9.4.1   INFRARED TECHNICAL FUNDAMENTALS

Infrared technology operates in that portion of the electromagnetic spectrum just below visible light. Infrared implementations support half-duplex or one-way and full-duplex or bi-directional data exchange at distances ranging from zero meters to one meter and higher. Infrared transmission is based on the use of light emissions from a Light Emitting Diode (LED) for establishing network connections. A flexible and reliable technology, infrared is integrated into wireless terminals and devices such as PDAs (Personal Digital Assistants), laptops, printers, digital cameras, overhead projectors, cellular telephones, portable scanners, credit card readers, headsets, game controls, fax (facsimile) equipment, and bank automated teller machines.

9.4.1.1   Serial Infrared (SIR), Fast Infrared (FIR), and Advanced Infrared (AIR)

Serial Infrared (SIR) networks transport data at 115 Kbps (Kilobits per second). By contrast, Fast Infrared (FIR) implementations foster transmission rates reaching 4 Mbps (Megabits per second) for supporting WLAN activities and services. Developed by IBM, Advanced Infrared (AIR) enables collaborative workgroup applications via infrared LAN configurations.

9.4.1.2 Diffuse Infrared, Direct Infrared, and Very Fast Infrared

Diffuse infrared platforms support multipoint-to-multipoint connections that are not dependent on a direct line-of-sight for transporting data within delimited locations such as classrooms. In contrast to diffuse infrared, direct infrared supports point-to-point connectivity for point-to-point data exchange and depends on direct line-of-sight for information transport. As with diffuse infrared, direct infrared is regarded as a secure medium for data reception and transmission.

In 1999, the Infrared Data Association (IrDA) approved a high-speed specification for VFIR (Very Fast IR). VFIR enables information transport via portable storage devices, desktop PCs, digital scanners, notebooks, and palmtops at rates reaching 16 Mbps.

9.4.2 Infrared Data Association (IrDA)

Organized in 1993, the Infrared Data Association (IrDA) develops standards for infrared networking solutions that facilitate point-to-point, point-to-multipoint, and multipoint-to-multipoint connections. The IrDA also supports an infrared connector specification for enabling persistent wireless links between peripheral devices such as cellular phones, digital scanners, digital cameras, and PDAs.

In addition, the IrDA publishes specifications clarifying the purpose and functions of IrLAN (Infrared LAN) protocols and technologies such as Serial Infrared (SIR) Link and the Infrared Link Management Protocol (IrLMP). These technologies and protocols enable two information appliances with standards-compliant infrared adapters to function as peer-to-peer WLAN nodes and exchange packets virtually as if they were attached via a single point-to-point wireline link.

The IrDA IrBus (Infrared Bus) specification enables in-room wireless use of up to eight peripheral devices such as gamepads, remote control units, joysticks, and keyboards.

In conjunction with the cellular phone and paging industries, the IrDA introduced the Infrared Mobile Communications (IrMC) standard for defining common data exchange formats. This specification also describes protocols such as IrOBEX (Infrared Object Exchange) for supporting the interoperability of mobile phones with IrDA-compliant devices to enable applications that include calendar synchronization and short messaging service (SMS). In 1999, the Bluetooth SIG (Special Interest Group) adopted the IrOBEX protocol for enabling interoperability between IrDA-compliant devices and Bluetooth network appliances.

9.4.3 Infrared Implementation Considerations

Although infrared signals cannot pass through walls, floors, or room partitions, infrared systems are easy to set up and install. These systems support short-haul, in-room or in-building applications such as data, file, e-mail, business card, and SMS exchange between IrDA-compliant devices.

9.5   LASER TECHNOLOGY

9.5.1   Laser Technical Fundamentals

Laser (Light Amplification by Stimulated Emission of Radiation) technology supports fast implementation of high-performance communications systems that operate in diverse environments. Lasers emit narrow light beams at precise wavelengths for enabling point-to-point connections to enable voice, video, and data transport. Laser configurations do not require rights-of-way permits or Federal Communications Commission (FCC) licensure for deployment. Laser technology fosters operations in the approximate infrared portion of the RF (Radio Frequency) spectrum. As a consequence, laser networks are also called laser infrared networks.

A laser network requires the use of laser transceivers (laser transmitters and laser receivers) for enabling direct line-of-sight, point-to-point transmissions and a tower to which this equipment is affixed. Natural and artificial obstructions such as foliage, dense fog, metal sheds, tall buildings, condensation, and ice accumulation block laser transmissions. Protective coatings, filters, and uniform air control systems safeguard laser equipment and transmissions from adverse weather conditions such as thunderstorms and snowstorms.

9.5.2   Free Space Optics (FSO) Laser Solutions (FSO)

Free Space Optics (FSO) network solutions employ lasers for transmitting voice, video, and data as optical signals through the air for enabling full-duplex transmissions between optical laser transceivers in license-exempt THz (Terahertz) spectral frequencies. Also called wireless optical implementations, Free Space Optics (FSO) configurations enable point-to-point transmissions over short distances for enabling last-mile or local loop connections.

Available since the 1980s, a basic FSO platform consists of two laser transceivers situated on rooftops or inside windows that maintain direct line-of-sight virtual connections. Depending on the weather and distance the between the laser transceivers, FSO installations facilitate rates between 10 Mbps and 2.488 Gbps (Gigabits per second or OC-48) over distances that range to 4 kilometers.

FSO broadband wireless implementations support building-to-building connections, disaster recovery operations, and emergency backup services; provide redundant links in case of disasters; and augment capabilities of LMDS (Local Multipoint Distribution System) networks. FSO configurations can be rapidly setup without permits or licenses in those areas where fiber optic landline connections are impractical or unavailable.

9.5.3   Laser Marketplace

9.5.3.1   Airlinx Communications

Airlinx Communications provisions laser solutions for high-speed, short-haul, point-to-point, and long-haul backbone applications and services in spectrum between the 5.725 and 5.825 GHz frequencies. Rates between 10 and 155.52 Mbps over distances that extend to 12 kilometers are supported. The Airlinx solution enables broadband FWA (fixed wireless access) services for enterprises situated at hazardous industrial locations.

9.5.3.2 AstroTerra TerraLink Solutions

Developed by AstroTerra Corporation, TerraLink free space optical (FSO) systems support voice, video, and data transmission at rates between 10 and 155.52 Mbps over distances that extend to 3.75 kilometers. Banks, hospitals, telecommunications companies, municipal governments, and military installations use TerraLink systems to foster communications in buildings separated by barriers such as highways. Although TerraLink solutions support operations in all types of weather, signal disruptions can occur if snow, smoke, fog, or other adverse atmospheric conditions block line-of-sight virtual connections between transceivers.

9.5.3.3 LSA Photonics SupraConnect Solutions

LSA Photonics supports implementation of SupraConnect laser communications networks that support short-range connections that extend to 400 meters and longer connections that extend to 4.5 kilometers. SupraConnect implementations employ transceivers mounted on building rooftops to provision direct line-of-sight links.

9.5.3.4 OrAccess

OrAccess supports implementation of FSO networks for eliminating network gridlock between long-haul optical fiber networks and the corporate premise. These WLL (Wireless Local Loop) solutions interoperate with Gigabit Ethernet technology; support optical protocols such as WDM (Wavelength Division Multiplexing) and DWDM (Dense WDM); provision reliable and dependable access to high-speed, high-capacity multimedia services; and can be rapidly implemented without advance notice.

9.6 SPREAD SPECTRUM TECHNOLOGY

9.6.1 SPREAD SPECTRUM TECHNICAL FUNDAMENTALS

A spread spectrum radio infrastructure consists of transmission towers or radio transceivers mounted on buildings, poles, or streetlights that are equipped with directional antennas for transporting signals via narrow beams directly to destination sites. Directional antennas are large outdoor devices that support direct line-of-sight transmissions over long distances. In contrast to directional antennas, omnidirectional antennas are small snap-on devices that enable data transmission in all directions to support short-range indoor operations.

9.6.2 SPREAD SPECTRUM FOUNDATIONS

The term "spread spectrum" refers to the broad spectral shape of the transmitted signal. Co-developed and patented by screen actress Hedy Lamarr, spread spectrum technology was classified by the U.S. Army for enabling secure transmissions during World War II. Following declassification of this technology in the early 1980s, spread spectrum transmission solutions became publicly available.

9.6.3 SPREAD SPECTRUM OPERATIONS

Spread spectrum technology distributes a transmitted signal over a much wider frequency than the minimum bandwidth necessary for signal transport. As noted in Chapter 8, spread spectrum signals penetrate walls and other obstructions and are immune to adverse weather conditions. However, spread spectrum transmissions are susceptible to interference generated by authorized radio transmitters that malfunction or are incorrectly installed. PDAs (Personal Digital Assistants), garage door openers, radios, and palmtops are examples of devices that employ spread spectrum technology.

9.6.3.1   3Com Palm Communicators

A representative example of a multifunctional, multipurpose wireless communicator, the Palm VII connected organizer is available from 3Com. Palm VII provisions two-way wireless data communications services via PalmNET. PalmNET supports operations throughout the United States and provisions connectivity to local and wider area networking applications.

Palm VII users access PalmNET by subscribing to vendor services and obtaining access to delimited wireless Web content in fields that includes finance, news, weather, travel, and entertainment. Palm VII devices also enable transmission and reception of short messages and E-commerce transactions.

9.6.4 SPREAD SPECTRUM ALLOCATIONS

9.6.4.1   U.S. Federal Communications Commission (FCC)

The U.S. Federal Communications Commission (FCC) monitors spread spectrum licensing and allocates spectrum between the 902 and 928 MHz frequencies and the 2.400 and 2.483 GHz frequencies for spread spectrum operations. In the United States, spread spectrum bands support everyday applications and critical government services.

9.6.4.1.1   FCC Web Survey

The FCC sponsors an interactive Web survey on wireless broadband development activities. Survey results enable educational institutions, medical centers, government agencies, and local communities to share information on their utilization of wireless broadband applications and contribute to the formation of a nationwide database that will be freely available.

9.6.4.2   United Kingdom Radio Communications Agency

The Radio Communications Agency in the United Kingdom manages nonmilitary spread spectrum allocations and monitors spread spectrum use to prevent signal interference. This Agency is responsible for frequency planning, spectrum fees and assignments, and license distribution.

9.6.5 DIRECT-SEQUENCE SPREAD SPECTRUM (DSSS) AND FREQUENCY-HOPPING SPREAD SPECTRUM (FHSS) SOLUTIONS

Spread spectrum networks employ DSSS (Direct-Sequence Spread Spectrum) and FHSS (Frequency-Hopping Spread Spectrum) technologies for enabling wireless networking operations. Generally, wireless home networks (WHNs) employ DSSS or FHSS solutions.

At present, DSSS and FHSS technologies are not interoperable. Transceivers enabling DSSS or FHSS implementations must be synchronized in order to enable spread spectrum services.

9.6.5.1 Frequency-Hopping Spread Spectrum (FHSS)

In FHSS (Frequency-Hopping Spread Spectrum) systems, a data signal is modulated with a carrier signal. This signal then hops from one frequency to another in spectrum between the 2.400 and 2.483 GHz frequencies. FHSS technology safeguards information transport by employing pseudo-random algorithms that enable signals to jump or hop from frequency-to-frequency. As a consequence, the carrier frequency changes periodically in order to reduce aggregate signal interference and the adverse impact of signals transmitted by other network stations or nodes operating concurrently in the same frequency band.

Identical hopping codes establish radio wave frequencies that are used for FHSS transmission by transceivers. In cases of interference, signals are retransmitted at different frequencies during subsequent hops. FHSS technology supports transmissions at rates reaching 2 Mbps. According to FCC regulations, FHSS systems utilizing spectrum between the 2.400 and 2.483 GHz RF bands must employ a minimum of 75 hopping frequencies.

9.6.5.2 Direct-Sequence Spread Spectrum (DSSS)

DSSS (Direct-Sequence Spread Spectrum) implementations support broadband applications and services at rates reaching 11 Mbps. With DSSS transmission, a transceiver or sending station transmits a data signal in conjunction with a higher data rate bit sequence to counteract signal interference. However, at high bit rates, DSSS transmissions are subjected to increased interference from other radio sources. For example, noise generated by cordless phones, microwave ovens, and portable bar code scanners that also operate in the same frequency bands as DSSS implementations corrupt the integrity of the data transmitted.

9.6.5.3 Spread Spectrum Services and Applications

Spread spectrum technology supports a diverse array of applications, including television broadcasts, air traffic control, CB (Citizens Band) radio programs, radar operations, and freight transport. In addition, spread spectrum networks enable

digital cellular communications services, Web browsing, emergency medical assistance, critical government functions, SMS, and e-mail relay. Spread spectrum technology also facilitates network interconnections between adjacent buildings, supports operations of satellite navigation systems, enables migration to third-generation
cellular networking solutions, and provisions fast access to broadband wireline
network applications.

9.6.6 SPREAD SPECTRUM MARKETPLACE

9.6.6.1 Bell Mobility ARDIS Network

The Bell Mobility ARDIS Network employs spread spectrum technology for
enabling transmission of voice and data at rates ranging from 4.8 to 19.2 Kbps in
cars, classrooms, and office buildings, as well as in metropolitan and wider area
networking configurations. As with cellular networks, the ARDIS Network consists
of large numbers of radio cells that support links between palmtops, laptops, and
notebooks and base stations, and facilitates seamless handovers and frequency reuse. To ensure trouble-free communications services, the ARDIS Network employs
error checking protocols, including cyclic redundancy check and supports frequency-
hopping modulation. Moreover, the ARDIS Network enables operations in the 800
MHz frequency block, routes transmissions via the strongest available radio air-link,
supports wide area communications coverage by switching between terrestrial and
satellite connections, and enables links to wireline networks.

9.6.6.2 Cingular Interactive Intelligent Wireless Network

The Cingular Interactive Intelligent Wireless Network (Cingular Interactive) employs
an Ericsson Mobitex spread spectrum, packet-switched network solution as its core
network technology. Cingular Interactive supports operations in conjunction with
America Online (AOL) to enable Web browsing, BlackBerry to support e-mail
exchange, and Fidelity InstantBroker to facilitate E-commerce transactions. Cingular
Interactive also enables subscribers with 3Com Palm VII connected organizers to
access PalmNET, a wide area packet data network developed by 3Com.

9.6.6.3 Ericsson Mobitex Solutions

Developed by Ericsson Telecommunications, Mobitex is a *de facto* international
standard for spread spectrum packet data networks. Information in a Mobitex network is transmitted in small individual packets via spread spectrum technology. The
Mobitex platform supports e-mail exchange, short messaging services (SMS), Web
browsing, touch-screen applications, dependable data transmissions, wireline and
wireless network connections, and always-on service.

Mobitex implementations interoperate with second-generation (2G) GSM (Global System for Mobile Communications) configurations. In addition, Mobitex solutions also interwork with third-generation (3G) UMTS (Universal Mobile Telecommunications Systems) and 3GSM (Third-Generation GSM) deployments. Mobitex
supports operations in spectrum between the 3.4 and 3.6 GHz frequencies.

In a Mobitex network, mobile devices such as bar-code readers, pagers, printers, handheld devices, and cellular phones communicate with the closest base station. Switches route traffic to and from base stations. Each base station serves a single Mobitex radio cell and provides communications coverage in an area extending to 30 kilometers. A Mobitex network can take the form of a public mobile network that provides nationwide coverage via hundreds of base stations, local and regional switches, and a national network management center. By contrast, a basic Mobitex system can be configured as a small, privately owned and privately administered local network for supporting several applications with only a few base stations and a single switch.

9.6.6.4   Lucent Technologies WaveLAN

Developed by Lucent Technologies, WaveLAN is a family of IEEE 802.11 spread spectrum radio devices that support WLAN implementations and enable WLAN interconnectivity with fixed wireline networks. The WaveLAN system employs a collision avoidance (CA) protocol so that each full-service wireless device supports rates at 2 Mbps. WaveLAN wireless networks function in temporary and frequently changing venues and provide Web service in difficult-to-wire buildings such as landmarks and historic structures.

9.6.6.5   NTT i-Mode

Available from NTT, i-Mode is a cellular communicator that employs spread spectrum technology for enabling data transmission at rates reaching 9.6 Kbps. i-Mode enables e-mail relay, voice calls, and short messaging service (SMS), and provisions access to i-Mode Web sites. Designed for i-Mode subscribers, these Web sites feature national and international news, foreign exchange rates, stock market updates, restaurant guides, and basic Japanese recipes. In addition, i-Mode Web sites provision access to health insurance applications, fortune-telling services, music clips, games, M-commerce (mobile commerce) activities, weather forecasts, tickets for concerts and theater performances, and travel reservations. An i-Mode communicator weighs approximately 3.6 ounces, features a large liquid crystal display that is easy to read and a command navigation button that moves a pointer on the display, and supports Java and Jini applications.

9.6.6.6   Metricom Ricochet Network

Operational until August 2, 2001, when Metricom filed for bankruptcy, the Metricom Ricochet Network employed FHSS packet-switched technology for enabling broadband FWA (fixed wireless access) local loop solutions. This network supported wireless applications via small-size external or internal radio modems that were attached to an RS-232 serial port or a USB (Universal Serial Bus) port on notebooks, laptops, and handheld computers. A Ricochet network consisted of clusters of radio transceivers or microcell radios that enabled transmissions between any two transceivers at distances between 1,000 and 1,400 feet. Radio transceivers were mounted on poles, tops of buildings, or street lights every quarter to one-half mile in a mesh topology in the coverage area. Wireline access points collected and transformed RF packets into formats for transport via T-1 or Frame Relay connections to a wireline IP backbone network.

The Metricom Ricochet Network enabled operations in license-exempt ISM (Industrial, Scientific, and Medical) bands between the 902 and 928 MHz frequencies. In Metricom Ricochet implementations, ISM bands were divided into approximately 162 channels that were 160 kHz (Kilohertz) wide. Metricom Ricochet networks supported transmissions between 28 and 128 Kbps, and enabled shared bandwidth services. As a consequence, network performance degraded when multiple users attempted to access the network concurrently. Ricochet service was available in metropolitan areas such as Atlanta, Seattle, Baltimore, and Denver; at major airports and selected corporate sites; and on university campuses.

9.6.6.7   Symbol Wireless Network Solutions

Symbol wireless networks are based on open bridge architecture for enabling easy installation, expandable capacity, interference immunity, and ready integration into wireline network environments. Symbol employs FHSS technology for enabling worldwide operations in spectrum between the 2.400 and 2.483 GHz frequency bands, and supports IP (Internet Protocol) packet transmission and voice-over-IP (VoIP) or Internet telephony services. Moreover, Symbol provisions high-rate WLAN services via an 11 Mbps Ethernet LAN solution that is compliant with the IEEE 802.11b specification. Symbol also participates in the Wireless Ethernet Compatibility Alliance (WECA).

9.7   MICROWAVE TECHNOLOGY

9.7.1   Microwave Technical Fundamentals

A derivative of radar technology developed in World War II, microwave systems employ short RF waves in the upper range of the electromagnetic spectrum for enabling high-bandwidth applications. In the present-day environment, microwave systems enable voice, data, video, and fax transmission; interlink cellular sites to the PSTN (Public Switched Telephone Network); and interconnect isolated LANs in different buildings separated by open spaces such as highways and bodies of water. Long-haul microwave configurations enable network backbone operations.

Common carrier microwave services typically support point-to-point configurations. Broadcast auxiliary microwave service relays television signals from a remote location back to a television studio. Private microwave service enables corporations to remotely control equipment, monitor operations of gas and oil pipelines at distant sites, and gather data from moving vehicles.

A basic microwave configuration includes a pair of towers to which microwave dishes are affixed. These towers are taller than nearby structures and trees. Each microwave dish is equipped with directional antennas that are aimed at each other for line-of-sight signal reception and transmission. Microwave network deployment requires the use of specialized equipment and devices such as terrestrial repeater stations that regenerate microwave signals for enabling signal robustness in long-haul networks that extend to 100 kilometers.

Microwave configurations support transmissions in the U-NNI (Unlicensed-National Network Infrastructure) spectrum between the 5.15 and 5.25 GHz frequencies and the 5.725 and 5.825 GHz frequencies for enabling high-speed data transmission and voice telephony service. Microwave networks also support information transport in UHF (Ultra-High Frequency) spectrum in frequencies above 40 GHz.

As with laser communications, microwave transmissions are disrupted by climate changes. Microwave signals carried in spectral bands below 10 GHz are also susceptible to multipath distortion caused by impenetrable barriers such as multistory metal office buildings in the transmission pathway. Heavy rains and snowstorms adversely affect microwave transmissions in the higher spectral ranges as well.

In the 1990s, the FCC established spectral allocations for PCS (Personal Communications Service) between the 1.950 and 1.990 GHz frequencies. These bands also support point-to-point microwave services. As a consequence of an FCC ruling, each PCS operator must provide interference protection to avoid disrupting the integrity of voice, video, and data transiting the in-place microwave network.

9.7.2 MICROWAVE SPECTRUM FREQUENCY ALLOCATIONS

9.7.2.1 U.S. Federal Communications Commission (FCC)

In 1972, the FCC made spectrum between the 38.6 and 40 GHz frequencies available to communications carriers implementing fixed microwave radio systems. Also called 39 GHz installations, these systems currently support point-to-point connections in WLL (Wireless Local Loop) implementations that support links from the customer premise to wireline networks such as corporate intranets and the Web. Moreover, 39 GHz installations optimally enable transmission rates at 1.544 Mbps (T-1) over distances that extend to 2 kilometers.

9.7.3 MICROWAVE MARKETPLACE

9.7.3.1 Nortel Networks

Nortel Networks supports broadband fixed wireless access (FWA) solutions that employ microwave technology for enabling wireline-equivalent voice services, always-on Web access, and high-speed packet data transport at 153 Kbps for residential subscribers and 326 Kbps for business subscribers. The network infrastructure consists of radio base stations that are interconnected via microwave links to switches at the local telephone exchange and to customer premise equipment (CPE). The CPE unit employs a small rooftop directional antenna pointed to the base station. This system supports operations in spectrum between the 3.4 and 3.6 GHz frequencies.

9.7.4 MICROWAVE IMPLEMENTATION CONSIDERATIONS

Microwave technology generally fosters information transmission at 1.544 Mbps (T-1) or 2.048 Mbps (E-1) rates. In addition, rates as high as155.52 Mbps (OC-3) are also supported. Because microwave systems employ licensed frequencies, co-channel interference is virtually eliminated.

Implementation of microwave systems typically requires real estate acquisition to ensure a line-of-sight path for signal transmission, construction permits for microwave facilities, and FCC licensing for microwave radio operations. The FCC can delay deployment of microwave solutions or prohibit their use in those areas where frequencies are overly congested. In addition, communities also block microwave system installations in response to health and safety concerns or for aesthetic reasons.

The initial specification for protecting the public from exposure to potentially harmful radiation emitted by microwave was adopted by the U.S. Congress in 1968. Formally known as the Radiation Control for Health and Safety Standard, this specification remains in place today.

9.8   SATELLITE TECHNOLOGY

9.8.1   SATELLITE BASICS

Satellite services employ microwave technology at very high frequencies for enabling narrowband and broadband implementations. Satellite configurations contain multiple pairs of receivers and transmitters (also known as transceivers) for signal reception and transmission, and enable a diverse array of communications solutions. Satellite technology features and functions are examined in Chapter 10.

9.8.1.1   Cornerstone Technologies, Inc. (CTI)

CTI (Cornerstone Technologies, Inc.) supports international satellite operations in Ku-band, Ka-band, and C-band frequencies for fostering high-speed transport of video, voice, and data, and for supporting high-quality applications that include videoconferencing, voice telephony, videotelephony, and Web browsing. CTI operates the West Coast Teleport in California, the East Coast Teleport in Connecticut, and the Asia-Pacific Teleport. These teleports or earth stations enable international communications coverage and support roaming in a broad coverage area. Additional satellite solutions are examined in Chapter 10.

9.9   STANDARDS ORGANIZATIONS AND ACTIVITIES

Diverse standards groups develop specifications for the design of interoperable broadband FWA networks that enable a variety of applications. The Federal Communications Commission, the European Telecommunications Satellite Organization (EUTELSAT), the International Telecommunication Union (ITU), the Institute of Electrical and Electronics Engineers (IEEE), and the European Telecommunications Standards Institute (ETSI) are active in all phases of wireless networking activities and support numerous forums, committees, and working groups for standards development. This section features an overview of major standards groups that contribute to the development of specifications for WPAN, WLANs, WMANs, and WWANs.

In addition, contributions of important groups such as the IEEE in the wireless networking area are also examined throughout this chapter

9.9.1 ALLIANCE FOR TELECOMMUNICATIONS INDUSTRY SOLUTIONS (ATIS) T-141.P WORKING GROUP

Sponsored by the Alliance for Telecommunications Industry Solutions (ATIS), the T-141.P Working Group promotes standardization of fixed wireless access point-to-multipoint systems that operate in unlicensed and licensed spectral bands. For these systems, this Working Group defines a common air interface and establishes minimum performance requirements for base stations and subscriber transceivers. The T-141.P Working Group also establishes procedures for implementation of fixed wireless access solutions that interface with wireline networks such as Public Packet Data Networks (PPDNs), Public Switched Telephone Networks (PSTNs), T-1 (1.544 Mbps) leased lines, and ISDN (Integrated Services Digital Network) configurations, and identifies procedures for establishing Web interconnections.

9.9.2 U.S. FEDERAL COMMUNICATIONS COMMISSION (FCC)

The FCC allocates radio spectrum for wireless applications and services between the 9 kHz and 300 GHz frequencies. An independent regulatory agency, the FCC is responsible for spectrum that supports non-federal government applications.

The FCC establishes spectral bands for wireless services that include PCS (Personal Communications Services), CMRS (Commercial Mobile Radio Services), broadband FWA (Fixed Wireless Access) systems, and analog and digital cellular telephony. The FCC also allocates spectral frequencies for applications in the fields of education, industry, science, and medicine and services associated with next-generation networking initiatives.

In 2000, the FCC allocated spectrum in the W-band between the 59 and 64 GHz frequencies for enabling license-exempt broadband FWA LAN services. In the European Union, approaches for implementation of the Mobile Broadband System (MBS) cellular network in the W-band are under consideration. Originally, the W-band supported communications between military satellites.

9.9.2.1 Unlicensed-National Information Infrastructure (U-NII) Spectral Bands

In 1997, the FCC designated 300 MHz of spectrum in three Unlicensed-National Network Infrastructure (U-NNI) spectral bands for enabling narrowband and broadband wireless networking services. The U-NNI spectrum between the 5.15 and 5.25 GHz frequencies supports short-range, room-to-room WLAN applications.

The U-NNI spectral block between the 5.35 and 5.725 GHz frequencies enables building-to-building WLAN connections. U-NNI spectrum between the 5.725 and 5.825 GHz frequencies facilitates network operations over a range of several kilometers in rural communities that lack an in-place wireline telecommunications infrastructure. This U-NNI spectral allocation also supports implementation of low-cost wireless networks in classrooms, on-site telemedicine emergency services, and teleradiology consultations.

9.9.2.2   Industrial, Scientific, and Medical (ISM) Spectral Bands

The FCC sets aside spectrum for wireless devices operating in license-exempt spectrum between the 2.400 and 2.483 GHz ISM (Industrial, Scientific, and Medical) bands. ISM frequencies also support the provision of telecommunications services such as DECT (Digital Enhanced Cordless Telecommunications), utilization of home appliances including microwave ovens, and implementation of FHSS (Frequency-Hopping Spread Spectrum) and DHSS (Digital-Hopping Spread Spectrum) solutions.

9.9.2.3   Fixed Narrowband and Broadband Services Spectrum

The FCC provisions licenses for spectrum between the 24.25 and 24.45 GHz frequencies and between the 25.05 and 25.25 GHz frequencies to support fixed wireless access services. Licenses are allocated in 172 economic areas throughout the United States. Licenses are also available in U.S. possessions and territories, including the U.S. Virgin Islands, American Samoa, Northern Mariana Islands, and the Gulf of Mexico.

9.9.3   UNITED KINGDOM (U.K.) RADIO COMMUNICATIONS AGENCY

The U.K. (United Kingdom) Radio Communications Agency auctions Wireless Telegraphy Act licenses for provisioning broadband fixed wireless access (FWA) services in spectrum between the 26 and 28 GHz frequencies. These RF bands support always-on connections, high-speed data transmission, video-on-demand (VOD), near video-on-demand (NVOD), videoconferencing, tele-entertainment services, teleshopping, E-commerce transactions, Web browsing, and LAN-to-LAN interconnectivity.

9.9.4   EUROPEAN CONFERENCE OF POSTAL AND TELECOMMUNICATIONS ADMINISTRATION (CEPT)

In 2000, the European Conference of Postal and Telecommunications Administration (CEPT) authorized the use of spectrum between the 40.5 and 43.6 GHz frequencies for MWA (Multimedia Wireless Access) services. MWA implementations provision asymmetric and/or symmetric transport of residential broadband fixed wireless access (FWA) applications.

CEPT allocates MWA frequencies in blocks of 1 or 2 MHz, and CEPT also defines approaches for accommodating legacy systems, coordinating MWA allocations with adjacent satellite frequencies, and enabling FDD (Frequency-Division Duplexing) and TDD (Time-Division Duplexing) operations. TDD solutions use the same channel for sending and receiving data. By contrast, FDD solutions employ separate channels for data transmission and reception.

9.9.5 ITU-Radio Communications Sector (ITU-R)

9.9.5.1 Broadband FWA Solutions and Radio Local Area Networks (RLANs)

The International Telecommunication Union-Radio Communications Sector (ITU-R) develops specifications for point-to-multipoint radio systems and utilization of IMT-2000 (International Mobile Telecommunications-Year 2000) technologies for enabling broadband fixed wireless access networking applications and services. In 2000, the ITU-R approved Recommendations for enabling an RLAN (Radio LAN) infrastructure to support broadband FWA network solutions. ITU-R allocates spectrum between the 1.350 and 2.690 GHz frequencies for broadband FWA networking operations. Emergent broadband RLAN specifications operate as extensions to wireline LANs and support ATM (Asynchronous Transfer Mode) and TCP/IP (Transmission Control Protocol/Internet Protocol) implementations that enable rates at 20 Mbps.

9.9.5.2 Broadband FWA Systems and Third-Generation (3G) Technologies

The International Telecommunications Union-Radio Communications Sector (ITU-R) supports development of a set of specifications for enabling broadband FWA systems to interwork with 3G (third-generation) technologies such as PCS (Personal Communications Services), W-CDMA (Wideband-Code-Division Multiplexing), cdma2000, and UMTS (Universal Mobile Telecommunications Service). Moreover, the ITU-R establishes procedures for interlinking broadband fixed wireless access systems with satellite and terrestrial networks and clarifies basic topologies, operations, and performance requirements for broadband FWA implementations.

9.9.6 International Telecommunications Commission-Telecommunications Standards Sector (ITU-T)

The ITU-T establishes frequency bands and operational requirements for wireless local loop (WWL) systems. Also called last mile, radio in the loop (RTIL), first mile, and broadband fixed wireless access (FWA) systems, WLL implementations enable operations in spectrum between the 3.5 and 10.5 GHz frequencies originally allocated for voiceband services.

9.9.7 National Telecommunications and Information Administration (NTIA)

The National Telecommunications and Information Administration (NTIA) monitors and administers spectrum for federal government utilization. In addition, the NTIA and the FCC assist the U.S. Department of State in coordinating radio frequency spectral allocations with international agencies such as the ITU (International Telecommunications Union) and countries that include Canada and Mexico.

9.10 WIRELESS NETWORK PROTOCOLS

9.10.1 Orthogonal Frequency-Division Multiplexing (OFDM)

As with other wireless transmission modulation techniques, including CDMA (Code-Division Multiplexing Access) and CDPD (Cellular Digital Packet Data), OFDM

(Orthogonal Frequency-Division Multiplexing Access) encodes data onto radio-frequency signals for enabling wireless transmissions. With OFDM modulation, signals are divided into narrowband channels or circuits for enabling effective bandwidth use in WLANs and cellular telephony applications. OFDM technology also supports HDTV (High-Definition Television) and digital audio broadcasts throughout the European Union.

9.10.1.1 Iospan

Developed by Iospan, AirBurst technology enables MIMO (Multiple Input and Multiple Output) OFDM implementations that eliminate line-of-sight requirements for wireless transmissions.

9.10.2 CODE-ORTHOGONAL FREQUENCY-DIVISION MULTIPLEXING

A variant of OFDM, Code-Orthogonal Frequency-Division Multiplexing (C-OFDM) enables seamless indoor IEEE 802.11a WLAN operations by dividing high-rate 20 MHz channels into lower rate 52 subchannels that are 300 kHz wide. Four subchannels carry out error-correction functions. The 48 remaining subchannels enable data transport at rates reaching 6 Mbps. The rate of data transported over the network can be increased from 6 Mbps to 12 Mbps or 24 Mbps and higher speeds with QAM (Quadrature Amplitude Modulation) and QPSK (Quadrature Phase Shift Key) modulation encoding schemes.

9.10.3 VECTOR-ORTHOGONAL FREQUENCY-DIVISION MULTIPLEXING

Vector-Orthogonal Frequency-Division Multiplexing (V-OFDM) employs multipath signal transmission and advanced multiplexing services for optimizing system performance and fostering dependable packet data throughput in residential broadband FWA networks operating in noisy environments.

9.10.3.1 V-OFDM Forum

Endorsed by the V-OFDM Forum, V-OFDM technology supports broadband FWA networking functions and services in the 5 GHz frequency block. V-OFDM also interworks with Ethernet technologies for enabling V-OFDM MAC (Media Access Control) Layer operations at Layer 2 or the Data-Link Layer of the Open Systems Interconnection (OSI) Reference Model. VOFDM Forum participants include Cisco Systems, Motorola, Broadcom, Bechtel, and Samsung.

9.10.4 WIDEBAND-ORTHOGONAL FREQUENCY-DIVISION MULTIPLEXING

Based on OFDM technology, W-OFDM (Wideband-Orthogonal Frequency-Division Multiplexing) is a bandwidth-efficient wireless transmission solution that enables high-speed rates to facilitate wireless in-home multimedia networking operations. Developed and patented by Wi-LAN, W-OFDM works in concert with the PCI (Peripheral Component Interconnect) IEEE 1394 High-Performance Serial Bus specification.

9.11 IEEE 802.11 SPECIFICATION

The IEEE 802.11 specification and its extensions, specifically IEEE 802.11a, IEEE 802.11b, and IEEE 802.11g, establish the framework and foundation for broadband fixed wireless access (FWA) LAN implementations. Adopted in 1997, this standard defines WLAN operations at the Physical Layer or Layer 1 and the MAC (Media Access Control) Layer or Layer 2 of the OSI Reference Model. This standard also describes wireless local networking capabilities in enabling transmission rates reaching 2 Mbps. In addition, approaches for enabling continuity of service between WLANs and wireline LANs employing Ethernet technologies are clarified.

The IEEE 802.11 standard supports seamless interoperability between WLAN equipment in multivendor environments and describes an open architecture. Moreover, the IEEE 802.11 specification indicates methods for using direct-sequence spread spectrum (DSSS) and frequency-hopping spread spectrum (FHSS) technologies in wireless local networks operating in spectrum between the 2.400 and 2.483 GHz frequencies.

9.11.1 HIGH-RATE LANS

9.11.1.1 IEEE 802.11a Extension

The IEEE 802.11a FWA LAN Extension defines capabilities of broadband fixed wireless access (FWA) LANs operating in the license-exempt U-NNI (Unlicensed-National Network Infrastructure) spectrum between the 5.15 and 5.25 GHz frequencies and the 5.725 and 5.825 GHz frequencies.

IEEE 802.11a broadband FWA LAN solutions enable transmission of bandwidth-intensive voice, video, and data such as terrestrial digital video broadcasts at rates ranging from 6 to 54 Mbps. IEEE 802.11a also supports implementation of next-generation broadband FWA LAN solutions with optimal rates at 100 Mbps.

As with HiperLAN-2 (High-Performance Radio Local Area Network-Type 2), IEEE 802.11a FWA LANs use the OFDM protocol to overcome multipath interference. Ethernet technology supports the IEEE 802.11a MAC (Media Access Control) Layer or Layer 2 operations in IEEE 802.11a FWA LAN implementations. WLANs are implemented throughout the United States and Canada. By contrast, ATM technology provisions MAC (Medium Access Control) Layer or Layer 2 services for HiperLAN-2 implementations in the European Union.

9.11.1.2 IEEE 802.11b Extension

Endorsed in 1999, the IEEE 802.11b extension establishes standards for broadband FWA Ethernet LANs. Standards-compliant IEEE 802.11b FWA Ethernet LANs support voice, video, and data transmission at rates reaching 11 Mbps and use spread spectrum technology for enabling license-exempt wireless operations in ISM (Industrial, Scientific, and Medical) RF bands between the 2.400 and 2.483 GHz frequencies. These systems support operations in fields that include business, education, medicine and government; extend the reach of wireline configurations; and employ WEP (Wired Equivalent Privacy) for safeguarding the integrity of transmissions. In addition, IEEE 802.11b systems are implemented in SOHO venues for supporting WPAN (Wireless Personal Area Network) applications and WHN (Wireless Home Network) operations. Approaches for enabling IEEE 802.11b FWA Ethernet LANs to support transmissions at rates reaching 54 Mbps are under consideration.

9.11.1.3 IEEE 802.11g Extension

Subsequent to the passage of the IEEE 802.11a and IEEE 802.11b Extensions, the IEEE Working Group established the IEEE 802.11g Task Force to develop parameters for enabling WLAN transmissions at rates greater than 20 Mbps.

9.12 BROADBAND FWA ETHERNET LANS

9.12.1 BROADBAND FWA ETHERNET LAN FUNDAMENTALS

Advances in communications technologies drive the popularity of broadband FWA Ethernet LANs. From the viewpoint of the communications carrier, a wireline network extending from the subscriber site to the local telephone exchange can be expensive to implement and difficult to maintain. By contrast, broadband FWA Ethernet LANs cost-effectively enable high-speed, high-capacity connections to wireline backbone networks. Broadband FWA Ethernet LAN configurations employ microwave technology to provision networking services over the first mile between the customer premise and the local telephone exchange, thereby eliminating the need to install an in-ground or pole-based wireline infrastructure.

9.12.2 BROADBAND FWA ETHERNET LAN TRANSMISSION ESSENTIALS

Also called radio in the loop (RITL), wireless local loop (WLL), and fixed radio access solutions, broadband FWA Ethernet LANs consist of mobile nodes that communicate via radio signals with fixed wireless access points (WAPs) or base stations. Basic subscriber equipment includes notebooks equipped with DSSS (Direct-Sequence Spread Spectrum) PC (PCMCIA or Personal Computer Memory Card International Association) Cards. When a wireless terminal or user moves out of range of one base station and into the range of another, handovers enable the provision of seamless communications services.

Broadband FWA Ethernet LAN configurations facilitate point-to-multipoint connections and enable operations in the higher radio frequencies of the electromagnetic spectrum that became available in the 1990s. Broadband FWA Ethernet LANs support services in an area that extends to 40 kilometers.

9.12.3 BROADBAND FWA ETHERNET LAN COMPONENTS

9.12.3.1 Personal Computer Memory Card International Association (PCMCIA)

PCMCIA (PC) Cards support mobile communications services for modular, peripheral digital audio and video equipment; Personal Digital Assistants (PDAs); laptop and notebook computers; smart cell phones; digital cameras; and cable television set-top boxes (STBs). PC Cards plug into PCMCIA slots for sending multimedia transmissions to a destination address such as a base station or another wireless device equipped with a compatible adapter. Connections remain active as long as users are within specified areas of cellular coverage. The PC Card also works in conjunction with the PCI (Peripheral Component Interconnect) IEEE 1394 High Performance Serial Bus specification and wireline Ethernet and Fast Ethernet networking solutions.

The Personal Computer Memory Card International Association (PCMCIA) designates electrical, physical, and software specifications for PCMCIA (PC) Card technology and establishes specifications for the PC Smart Card, Miniature Card, and SmartMedia Card.

9.12.3.2 Peripheral Component Interconnect Special Interest Group (PCI SIG)

The Peripheral Component Interconnect Special Interest Group (PCI SIG) supports the universal connection of all types of low-cost PC peripherals via the PCI (Peripheral Component Interconnect) standard. The Mini PCI specification fosters integration of communications peripherals with mobile computing devices. PCI SIG also endorses the PCI Local Bus specification for enabling easy removal or replacement of adapter cards and PCI-X, a high-performance extension to the PCI Local Bus that fosters interoperability between mobile devices and landline networks employing wireline and/or wireless Ethernet technologies.

9.12.4 BROADBAND FWA ETHERNET IMPLEMENTATION CONSIDERATIONS

A cost-effective approach for extending the reach of wireline LANs, broadband FWA Ethernet LAN installations support networking operations in older buildings, hard-to-wire and historic structures, hospitals, warehouses, libraries, homes, and office complexes. In school and university venues, broadband FWA Ethernet LANs enable extension of landline network services to chemistry labs, libraries, classrooms, student centers, cafeterias, and dormitories without new construction. Government agencies employ broadband FWA Ethernet LANs to manage crisis situations in the aftermath of natural and artificial disasters.

The use of radio frequency (RF) channels by broadband FWA Ethernet LANs places fundamental limitations on network operations. Wireless transmission rates are generally slower than rates enabled by direct wireline connections. Acceptable levels of signal synchronization can be difficult to achieve and maintain and connectivity to landline networks can also be sporadic. (See Figure 9.1.)

9.12.5 BROADBAND FWA ETHERNET LAN MARKETPLACE

Broadband FWA Ethernet LANs that conform to IEEE 802.11b requirements enable rates reaching 11 Mbps. Vendors active in the broadband FWA Ethernet marketplace include 3Com, Symbol, Apple, Extreme Networks, Ericsson, Fujitsu, Dell, Compaq, Sony Corporation, Siemens, and Nokia.

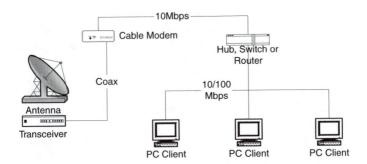

FIGURE 9.1 A wireless office network.

9.12.5.1 BreezeCom

BreezeCom products use direct-sequence spread spectrum (DSSS) technology for enabling broadband FWA Ethernet LAN operations in the license-exempt ISM spectrum between the 2.400 and 2.483 GHz frequencies. BreezeCom enables indoor and outdoor temporary networks for provisioning communications services at one-time events such as outdoor concerts and tennis matches. These solutions also provision links to remote office networks and connections between buildings in a campus environment at 11 Mbps.

9.12.5.2 Cisco Systems Wi-Fi AirConnect Solution

Available from Cisco Systems, Wi-Fi AirConnect broadband FWA Ethernet LAN solutions enable ad hoc and peer-to-peer WLAN topologies, seamless in-building roaming, streaming video, Internet telephony or voice-over-IP, connectivity to wireline LANs, and transmission rates at 11 Mbps. This IEEE 802.11b standards-compliant wireless Ethernet solution supports utilization of the WEP (Wired Equivalent Privacy) protocol and interlinks network segments that are up to 25 meters apart

9.12.5.3 Lucent Technologies Wi-Fi Network Solution

The Lucent Technologies Wi-Fi Network supports broadband FWA Ethernet LAN deployments for enabling services in a 25-meter area of coverage. This license-exempt solution supports the WEP (Wired Equivalent Privacy) protocol and transmission rates at 1, 2, 5.5, and 11 Mbps depending on subscriber requirements.

9.13 BROADBAND FWA ETHERNET LANS STANDARDS ORGANIZATIONS AND ACTIVITIES

9.13.1 WIRELESS ETHERNET COMPATIBILITY ALLIANCE (WECA)

Established in 1999, the Wireless Ethernet Compatibility Alliance (WECA) certifies the interoperability of broadband FWA Ethernet LAN products based on the IEEE

802.11b specification. WECA also certifies the interoperability of high-rate (HR) WLANs that operate in U-NNI license-exempt ISM (Industrial, Scientific, and Medical) spectrum between the 2.400 and 2.483 GHz frequencies. In addition, WECA certifies the performance of wireless Ethernet products such as PC (PCM-CIA) cards for notebook computers, PCI (Peripheral Component Interconnect) cards for desktop computers, and USB (Universal Serial Bus) ports that work in concert with the IEEE 802.11b specification.

Certified WECA products feature Wi-Fi (Wireless-Fidelity) logos, enable message exchange with 40-bit encryption, and support information transport at rates ranging from 1 to 11 Mbps. Tests benchmarking the conformance of broadband FWA Ethernet LAN equipment with Wi-Fi specifications are conducted at the Silicon Valley Networking Lab (SVNL). WECA also supports development of Wi-Fi networks that enable transmission at rates between 22 and 54 Mbps.

The Interoperability Lab (IOL) at the University of New Hampshire also tests the compatibility of broadband FWA Ethernet LAN devices with the IEEE 802.11b specification. However, only WECA provisions an internationally recognized Wi-Fi logo. WECA also endorses the use of Wi-Fi as a replacement for IEEE 802.11b. WECA-certified multifunctional equipment enables wireless networking operations in diverse settings including hotel rooms, train stations, bus and airport terminals, SOHO (Small Office/Home Office) venues, classrooms, museums, parks, and business offices.

9.14 HIPERLAN-1 (HIGH PERFORMANCE RADIO LOCAL AREA NETWORK-TYPE 1) AND HIPERLAN-2 (HIPERLAN-TYPE 2)

9.14.1 HiperLAN-1 and HiperLAN-2 Specifications

Developed and endorsed by the European Telecommunications Standards Institute (ETSI), HiperLAN-1 (High-Performance Radio Local Area Network-Type 1) and HiperLAN-2 (HiperLAN-Type 2) specifications define capabilities, operations, and services enabled by radio LAN (RLAN) solutions. HiperLAN-1 and HiperLAN-2 support wireless connectivity between network nodes such as workstations, PCs, laptops, servers, printers, and consumer electronic equipment in a diverse range of settings including SOHO environments.

9.14.2 HiperLAN-1 and HiperLAN-2 Capabilities

With HiperLAN-1 and HiperLAN-2 implementations, physical cables for connection of data networks are no longer necessary. In contrast to wireline LANs, radio LANs (RLANs) such as HiperLAN-1 and HiperLAN-2 enable a more flexible approach to reconfiguration, installation, and network use, thereby minimizing operating costs associated with network wiring modifications and upgrades. HiperLAN-1 and HiperLAN-2 support wireless operations and communications services in ad hoc and infrastructure-based configurations.

9.14.2.1   HiperLAN-2 Features and Functions

HiperLAN-2 enhances the capabilities of HiperLAN-1 and enables RLAN (Radio Local Area Network) access to broadband networks. This short-range RLAN supports information transport at rates reaching 25 Mbps. A connection-oriented technology, HiperLAN-2 employs TDM (Time-Division Multiplexing) modulation and facilitates high-speed connectivity to wireline and wireless networks. HiperLAN-2 interworks with ATM, Ethernet, Fast Ethernet, 3GSM, UMTS, and IP technologies and fosters transmission of voice and video applications with QoS (Quality of Service) assurances.

HiperLAN-2 network implementations provide high-capacity services at venues that include offices, homes, classrooms, school buildings, exhibition halls, theaters, arenas, hospitals, factories, and wherever else radio transmission is a viable alternative to wireline networking technology. A versatile and flexible high-speed broadband FWA local networking solution, HiperLAN-2 supports secure and interoperable applications; enables unicast, multicast, and broadcast transmissions; and works in concert with second-generation and third-generation cellular networks. HiperLAN-2 supports operations in license-exempt spectrum between the 5.15 and 5.25 GHz frequencies, the 5.25 and 5.35 GHz frequencies, and the 5.725 and 5.825 GHz frequencies.

9.14.2.2   HiperLAN-2 Global Forum

The HiperLAN-2 Global Forum promotes adoption of HiperLAN-2 as the basis for worldwide RLAN implementations and supports interoperability between Hiper-LAN-2 and BRAN (Broadband Radio Access Network). The HiperLAN-2 Global Forum also sponsors tests to ensure the conformance of networking devices to the HiperLAN-2 specification.

HiperLAN-2 Global Forum participants include Dell, Nokia, Telia, Bosch, and Texas Instruments. In addition, Alcatel, Matsushita Communications, Motorola, and NTT take part in the HiperLAN-2 Global Forum as well.

9.15   WIRELESS PERSONAL AREA NETWORKS (WPANS)

9.15.1   WPAN TECHNICAL FEATURES

Affordable low-power, license-exempt, streamlined networking solutions, WPANs (Wireless Personal Area Networks) support network operations at rates reaching 1 Mbps and provision access to Web services in and around the home. In addition, WPANs enable meter reading, medical monitoring, home automation, and activation of home electronic and security systems.

9.15.2   WPAN DEVICES

WPAN devices include unobtrusive and inexpensive mobile and portable computing peripherals and modules such as smart tags and badges, interactive toys, remote controls, portable bar-code readers, PCs, printers, PDAs, and laptop computers.

These devices can be worn or carried and support interoperability in a personal operating space (POS). According to the IEEE 802.15 WPAN Working Group, a POS (Personal Operating Space) is an area that extends to 10 meters in all directions around a stationary or mobile network user.

9.15.3   IEEE 802.15 WPAN Working Group

The IEEE 802.15 WPAN Working Group develops specifications for WPAN solutions and procedures for improving the operations, capabilities, and performance of WPAN implementations. This Working Group clarifies procedures for the establishment of wireless connections by moving, stationary, or portable devices that enter or operate within a POS and delineates recommended practices for enabling the coexistence of WLANs and WPANs operating in the license-exempt spectrum between the 2.400 and 2.485 GHz frequencies.

9.15.4   IEEE 802.15 High-Rate Study Group

Specifications for short-range, high-rate WPANs that enable voice, video, and data transmission at rates reaching 20 Mbps are under consideration by the IEEE 802.15 High-Rate Study Group. In addition, specifications for implementation of interoperable WPAN devices such as printers, laptops, PCs, and personal digital assistants (PDAs) are in development. Procedures for enabling interactive toys to function at 200 Kbps are also examined. Participants in the IEEE 802.15 High Rate Study Group include Motorola, Eastman Kodak, and Cisco Systems.

9.15.5   IEEE 802.15 Task Groups

9.15.5.1   IEEE 802.15 Task Group 1 (TG1)

The IEEE 802.15 Task Group 1 (TG1) endorses the utilization of the 1 Mbps specification established by the Bluetooth Special Interest Group (BSIG) and Home RF Working Group for enabling seamless transmission in WPAN configurations. Formally called IEEE 802.15.1, this specification defines Physical Layer or Layer 1 and MAC Layer or Layer 2 specifications for provisioning wireless services with wearable or handheld communicators and mobile and fixed devices operating within a POS. Moreover, the TG1 clarifies approaches for enabling interoperability between WPANs and IEEE 802.11-compliant BFA LAN devices and defines procedures for supporting seamless operations of WPAN and IEEE 802.11 broadband WLAN solutions functioning in the license-exempt U-NNI spectrum between the 2.400 and 2.485 GHz frequencies.

9.15.5.2   IEEE 802.15 Coexistence Task Group (TG2)

The IEEE 802.15 Coexistence Task Group (TG2) is charged with resolving potential interference problems between WPANs and WLANs operating in the same or adjacent frequencies.

9.15.5.3   IEEE 802.15 Task Group 3 (TG3)

The IEEE 802.15 Task Group 3 (TG3) promotes development of affordable high-rate WPAN configurations and protocols that support Physical Layer or Layer 1 and MAC Layer or Layer 2 operations at 20 Mbps and faster rates.

9.16   BLUETOOTH TECHNOLOGY

9.16.1   BLUETOOTH TECHNICAL FUNDAMENTALS

The rapid proliferation of mobile devices that support diverse applications such as personal finance, education, and entertainment, and the demand for wireless access to multimedia applications and spontaneous interoperability between wireless devices contribute to the implementation of Bluetooth WPANs. Bluetooth WPANs are ad hoc and freestanding network configurations that operate without a fixed wireline infrastructure. Wireless terminals acting as routers establish links for information exchange in a peer-to-peer topology.

Ad hoc Bluetooth WPANs are rapidly deployed. These networks dynamically establish communications links that are seamlessly reconfigured to accommodate personal and terminal mobility. Bluetooth technology also enables co-located services to be provisioned by WPANs that operate in the same physical space. Moreover, Bluetooth solutions support multipoint-to-multipoint interactivity within groups of compatible mobile hosts. Bluetooth systems operate globally in the 2.400 GHz license-exempt ISM band and provision service in the United States, the European Union, and Japan.

9.16.2   BLUETOOTH TRANSMISSION BASICS

An evolving third-generation specification, Bluetooth technology utilizes omnidirectional radio signals for enabling transmission between portable communicators within range of each other, thereby eliminating the need for wireline connections and an established wireline network infrastructure. In temporary next-generation mobile computing environments, WPANs also eliminate the need for users to maintain a fixed and stationary position in relationship to network implementations.

Designed as a universal radio interface technology, Bluetooth facilitates operations in license-exempt and noisy RF environments. Bluetooth technology transmits information through non-metal solid objects at distances ranging from 10 centimeters to 10 meters. The area of coverage provisioned by a Bluetooth WPAN can be extended to 100 meters by increasing the transmission power of wireless devices.

For enabling seamless operations, Bluetooth WPANs employ forward error correction (FEC) to limit the effects of random noise on transmission, frequency-hopping spread spectrum (FHSS) technology to eliminate signal interference, and time-division duplexing (TDD) technology for modulation. Bluetooth WPANs support asynchronous data and synchronous voice channels for enabling voice and data services. Microchips are embedded into electronic devices to facilitate Bluetooth services.

9.16.3   BLUETOOTH APPLICATIONS AND SERVICES

Bluetooth WPANs enable IP packet transmission, feature built-in encryption and authentication mechanisms for security, and interoperate with DECT (Digital Enhanced Cordless Telecommunications) services. Bluetooth WPANs are flexible and extendible. These networks support interoperability of wireless devices such as cellular phones, PDAs, digital cameras, digital scanners, printers, notebook computers, joysticks, and handheld communicators that are within range of each other and facilitate dependable access to broadband fixed wireless access (FWA) LANs.

Bluetooth WPANs enable full-duplex data transmissions at rates ranging from 720 Kbps to 1 Mbps. Using a multiservice, multifunctional platform, Bluetooth implementations foster applications that include e-mail, short messaging services (SMS), data transport, voice telephony, and Web browsing. (See Figure 9.2.)

9.16.4   BLUETOOTH PICONETS

Bluetooth technology fosters implementation of ad hoc networks that operate within a ten-meter radius in license-exempt spectrum between the 2.400 and 2.483 GHz frequencies in piconet configurations. Piconets support point-to-point and point-to-multipoint transmissions via two to eight portable communicators. In a piconet, one communicator functions as a master. The remaining Bluetooth devices function as slaves throughout the duration of the connection. A collection of independent piconets forms a scatternet. A scatternet enables data and voice transmission from one piconet to another piconet.

9.16.5   BLUETOOTH AND IEEE 802.11B FWA ETHERNET LAN OPERATIONS

IEEE 802.11b FWA Ethernet LANs and Bluetooth radios share common spectrum between the 2.400 and 2.483 GHz ISM RF bands. Approaches for enabling the co-existence of Bluetooth technology and broadband FWA Ethernet LANs are examined by the IEEE 802.11b Task Forces.

9.16.6   BLUETOOTH SPECIAL INTEREST GROUP (BLUETOOTH SIG)

The Bluetooth Special Interest Group (SIG) promotes development of interoperable Bluetooth devices and products and implementation of Bluetooth services for business travelers and telecommuters. Bluetooth SIG participants include Intel, Nokia, IBM, and Toshiba.

9.17   WIRELESS HOME NETWORKS (WHNS)

9.17.1   WHN FEATURES AND FUNCTIONS

Developed by the HomeRF (Radio Frequency) Working Group, wireless home networks (WHNs) enable operations in license-exempt spectrum between the 2.400 and 2.483 GHz frequencies. WHN solutions are based on the SWAP (Shared Wireless Access Protocol) specification. Also developed by the HomeRF Working Group,

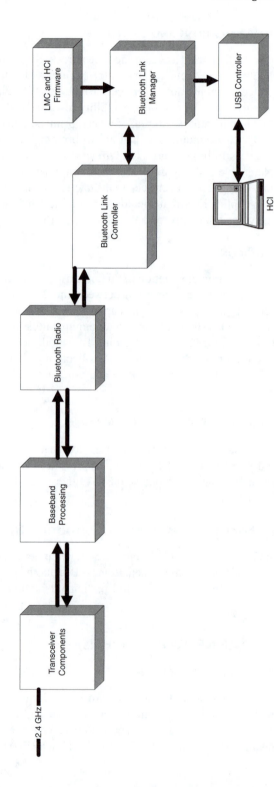

FIGURE 9.2 Processes supported by a Bluetooth first-generation USB configuration that is also compliant with HCI (Human-Computer Interface) requirements.

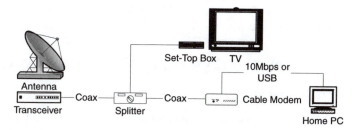

FIGURE 9.3 A wireless home network (WHN).

SWAP defines a common interface for supporting networking operations inside a home and the grounds immediately outside. SWAP is based on the Digital Enhanced Cordless Telecommunications (DECT) standard and the IEEE 802.11b broadband FWA Ethernet LAN specifications

9.17.2 WHN Configurations

A WHN configuration supports up to 127 network nodes or stations and employs frequency-hopping spread spectrum (FHSS) technology for enabling data and voice transmission. SWAP-compliant devices such as notebook and laptop computers, wireless terminals, digital cordless phones, and PDAs exchange information in configurations that include ad hoc WHNs as well as WPANs and WLANs.

9.17.3 WHN Applications

WHNs enable applications that include voice conversations, data exchange, file and equipment sharing, e-mail, and control of air conditioning, heating, and electrical systems by voice commands at rates reaching 10 Mbps and at distances up to 10 meters. WHN implementations complement services supported by the in-place wireline network. (See Figure 9.3.)

9.17.4 HomeRF (Radio Frequency) Working Group

The HomeRF Working Group (HomeRF WG) ratified the SWAP (Shared Wireless Access Protocol) 1.0 specification in 1999. In addition to supporting the WHN initiative, the HomeRF Working Group also fosters development of wireless products that support delivery of streaming video and audio and broadband wireless Web connections outside the home. Participants in the HomeRF Working Group include 3Com, Compaq, Microsoft, Samsung, Motorola, Apple Computer Inc., Broadcom Corporation, France Telecom, Global Convergence Technologies, IBM, Intel Corporation, Microsoft, Siemens, Sprint, and Texas Instruments.

9.17.4.1 SWAP-CA (Shared Wireless Access Protocol-Cordless Access) Forum

The SWAP-CA (Shared Wireless Access Protocol-Cordless Access) Forum promotes the utilization of standardized technologies for SWAP services such as FHSS and

DECT. Participants in this forum include 3Com, Texas Instruments, IBM, Nortel, Intel, Compaq, Ericsson, and Microsoft.

9.18   IEEE 802.16 BROADBAND WIRELESS ACCESS (BWA) SYSTEMS

9.18.1   BWA Technical Features and Functions

The IEEE 802.16 specification on broadband wireless access systems facilitates implementation of interoperable fixed point-to-multipoint broadband wireless access networking solutions. These systems transport voice, video, and data in spectrum between the 2 and 11 GHz frequencies and in the 10 and 66 GHz spectral bands.

Broadband wireless access (BWA) systems support high-capacity, high-performance interactive multimedia applications in fields that include E-commerce, tele-education, telemedicine, and E-government and enable multimedia videoconferencing, video multicasts, and voice calls. These systems also enable transmission of variable-length IP datagrams, work in concert with IPv4 (Internet Protocol version 4) and IPv6 (Internet Protocol version 6), and provide long-haul services for wireless and cellular networks. Broadband wireless access metropolitan networks interoperate with high-speed broadband wireline networks and provision equivalent services.

9.18.2   BWA Transmission Basics

Broadband wireless access implementations support standardized point-to-multipoint radio-air interfaces and enable functions in wireless metropolitan networks and wider area wireless network configurations. Network equipment includes integrated antennas and broadband transceivers that are attached to poles, rooftops, and towers. Broadband transceivers employ Quadrature Phase Shift Key (QPSK) modulation that operates with Time-Division Multiple Access (TDMA) channels and Quadrature Amplitude Modulation (QAM) modulation that operates with FDMA (Frequency-Division Multiple Access) channels to provision increased channel capacity and enable robust signal transmissions.

The Network Interface Unit (NIU) at the customer premise generates and terminates network transmissions. The NIU works in concert with an array of technologies, including Ethernet and ATM. Broadband wireless access metropolitan networks co-exist with or take the place of DSL (Digital Subscriber Line), LMDS (Local Multipoint Distribution System) and MMDS (Multichannel Multipoint Distribution System) implementations.

The quality, reliability, and availability of network connections in broadband wireless access metropolitan networking implementations change frequently and unexpectedly. Factors that contribute to variations in network performance include the complexity of the infrastructure, extent of geographic coverage, population density of the service area, application requirements, and available resources. Moreover, rainstorms, thunderstorms, and environmental noise are among the variables that also affect network operations.

The IEEE 802.16 broadband wireless access specification also clarifies Physical Layer or Layer 1 and MAC (Medium Access Control) Layer or Layer 2 functions.

Procedures for providing QoS (Quality of Service) guarantees and CoS (Classes of Service) assurances are in development.

9.18.3 IEEE 802.16 WORKING GROUP ON BROADBAND WIRELESS ACCESS SYSTEMS

The IEEE 802.16 Working Group on Broadband Wireless Access Systems defines the radio-air interface for broadband wireless access configurations that operate in the licensed spectrum between the 2 and 11 GHz frequencies. This Working Group also delineates the radio-air interface for broadband wireless systems that operate in spectrum between the 10 and 66 GHz frequencies. These frequencies also support U-NNI operations. Approaches for enabling the co-existence of broadband wireless access metropolitan networks and LMDS (Local Multipoint Distribution System) configurations that operate in the 23.5 and 43.5 GHz frequencies are also explored.

IEEE 802.16 specifications conform to IEEE 802 standards governing LAN and MAN operations endorsed by the IEEE in 1990. As opposed to IEEE 802.11 that focuses on WLAN implementations and applications, the IEEE 802.16 Working Group develops affordable point-to-multipoint broadband wireless access configurations that enable multimedia services in MAN and WAN environments. The IEEE 802.16 Working Group works with the Broadband Radio Access Network (BRAN) Committee sponsored by the European Telecommunications Standards Institute (ETSI) in developing compatible broadband wireless access system specifications.

9.18.4 IEEE 802.16A EXTENSION

The IEEE 802.16a Extension clarifies operations, services, and the radio-air interface for broadband wireless access networks that operate in spectrum between the 2 and 11 GHz frequencies.

9.18.5 IEEE 802.16B EXTENSION: THE WIRELESSHUMAN (HIGH-DATA RATE UNLICENSED METROPOLITAN AREA NETWORK) INITIATIVE

Sponsored by the IEEE 802.11 Working Group on broadband wireless access systems, the WirelessHUMAN (High-Data Rate Unlicensed Metropolitan Area Network) initiative fosters implementation of broadband wireless access metropolitan network specifications. These specifications provide the foundation and framework for the IEEE 802.16b Extension.

Approved in 2001, the IEEE 802.16b Extension clarifies broadband wireless access metropolitan network functions and capabilities of the radio-air interface. Licensed-exempt BWA metropolitan networks support multimedia services in license-exempt spectrum between the 5.15 and 5.25 GHz frequencies, between the 5.25 and 5.35 GHz frequencies, and between the 5.725 and 5.825 GHz frequencies. To facilitate the standardization process, the WirelessHUMAN initiative supports utilization of the Physical Layer OFDM protocol defined in the IEEE 802.11a Extension and MAC (Medium Access Control) Layer operations defined in the IEEE 802.16 standard.

9.18.6   NATIONAL-WIRELESS ELECTRONIC SYSTEMS TESTBED (N-WEST)

Sponsored by the IEEE 802.16 Working Group, the National-Wireless Electronic Systems Testbed (N-WEST) promotes the implementation of interoperable broadband wireless access metropolitan network products by carrying out test and measurements at the system and component levels. These tests and measurements also contribute to development of operational standards for broadband wireless access metropolitan networks and LMDS (Local Multipoint Distribution System) configurations. The U.S. Department of Commerce, the National Institute of Standards and Technology (NIST), and the National Telecommunications and Information Administration (NTIA) support N-WEST operations.

9.19   BROADBAND WIRELESS INTERNET FORUM

Sponsored by the IEEE Industry Standards and Technology Organization (IEEE ISTO), the multivendor Broadband Wireless Internet Forum (BWIF) supports implementation of standardized broadband wireless access networks and interoperable broadband wireless access products. Participants in BWIF include Cisco Systems, Broadcom, Toshiba, Wireless Facilities, Inc. (WFI), Bechtel, and Pace Micro.

9.20   BROADBAND RADIO ACCESS NETWORK (BRAN)

9.20.1   BRAN CAPABILITIES

Developed by the ETSI (European Telecommunications Standards Institute), the Broadband Radio Access Network (BRAN) defines a highly flexible radio system that provides indoor fixed wireless and cordless access to a diverse array of core networks and multimedia services. Distinguished by its flexibility, BRAN also interworks with wireless and wireline ATM networks, IP configurations, and third-generation cellular solutions such as UMTS (Universal Mobile Telecommunications System).

The Broadband Radio Access Network (BRAN) enables rates ranging from 25 to 155.52 Mbps (OC-3) over distances that extend from 50 meters to 5 kilometers. In addition, BRAN fosters high-speed symmetrical connections at rates reaching 25 Mbps in each direction and can be rapidly deployed. BRAN research contributes to the development of Physical Layer or Layer 1 and MAC Layer or Layer 2 operations for HiperLAN-2 and the provision of QoS guarantees in broadband FWA wider area networks.

9.20.2   BRAN PARTICIPANTS

The ETSI Signaling Protocols and Switching (SPS) Technical Committee, the ATM Forum, and the Internet Engineering Task Force (IETF) support the BRAN initiative and develop specifications for radio access systems, network management functions, and internetwork interoperability. In addition, BRAN participants coordinate activities with the Mobile Multimedia Access Communications Council (MMAC), the

IEEE 802.11 and IEEE 802.16 Working Groups, and the European Commission Advanced Communications Technologies and Services (EC-ACTS) Program.

9.20.3   BROADBAND FWA NETWORK MARKETPLACE

Applications and services provided by communications carriers in the broadband FWA (fixed wireless access) LAN, MAN, and WAN marketplace are highlighted in the material that follows.

9.20.3.1   Aerial Telecom

Aerial Telecom supports implementation of broadband wireless access solutions built as overlays to Internet-based PoPs (Points of Presence) for enabling direct links to fiber optic backbone networks such as UUNet and Genuity.

9.20.3.2   Aperto Networks

Aperto Networks implements multiservice broadband FWA networks that employ PacketWave products for eliminating bottlenecks at the local loop. These point-to-multipoint network solutions operate in spectrum that includes the 2.500, 3.500, 5.300, and 5.800 GHz frequencies. PacketWave products support line-of-sight (LOS), non-line-of-sight (NLOS), and obstructed line-of-sight (OLOS) operations and IP packet delivery. TDD utilization enables the flexible implementation of a broadband channel for facilitating multimedia transport at 20 Mbps. Aperto Networks solutions dynamically allocate bandwidth on-demand for accommodating changes in traffic patterns and provide QoS assurances for delivery of time-sensitive voice and video applications.

9.20.3.3   Harmonix

In 2000, the FCC allocated W-band spectrum between the 59 and 64 GHz frequencies for enabling license-exempt broadband FWA LAN services. Developed by Harmonix, GigaLink supports implementation of broadband point-to-point broadband FWA LANs that operate in the spectrum between the 59 GHz and 64 GHz frequencies. GigaLink wireless local loop solutions interoperate with Gigabit Ethernet wireline networks and enable transmission rates reaching 622.08 Mbps (OC-12). Broadband wireless access MANs can be rapidly implemented in sparsely populated rural settings, as well as in large cities, without fixed wireline infrastructures. These versatile and flexible networks complement services supported by in-place wireline infrastructures, regardless of geographic and artificial constraints such as canyons, rivers, lakes, railroad lines, asbestos hazards, and international borders.

9.21   IMPLEMENTATION CONSIDERATIONS

New generations of broadband wireless access networking solutions support network management and security protocols such as SNMP (Simple Network Management Protocol) and DES (Digital Encryption Standard) for enabling dependable voice, video, and data delivery to stationary and mobile subscribers with the reliability of wireline networking configurations. Moreover, broadband wireless access MANs interoperate with legacy network architectures and overcome local loop constraints in densely populated cities where DSL and cable modem services are not available.

Web popularity fosters the growth in wireless communicating devices and technologies. However, Web applications are designed for utilization in traditional wireline environments. As a consequence, broadband wireless access network performance can also be unreliable. In addition, wireless signals are subject to attenuation and degradation. Poor response time, latency, and dropped connections are not uncommon as a consequence of signal interference. Obstructions such as hills, tall buildings, and foliage that block the line-of-sight transmission path between a wireless transceiver and base station disrupt signal integrity. In addition, reflections off of multistory hotels, office buildings, and apartment complexes trigger rapid changes in signal amplitudes that result in multipath signal fading. Signal strength and robustness also grow weaker as the distances between wireless transmitters and base stations increase.

Moreover, broadband wireless access MANs that share frequency bands with radar services provisioning air and sea traffic control, weather and flood forecasting, and airport and harbor surveillance can experience unacceptable levels of radio interference. The ITU addresses this problem through the development of recommendations on preferred frequency bands for broadband wireless access MAN installations. In addition, the National Telecommunications and Information Administration (NTIA) sponsors a comprehensive survey of adjacent band interference. This survey indicates spectral bands used by high-powered transmitters such as radar and identifies technologies that support increased immunity to in-band and co-located band noise. Findings are expected to contribute to technical and regulatory solutions for controlling transmitter emissions and in-band and adjacent band interference.

9.22 WIRELESS NETWORK SECURITY

The use of shared airwaves in wireless transmissions puts information at risk unless policies and techniques enabling network security, privacy, accountability, integrity, and authentication are put into place. As with their wireline counterparts, broadband wireless networks are susceptible to attack by intruders seeking free and anonymous access to network assets. Even with proper safeguards, wireless transmissions are susceptible to interception that can compromise confidentiality and privacy of network users and disrupt network services.

As an example, the IEEE 802.11b Extension supports the use of WEP (Wired Equivalent Privacy) protocol for enabling secure wireless transmissions. In 2001, investigators at the University of California at Berkeley identified a security flaw in WEP that enables intruders in the vicinity of a broadband FWA (Fixed Wireless Access) LAN to intercept transmissions without detection.

Wireless computing is experiencing remarkable expansion in the number of users. The quantity of data transported across wireless networks is on the upswing as well. In addition to the increasing numbers of trusted users, wireless computing is also sparking an expansion in the number of rogue users or hackers.

It is important to note that, in general, the risks associated with wireless networks are not any greater than the risks associated with wireline networks. Passwords, firewalls, smart cards, biometrics, digital signatures, and encryption are among the mechanisms that ensure information integrity and safeguard network resources in wireless and wireline network configurations from incursion.

9.23   U.S. TELE-EDUCATION WIRELESS SOLUTIONS

Wireless network solutions in the tele-education domain foster access to information resources in campus buildings such as libraries, classrooms, and dormitories. In the 1990s, Metricom supported wireless infrastructure installation for mobile users on college campuses that resulted in the deployment of Ricochet wireless configurations at the University of Miami, Oregon State University, and Austin College in Texas. On these campuses, the Metricom Ricochet wireless network fostered access to computing services from any point on campus, including faculty apartments and campus dormitories, where optical fiber was not yet installed or where wireline plants did not support robust connections to networking services.

Wireless networks feature an array of technologies, services, and architectures for fostering access to a rich array of Web resources inside or outside a classroom and enable applications ranging from team teaching to scientific experiments. For example, with WLAN implementation, faculty can enter student grades and attendance records and review campus news and events in school intranets from any point on campus. Administrative staff can communicate with teachers throughout a school district via e-mail and SMS. Students work collaboratively, exchange ideas, and participate interactively with their peers in classroom activities. School psychologists, social workers, and guidance counselors readily update student records and monitor student progress.

9.23.1   CALIFORNIA

9.23.1.1   California State University at Los Angeles (CSU Los Angeles)

California State University at Los Angeles utilizes a microwave ITFS (Instructional Televised Fixed Service) network for broadcasting courses leading to Bachelor of Science degrees in electrical engineering and fire protection administration to students in Orange, Los Angeles, Ventura, and Kern Counties.

9.23.1.2   San Francisco Public Library

The San Francisco Public Library employed Metricom Ricochet service in bookmobiles to determine book availability, check patron records, and facilitate book circulation. Wireless bookmobile configurations also enabled underserved children and senior patrons in nursing homes and convalescent facilities to browse the Web.

9.23.1.3   Stanford University

The Stanford Center for Professional Development (SCPD) at Stanford University supports delivery of graduate education courses and programs to more than 250 corporate sites in the 35-mile area from San Francisco to San Jose via a microwave network.

9.23.2   COLORADO

9.23.2.1   Old Colorado City Communications Initiative

The role of wireless connections in promoting collaboration, Web access, distance learning, home schooling, and interoperability with wireline networks is examined in Colorado as part of the Old Colorado City Communications Initiative sponsored by the National Science Foundation (NSF). Participants include school districts in Monte Vista, San Luis, and Colorado Springs.

In addition, the Old Colorado City Communications Initiative used spread spectrum technology for interconnecting educational institutions and the United States Embassy in Ulaan Baatar, the capital city of Mongolia, to the Internet. By employing satellite links, scientists in the United States could also communicate with their peers at the National University of Mongolia, the Mongolian Technical University, and the Mongolian Academy of Sciences.

9.23.3   FLORIDA

9.23.3.1   Nova Southeastern University (NSU)

The Shepard Broad Law School at Nova Southeastern University (NSU) employs wireless LAN technology for enabling connectivity via laptop computers to Web services, library resources, and legal documents, and for supporting links to e-mail from any point on campus. The infrastructure includes transceivers or access devices installed in drop ceilings above the law school library, cafeteria, and classrooms. This WLAN solution enables law students to form virtual study groups, access grades, view previous exams and syllabi, register for courses, post resumes, and take exams online.

9.23.3.2   University of Central Florida (UCF)

The University of Central Florida (UCF) utilizes a wireless Ethernet LAN in its College of Engineering building for enabling students and faculty to access the UCF wireline campus network from any inside location.

9.23.4   HAWAII

9.23.4.1   Hawaii Wide Area Integrated Information Access Network (HAWAIIAN)

The Hawaii Wide Area Integrated Information Access Network (HAWAIIAN) uses an intra-island SONET (Synchronous Optical Network) system that works in conjunction with an inter-island digital microwave backbone network for provisioning access to educational services, Web resources, and applications. This mixed-mode wireless and wireline network supports voice, data, and video transmission; videoconferences; and delivery of teleprograms and telecourses offered by the University of Hawaii System.

9.23.5 MICHIGAN

9.23.5.1 Battle Creek School System

Located in Troy, the Battle Creek School System employs broadband fixed wireless access IEEE 802.11b LANs operating in the 2.400 GHz frequency block for enabling voice, video, and data transmission at rates reaching 11 Mbps and supporting networking services in every district junior high school situated in historic brick buildings.

Broadband FWA Ethernet LANs support Web browsing and connections to a wireline school system WAN. Faculty and administrators employ broadband FWA Ethernet LAN connections for accessing e-mail, maintaining class attendance records, and setting-up meetings and conferences with educators and parents. Wireless Access Points (WAPs) are located in ceiling tiles in each junior high school.

9.23.5.2 Wayne State University

WLANs offer significant flexibility not possible with traditional wireline LANs. A wireless link based on microwave technology can provide immediate fixed network access for temporary or new users and reduce costs associated with cabling and equipment installation. For example, at Wayne State University, a microwave network eliminates barriers of distance by interlinking stand-alone networks in the medical school and in the medical library.

9.23.6 MISSOURI

9.23.6.1 Independence Public School System

The Independence Public School System utilizes a systemwide wireless IEEE 802.11b Ethernet infrastructure implemented in 22 schools and 3 administration buildings. This WLAN configuration enables a broad array of educational, staff, and administrative applications and services and saves the district several thousand dollars each month in leased line fees.

9.23.7 NEW YORK

9.23.7.1 Smithfield School District

In the Smithfield School District, the Symbol Technologies Palm Computer Project facilitates utilization of bar-coded student identification cards with Palm handheld communicators to track textbooks, library resources, and lunch debits. Teachers, administrators, and security personnel utilize Palm handheld devices to access student grades, schedules, attendance records, and discipline records via the districtwide wireline network.

9.23.7.2   State University of New York (SUNY) at Potsdam and Potsdam Central Schools

The State University of New York (SUNY) at Potsdam and Potsdam Central Schools employ spread spectrum technology for implementing 10BASE-T wireless Ethernet LANs on the university campus and in the Potsdam Central Schools. These economical solutions support building-to-building connections over distances of less than one mile, foster links to wireline networks, enable Web browsing, and eliminate the need for leased T-1 lines. The license-exempt wireless Ethernet LANs at university and school sites support data and voice transmission at rates reaching 2 Mbps.

9.23.8   TEXAS

9.23.8.1   Canyon Independent School District (CISD)

Canyon Independent School District (CISD) in Amarillo employs microwave technology for interlinking schools throughout the district and two campuses on a canyon plateau in a regional network configuration. This wireless configuration enables line-of-sight transmissions at 10 Mbps over relatively flat terrain; supports intercampus wireless connections for e-mail exchange; and facilitates access to student grades, attendance records, and accounting and payroll data on the district network. A firewall helps to ensure information security. The CISD also supports classes on Web utilization that are available to teachers, students and their parents or caregivers, and members of the local community.

9.23.8.2   Lamesa School District

The Lamesa School District employs a wireless Ethernet configuration that enables Web browsing and access to educational resources via the wireline network at Texas Tech University. The Lamesa School District network also supports access to grading and attendance records, class schedules, library resources, and educational CD-ROMs. Transmission rates at 20 Mbps are supported. Plans by the Lamesa School District to wirelessly extend network services to outlying school districts through the use of microwave links and roof-mounted antennas are in development.

9.23.9   WASHINGTON, D.C.

9.23.9.1   National Defense University (NDU)

At National Defense University (NDU), a wireless LAN extends the capabilities of the university wireline LAN so that faculty and students can wirelessly access coursework, data files, e-mail, and document printing capabilities from any point on campus. Moreover, faculty can utilize WLANs for conducting group decision-support demonstrations and seminars at off-campus locations.

9.23.10 U.S. Virgin Islands

9.23.10.1 Department of Education

After a hurricane destroyed the U.S. Virgin Islands Department of Education wireline communications network, administrators moved forward with wireless Ethernet network implementations that were rapidly deployed. These wireless network configurations interlink schools in districtwide networks situated on three islands.

9.24 INTERNATIONAL TELE-EDUCATION INITIATIVES

9.24.1 Australia

9.24.1.1 Perth Academic Regional Network (PARNET)

The Perth Academic Regional Network (PARNET) in Western Australia employs ATM and microwave technologies for interlinking institutions that include the University of Western Australia, the Curtin University of Technology, and Edith Cowan and Murdoch Universities. In addition, PARNET supports diverse tele-education applications for the Western Australian Regional Network Organization (WARNO). WARNO also participates in the Australian Academic Research Network (AARNET).

9.24.2 Canada

9.24.2.1 Medicine Hat, Alberta, WWAN

Spread spectrum technology links approximately 20 elementary schools, high schools, and libraries in a WWAN configuration in Medicine Hat, Alberta. Capable of supporting rates at 1.5 Mbps, this technology also enables access to Web resources.

9.24.2.2 Peel School District, Ontario

Located in the municipalities of Missassauga, Brampton, and Caledon in the province of Ontario, the Peel School District employs a wireless WAN backbone in concert with wireless Ethernet bridges for interlinking 28 secondary schools. This WWAN configuration interoperates with wireline Ethernet networks, enables students to access Web services and online databases for educational research, and supports utilization of tele-education courses and videoconferencing for curriculum enhancement.

9.24.2.3 University of British Columbia

At the University of British Columbia, a license-exempt Ethernet WLAN solution operates in the 2.400 GHz RF band. This WLAN supports short-distance point-to-point links and accommodates WLAN requirements for small groups of students. Approaches for delivering WLAN services in an extended area of coverage are under consideration.

9.24.2.4   University of Toronto and Center for Wireless Communications

The University of Toronto and the Center for Wireless Communications support implementation of a wireless ATM (WATM) LAN capable of sustaining data rates of 10 Mbps. This WATM LAN enables operations in spectrum between the 5.15 and 5.25 GHz frequencies. The capabilities of this WATM LAN in promoting the transparent delivery of voice, video, and data transmission to mobile users are evaluated. The National Science and Technology Board of Singapore and the Government of Ontario also participate in this initiative.

9.24.2.5   University of Waterloo in Ontario

Investigators at the University of Waterloo in Ontario explore issues associated with the provision of QoS measurements and mobility management services and develop distributed multimedia applications that operate in mobile network environments.

9.24.3   LATVIA

9.24.3.1   LATNET (Latvian Academic and Research Network)

Developed by the University of Latvia and Riga Technical University, LATNET (Latvian Academic and Research Network) employs spread spectrum technology for provisioning telephony services, Internet connectivity, and data transport in Riga and its suburbs. This wireless network employs DSSS (Direct Sequence Spread Spectrum) technology for enabling point-to-multipoint links. Rates up to 2 Mbps are supported at distances extending to 7 kilometers in an area of coverage that includes the Hotel Latvia and the nearby television station.

LATNET employs standard WLAN equipment and supports operations in the 900 MHz and 2.400 GHz frequencies. Marginal radio signal quality and heavy traffic loads significantly impair network performance. Government licenses are not required for WLAN implementations supporting wireless information transport over short distances. LATNET also provisions satellite links to the Norwegian TAIDE network.

9.24.4   NORDIC COUNTRIES

9.24.4.1   NORDUnet (Nordic Countries Network)

A network initiative sponsored by Denmark, Finland, Iceland, Norway, and Sweden, NORDUnet (Nordic Countries Network) employs microwave connections that interlink the NORDUnet Point of Presence (PoP) at the Technical University of Denmark in Lyngby with the University of Lund in Sweden. These connections foster telecollaborative activities between educational institutions and research centers in Denmark and Sweden, and enable creation of the Virtual Oresund University.

9.24.5   NORWAY

9.24.5.1   Wireless Experimental Metropolitan Area Network (WEMAN)

Sponsored by the Norwegian Research Council, the Norwegian Academic Network for Research and Education (Uninett), and the Institute for Informatics (IFI) at the

University of Oslo, WEMAN (Wireless Experimental Metropolitan Area Network) provides IPv6 services and conforms to the IEEE 802.11 standard. Low noise amplifiers and specially designed receive-only antennas extend the reach of WEMAN to sparsely populated locations. Rates up to 3 Mbps are supported in the license-exempt spectrum between the 2.400 and 2.483 GHz frequencies. In addition, WEMAN demonstrates the feasibility of using WLAN technology in a WMAN environment.

Problems associated with achieving wireless network scalability, reliability, and stability in WEMAN-based research are explored. In addition, the impact of meteorological conditions on wireless networking operations and the degree of interference on adjacent nodes generated by frequency-hopping spread spectrum (FHSS) technology are examined. WEMAN supports continuous Internet connectivity and applications such as telecommuting and teleworking and provisions a framework for accessing wireless Internet2 applications.

9.25 EUROPEAN COMMISSION TELEMATICS APPLICATIONS (EC-TAP) PROGRAM

9.25.1 EDUCATION AND TRAINING SERVICES FOR HOME LEARNERS (TOPILOT)

A wireless training and education initiative, the TOPILOT was designed for young students traveling with their families and, therefore, unable to attend traditional schools. TOPILOT tutors monitored and supported learners at a distance through the use of tools and technologies that included CD-I (Compact Disk-Interactive) players and disks; dial-up access connections; 2G (second-generation) GSM communicators. Children from three to nine years old learned basic counting, reading, and social skills. For students over age 14, TOPILOT provided access to vocational teletraining and tele-education services.

9.26 BUSINESS SECTOR WIRELESS INITIATIVES

Wireless networks in the business sector optimize information access and telecollaboration, thereby increasing on-the-job productivity and optimizing customer service. Wireless configurations enable diverse tele-operations at business venues that include specialty shops, corporate offices, grocery stores, banks, stock brokerage firms, insurance agencies, hotels, and conference facilities, and teleservices such as package delivery and E-commerce applications.

Wireless networks support real-time streaming of music videos and ticket selling services. Repair technicians consult wireless networks for obtaining technical and inventory data and scheduling maintenance jobs. Landlords equip buildings with antennas for supporting access to Web sites to attract potential tenants.

Business travelers on the move employ wireless networks for accessing the Internet, groupware applications, virtual offices, and corporate intranets and extranets. In the financial sector, wireless network solutions enable connections to bank networks featuring information on accounts, stock quotes, credit card balances, and interest rates. In the retail sector, wireless configurations enable sales associates and their supervisors to access gift registries, verify payments and orders, track inventory status, reorder merchandise, check prices, and process automatic markdowns,

9.27 PUBLIC SECTOR WIRELESS INITIATIVES

Wireless and mixed-mode wireline and wireless systems are viable alternatives to conventional twisted copper pair telecommunications networks for supplementing access to public sector services and provisioning voice and data coverage in areas that lack a fully developed POTS (Plain Old Telephone Service) infrastructure. These configurations enable fire departments and law enforcement agencies to deliver high-quality services in life-threatening situations and support operations in the office and in the field for problem resolution. Security mechanisms such as encryption, authentication, and compression foster secure and fast information transport.

Police departments utilize wireless networks for secure two-way messaging services at accident and crime scenes, filing reports, and tracking stolen vehicles. These wireless networks also foster access to vital data, location information, and incident log updates in local, state, and national law enforcement networks. In addition, wireless networks enable transmission of video images from remote digital cameras at disaster sites such as the Pentagon and the World Trade Center for determining the need for additional personnel and resources.

Wireless transportation networks provide motorists with real-time information on weather conditions, traffic delays, and road construction. Moreover, these configurations support access to municipal services in rural and isolated communities as well as in public housing projects, correctional facilities, courthouses, and public libraries.

9.27.1 PUERTO RICO

9.27.1.1 PRStar.net

A government-sponsored initiative, PRStar.net consists of an ATM backbone network supporting administrative services at rates reaching OC-3 (155.52 Mbps). PRStar.net interoperates with broadband fixed wireless access (FWA) Ethernet LANs that support information transport at 11 Mbps. FWA Ethernet LANs are implemented in public schools, courts, correctional facilities, and public libraries.

9.27.2 SWEDEN

9.27.2.1 CityNet

Wireless networks are popular solutions in those municipalities, regions, and countries lacking a fixed wireline infrastructure. In Stockholm, Malmo, and Gothenburg, a wireless Internet service called CityNet, developed by Telia and Wi-LAN, supports high-speed wireless Internet, intranet, and extranet access and networking applications sponsored by government agencies. Designed for SOHO and business users, CityNet fosters rates reaching 2 Mbps. Future implementations are expected to incorporate W-OFDM technology developed by Wi-LAN for enabling transmission rates reaching 20 Mbps.

9.28 REMOTE MONITORING WIRELESS NETWORKING SERVICES

Wireless communications networks enable remote monitoring, surveillance, and search-and-rescue activities. For example, remote monitoring supports inventory control of vending machines, verification of package deliveries, and security of commercial and residential buildings and neighborhoods. In addition, wireless networks enable automated meter readings for determining gas, water, and electric charges and management of public facilities such as irrigation systems and storage tanks at remote locations.

9.29 WIRELESS TELEMEDICINE INITIATIVES

Wireless hospital LANs extend the reach and capabilities of fixed wireline LANs by bringing computing services directly to a patient's bedside. As an example, physicians use wireless communicators to access patient records, manage medication, obtain lab test results in real-time, and confirm diagnoses. Nurses use WLANs to place medication orders directly with the hospital pharmacy, monitor drug interactions, and check vital signs.

9.29.1 CALIFORNIA

9.29.1.1 Cedars-Sinai Medical Center

Clinicians at Cedars-Sinai Medical Center access surgical reports, physical evaluations, consultations, medical records, and admitting histories maintained on the wireline hospital network. This system also provides access to e-mail, processes book and article requests for the medical library, and features sophisticated security mechanisms for maintaining medical information confidentiality.

9.29.2 NORTH CAROLINA

9.29.2.1 Duke University Medical Center

At Duke University Medical Center, physicians and nurses access patient records and laboratory data in real-time at the point of care via an 11 Mbps broadband WFA Ethernet LAN.

9.29.3 OHIO

9.29.3.1 Good Samaritan Hospital

Physicians at Good Samaritan Hospital employ a broadband FWA Ethernet LAN for accessing records of patients in cardiac and surgical units, intensive care recovery rooms, a birthing center, and hospital clinics. In addition, physicians also monitor patient medications, access intake orders, and review treatment protocols in real-time at the patient's bedside via the WLAN solution.

9.29.4   SOUTH CAROLINA

9.29.4   Medical University of South Carolina

At the Medical University of South Carolina Albert Florens Storm Eye Institute, ophthalmology residents employ handheld wireless devices for completing patient history forms in real-time.

9.30   WIRELESS NETWORK CHALLENGES

Advances reflected in the aforementioned wireless initiatives promote increased opportunities for instruction, collaboration, and interaction. Faculty members use WLAN services to stay current in best practices in their field and access the Internet, campus intranets, and community networks. Wireless networks that interoperate with landline networks take advantage of existing applications and devices while providing flexibility for diverse activities.

Despite benefits such as lowered costs for laptops, notebooks, and pen-based portables, and the diversity of wireless services in the wireless marketplace, deliberation continues, particularly in academic circles, about the feasibility of implementing wireless communications services and integrating wireless technologies into wirebound network infrastructures.

Generally, wireless networks are easier to install and configure than fixed wireline networks. Wireless networks support data integrity through utilization of security measures such as encryption and protocols such as WEP.

The emergence of small, lightweight, lower-power, inexpensive wireless terminals contributes to widespread interest in the wireless domain. These devices include palm-sized PCs such as the Palm Pilot VII connected organizer and handheld devices with miniature keyboards that use the Windows CE operating platform. Personal Digital Assistants (PDAs), bar-code scanners, digital cameras, pagers, printers, mobile phones, and image capture devices are also examples of wireless information communicators. By supporting links to wireless and/or wireline networks, these versatile and affordable devices eliminate unnecessary paperwork; optimize user productivity; and bring products, services and transaction points directly to the user.

Nonetheless, there are still challenges in achieving high performance rates for file transfer and Web access and problems associated with high and bursty data losses and network congestion. In comparison to wireline networks, wireless networks operate in a constrained communications environment. Connectivity to a fixed wireline network is not always reliable, stable, or secure. Problems associated with wireless ATM implementations that interwork with broadband wireless network configurations persist as well.

As a consequence of battery limitations, wireless networking devices are less powerful than conventional desktop computers. Moreover, despite widespread advances in broadband FWA technology, wireless networks support slower transmission rates and are more susceptible to security intrusions than wireline configurations. In addition, wireless devices generally have smaller displays, employ nontraditional input devices ranging from a stylus to an abbreviated keyboard, and lack the functional flexibility and performance capabilities of their conventional wireline counterparts. As with twisted copper pair POTS configurations, wireless systems can be limited in supporting fast voice and data transmission rates. Any decision to deploy wireless networks must take these aforementioned constraints into account.

Like wireline network counterparts, wireless networks enable diverse applications, architectures, and configurations. Successful network implementation in the mobile communications arena involves resolving problems associated with handling heterogeneous teletraffic and terminal and personal mobility. Approaches for supporting multimedia teleservices with Quality of Service (QoS) guarantees or Class of Service (CoS) assurances; provisioning on-demand bandwidth; and fostering real-time access to voice, video, and data resources must be clarified as well.

In parallel with wireline counterparts, wireless networks are subject to operational problems that impact their reliability. Typically, wireless configurations are limited in supporting fast transmissions. Continuous changes in wireless devices that operate in mobile environments contribute to problems in network monitoring, administration, and management. Sources of interference such as environmental noise, thunderstorms, blizzards, and line-of-sight obstructions disrupt the integrity of wireless networking operations.

Moreover, wireless transmissions are subject to fading, high-bit error rates, and sporadic connectivity. Ultimately, successful wireless network implementation involves selection of software, hardware, and transmission techniques for accommodating current and projected institutional goals, objectives, and requirements.

9.31   U.S. WIRELESS RESEARCH INITIATIVES

The DARPA ITO (Information Technology Office) sponsors development of a wireless infrastructure that supports data access independent of user location. This extendible and flexible infrastructure scales to accommodate computing requirements of thousands of users, accommodates operations of heterogeneous devices, and enables virtual interactions. University-sponsored wireless initiatives define a technical infrastructure for mobile computing, foster development of advanced packet radio networking protocols and adaptive network algorithms, contribute to implementation of sophisticated wireless protocols, and facilitate practical deployment of wireless network solutions and access to mobile computing resources in diverse networking environments.

9.31.1   CALIFORNIA

9.31.1.1   University of California at Berkeley

Researchers at the University of California at Berkeley use computer-aided design (CAD) techniques for design of high-performance wireless network systems enabling multiple media transport. Strategies for the design of an adaptable and reconfigurable wireless network employing spread spectrum radios and procedures for implementation of an instant infrastructure supporting voice, data, and video networking with PC notebooks are examined as well. Findings contribute to the development of HDTV (High Definition Television) systems, CDPD (Cellular Digital Packet Data) modems, and handheld devices for personal communications systems and environmental monitoring.

9.31.1.2.1 Berkeley Wireless Research Center

The Berkeley Wireless Research Center and the Wireless Information Network Laboratory (WinLAB) at Rutgers University sponsor development of smart plug-and-play radio networks that operate in license-exempt spectrum. These radio networks interlink phones, smart appliances, printers, cameras, scanners, and desktop, wearable, and palmtop computers.

9.31.1.3 InfoPad Project

Innovative solutions accelerating the use of portable and mobile computers and wireless communications systems to support ubiquitous computing, traffic integration, and seamless network connectivity for instructional enrichment and tele-education are reflected in the InfoPad initiative developed at the University of California at Berkeley. InfoPad is a lightweight wireless portable terminal featuring pen input and handwriting and speech recognition capabilities. With InfoPad deployment, groups of students will be capable of simultaneously receiving streams of text, graphics, audio, and video in an indoor classroom setting; retrieving data from campus intranets; and browsing the Internet in real-time.

9.31.1.4 Ninja Initiative

The Ninja initiative conducted at the University of California at Berkeley fosters development and deployment of a software infrastructure for enabling next-generation Internet applications that work in concert with diverse information appliances, including workstations, personal computers, laptops, cellular telephones, and Personal Digital Assistants (PDAs). Ninja architecture is in development. Upon completion, Ninja is expected to provision ubiquitous access to e-mail, pages, real-time streaming audio services, and calendar data at anytime and from anywhere.

9.31.2 INDIANA

9.31.2.1 Purdue University

The Crosspoint Project at Purdue University is exploring the feasibility of combining spread spectrum and ATM (Asynchronous Transfer Mode) technologies for creating a seamless network that accommodates multimedia computing requirements of each individual on campus. Also a Purdue wireless initiative, the MediPad Project involves the creation of a prototype telemedicine environment for supporting distance education as well as remote diagnosis, supervision, and patient monitoring. Approaches for enabling mobile computing environments for telemedicine applications and techniques for data management in mobile distributed environments are explored. A MediPad prototype is implemented in the Indiana Rural Medical Network (IRMN).

9.31.3 NEW JERSEY

9.31.3.1 Rutgers University WinLAB

To support seamless wireless connectivity, the Wireless Information Network Laboratory (WinLAB) is testing the effectiveness of a new geographic routing protocol for transporting data packets through an internetwork of computers. Geographic routing is based on the utilization of arbitrary geographic regions denoted by longitude and latitude rather than logical computer addresses as criteria for packet routing. Tests promote an understanding of trade-offs associated with geographical routing in actual situations.

9.31.4 PENNSYLVANIA

9.31.4.1 Carnegie Mellon University (CMU)

9.31.4.1.1 Monarch Project

The Carnegie Mellon University School of Computer Science sponsors the Monarch Project for enabling seamless and transparent wireless networking services and applications. Operational since 1992, the Monarch Project develops methods and techniques for enabling users on the move to communicate with one another and achieve network connections in ad hoc networking environments. Protocols that operate at the Data Link Layer or Layer 2 through the Presentation Layer or Layer 6 of the OSI Reference Model are also in development.

9.31.4.2 Spot Initiative

Developed by the Wearable Group at CMU, the Spot initiative supports research in the field of wearable computing systems. This initiative clarifies interactive and architectural requirements for enabling effective interaction via wearable devices and promotes development of customizable design elements such as digital visual interfaces that contribute to the field of wearable research.

9.31.5 VIRGINIA

9.31.5.1 Virginia Polytechnic Institute and State University (Virginia Tech) Mobile and Portable Radio Research Group (MPRG)

The Mobile and Portable Radio Research Group (MPRG) at Virginia Tech designs next-generation wireless systems and develops products for use in the commercial sector. Areas of research include wireless LAN deployment, smart antenna functionality, and microwave broadband channel capabilities, simulation and design of communications systems, and intelligent communications systems and solutions. The

MPRG also provides design and analysis methods, tools, and techniques for regulatory agencies, manufacturers, and government consumer service providers. MPRG industrial affiliate members include Ericsson, Lucent Technologies, Nokia, Nortel, Qualcomm, AT&T, and Hughes Electronics.

9.32 DARPA GLO-MO
(GLOBAL MOBILE INFORMATION SYSTEMS) INITIATIVES

The Defense Advanced Research Projects Agency is the key research and development organization for the U.S. Department of Defense. DARPA directs and supports research in the wireless arena that advances military objectives and contributes to public infrastructure enhancements. In the wireless domain, DARPA-sponsored research includes development and deployment of network paradigms, protocols, architecture, and mechanisms for supporting real-time communications services and dependable transport of multiple media applications.

DARPA initiatives in this section demonstrate the feasibility of provisioning wireless support that accommodates sporadic and changing connectivity and movement within and between networks. State-of-the art wireless networks, services, applications, and techniques for accommodating DARPA objectives also promote wireless capabilities and innovative wireless solutions in academic, businesses, and local and municipal agencies.

Network technology implementations and innovations such as hybrid wireless and ATM configurations, and powerful low-cost handheld wireless devices that support military needs also accommodate business and educational requirements.

GloMo (Global Mobile Information Systems) initiatives support utilization of the Web, videoconferencing, electronic whiteboards, voice communications, e-mail, and MBone capabilities by users on the move. These initiatives also support development of application program interfaces and innovative network techniques that enable fast implementation of wireless services in dynamic and hostile environments.

9.32.1 CALIFORNIA

9.32.1.1 STANFORD UNIVERSITY

9.32.1.1.1 MosquitoNet Project
The MosquitoNet Project supports development of a wireless communications infrastructure that includes multiple low-power radio transceivers, GPS receivers, and amplifiers that enable reliable operations in a distributed mobile computing network. The MosquitoNet Project also identifies methods for enabling continuous virtual connections between mobile computers and the Web that enables network users to switch seamlessly between different types of wireline and wireless network services.

Stanford University investigates approaches for accommodating secure, reliable, and always-available defense mobile communications requirements. Procedures for enabling development of a reconfigurable multimode multiband radio and architecture that adapts to changing requirements and environmental conditions are evaluated.

9.32.2 New Jersey

9.32.2.1 Rutgers University Wireless Information Network Laboratory (WinLAB)

Sponsored by the Computer Science Department at Rutgers University, the Wireless Information Network Laboratory (WinLAB) supports development of next-generation communicators that facilitate dependable wireless access to voice and non-voice services. Each WinLAB communicator features an integrated platform to support applications and services that previously required the use of mobile computing devices, pagers, or portable phones.

9.33 EUROPEAN COMMISSION ADVANCED (EC-ACTS) WATM (WIRELESS ATM) PROJECTS

Research institutions worldwide are investigating problems associated with wireless functionality and benchmarking mobile computing performance in simulations, experimental prototypes, and actual situations. Approaches facilitating data consistency, indoor signal quality prediction, traffic modeling, security, and provision of capabilities supporting users on the move while others remain in a fixed location are in development. EC-ACTS initiatives in the wireless domain that focused on provisioning WATM services are highlighted in this section.

9.33.1 Magic Wireless ATM Network Demonstrator (WAND)

The Magic WAND project demonstrated the technical feasibility of providing real-time wireless multimedia services to users on the move via a WATM infrastructure that supports transmission at rates reaching 20 Gbps in the 5 GHz spectral block. Test efforts focused on the use of mobile terminals for supporting information access, x-ray transmission, and teleconferencing at rates and service levels that were equivalent to those of the wireline ATM world. Approaches for counteracting signal fading on the wireless network component were also examined.

9.33.2 Wireless Broadband CPN/LAN for Professional and Residential Multimedia Applications (MEDIAN)

The MEDIAN project supported deployment of a high-speed, wireless, customer premise, local area network (WCPN/WLAN) configuration for enabling transparent access to fixed ATM multimedia applications. MEDIAN employed an ATM interface and utilized the 60 GHz frequency band for enabling transmission at rates reaching 155.52 Mbps (OC-3). MEDIAN contributed to the refinement of specifications associated with DECT, GSM, UMTS, and HiperLAN.

9.34 PLANNING GUIDELINES

Wireless network innovations are underway to improve the usability and performance of mobile computing services and optimize connectivity to Web applications. Broadband fixed wireless access (FWA) LANs, MANs, and WANs provide individuals with a rich set of communications capabilities in an integrated and easy to use format as they move from place to place and enable connectivity to the Internet, intranets, and extranets. Ideally, these configurations feature integrated hardware and software systems, encryption mechanisms, support for multimedia applications, and fixed costs.

Design decisions for building wireless networks in the educational environment are challenging. Modems range from built-in and stand-alone units to removable PC Cards. Wireless terminals include portable devices such as notebooks, laptops, palmtops, wearable computers, and PDAs. Effective wireless network deployment depends on such dynamics as the user interface, network log-on rate, response time, throughput, reliable data delivery, and features and functions enabled. Technical support is critical in ensuring trouble-free wireless implementation, management, and maintenance.

In evaluating the suitability of wireless communications for the learning environment, the need for wireless technology and its priority in relation to other computer communications requirements should be addressed initially. The system design must take into account the selection of appropriate portable data terminals or mobile devices, radio modems, and gateway software connectivity requirements. Transmission methods between the host computer and mobile or portable data terminals, and safeguards ensuring data integrity and network reliability and availability, must also be put into place. A site survey ensures that signal frequencies can support wireless communications as mobile workers move around the wireless network venue. This survey demonstrates the potential impact of a WLAN or WWAN on overall network applications and establishes with greater accuracy the likely success of wireless network deployment. In addition, the site survey identifies wireless costs, requirements for teletraffic support, wireless communications constraints, and operational considerations.

Ideally, effective broadband FWA LANs or wireless wider area networks provide thorough coverage in a specified area; handle extensive traffic; foster simultaneous voice and data transmission at high-rates, and enable continuous, clear, and reliable signals. This survey provides a comprehensive guide to network design and installation. The physical area, the placement of inside and/or outside wireless access points, antenna performance requirements, and spatial characteristics affecting wireless coverage are defined. In addition, potential sources of interference from airplanes, airports, and heavy trucks are identified. A common platform that enables interconnectivity of wireless and wireline institutional networks is defined.

At this point, a decision can be made concerning the viability of wireless use in terms of institutional goals and objectives. Applications selected for wireless implementation should accommodate user requirements, justify the expense of investment in establishing the wireless infrastructure, and facilitate collaboration and enhance productivity among faculty, administrators, students, and staff in specified user communities.

Guidelines and procedures for connectivity, integration with the fixed wireline infrastructure, acceptable performance levels, user training, information security, and monitoring and managing the wireless configuration under various conditions should be clearly defined to support seamless migration to a wireless networking infrastructure. A timetable for wireless networking implementation should also be established.

Wireless solutions can be developed in-house or outsourced to communications carriers and equipment providers. Communications carriers active in this domain include GTE, WorldCom, and McCaw Cellular Communications. Additional communications carriers supporting wireless network deployment include AT&T, Motorola, American Personal Communications, Western Wireless Corporation, and Verizon. Wireless equipment providers in this fast developing field include Digital Ocean, IBM, Solectek, Motorola, Proxim, and Nokia.

9.35 SUMMARY

Wireless networks vary in scope and size from ad hoc and in-room infrared installations to metropolitan cellular networks and global satellite configurations. Recent advances in wireless technologies have contributed to making the utilization of wireless networks in the home, school, and university environments economically viable and technically feasible. Approaches for enabling wireless network to support Quality of Service (QoS) requirements for provisioning effective multimedia applications in multi-user multiservice wireless networks are in development.

The rapid implementation of wireless network and associated portable computing and communications devices enable users on the move to access voice, video, and data resources at anytime and from anywhere. Wireless networks are distinguished by their support of flexible and instantaneous access to intranets, extranets, and the Internet, regardless of the user's location or mobility.

In this chapter, trends and developments in wireless network technologies and distinctive characteristics of wireless network configurations are examined. Operational considerations associated with wireless networking implementation and standards activities in the wireless networking domain are explored. Guidelines for planning a wireless network implementation are reviewed. Emergent technologies supporting wireless connectivity in the home and in the workplace are described.

Wireless initiatives employ innovative techniques for provisioning acceptable performance levels and cost-effective operations. Lessons learned from these projects contribute to the development of wireless terminals, interface techniques, and standards for supporting wireless computing in heterogeneous real-world environments. Procedures for integrating multiple wireless networks that can interoperate with a wide array of wireless information appliances are under investigation as well.

Building a wireless network that supports reliable performance, links to multiple systems, and integration of current and next-generation technologies involves careful planning and execution. Mobile users are nomadic and the communications infrastructure for accommodating real-time traffic requirements is unstable. Applications supported remotely by a wireless network infrastructure, such as multimedia e-mail and seamless connections to global networks and imaging databases, require the same computing power and communications support now available on desktop PCs and the local and wide area landline networks to which they are joined.

Capabilities of a wireless network should be carefully evaluated in pilot tests prior to full-scale implementation. Variables affecting the structure and format of a wireless network include costs, legal regulations, transmission capabilities, and configuration requirements. Factors affecting wireless network performance include mobility, immunity to interference, power consumption, performance consistency, security, and network capacity, scalability, and reliability. Despite significant technical advances and recent innovations, it is important to note that wireless information appliances are still limited in effectively providing access to Web resources and e-mail as a consequence of such constraints as battery life and processing power.

Effective wireless network implementations overcome technical, architectural, and operational challenges as well as accommodate institutional mission, goals, and constraints; user objectives; and budgetary requirements. Wireless networks that support multimedia services enable individuals to utilize mobile terminals for seamlessly accessing multiple databases at diverse locations on-demand. This capability also supports navigation through resources in digital libraries and participation in joint projects on curricular design and development with administrators using stationary terminals in fixed locations. Widespread user acceptance and effective deployment of wireless networks in the learning environment ultimately depends on the ability of the wireless configurations to provide the same level of performance and quality of service (QoS) guarantees as their wireline counterparts.

Challenges associated with wireless network implementation include providing on-demand bandwidth allocation to support a broad spectrum of applications. Mobility management issues associated with wireless ATM, mobile IP, and next-generation wireless networks must also be addressed.

Wireless network technologies work in concert with diverse narrowband and broadband wireline solutions. Third-generation cellular communicators enable increasingly sophisticated music, voice, video, imaging, and data services with QoS guarantees and provision access to networking applications in densely populated urban communities as well as in rural and isolated locations. Third-generation network solutions such as Bluetooth and UMTS are expected to deliver mobile IP services at rates ranging from 144 and 384 Kbps for users on the move to 2 Mbps for users at fixed locations.

Research in wireless communications is progressing in many directions. Next-generation wireless networks are designed to provide seamless access to local and wider area applications via compact wireless appliances and enable dependable and reliable voice, video, and data transmissions at rates reaching 155.52 Mbps (OC-3). Accelerating demand for ubiquitous access to communications resources and wireline and/or wireless networks, regardless of user location or mobility, drives the development and implementation of scalable and extendible next-generation wireless network solutions.

9.36   SELECTED WEB SITES

Alcatel. Broadband Wireless Access Solutions
    Available: http://www.cid.alcatel.com/solutions/
Bluetooth. The Official Bluetooth Website.
    Available: http://www.bluetooth.com/

Carnegie Mellon University School of Computer Science. Monarch Project. Research Papers. Last modified on April 25, 2000.
Available: http://www.monarch.cs.cmu.edu/

Carnegie Mellon University Wearable Group. Spot. Last modified on September 5, 2001.
Available: http://www.wearablegroup.org/hardware/spot/

Federal Communications Commission (FCC). FCC National Regulatory Research Institute. Community Broadband Deployment Database.
Available: http://www.nrri.ohio-state.edu/programs/telcom/broadbandquery.php

European Telecommunications Standards Institute (ETSI). HiperLAN2 Technical Overview.
Available: http://www.etsi.org/technicalactiv/hiperlan/hiperlan2tech.htm

Infrared Data Association (IrDA). The Association for Defining Infrared Standards. Welcome to IrDA. The IrDA Home Page.
Available: http://www.irda.org/

Personal Communications and Industry Association (PCIA). About PCIA.
Available: http://www.pcia.com/industryconnect/index.htm

Rutgers University. WinLAB. Research Areas.
Available: http://www.winlab.rutgers.edu/pub/Index.html

Wireless Ethernet Compatibility Alliance (WECA). Articles and News.
Available: http://www.wi-fi.com/articles.asp

Wireless LAN Association (WLANA). The Wireless Networking Industry's Information Source. Home Page.
Available: http://www.wlana.org/index.html

U.S. Department of Defense Advanced Research Projects Agency (DARPA) Information Technology Office (ITO). Global Mobile Information Systems (GloMo). Robert Ruth, Program Manager.
Available:
http://www.darpa.mil/ito/research/pdf_files/glomo_approved.pdf

U.S. Department of Defense Advanced Research Projects Agency (DARPA) Information Technology Office (ITO). NGI (Next-Generation Internet) Project List.
Available: http://www.darpa.mil/ito/research/ngi/projlist.html

U.S. Department of Defense Advanced Research Projects Agency(DARPA) Information Technology Office (ITO). Ubiquitous Computing. Goals. Dr. Jean Scholtz, Program Manager.
Available: http://www.darpa.mil/ito/research/uc/goals.html

10 Satellite Networks

10.1 INTRODUCTION

This chapter presents an examination of satellite network capabilities, technical attributes, and implementations. Satellite systems employ radio waves in the super-high and extremely high RF (radio frequency) bands of the electromagnetic spectrum for enabling dependable transport of voice, video, data, and still images. Satellite configurations utilize state-of-the-art technologies for facilitating high-speed access to bandwidth-intensive resources and time-critical data. Rapidly evolving satellite networks are further distinguished by their provision of on-demand seamless mobile communication services at anytime and in every place and delivery of broadband multimedia applications to subscribers at rural locations.

10.2 PURPOSE

The growing popularity of wireless communications and mobile computing inten-sifies interest in satellite technologies, services, and networks. Satellite communi-cation solutions support a broad array of tele-applications that include high-definition television (HDTV) broadcasts, video-on-demand (VOD), videoconferencing, tele-instruction, and Web browsing. In this chapter, satellite technical features, functions, regulations, standards, and operational considerations are examined. Challenges and prospects associated with utilization of satellite configurations for enabling routine access to present-day and next-generation networks and narrowband and broadband services are reviewed.

Satellites are either natural or artificial. A natural satellite is a celestial body that revolves around a large-size planet such as the Earth. By contrast, an artificial satellite is an object placed into orbit around the Earth. Artificial satellites enable services that include weather forecasting, forest fire detection, scientific research, navigation, E-commerce (electronic commerce) transactions, photographic surveillance, and detection of nuclear explosions. In this chapter, the capabilities of artificial satellite constellations in enabling tele-education and telemedicine services are explored.

10.3 FOUNDATIONS

First proposed in 1945 by the British science fiction writer Arthur C. Clarke, satellite communications became a reality with the launching of the Russian satellite Sputnik I in 1957 and the American satellite Explorer I by the National Aeronautics and Space Administration (NASA) in 1958.

The first National Oceanic and Atmospheric Administration (NOAA) series of Polar-Orbiting Environmental Satellites (POES) commenced in 1960 with the launch of the Television Infrared Operational Satellite (TIROS). The POES fleet provides global meteorological data for research and geographic and atmospheric applications.

In the 1970s, NASA launched the first geostationary (GEO) Applications Technology Satellite (ATS-1) with the capabilities to record weather systems in motion and SEASAT. SEASAT was the first earth orbiting satellite designed specifically to monitor oceanic activities remotely. ATS-1 and SEASAT also provided critical data on adverse weather conditions to NOAA scientists for enabling accurate weather forecasts.

The first Synchronous Meteorological Satellite (SMS-1) was placed into orbit during the 1970s. This satellite served as the prototype for the Geostationary Operational Environmental Satellite (GOES) system. The GOES system monitors the Earth's atmosphere and surface in the Western Hemisphere. GOES imagery enables scientists to predict the amount of rainfall during hurricanes.

10.3.1 THE U.S. DEPARTMENT OF DEFENSE GLOBAL POSITIONING SYSTEM (GPS)

10.3.1.1 GPS Features and Functions

A U.S. Department of Defense (DoD) initiative, the Global Positioning System (GPS) uses a fleet of 24 satellites that were launched between 1978 and 1994 for transmitting extremely precise positioning information to GPS receivers. GPS applications were developed initially by the U.S. Department of Defense to accommodate military needs and support military operations. A satellite-based radio positioning system, GPS enables applications that include maritime navigation and messaging services.

Satellite configurations consist of space, ground, and control segments. In terms of GPS, the space segment consists of a fleet of 24 satellites placed into orbit between 1978 and 1994. These satellites orbit the Earth every 12 hours. The ground segment is comprised of receivers that are handheld or mounted on aircraft, ships, trucks, tanks, and cars. Generally, GPS handheld receivers are about the same size as cellular phones. The control segment includes ground stations that monitor satellite operations and functions.

GPS supports information transport between the 1.559 and 1.610 GHz (Gigahertz) frequency bands. GPS also enables signal transmission in the 1.227 GHz spectral block for non-safety critical applications and in the 1.176 GHz spectral block for life-threatening emergencies. Additional systems that supplement GPS services are expected to operate in spectrum between the 1.559 and 1.610 GHz frequency bands.

GPS receivers process satellite signals to compute velocity, position, and time and generate accurate location and navigational data. The Global Positioning System supports Standard Positioning Service (SPS) and Precise Positioning Service (PPS). GPS services are used by surveyors, mapmakers, and motorists to determine locations, and by police officers and firefighters for vehicle dispatch. In addition, GPS services enable railway engineers to monitor train movements and naval officers on ships at sea to utilize GPS data for navigation.

GPS data resulted in predictions of Hurricane Andrew in 1992, the floods in the Midwest in 1993, and the North Ridge Earthquake in California in 1994. During Operation Desert Storm, GPS data contributed to the execution of quick and successful air and ground attacks by American troops. The Southern California Integrated GPS Network (SCIGN) uses GPS technology for forecasting earthquakes.

In addition, GPS supports diverse initiatives that demonstrate the application of space technologies in public health, travel, engineering, and public transportation. During the construction of the tunnel under the English Channel for linking Dover, England and Calais, France, construction crews used GPS communicators to check positions. The Global Positioning System also tracks oil spills, volcanic eruptions, droughts, and nuclear accidents to mitigate the effect of environmental disaster. Data derived from GPS satellites are maintained in the Global Disaster Information Network (GDIN).

10.3.2 NASA Advanced Communications Technology Satellite (ACTS) Experiment Program

NASA launched the Advanced Communications Technology Satellite (ACTS) aboard the space shuttle Discovery on September 12, 1993. Simultaneously with the ACTS launch, NASA initiated the NASA ACTS Experiment Program. The NASA ACTS Experiment Program supported innovative satellite applications in the educational domain and contributed to the development of the global information infrastructure. In addition, capabilities of the Ka-band in enabling satellite-based voice, video, and data transmission were verified.

With the initiation of the ACTS Experiment Program, NASA also began development of the National Research and Education Network (NASA-NREN). During the course of the NASA ACTS Experiment Program, organizations representing industry, academia, and government conducted approximately 100 research projects and education initiatives. On May 31, 2000, the NASA ACTS Experiment Program came to a close. At that time, NASA predicted that the Advanced Communications Technology Satellite (ACTS) still had from two to four years of useful life.

10.3.3 Ohio Consortium for Advanced Communications Technology (OCACT)

Following the close of the NASA ACTS Experiment Program, the Ohio Consortium for Advanced Communications Technology (OCACT) under the leadership of Ohio University announced plans to take over continued operation of the NASA Advanced Technology Satellite. Ohio University indicated that this satellite would serve as a laboratory for enabling next-generation research projects and academic initiatives in space science, satellite communications, satellite operations, and distance learning. In addition to Ohio University, OCACT participants include the NASA Glenn Research Center, Texas A&M University, the Ohio Board of Regents, the Naval

Postgraduate School, the Air Force Research Laboratory (Lab), and the Space and Naval Warfare Systems Center.

10.4   SATELLITE TECHNICAL FUNDAMENTALS

Satellites are Earth-orbiting spacecraft that employ microwave technology in the super high and extremely high frequencies for enabling wide area interactive communications and transmission of broadcast services to locations that cannot readily be served by terrestrial facilities. As noted in the GPS description, satellite configurations include three segments: space, ground, and control. The number of satellites in the space segment depends on service requirements and budget allocations.

Control segments include low-cost Very Small Aperture Terminals (VSATs) that are linked to end-user equipment, as well as technically complex facilities costing millions of dollars that track the accuracy of satellite operations. As with control segments, ground components support varying functions. For example, receive-only Earth stations serve as distribution sites for broadcast television (TV) programs. Earth stations that support dual-mode operations for enabling signal reception and signal transmission are typically situated at key research centers, university campuses, and corporate headquarters. Earth stations can also support command and control operations.

The pathway for transporting radiowaves from the satellite to the designated Earth station or reception site is called the downlink. The pathway for transmitting radiowaves or signals from the Earth station to the satellite is called the uplink. Transponders are combinations of receivers and transmitters carried by satellites for enabling signal reception on the downlink path, signal amplification, and signal retransmission on the uplink path.

The crispness and robustness of satellite signals or radiowaves depends on the condition of the uplink and downlink. To reduce the effects of signal attenuation, the downlink typically operates at lower frequency levels than the uplink.

The satellite, the transmitting Earth station, and the receiving Earth station form a basic three-node network with the satellite situated at the apex of the configuration. Because satellite signals cannot pass through the Earth's surface, satellite transmissions travel in straight line-of-sight (LOS) paths from one point to another.

Satellite systems are used extensively for enabling long-haul and transoceanic information distribution. Landline links such as coaxial cable, optical fiber, and twisted copper pair are used in conjunction with satellite networks for connecting residential, educational, medical, and corporate networks with Earth or ground stations.

10.4.1   ANTENNAS

An antenna is an aerial communications device that receives and radiates radiowaves through free space to remote locations. An antenna enables a satellite to communicate with a ground or Earth station that can also provide command and control services. Conventional antennas for enabling wide area television broadcasts are omni-directional. By contrast, higher gain directional antennas employ spot beam technology to support full-duplex point-to-point transmissions. With spot-beam technology, an antenna system divides a single footprint or coverage area into smaller coverage areas or sub-footprints for supporting lower power transmissions from and to multiple subscriber sites.

Antenna shape and size depend on the wavelength of radiowaves that are transmitted or received. As an example, dish antennas work in concert with satellite signals that are transported in extremely high frequencies and feature very short wavelengths.

The Federal Communications Commission (FCC) adopted rules governing antenna placement for receiving Direct Broadcast Satellite (DBS), Multichannel Multipoint Distribution System (MMDS), Local Multipoint Distribution System (LMDS), and Television Broadcast Station (TVBS) transmissions. These rules are included in the Telecommunications Act of 1996. Fire codes prevent antenna installations on fire escapes, and local codes require antenna placement at specified distances from powerlines.

10.4.2 SATELLITE PROTOCOLS

As with cellular technologies and communications services, protocols enabling satellite transmissions include Time-Division Multiple Access (TDMA) and Code-Division Multiple Access (CDMA). TDMA divides each satellite channel into timeslots in order to increase the quantity of data that are transported. Using TDMA, satellite networks support access to Web applications, Public Switched Telephone Network (PSTN) services, wireline network links, and MPEG-2 (Moving Pictures Experts Group-2) videoconferences. For enabling mobile satellite communications services, CDMA assigns a unique sequence code to each call so that multiple calls can then be overlaid on each other to optimize utilization of the available bandwidth.

10.5 SATELLITE FREQUENCY BANDS

Satellite communications generally take place in that part of the radio spectrum that ranges between 1 and 30 GHz. In the United States, frequency bands typically employed for satellite communications include the C-band, Ku-band, and Ka-band. The letters designating the names of the frequency bands are randomly selected.

10.5.1 C-BAND TRANSMISSIONS

Operating at the lower end of the spectrum, the C-band supports uplink operations between the 5.9 and 6.4 GHz frequency bands. For the downlink, frequencies ranging from 3.7 to 4.2 GHz are employed. C-band transmissions are fairly immune to atmospheric disturbances.

Fixed satellite services supporting VSAT operations typically operate in the C-band and the Ku-band. In the United States, the C-band operates within the same portion of the frequency spectrum as terrestrial microwave. As a consequence, C-band transmissions are susceptible to radio frequency interference (RFI) from microwave signals.

10.5.2 KU-BAND TRANSMISSIONS

The higher power Ku-band supports communications services in densely populated urban areas. Ku-band uplinks support operations between the 14 and 14.5 GHz. frequencies and between the 17.3 and 17.8 GHz frequencies. Ku-band downlinks operate between the 11.7 and 12.7 GHz frequencies.

10.5.3 KA-BAND TRANSMISSIONS

With the launching of the NASA Advanced Communications Technology Satellite (ACTS) spacecraft in 1993, the Ka-band portion of the RF spectrum in the 20 to 30 GHz frequency block became available. A GEO (Geosynchronous Earth-Orbit or Geostationary Earth-Orbit) spacecraft, ACTS enabled uplink operations in the 30 GHz frequency block and downlink operations in the 20 GHz frequency block. Currently, GEO and LEO (Low-Earth Orbit) constellations utilize Ka-band and Ku-band frequencies for enabling high-speed, bandwidth-intensive Web services.

10.5.4 L-BAND AND S-BAND TRANSMISSIONS

L-band operations employ the spectral block between the 1.5 and 2.7 GHz frequencies. S-band operations utilize spectrum between the 2.7 and 3.5 GHz frequencies. Satellite constellations such as INMARSAT (International Maritime Satellite) and Globalstar employ the L-band and the S-band for enabling paging services, cellular telephony, and data transmissions.

10.6 GSM, 3GSM, AND S-UMTS

The satellite components of next-generation mobile communications networks interoperate with terrestrial technologies such as IP (Internet Protocol), Frame Relay, ISDN (Integrated Services Digital Network), and ATM (Asynchronous Transfer Mode). In addition, satellite components also work in concert with second-generation GSM (Global System for Mobile Communications) networks, and third-generation 3GSM (Third-Generation GSM), S-UMTS (Satellite-Universal Mobile Telecommunications Service), and GMPCS (Global Mobile Personal Communications by Satellite) configurations.

10.6.1 GLOBAL SYSTEM FOR MOBILE COMMUNICATIONS (GSM) AND 3GSM (THIRD-GENERATION GSM)

The success of GSM mobile radio systems has contributed to research initiatives that explore methods for augmenting the capacity of GSM and 3GSM networks. For example, Ericsson developed ACeS (Asia Cellular Satellite System), a combined satellite and cellular telephony system that provisions digital services to computer and mobile phone users throughout the Asia-Pacific service area bounded by Indonesia, Japan, Pakistan, and New Guinea. By merging satellite communications capabilities with GSM and 3GSM technologies, ACeS enables subscribers using Ericsson dual-mode portable communicators to switch between local cellular service and satellite service as they roam throughout the service region. ACeS subscribers employ GSM SIMs (Subscriber Identity Modules) or 3GSM SIMs and network codes to access ACeS services when outside the ACeS coverage area.

10.6.2 SATELLITE-UNIVERSAL MOBILE TELECOMMUNICATIONS SERVICE (S-UMTS)

Distinguished by its provision of global coverage, satellite technology is viewed as a key enabler for global operations. S-UMTS supports seamless roaming and transparent handovers between terrestrial and satellite networks. S-UMTS operates in spectrum between the 1.980 and 2.010 GHz frequencies and between the 2.170 and 2.200 GHz frequencies. It is interesting to note that conventional mobile satellite services support operations in two 30 MHz (Megahertz) channels in the 2 GHz frequency bands.

Low-Earth Orbit (LEO), Mid-Earth Orbit (MEO), and Geosynchronous or Geostationary-Earth Orbit (GEO) constellations support S-UMTS services. S-UMTS is designed to work in conjunction with the wireline Public Switched Telephone Network (PSTN) and technologies that include cable modem, DSL (Digital Subscriber Line), ISDN (Integrated Services Digital Network), ATM (Asynchronous Transfer Mode), and Gigabit Ethernet.

A core IMT-2000 (International Mobile Telecommunications-2000) technology, S-UMTS supports interactive videoconferences that are compliant with the MPEG-4 specification and the ITU-T H.323 Recommendation. Advanced S-UMTS services are expected to provision bandwidth on-demand for accommodating multimedia requirements. The S-UMTS (Satellite-UMTS) Forum supports the allocation of a 50 MHz spectral block in 2005 and a 90 MHz spectral block in 2010 for supporting universal broadband S-UMTS applications that are accessible at anytime and from every place.

10.6.3 GLOBAL MOBILE PERSONAL COMMUNICATIONS BY SATELLITE (GMPCS) SPECIFICATIONS AND SERVICES

10.6.3.1 World Telecommunications Policy Forum and ITU

The World Telecommunication Policy Forum (WTPF) develops regulatory guidelines for the deployment of third-generation GMPCS (Global Mobile Personal Communications by Satellite) systems. A core IMT-2000 (International Mobile Telecommunications-2000) technology, GMPCS systems provision global, regional, and transborder personal telephony services, and enable access to narrowband and broadband applications via GEO and LEO satellite constellations.

Building upon the framework established by the WTPF, the ITU-T finalized GMPCS specifications by working with an informal group of GMPCS service providers, operators, and terminal manufacturers. This group approved procedures for GMPCS implementations and developed a Memorandum of Understanding (MoU) for provisioning GMPCS services.

In conformance with this MoU, signatories such as Motorola agree to support the utilization of interoperable and compact GMPCS communicators so that GMPCS subscribers can use GMPCS-compliant devices, regardless of the manufacturer, in each member state in the European Union in which GMPCS services are licensed. GMPCS MoU signatories also develop the framework for GMPCS maritime, aeronautical, and terrestrial services and clarify system architecture for vehicular mobile Earth stations.

10.7 NATIONAL AND INTERNATIONAL STANDARDS ORGANIZATIONS AND ACTIVITIES

Satellite service involves the use of radiowaves in the super high and extremely high frequency bands of the RF spectrum for enabling utilization of voice, video, images, and data services by educational military, government, and corporate entities. National administrations such as the Federal Communications Commission (FCC), the Canadian Department of Communications (DOC), and the European Technical Standards Institute (ETSI) establish procedures for frequency band allocations. The ITU (International Telecommunications Union) coordinates policies for spectral allocations internationally.

10.7.1 INTERNATIONAL TELECOMMUNICATIONS UNION-TELECOMMUNICATIONS SECTOR (ITU-T)

The ITU-T (International Telecommunications Union-Telecommunications Standards Sector) defines specifications for satellite transmission of digital television programs and supports utilization of MPEG-2-compliant applications for enabling uniform services in the ITU-T J.84 Recommendation. In addition, this Recommendation clarifies procedures for channel coding and modulation of voice, video, and audio signals received from satellite systems and distributed by Satellite Master Antenna Television (SMATV) networks.

10.7.2 INTERNATIONAL TELECOMMUNICATIONS UNION-TELECOMMUNICATIONS DEVELOPMENT SECTOR (ITU-D)

The International Telecommunications Union-Telecommunications Development Sector (ITU-D) supports programs that provision affordable access to wireless applications in rural and remote communities, and promotes implementation of satellite communications services in developing countries with a severe shortage of phonelines. ITU-D initiatives are carried out in the African countries of Benin, Tanzania, Mali, and Uganda, as well as in India, Romania, and Vietnam.

10.7.3 DIGITAL AUDIO VISUAL COUNCIL (DAVIC)

Based in Geneva, Switzerland, the ETSI-sponsored Digital AudioVisual Council (DAVIC) developed specifications for the design, implementation, and maintenance of satellite broadband video networks. To facilitate cost-effective transport and distribution of digital television services, DAVIC established agreed-upon methods for equipment interoperability.

DAVIC specifications for supporting applications such as near video on-demand (NVOD), video on-demand (VOD), and television distribution were endorsed by the ISO (International Standards Organization). DAVIC also defined DTV (Digital Television) formats and systems, developed tools for Web compatibility, and promoted utilization of MMDS and LMDS networks and interoperable set-top boxes (STBs). DAVIC operations were carried out between 1994 and 1999.

10.7.4 DIGITAL VIDEO BROADCASTING (DVB) PROJECT

Sponsored by an ETSI Consortium consisting of manufacturers, broadcasters, regulatory bodies, and network operators, the Digital Video Broadcasting (DVB) project supported the development of a global set of standards and specifications for enabling trouble-free and seamless delivery of digital broadcast services and digital television programming. As with the DAVIC initiative, the DVB project was operational between 1994 and 1999.

10.7.5 WORLD RADIOCOMMUNICATIONS CONFERENCE (WRC)

At WRC-2000 (World Radiocommunications Conference-Year 2000), a resolution supporting operations of non-geostationary satellite systems in the Ku-band was endorsed and existing GSO (Global Satellite Orbital) systems were protected. Moreover, the United States obtained a 2 GHz frequency block of spectrum between the 40 and 52 GHz frequencies for implementation of fixed satellite services in the V-band and secured additional spectrum for GPS (Global Positioning System) applications.

10.8 GEOSTATIONARY-EARTH ORBIT OR GEOSYNCHRONOUS-EARTH ORBIT (GEO), MID-EARTH ORBIT (MEO), AND LOW-EARTH ORBIT (LEO) SATELLITE SYSTEMS

Depending on the orbits employed, satellite systems are broadly categorized as Geostationary-Earth Orbit or Geosynchronous-Earth Orbit (GEO), Mid-Earth Orbit (MEO), or Low-Earth Orbit (LEO) systems. Positions of communications satellite fleets in relationship to their service areas affect transmission quality, usage fees, and the structure and complexity of satellite network configurations.

10.8.1 GEO SATELLITE SYSTEMS

Geostationary-Earth Orbit or Geosynchronous-Earth Orbit (GEO) satellites orbit the earth at an altitude of 35,800 kilometers or 22,282 miles above the Earth's surface. GEO satellites appear stationary because they revolve at the same rotational speed as the Earth. Digital satellite systems employ GEO configurations for multisite broadcast transmissions. GEO satellites provide communications services such as long distance and international telephony, HDTV (High-Definition Television) programming, and broadband applications.

In 2000, the ITU established power limits for GEO, MEO, and LEO configurations and procedures for sharing spectrum between the 10 and 18 GHz frequency bands in the Ku-band. These controls enable the coexistence of GEO, MEO, and LEO fleets and facilitates dependable delivery of advanced multimedia services to subscribers. Regional coverage provided by GEO satellites is limited. As a consequence, multiregional mesh satellite configurations are established by using inter-satellite links that enable interoperable transregional satellite communications and services.

10.8.1.1   Cyberstar

Cyberstar provisions access to IP applications and multicast communications service via mixed-mode satellite and wireline optical fiber networks. The terrestrial component employs technologies such as Frame Relay and ATM. The space component consists of three satellites. Cyberstar supports voice, video, and data transmissions at rates reaching 30 Mbps in the Ku-band and the Ka-band. In addition, the space component employs MMDS (Multichannel Multipoint Distribution System) spectrum for delivery of IP multicasts, news feeds, and distance education programs. Fast file transfer, teleseminars, teleworkshops, virtual classrooms, and streaming media services to the desktop are also supported. A Loral Space and Communications initiative, Cyberstar provisions network services in Slovakia, El Salvador, Gibraltar, Latvia, Vietnam, Uzbekistan, Indonesia, Belarus, and Estonia.

10.8.1.2   EuroSkyWay (ESW)

The EuroSkyWay (ESW) constellation employs three GEO satellites in transregional configurations for supporting voice, video, and data delivery to low-cost portable, mobile, and fixed communicators. The ESW platform fosters access to multimedia applications throughout the European Union and the Mediterranean basin. In addition, ESW supports development of a Global Multimedia Satellite Network in partnership with agencies that include NASA. Approaches for utilizing the ESW satellite component in conjunction with wireline ATM and IP services are also examined.

10.8.1.3   Spaceway

Developed by Hughes Network Systems, the Spaceway broadband GEO satellite network is slated to become operational in 2002. This network will enable voice, video, and data transmission in Ka-band frequencies at rates reaching 16 Mbps on the downlink path and 400 Mbps on the uplink path.

10.8.2   Mid-Earth Orbit (MEO) Satellite Systems

Mid-Earth Orbit (MEO) satellites orbit the earth at altitudes ranging from 6,000 to 12,000 miles. INMARSAT (International Maritime Satellite) and the Global Positioning System (GPS) are examples of MEO satellite fleets.

10.8.2.1   Constellation

Currently in development, the Constellation satellite initiative will support voice, data, and fax applications and CDMA (Code-Division Multiple Access) technology for enabling subscriber access to broadband applications. The Constellation MEO configuration consists of 12 satellites in orbit around the equator at an altitude of 2,000 kilometers or 1,240 miles. Each satellite will support 24 antenna beams.

On the downlink path, satellite transmissions from the Constellation fleet will be sent to mobile terminals such as handheld and compact communicators and vehicle-mounted devices for trucks and cars and fixed-site terminals at SOHO (Small Office/Home Office) sites and public telephone booths. The Constellation fleet will enable operations between the 24.83 and 25.00 GHz frequency bands.

10.8.3   Low-Earth Orbit (LEO) Satellite Systems

Low-Earth Orbit (LEO) satellites maintain orbits at altitudes ranging from 500 to 900 kilometers above the surface of the Earth. LEO configurations employ multiple satellites to provision wide area communications coverage. LEO constellations consist of Big LEO fleets that mainly support global telephone services and Little LEO fleets that primarily provide mobile data services.

10.8.3.1   Globalstar

Developed by Loral Space and Communications, Globalstar is a Big LEO satellite initiative that links present-day GSM networks in 33 countries via a constellation of 48 satellites for provisioning GSM cellular services to subscribers with compact Globalstar phones, regardless of location and mobility. Ground segments include control centers and gateways. Global Touch Communications and Mobile Telecommunications Africa are distributors of mobile satellite phones that work in concert with Globalstar.

Globalstar enables global roaming or the ability of global travelers to move seamlessly around a cluster of access points in an area of coverage without losing telephony services. Globalstar subscribers use the less-expensive GSM-mode when they are in a GSM coverage area and then automatically transfer to the satellite-mode when they roam outside the GSM coverage area without dropping the telephone connection. Moreover, the Globalstar constellation also facilitates links to narrowband and broadband applications; supports fax, paging, and GPS services; and accommodates communications requirements of individuals in underserved communities without fixed wireline infrastructures that use Globalstar fixed-site phones primarily for basic telephone calls.

10.8.3.2   Teledesic

Slated to become operational in 2003, the Teledesic initiative, also known as the Internet-in the-Sky, will support broadband operations via a Little LEO constellation that is expected to consist of approximately 288 satellites. The Teledesic configuration will enable bi-directional high-speed, location-insensitive connections to the Web and Virtual Private Networks (VPNs) in real-time and will provide on-demand gigalink services for multimedia transport to accommodate communications requirements of individuals anywhere on the planet at affordable prices. Teledesic portable communicators are also expected to provide links to GPS services.

Teledesic will use Ka-band frequency bands for communications services. Spectrum between the 18.8 GHz and 19.3 GHz frequency bands will be employed for downlink transmissions, and rates at 64 Mbps will be supported. The spectral block between the 28.6 and 29.1 GHz frequency bands will be used for uplink transmissions rates reaching 2 Mbps. In 1998, a demonstration Teledesic satellite called T-1 was launched into orbit from the Vandenberg Air Force Base.

10.9   VERY SMALL APERTURE TERMINAL (VSAT) SOLUTIONS

Very Small Aperture Terminals (VSATs) are small-sized fixed Earth stations or terminals that handle high throughput rates and enable one-way or receive-only and bi-directional or two-way interactive communications. The acronym "VSAT" refers to the size of the satellite terminal or earth station. The term "aperture" refers to the portion of the antenna exposed to satellite signaling. A VSAT network includes a master Earth station that serves as the communications hub for the rest of the configuration and a variable number of remote VSAT ground stations. Companies that hold FCC-designated orbital locations and Ka-band licenses for VSAT transmissions include PanAmSat, Loral, Echo Star, Hughes Network Systems, and Motorola.

10.9.1   VSAT OPERATIONS

Generally VSAT services employ Single Channel per Carrier (SCPC) and Time-Division Multiple Access (TDMA) protocols. A point-to-point networking technology, SCPC enables multiple VSAT-equipped sites to directly communicate via dedicated satellite channels with a hub or Earth station maintained by the communications provider. Designed for SOHO (Small Office/Home Office) venues, VSAT service enables rates reaching 2 Mbps on the downlink and 19.2 Kbps on the uplink.

10.9.2   VSAT CAPABILITIES

Scalable and affordable VSAT installations support broadcast and cable television distribution in hotels, apartment complexes, public school districts, colleges, and universities. VSAT configurations also facilitate LAN interconnectivity, Web browsing, and dependable access to multimedia applications. In the corporate sector, private networks supporting voice, video, and data are usually based on VSAT solutions.

VSAT videoconferencing solutions foster receive-only video transmission on the downlink. Portable communicators and PCs enable operations on the uplink. VSATs extend the reach of corporate and educational networks to distant locations and support reliable connections that are more dependable than traditional leased line service.

Moreover, VSAT technology provisions links to locations where the conventional telecommunications infrastructure is nonexistent or obtaining land telephony service is problematic. Generally, prices for VSAT and POTS services are equivalent. Advanced VSAT configurations are expected to support transmission reaching 155.52 Mbps (OC-3).

10.9.3   Direct Broadcast Satellite (DBS) Service

VSAT technology is a popular enabler of Direct Broadcast Satellite (DBS) service, also known as Direct Satellite Service (DSS). With DBS service, television programs are transmitted to the satellite and then rebroadcast directly to individual viewers at a variety of venues, including homes, offices, K–12 schools, community colleges, universities, and corporations.

In addition to providing cable-like television programming directly to subscribers, VSAT implementations also foster distribution of multimedia applications and enable fast connectivity to the Web, intranets, and extranets. A typical VSAT reception system at the subscriber premise includes a satellite dish mounted outdoors that measures between 18 and 21 inches in diameter. This dish is linked via cabling to a digital decoder. The digital decoder is usually in the form of a set-top box (STB) for the television set or a PCMCIA (PC) card for the computer.

The Federal Communications Commission (FCC) allocates eight orbital positions at the equator for DBS service. At each orbital position, the FCC permits a maximum of 32 broadcast frequency bands.

DBS downlinks support transmissions in spectrum between the 11.7 and 12.2 GHz frequencies. DBS uplinks transport transmissions between the 14 and 14.5 GHz frequencies. DBS transmissions are affected by the weather. Thunderstorms and snow build-up interrupt signal reception. System outages also occur during thunderstorms. Trees and buildings interfere with direct line-of-sight transmission requirements and also block signal reception. (See Figure 10.1.)

10.9.4   Global VSAT Forum

The Global VSAT Forum encourages development of uniform VSAT regulatory standards, elimination of barriers to transborder broadcasts, and a framework for streamlining the VSAT Earth station approval procedures. In addition, this Forum supports approaches for eliminating VSAT implementation barriers, including costly licensing fees and custom duties.

The VSAT Forum also participates in the Global Mobile Personal Communications by Satellite (GMPCS) initiative and works in concert with ETSI in developing specifications for provisioning uniform satellite broadband interactive multimedia services in the Ku-band and the Ka-band. Participants in the Global VSAT Forum include satellite system integrators, equipment manufacturers, and communications operators such as Belgacom, British Telecom, EUTELSAT, Gilat Satellite Networks, and Hughes Network Systems. In addition, IBM Global Network, INTELSAT, WorldCom, PanAmSat, Singapore Telecom, NEC, and NORSAT participate in Global VSAT Forum activities.

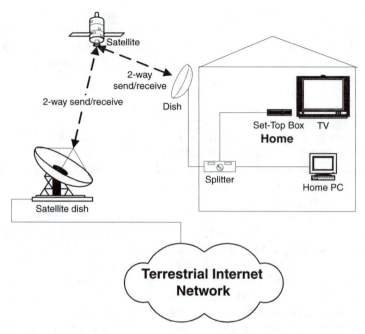

FIGURE 10.1 A VSAT (Very Small Aperture Terminal) Network.

10.9.5 EUROPEAN TELECOMMUNICATIONS STANDARDS INSTITUTE (ETSI)

The ETSI develops specifications in the VSAT domain for enabling seamless reception and transmission of broadband interactive multimedia services. Moreover, the ETSI clarifies operations of Satellite Interactive Terminals (SITs) that work in concert with GEO satellite configurations that support downlink transmissions between the 11 and 12 GHz frequency bands and uplink transmissions between the 29 and 30 GHz frequency bands. In addition, ETSI defines capabilities of SUTs (Satellite User Terminals). SUTs work in conjunction with GEO satellite configurations that support downlink transmissions between the 19.7 and 20.2 GHz frequency bands and uplink transmissions between the 29.5 and 30 GHz frequency bands.

Moreover, ETSI defines capabilities of SUTs (Satellite User Terminals). SUTs work in conjunction with GEO satellite configurations in enabling reception of downlink transmissions in the FSS spectrum between the 19.7 and the 20.2 GHz frequency bands and uplink transmissions between in the FSS spectrum between the 29.5 GHz and the 30.00 GHz frequency bands.

SUTs support reception of data signals on the downlink path. SITs enable data and audiovisual signal reception on the downlink path and provision a return channel in support of real-time interactive satellite services on the uplink path.

10.9.6 VSAT INITIATIVES

10.9.6.1 ASTRA Satellite Services

In 2000, EMS Technologies, the Luxembourg operator of ASTRA Satellite Services, demonstrated the capabilities of DVB (Digital Video Broadcasting) RTC (Return

Channel via Satellite) installations in a series of pilot implementations. This DTTH (Direct-to-the-Home) system transports voice, video, and data and radio and television broadcasts directly to the subscriber site at 40 Mbps on the downlink path. On the uplink path, transmissions at rates of 2 Mbps are enabled. The ASTRA Broadband Interactive System provisions high-speed multimedia delivery via satellites in Central and Western Europe.

10.9.6.1.1 Europe Online

Europe Online uses transponder capacity on the ASTRA Satellite System for supporting voice, video, and data transmission, and providing connectivity to the Web and narrowband and broadband public and private networks. Reception sites feature VSAT dishes that are linked directly to digital decoders such as STBs. Europe Online maintains a Point of Presence (PoP) site in Betzdorf that supports information transport on the uplink. Designed for subscribers throughout the European Union, this initiative enables operations that conform to the DVB-IP (Digital Video Broadcast-Internet Protocol) specification.

10.9.6.2 Hughes Network Systems (HNS) DirecDuo Service

Hughes Network Systems (HNS) DirecDuo service features an amalgam of Digital Satellite Service (DSS) and DirecPC technologies that enable transmission rates downstream at 400 Kbps as well as connectivity to more than 100 television channels. DirecDuo service supports satellite television broadcasts, telecourse delivery, and Web TV programming. The reception site features VSAT dishes and digital decoders that support interactive functions such as electronic shopping. Web service providers and communications carriers support information transport on the uplink path.

10.9.6.3 MedwebSAT

MedwebSAT provisions low-cost VSAT services across the United States, Mexico, the Virgin Islands, the European Union, and Southeast Asia to rural clinics and healthcare centers situated in communities without access to a wireline infrastructure. MedwebSAT also supports delivery of continuing medical education courses to medical practitioners and clinical personnel at remote and isolated locations.

10.9.6.4 NORSAT VSAT Services

NORSAT VSAT services enable narrowband and broadband video, voice, and data transmission at rates ranging from 100 Kbps to 52 Mbps. The NORSAT VSAT platform also supports applications that include Web browsing, VoIP (voice-over-IP), VoFR (voice-over-Frame Relay), teletraining programs, telecourses, videoconferences, LAN interconnections, and television broadcasts. In addition, this platform provides Internet, intranet, and extranet connectivity; security and surveillance applications; and electronic commerce (E-commerce) services.

The NORSAT VSAT network configuration includes earth stations that serve as transfer points between satellite and subscriber sites. The earth station features uplink and downlink equipment for signal transmission and reception. Each subscriber site is equipped with a VSAT terminal that consists of an outdoor antenna for signal transmission and reception and indoor equipment for enabling connectivity to an information appliance such as a laptop or PC (Personal Computer).

NORSAT also provisions Ku-band and C-band satellite communications services to drilling rigs, submarine vessels, offshore oil production units, ships in international waters, terrestrial vehicles, supply vessels, and airplanes. The NORSAT satellite tracking system monitors locations of fishing vessels, trains, ships, yachts, containers, and trucks. This service is available in the United Kingdom, Asia, and Africa, as well as in Central, Eastern, Northern, and Western Europe.

10.9.6.5 Telesat Canada

Telesat Canada provides VSAT Direct PC services via the Nimiq, Canada's first broadcast satellite, for residential users that include news clips, live video feeds, and broadcast programming. Communications service providers include the Canadian Broadcasting Corporation, Canadian Satellite Communications, the CanWest-Global Network, and Bell Express Vu. In addition, Telesat Canada also allows North American broadcasters to access the space segment in order to televise special events and news bulletins from mobile units.

10.9.7   VSAT RESEARCH INITIATIVES

Recent research initiatives in the VSAT arena are now examined. As indicated in these cases, the explosive growth of Web services has increased demand for delivery of low-cost telecommunications services and applications via VSAT technology.

10.9.7.1   GloMo Program

Sponsored by DARPA (U.S. Department of Defense Research Agency), the GloMo (Global Mobile Information Systems) Program supports initiatives that contribute to the integration of Direct Broadcast Satellite (DBS) technology into the digital battlefield for enabling military personnel to maintain a consistent tactical picture in the field. In addition, the GloMo Program fosters development of network protocols and management algorithms for enabling fast transmission rates and dependable delivery of voice, video, and data services. Approaches for facilitating instantaneous access to Web resources are also explored.

10.9.7.2   European Space Agency (ESA)

The European Space Agency (ESA) supports an array of services, ranging from spacecraft operations to educational outreach programs that promote the inclusion of space-related topics in K–12 grades. The ESA sponsors the ARTES (Advanced Research in Telecommunications System) Program to facilitate application development in multimedia services, tele-healthcare, and tele-education. The ARTES Program also enables utilization of broadband mobile services that work in conjunction with VSAT networks. To support the ARTES Program, the European Space

Agency (ECA) establishes partnerships with the European Commission, member states in the European Union, government agencies, and the corporate sector.

10.9.7.2.1 ESPRESSO for Schools

The ESPRESSO for Schools project involves the participation of 200 schools in the United Kingdom in a tele-education pilot project featuring the use of satellite delivery for providing access to high-bandwidth applications in the classroom. Course modules support curricular enhancement.

10.9.7.2.2 EUROMEDNET

The EUROMEDNET initiative supports pilot tests and experimental activities in the Mediterranean Region to validate satellite capabilities in provisioning telemedicine. Clinical sites are part of the SHARED (Satellite Health Access for Remote Environment Demonstrator) network. VSAT teleconsultations and continuing medical education courses are provided. VSAT services also support desktop videoconferencing at rates of 384 Kbps. Medical institutions taking part in EUROMEDNET include the Habib Thameur Hospital in Tunis, the Italian Field Hospital in Durazzo, the San Raffaele Hospital in Rome, the V. Babes Hospital in Bucharest, and the Garibaldi Hospital in Palermo.

10.9.7.2.3 European Medical Network (EMN)

The EMN (European Medical Network) initiative supports distance learning activities and applications for medical professionals throughout the European Union. The infrastructure for tele-education delivery includes menu-driven digital satellite broadcasts that work in combination with Web service. Programs on cardiology and bone nailing and plating techniques are distributed to 50 downlink sites in a series of trials in Germany, Switzerland, France, the United Kingdom, and the Netherlands. Participants include the European Union of Medical Specialists, Deutsche Telekom, and IBM Global Information Services.

10.9.7.2.4 TRAPEZE

The TRAPEZE initiative fosters satellite tele-education delivery to K–12 students in circus, carnival, and fairground communities that do not have access to wireline communications services. The TRAPEZE infrastructure enables one-way broadband communications on the downlink; dial-up modem connections are used for the uplink. The West Midlands Consortium and the Stichting Rijdende School sponsor this initiative.

10.10 SATELLITE BROADCASTS IN THE TELE-EDUCATION DOMAIN

Satellite broadcasts support diverse applications, including military operations, environmental monitoring, and disaster response. Satellite technology cost-effectively delivers large quantities of information to broadly distributed reception sites. In conventional satellite networks, numerous programs are transmitted to the satellite and then rebroadcast to specified earth or ground stations or operators that then in turn distribute the programs to individual viewers. For example, PBS (Public Broadcasting

System) distributes television programming exclusively via satellite channels. Cable television systems receive programming from satellites as well.

Satellite communications networks also support delivery of television broadcasts for curricular enrichment in public and private schools and universities and enable access to movies, Broadway shows, opera performances, and symphonic concerts.

Distinctive attributes of satellite television broadcasts in the tele-education arena are highlighted in the material that follows. Broadcasts supported by national educational broadcasting agencies and satellite program providers are initially explored.

10.10.1 SATELLITE PROGRAM PROVIDERS IN THE UNITED STATES

Satellite program providers in the United States include the American Law Network, the Healthcare Informatics Telecom Network, and Knowledge Online. Additional satellite program providers active in this arena also include the Public Health Training Network, America's Continuing Education Satellite (ACE) Network, and A*DEC, a national consortium of land grant institutions and state universities.

In an effort to create new tele-educational opportunities, educational broadcasting agencies such as the Public Broadcasting System (PBS), American Public Television Stations (APTS), the Corporation for Public Broadcasting (CPB), and the National Educational Telecommunications Association (NETA) offer instructional, informational, and cultural programming rather than conventional entertainment broadcasts.

Public education satellite telecourses cover topics ranging from computers and technology, human development, and women's studies to philosophy and ethics, political science, and anthropology. In the material that follows, teleprograms developed by state consortia and public and private partnerships are noted.

10.10.1.2 The PBS Adult Learning Service

Developed by PBS, the Adult Learning Service (ALS) supports satellite delivery of college credit courses and lifelong learning programs that accommodate the needs and objectives of adult learners. ALS educational programming is delivered to universities, colleges, libraries, and schools nationwide in analog and digital format. Joint Annenberg and CPB (Corporation for Public Broadcasting) initiatives that promote the utilization of satellite television broadcasts delivering course modules in math and science to students in primary and secondary schools are distributed by the PBS Adult Learning Service as well. Public television stations are required to support HDTV broadcasts by the spring of 2003.

10.10.1.3 Air Pollution Distance Learning Network (APDLN)

The Air Pollution Distance Learning Network (APDLN) is an educational satellite broadcast network with more than 100 university and governmental affiliates across the United States. Developed by a consortium that includes the U.S. Environmental Protection Agency (U.S. EPA), state and local air government agencies, and North Carolina State University, the APDLN broadcasts teleseminars and telecourses on air quality and pollution control.

10.10.1.4 Executive Education Network (EXEN)

The Executive Education Network (EXEN) delivers interactive management programs via satellite to downlinks at corporate locations to facilitate executive training. Babson College; the Harvard Business School; Carnegie Mellon, Pennsylvania State, and Southern Methodist Universities; and the Universities of North Carolina, Pennsylvania, Notre Dame, and Texas develop EXEN tele-instructional programs for distribution.

10.10.1.5 Hughes Network Systems (HNS)

Satellite operators in the educational sector such as Hughes Network Systems (HNS) form strategic partnerships with educational institutions, government agencies, professional associations, and nonprofit groups to facilitate production and distribution of tele-education credit and non-credit courses relevant to school and university curricula.

HNS supports satellite broadcasts specifically designed for the Indiana Higher Education Telecommunications System (IHETS) and Oklahoma State and California State Universities. HNS also provides communications capabilities for interactive distance learning programs implemented by the Universities of Alabama, Florida, Maryland, and Tennessee; Northern Virginia Community College; the Western Kentucky Educational Technology Corporation; and the Nebraska Educational Telecommunications Network.

10.11 COMMERCIAL SATELLITE OPERATORS

With the popularity of satellite distance education programs, competition escalates worldwide among commercial satellite operators interested in selling or leasing space segments and offering partial, value-added, or full network services. Satellite communications carriers active in the commercial satellite arena include WorldCom, AT&T, Sprint, and Loral Corporation. Commercial satellite operators in the Asia and Pacific Region include Korea Telecom and ChinaSat. Turksat is the commercial satellite provider in Turkey. Deutsche Telekom offers commercial satellite services in Germany.

10.11.1 CANAL SATELLITE (CANALSAT)

In France, Canal Satellite (CANALSAT) broadcasts interactive DTV (Digital Television) programs and provisions access to local news, special events, weather forecasts, teleshopping activities, e-mail, home banking services, and tele-entertainment.

10.11.2 EUROPEAN TELECOMMUNICATIONS SATELLITE ORGANIZATION (EUTELSAT)

EUTELSAT provides digital and analog television programs, radio broadcasts, and multimedia services to more than 70 million homes connected to cable networks or equipped with VSATs for DBS reception in the European Union, the Middle East, Africa, and Asia. EUTELSAT also fosters development of multimedia digital platforms for provisioning Web services and streaming video and audio via satellite links.

Equipment that works in concert with the EUTELSAT DVB (Digital Video Broadcast) platform at the customer premise includes a small satellite dish, a DVB/MPEG-2 card for a desktop computer, and a modem for transmitting information via a PSTN or ISDN connection on the uplink.

To accommodate growing demands for satellite service between the Mediterranean basin, the Europe Union, Central Asia, Russia, and Africa, EUTELSAT employs W-series satellites for provisioning access to narrowband and broadband applications and services in DTTH (Direct-to-the-Home) configurations. In addition, W-series satellites provide access to mobile communications services, support Web browsing, and enable high-speed multimedia distribution.

EUTELSAT also uses the SESAT satellite as part of its fleet for delivering a wide array of multimedia and telecommunications services in an area of coverage that extends from Russia to the United Kingdom. EUTELSAT satellite services are also available in Saudi Arabia and North Africa. These satellite networks enable high-speed Web connectivity, rural telephony service, news distribution, and DTTH programs. Television broadcasts are distributed to MMDS providers for retransmission to subscriber sites.

EUTELSAT has used GEO satellites for supporting cellular telephony services throughout the European Union since the 1980s. EUTELSAT currently supports delivery of standardized Web services via digital television broadcasts that are compliant with MPEG-2 DVB (Moving Picture Experts Group-2 Digital Video Broadcast) specifications.

EUTELSAT also plans to deploy a new satellite called Atlantic Bird 1 for supporting interconnectivity to wireline and wireless intra-European and trans-Atlantic networks. Atlantic Bird 1 will support operations in the Ku-band and broadcasts programs in a coverage area that includes the European Union, North Africa, North America, South America, and the Middle East. EUTELSAT service providers include Antenna Hungaria, Telecom Italia, British Telecom, and Easynet.

10.11.3 TELENOR BROADBAND SATELLITE SERVICES

A major satellite broadcasting service provider in Northern Europe, Telenor uses THOR satellites and leased capacity on INTELSAT for distribution of digital video broadcasts (DVBs). This configuration fosters transmission of sports events, news, and instructional and entertainment programming to broadcasters and cable network operators in Scandinavia, Eastern Europe, Africa, and the United Kingdom. The Nittedal Earth Station near Oslo enables network management operations.

10.11.3.1 Taide Network

A Telenor subsidiary, the Taide Network provisions connectivity to major network implementations in the United States and the European Union and delivers global satellite services to the customer premise. The Taide infrastructure supports transmission at rates reaching 155.52 Mbps (OC-3), facilitates dependable reliable voice, video, and data distribution, and enables IP telephony, and VPN implementations.

Telenor operates the Svalbard Satellite (SvalSat) earth station situated at Plataaberget on the Arctic Island of Svalbard. The SvalSat earth station is the northernmost earth station on the planet. This earth station retrieves and processes environmental and meteorological data from Polar-Orbiting Environmental Satellites (POES) and then transmits this data to NASA for utilization in the Mission to Planet Earth Program.

10.12 SATELLITE NETWORK CONFIGURATIONS

The popularity of Web services drives demand for faster and more efficient access to the Internet from any location. This demand contributes to the emergence of mixed-mode configurations featuring satellite networks that operate in conjunction with cable, Asynchronous Digital Subscriber Line (ADSL), Integrated Services Digital Network (ISDN), Asynchronous Transfer Mode (ATM), and Synchronous Optical Network (SONET) technologies and configurations.

These hybrid network configurations distribute high-speed, high-performance, data-intensive applications, and support instantaneous and ubiquitous access to Web resources.

As illustrated in the following initiatives sponsored by NASA (National Aeronautics and Space Administration), satellite configurations provide services where terrestrial networks cannot be deployed and extend coverage of in-place landline networks. These initiatives also demonstrate the viability of satellite configurations in supporting point-to-point and point-to-multipoint connections, fostering communications in local and wider area environments, and enabling dependable access to present-day and emergent multimedia services and Web resources.

10.12.1 NASA Tele-education Projects

NASA sponsors an array of tele-instruction projects that demonstrate benefits of satellite technologies in the tele-education domain. NASA satellite initiatives in the educational sector involve design, development, and implementation of models, prototypes, experiments, pilot tests, and full-scale programs.

NASA also provides educational content for video and real-time educational course; develops classroom materials for content areas in mathematics, science, and technologies; and sponsors multidisciplinary workshops and professional development courses for teachers in public, private, charter, and home schools. In addition, NASA posts curricular materials, posters, brochures, fact sheets, and educational resources for K–12 educators, and develops tele-instructional modules on topics such as the impact of the atmosphere and the greenhouse effect on surface temperatures to complement classroom activities. Post-secondary institutions that take part in NASA educational enrichment teleprograms and tele-activities include La Guardia and Queensborough Community Colleges; Marymount and Queens Colleges; the University of Miami; the Florida Agricultural and Mechanical University (FAMU); and Georgetown, Dartmouth, Brown, and Southern Connecticut State Universities.

10.12.1.1 NASA Jason Project

The NASA Jason Project enables middle school and high school students to explore NASA expedition sites in real-time, interact with NASA scientists, and control live-feed video cameras. Data, video, and audio signals for Jason expeditions originate from simultaneous live satellite broadcasts. These broadcasts are then downlinked to primary interactive network sites (PINS) throughout the United States, Bermuda, Mexico, and the United Kingdom. The Jason Project also sponsors Web sites that enable students and their teachers to track NASA discoveries and participate in online discussion groups.

10.12.1.2 NASA Passport to Knowledge Program

The NASA Passport to Knowledge program includes satellite broadcasts about Mars, Antarctica, and the stratosphere. Designed for secondary school students and their teachers, these programs foster telecollaboration with NASA researchers and scientists and enable students, regardless of location to participate virtually in space missions.

10.12.2 NASA RESEARCH AND EDUCATION NETWORK (NASA-NREN)

The NASA Research and Education Network (NASA-NREN) is a high-speed, high-performance communications network testbed that fosters advanced research, next-generation applications, and innovations in network design. The NASA-NREN testbed enables researchers to validate the capabilities of high-definition television broadcasts and IPv6 services; verify QoS guarantees for transport of audio and bandwidth-intensive still images and video-over-gigabit networks, and analyze massive datasets generated by satellite systems. In addition, the outcomes of NASA-NREN trials and experiments in provisioning fast, dependable, and reliable throughput; interoperable networking solutions; and access to data-intensive resources contribute to development of the Next-Generation Internet (NGI) and Internet2. NASA-NREN research findings are also available to researchers using these peer-level networks via NGI exchanges (NGIXs) and I2 GigaPoPs (Gigabit Points of Presence).

NASA-NREN activities were initiated with the launch of the NASA ACTS spacecraft in 1993. At the outset, the NASA-NREN testbed supported interconnectivity to the National Science Foundation Network (NSFnet) and transmission rates at 45 Mbps.

Currently enabling multimedia transport at 622.08 Mbps (OC-12), the NASA-NREN testbed facilitates development of middleware applications that gracefully adapt to bandwidth degradation and deployment of gigabit networks that provision a minimum of 500 Mbps for a single desktop application. Moreover, the NASA-NREN testbed enables utilization of IP multicasts to distribute massive datasets to multiple NASA sites for scientific analysis and fosters implementation of sophisticated hybrid satellite and terrestrial gigabit networks that contribute to innovations in earth science, space science, aeronautics, astrophysics, astrobiology, and geophysical studies. The NASA-NREN platform also supports access to still images and telemetry from robotic planetary excursions, virtual participation in remote geological experiments, and real-time transport of echocardiograms.

NASA-NREN initiatives such as the NASA ATM Research and Industrial Studies (ARIES) demonstrate capabilities of mixed-mode satellite and terrestrial networks in supporting emergency services in response to crisis situations and natural disasters. NASA-NREN teletraining programs help individuals adapt to remote environments such as those encountered in space flight. NASA-NREN pilot studies facilitate utilization of specialized hardware, metacomputing resources, and distributed tools to support visualizations of massive datasets generated by NASA research satellites.

In the field of earth sciences, the NASA OhioView project uses the NASA-NREN platform to facilitate applications in agriculture, cartography, geography, forestry, urban planning, and geology. Academic participants include Bowling Green, Kent State, Ohio State, and Ohio Universities

The NASA Space Communications Program (SCP) supports a Web distributed information system called the Space Internet. Every spacecraft and instrument serves as node on the Space Internet.

10.12.2.10   Ultra-Small Aperture Terminal (USAT) Systems

Developed by the NASA ACTS (Advanced Communications Telecommunications Satellite) Experiment Program, Ultra-Small Aperture Terminal (USAT) systems employ the Ka-band frequencies for data transport, Web browsing, distance learning, and interactive videoconferencing. In a series of healthcare experiments, USATs enabled teleconsultations between patients in remote locations and physicians at key medical centers and transmission of CAT and MRI scans via the NASA-NREN testbed for assessment by medical specialists. These experiments verified performance of mixed-mode satellite and terrestrial networks in provisioning access to affordable healthcare services to individuals in medically underserved rural communities.

10.13   U.S. DEPARTMENT OF DEFENSE SATELLITE TELEMEDICINE INITIATIVES

The U.S. Department of Defense sponsors telemedicine projects that support access to specialty healthcare services, decrease professional isolation of rural healthcare providers, and promote telehealthcare education and teleresearch. Sponsored by the U.S. Department of Defense, the Telemedicine and Advanced Technology Research Center (TATRC) provisions access to programs in telehealthcare, telemedicine, and medical informatics. TATRC also employs satellite services to support delivery of telemedicine treatment to U.S. military forces in Iraq, Somalia, Rwanda, Croatia, Macedonia, Haiti, Panama, Cuba, Egypt, Russia, Sweden, Kenya, and the United Kingdom.

The ongoing deployment of military forces contributes to the utilization of ATM and satellite technologies for provisioning access to remote telehealthcare treatment. As an example, medical specialists in Germany provision telemedicine services to armed forces in Croatia, Bosnia, and Hungary. In addition, TATRIC supports utilization of Personal Information Carriers (PICs), which are portable memory devices that contain medical information in small wearable computers used by soldiers in the field. The TATRIC Battlefield Medical Information Systems Technologies (BMIST) Project employs a mix of technologies, including satellite links and wireline connections for enabling medical specialists at major metropolitan medical centers to treat battlefield casualties.

10.14 SATELLITE INITIATIVES SPONSORED BY PUBLIC AND PRIVATE PARTNERSHIPS AND CONSORTIA IN THE TELE-EDUCATION DOMAIN

In addition to NASA and the U.S. Department of Defense, schools, school districts, community colleges, colleges, universities, and state education agencies are also active in the satellite networking domain. Satellite initiatives sponsored by public and private partnerships and consortia are examined in the material that follows.

10.14.1 COMMUNITY COLLEGE SATELLITE NETWORK

With satellite communications, barriers such as location and distance blocking access to a quality education are virtually eliminated. For example, the Community College Satellite Network employs satellite technology for broadcasting in-service and pre-service training programs and credit and non-credit courses to staff, students, and faculty across the United States.

10.14.2 STRATYS LEARNING SOLUTIONS

In 1998, NTU (National Technological University) and the Business and Technology Network formed Stratys Learning Solutions to provision access to satellite-based telecourses and teleprograms in the E-learning arena. An accredited degree-granting institution, NTU currently delivers graduate and undergraduate courses, master's programs, and non-credit short courses via the Business Technology Network, formerly a subsidiary of the Public Broadcasting System (PBS).

NTU telecourses and teleprograms are broadcast in real-time to reception sites that support MPEG-2 and DVB technology throughout North America and the Asia-Pacific region. Universities affiliated with Stratys Learning Solutions include Arizona State, Colorado State, North Carolina State, Clemson, Columbia, Iowa State, and Michigan State Universities. The Universities of Delaware, Florida, Idaho, Kentucky, Michigan, and Minnesota are also affiliated with Stratys Learning Solutions.

Stratys Learning Solutions has a client base of approximately 1,000 universities, government agencies, and other organizations. Participants include corporate managers, information technology professionals, scientists, engineers, and educators. These individuals represent universities, government agencies, research institutions, and corporations.

10.14.3   STAR SCHOOLS PROGRAM

Sponsored by the U.S. Department of Education, the Star Schools Program provisions tele-instruction in academic fields that include science, mathematics, and foreign languages, and in vocational education through the use of satellite links. Connections 2000, a Star Schools telelearning project, enables upper elementary and middle school students and their parents and teachers in 21 states, Washington, D.C., and the U.S. Virgin Islands to participate in interactive satellite-supported telecourses in social science and the language arts. A multistate consortium associated with the Star Schools Program, the Pacific Star School Partnership provides satellite telecourses for adult workers in a satellite GED (General Education Diploma) program.

10.14.3.1   SERC (Satellite Educational Resources Consortium)

Funded by the Star Schools Program, the Satellite Educational Resources Consortium (SERC) is a nonprofit organization that promotes access to satellite telecourses for student enrichment and staff development. SERC works in partnership with public television agencies, state departments of education, and the Distance Learning Resource Network (DLRN). DLRN monitors trends in distance education and provisions access to Web sites associated with the Star Schools program.

10.14.3.2   United Star Distance Learning Consortium (USDLC)

Initially funded through the Star Schools Program, the United Star Distance Learning Consortium (USDLC) is a partnership among schools, universities, school districts, and state departments of education in Florida, Illinois, New Mexico, North Carolina, and Texas.

The USDLC provides satellite-based teleprograms for enabling local school districts to improve student performance and school operations, and broadcasts teleworkshops and teleseminars on topics that include multicultural learning environments, technology in instruction, and instructional standards in science and mathematics. The USDLC also produces satellite-based staff development programs for teachers.

10.14.4   WESTERN GOVERNORS UNIVERSITY (WGU)

Western Governors University (WGU) is a virtual university developed by the Western Governors Association. WGU offers degree programs and certificates in fields that include network administration and learning and technology. Colleges, universities, and corporations worldwide develop distance learning courses for WGU. This virtual university has an online campus. Students select courses from an online catalog that features searchable course and program listings. Advisors and mentors provision online assistance in course and program selection. An electronic library featuring full-text and citation databases supports the curriculum. Courses offered through WGU are delivered via satellite broadcasts, the Internet, videotapes and audiotapes, closed-circuit television, e-mail, and the postal service.

10.15 SATELLITE TELE-EDUCATION INITIATIVES IN THE UNITED STATES

10.15.1 ALABAMA

10.15.1.1 University of Alabama

The University of Alabama Center for Communication and Educational Technology (CCET) works in conjunction with the University's Center for Public Television and Radio in designing and producing satellite telecourses in foreign languages and science. These video programs are distributed to secondary schools nationwide and in Canada.

10.15.2 ALASKA

10.15.2.1 University of Alaska

Sponsored by the University of Alaska Learning Cooperative, the LIVENET (Live Interactive Video Education Network) project broadcasts teleclasses that originate from university classrooms. Students view LIVENET programs from remote sites equipped with satellite downlink capabilities and communicate with instructors by e-mail, fax, land mail, and telephone. LIVENET tele-instruction appeals to working adults, single parents with small children, and the physically challenged.

10.15.3 ARIZONA

10.15.3.1 Northern Arizona University (NAU)

The satellite-based Northern Arizona University (NAU) Earth and Environmental Science Program is designed for traditionally underserved students in grades four through six attending Star Schools. In addition, the Northern Arizona University (NAU) Educational Systems Programming (ESP) Unit produces a national interactive satellite program featuring twice-weekly instruction in Spanish and culture for elementary school students at more than 285 schools in 26 states.

Moreover, the NAU ESP Unit works with the Northern Arizona University Learning Alliance in distributing satellite programming to approximately 100 sites throughout Arizona. The NAU ESP Unit also collaborates with the Museum of Northern Arizona and the Grand Canyon Association in developing in-service tele-programs and enrichment applications for satellite delivery

10.15.4 CALIFORNIA

10.15.4.1 California State University (CSU) at Chico

Sponsored by California State University (CSU) at Chico, CSUSAT is a satellite telelearning network that delivers a full range of upper-division courses in Liberal Studies, Social Science, Business Administration, Psychology, and Family Relations to regional reception sites. In addition, CSUSAT courses are delivered to an Indian

Reservation, two prisons, and a vocational facility operated by the Sheriff's Department. Remote students pay the same tuition fees and meet the same requirements as on-campus students.

CSU at Chico also takes part in 4Cnet, a consortium consisting of California Community Colleges. 4CNet institutions develop satellite tele-instruction courses and teleprograms that are broadcast from the CSU at Chico satellite facilities.

10.15.4.2   Los Angeles County Office of Education

The Los Angeles County Office of Education sponsors TEAMS, a distance learning program for K–8 teachers and students. Through the utilization of the Educational Telecommunications Network (ETN), TEAMS provisions access to professional development programs and nationally televised satellite broadcasts in mathematics, science, history, social science, and language arts. The TEAMS program is partially funded by the U.S. Department of Education.

10.15.4.3   University of California at Irvine

The University of California at Irvine sponsors a distance learning system called the Distributed and Interactive Courseware Environment (DICE). The DICE initiative supports delivery of bandwidth-intensive interactive multimedia courseware to K–12 schools via a high-speed, high-performance ATM network, and distribution of courseware via satellite to geographically dispersed schools.

10.15.4.4   University of Southern California (USC)

The University of Southern California (USC) Center for Educational Telecommunications, USC School of Education, and the Annenberg School of Communications produce satellite broadcasts featuring tele-education programs, continuing education courses, and teletraining sessions for the Educational Telecommunications Network (ETN).

10.15.5   Georgia

10.15.5.1   Interactive Teaching Network (ITN)

Georgia's statewide satellite Interactive Teaching Network (ITN) includes more than 2,200 downlinks. Each institution in the state university system has at least one digital dish that supports reception of C-band and Ku-band transmissions. The state of Georgia also owns a transponder on Telstar 401 that provides eight digital channels of broadcast capacity for academic programs and medical services. Georgia Public Broadcasting coordinates access to programming services. The University of Georgia and the Georgia Center for Continuing Education produce ITN interactive satellite programs for the economically disadvantaged and develop satellite broadcasts featuring in-service and pre-service training that are delivered to reception sites throughout the United States and Canada.

In addition, ITN delivers satellite teleworkshops and teleseminars for general and special education teachers, parents, school psychologists, school administrators, social workers, juvenile probation officers, and paraprofessionals to satellite downlink facilities in the United States, Canada, and the American territories of Samoa and Guam. These broadcasts explore topics that include attention deficit disorder, the role of technology in the classroom, teen parenting, special education, and dropout prevention. ITN program participants earn continuing education units through the University of Georgia and academic credits through the University of Colorado at Colorado Springs.

10.15.5.2 PeachStar Education Services

PeachStar Education Services produces distance learning programs delivered via ITN. Satellite reception sites include public elementary, middle, and high schools; colleges, universities, and technical institutes; and regional public libraries and educational agencies. PeachStar satellite broadcasts support delivery of credit, non-credit, and enrichment courses, and staff development programs for educators.

10.15.5.3 Valdosta State University

A participant in the University of Georgia System, Valdosta State University broadcasts satellite teleprograms produced at the university television production and satellite uplink facility. Satellite broadcasts include SAT (Scholastic Aptitude Test) preparation courses, writing projects and English as Second Language (ESOL) programs.

10.15.6 HAWAII

10.15.6.1 University of Hawaii

The University of Hawaii produces satellite broadcasts in science, geography, education, music, and the environment for delivery to elementary school students via I-Net (Institutional Network). In addition, the University of Hawaii delivers satellite courses via I-Net to high schools with small enrollments for enabling students to earn college credits and prepare for advanced placement tests. Satellite-based professional development sessions, teacher certification courses, and teleseminars on safety awareness and accident prevention for faculty are also broadcast via I-Net.

10.15.7 INDIANA

10.15.7.1 Indiana College Network (ICN)

Sponsored by the IHETS (Indiana Higher Education Telecommunications System), the Indiana College Network (ICN) broadcasts satellite-based undergraduate telecourses in degree programs developed by college and universities to local reception sites. Courses are also transmitted directly to student homes via public television or cable modem network systems. Additional telecourse delivery mechanisms include the Web, CD-ROMs, and computer disks. Participants include Purdue, Indiana State,

Indiana, and Vincennes Universities. The Independent Colleges of Indiana, Ivy Tech State College, and the University of Southern Indiana participate in ICN as well. Also an ICN participant, Ball State University produces satellite-based tele-education courses for K–12 curricular enrichment.

10.15.7.2   Indiana Higher Education Telecommunications System (IHETS)

The Indiana Higher Education Telecommunications System (IHETS) delivers satellite broadcasts to vocational schools, high schools, colleges, universities, corporations, and community centers in Indiana and to other reception sites throughout the United States, with the exception of Hawaii and Alaska. IHETS satellite teleprograms are also available in Mexico and Canada.

The IHETS satellite broadcasts feature credit and non-credit telecourses, teleseminars for professional and staff development, teleworkshops, and virtual field trips to zoos and museums.

In addition, the IHETS maintains collections of video clips at the IHETS-sponsored Web site. These video clips enable prospective distance education students to examine televised courses offered by IHETS member institutions. The IHETS Web site also provides technical information for coordinators at IHETS reception sites.

Programming for IHETS satellite delivery originates from campus studios at universities in Bloomington, Evansville, Indianapolis, Muncie, Lafayette, South Bend, Terre Haute, and Vincennes. Signals are transmitted from these locations to the GE-3 satellite. GE-3 is an American Communications satellite in geostationary orbit above the equator.

10.15.8   IOWA

10.15.8.1   Iowa Public Television

With the assistance of a U.S. Department of Education Star Schools grant, Iowa Public Television supports special educational satellite broadcasts for K–12 students. Educational programs feature virtual field trips to geological sites, living history farms, wildlife preserves, and art museums. Iowa Public Television coordinates classroom scheduling and manages operations in 800 distance learning classrooms statewide.

10.15.9   KANSAS

10.15.9.1   University of Kansas Medical Center

In Kansas, KUTMTV, a public television channel, delivers satellite broadcasts developed by the University of Kansas Medical Center. Satellite teleprograms enable physicians, nurses, and allied health personnel to earn continuing education credits. Moreover, KUTM-TV delivers patient education teleprograms on smoking and cancer management, and hospital in-service staff education teleprograms on fire safety, violence in the workplace, and radiation safety.

KUTM-TV also features satellite broadcasts for underserved school children and patients under hospice care.

10.15.10 MARYLAND

10.15.10.1 Baltimore City Public School System (BCPSS)

10.15.10.1.1 TEAMS Programming

The Baltimore City Public School System (BCPSS) provides every school in the district with free access to satellite classroom programs developed by TEAMS. Sponsored by the Los Angeles County Board of Education, TEAMS broadcasts lessons in science, mathematics, history, social studies, reading, and language arts. Designed for elementary and middle schools, TEAMS lessons are taught by a teacher in the studio and facilitated by a teacher in the classroom.

10.15.10.2 BCPSS Satellite Projects, Programs, and Services

In the Baltimore City Public School System (BCPSS), satellite broadcasts expand course offerings and provision upper-level courses that may involve only several students in a few schools. The BCPSS pays the licensing fees and tariffs for satellite programming and coordinates schedules for students participating in telecourses. Schools are kept open after-hours for delivering satellite programming to teachers, staff, and members of the community.

In addition, the BCPSS produces satellite broadcasts in partnership with Baltimore City Community College, Coppin State College, and Towson University for homebound students unable to participate in traditional classroom sessions. The BCPSS also provisions satellite teleprograms for the Maryland Interactive Distance Learning Network (MIDLN).

10.15.10.3 Hagerstown Community College

The Hagerstown Community College Distance Education Center produces and delivers satellite teleprograms and telecourses via the Maryland Instructional Television (ITV) Network to students that work full-time and therefore cannot take a full courseload or travel to campus. Courses for credit are available in Computer Science, Business and Management, and Electrical Engineering.

10.15.10.4 Maryland Instructional Television (ITV) Network

Participants in the Maryland Instructional Television (ITV) Network in the Baltimore and Washington, D.C. area include Hagerstown Community College, the University of Maryland at College Park, NASA, the United States Naval Research Lab, the National Science Foundation (NSF), and the Social Security Administration (SSA). The World Bank, the Bureau of the Census, the U.S. Department of Defense at Fort Meade, and the National Imagery Mapping Association (NIMA) produce satellite broadcasts for the Maryland ITV Network as well. These satellite broadcasts are delivered to colleges and universities, libraries, community centers, corporations, and industrial venues. In addition to provisioning courses for credit, satellite broadcasts on the Maryland ITV Network also enable electronic field trips, interactive town meetings, videoconferences with experts around the world, and staff development activities.

10.15.10.5 University of Maryland at College Park

The University of Maryland at College Park operates the Maryland Instructional Television (ITV) Network. The College of Engineering and the University of Maryland at College Park deliver undergraduate and graduate distance education courses and degree programs to adult learners via satellite broadcasts over the Maryland ITV Network. In addition, the Maryland ITV Network supports videoconferences and professional development programs, provisions NTU academic courses and noncredit short courses, and links students and faculty on-campus with off-campus students and subject experts at local and remote sites in virtual classroom environments.

10.15.10.6 Maryland Interactive Distance Learning Network (MIDLN)

The Maryland Interactive Distance Learning Network (MIDLN) is a two-way interactive video distance learning delivery system that consists of a terrestrial fiber-optic network and a satellite component. MIDLN distance education classrooms are situated in secondary schools, community colleges, colleges, universities, hospitals, museums, and research centers throughout the state. MIDLN faculty and administrators are required to participate in distance education technical training sessions. Technical support is available throughout the semester to eliminate class disruptions.

As with the Maryland ITV Network, the MIDLN facilitates access to educational courses and programs that otherwise would not be available because of small enrollment. The MIDLN and the Maryland ITV Network broadcast staff development programs, administrative meetings, workshops, and seminars, thereby eliminating the time and expense associated with traveling to remote sites.

10.15.10.7 Anne Arundel Community College

The Maryland Interactive Distance Learning Network (MIDLN) also accommodates diverse student requirements. As an example, at Anne Arundel Community College, homebound learners and students who work full-time can complete distance education courses via MIDLN for satisfying requirements for an Associate of Arts degree or certificate.

10.15.10.8 Towson University

Towson University operates two interactive video classrooms located in the Albert S. Cook Library. Both classrooms have similar capabilities but are linked to different networks. These networks include the University of Maryland College Park Interactive Video Network (IVN) and the Maryland Interactive Distance Learning Network (MIDLN).

Through satellite downlinks, Towson University provides a wide range of programs from regional, national, and international sources for curricular enrichment and professional development. In cooperation with the Maryland College of the Air Consortium, Towson University also broadcasts introductory courses over public and cable television networks.

10.15.11 Michigan

10.15.11.1 Michigan Information Technology Network (MITN)

The Michigan Information Technology Network (MITN) offers satellite broadcasts of short non-credit courses and teleseminars in healthcare, computer technology, engineering, and waste management. The MITN also facilitates access to satellite telecourses in emergency management, applied computing, telecommunications, business administration, and computer engineering. In addition, the MITN provisions satellite broadcasts of K–12 outreach activities and corporate training sessions. The University of Michigan and Michigan State, Wayne State, Western Michigan, and Michigan Technological Universities maintain uplink and downlink facilities and distribute MITN satellite broadcasts to K–12 schools, post-secondary institutions, corporations, and government agencies.

10.15.12 Mississippi

10.15.12.1 Mississippi Authority for Educational Television (MAETV)

The Mississippi Authority for Educational Television (MAETV) supports delivery of satellite tele-education broadcasts to rural school districts throughout the state. This satellite solution provides small and isolated K–12 schools and communities with access to interactive distance learning courses in foreign languages, science, and mathematics. Interactive satellite programs also enable individuals in rural communities to participate in personal and professional enrichment programs.

10.15.13 Missouri

10.15.13.1 Missouri School Boards Association (MSBA) Education Satellite Network

Sponsored by the Missouri School Boards Association (MSBA), the Education Satellite Network (ESN) presents staff development programs, telecourses in U.S. history and mathematics, and curricular enrichment activities such as virtual trips. Additionally, the MSBA ESN enables students, faculty, and staff to question leaders in national and state government through interactive videoconferences, participate in mock elections, and explore the effect of substance abuse with medical experts.

10.15.14 Montana

10.15.14.1 Montana Education Telecommunications Network (METNET)

Established by the State of Montana, METNET (Montana Education Telecommunications Network) delivers satellite broadcasts to K–12 schools, state agencies, colleges and universities, and nonprofit agencies. Moreover, METNET supports teletraining sessions and televideo classes and provisions access to interactive teleprograms on the Eastern Montana Telemedicine Network (EMTN) and at North Dakota State, Wyoming State, and South Dakota State Universities. The Burns Telecommunications Center, a department of Montana State University (MSU), manages satellite facilities and coordinates statewide and regional satellite videoconferences.

10.15.15 NORTH CAROLINA

10.15.15.1 East Carolina University (ECU) School of Medicine

The East Carolina University (ECU) School of Medicine provisions REACH-TV (Rural Eastern Carolina Health Television) satellite broadcasts, videoconferences, teleworkshops, and continuing medical education courses. Satellite services enable ECU medical professionals to participate in medical teletraining sessions, teleconsultations, and teleradiology programs, and provide emergency medical services across the state in real-time. In addition, university practitioners use satellite videoconferences to meet with prisoners at a distance in real-time and provision healthcare treatment.

The Center for Health Sciences Communication (CHSC) at ECU operates a local cable access channel that provides medical news and information throughout Eastern North Carolina. CHSC satellite downlinks capable of receiving C-band and Ku-band transmissions foster access to national teleconferences and tele-instructional programs. Microwave services supporting audio, video, and data transmissions provide full-duplex connectivity to sites throughout the state.

10.15.15.2 North Carolina School of Science and Mathematics (NCSSM)

The North Carolina School of Science and Mathematics (NCSSM) uses satellite broadcasts to provision access to distance education courses, special class programs, and teacher in-service training. Through NCSSM satellite programming, high school students and residents in Brunswick, Catawba, Durham, Dare, Guilford, Ash, Lincoln, Pender, and Tyrell Counties participate in enrichment programs and college credit courses.

10.15.16 OKLAHOMA

10.15.16.1 *OneNet*

Originally called the Arts and Science Teleconferencing Service (ASTS), OneNet is a statewide network infrastructure that provides citizens with connectivity to information resources and contributes to state economic growth. OneNet is a hybrid wireless and wireline network that consists of satellite and microwave technologies and an optical fiber terrestrial component for transmitting full-motion video, voice, and data. OneNet supports enrichment programs in rural high schools, connectivity to online materials such as law library resources and state statutes, and in-service training for administrators and teachers in vocational and technical schools. Moreover, OneNet enables access to regional and national networks such as Internet2.

The Oklahoma State University K–12 Distance Learning Academy uses OneNet to broadcast satellite telecourses in mathematics, German, and Spanish to students in Grades 3 to 12. In addition, the Oklahoma Telemedicine Network uses OneNet satellite services for enabling doctors at rural hospitals to participate in real-time videoconferences and teleconsultations with radiologists at regional healthcare facilities.

10.15.16.2   University of Oklahoma

The University of Oklahoma Center for Distance Education delivers satellite tele-courses in history, economics, and geography produced by JEC Knowledge Online. Students are required to visit the university campus only for orientation and exam-inations. In addition, JEC Knowledge Online produces satellite instruction programs for undergraduate and graduate students enrolled at the University of Colorado and George Washington, Regis, and California State Universities.

10.15.17   OREGON

10.15.17.1   Oregon State University (OSU)

Satellite technology extends the boundaries of classroom instruction and academic scholarship beyond the limitations of time and space. As an example, Oregon State University (OSU) provides satellite-supported distance education courses for stu-dents unable to participate in on-campus sessions because of work schedules, phys-ical disabilities, educational preference, or commuting distance. Formal admission to OSU is not required for enrollment. Courses can be taken as part of a formal degree program, for professional development, or personal enrichment. Assignments are submitted via e-mail or fax. Site coordinators at remote locations provide tech-nical assistance. In addition, OSU also utilizes satellite services for delivery of marine biology and oceanography courses for college credit to secondary school students throughout the United States.

10.15.18   SOUTH CAROLINA

10.15.18.1   South Carolina Department of Education

The South Carolina Department of Education delivers satellite broadcasts over the South Carolina Educational Television Network to public and independent colleges that are equipped with satellite downlink facilities. Satellite programs include under-graduate and graduate courses in business administration, agriculture, allied health, and world geography. Satellite broadcasts originate from Winthrop, Clemson, and South Carolina Universities and the College of Charleston.

10.15.19   SOUTH DAKOTA

10.15.19.1   South Dakota Rural Development Telecommunications (RDT) Network

The South Dakota Rural Development Telecommunications (RDT) Network provi-sions access to tele-education programs via a one-way video and two-way audio statewide satellite network. Participants in satellite-based college and high school courses include individuals interested in obtaining college credits while still in high school, completing a college degree, or earning a high school diploma.

10.15.20 TENNESSEE

10.15.20.1 Middle Tennessee State University (MTSU)

Middle Tennessee State University (MTSU) operates a satellite tele-education project in rural Franklin, Bedford, Cannon, Coffee, Moore, and Marshall Counties. This satellite-based distance learning initiative offers teacher-training courses and K–12 enrichment classes.

10.15.21 TEXAS

10.15.21.1 STARLINK (State of Texas Academic Resources Link)

Created to overcome barriers of time, cost, and geography, STARLINK (State of Texas Academic Resources Link) is a collaborative initiative sponsored by 64 Texas community and technical colleges and one university. Managed by Austin Community College and the Dallas County Community College District, STARLINK uses satellite technology for broadcasting teleconferences, interactive training sessions, and tele-education activities to business and professional organizations statewide.

10.15.21.2 University of Texas (UT) at Austin

The German program at the University of Texas at Austin helps business students gain fluency in German by using satellite broadcasts from German television stations. These satellite broadcasts enable students to develop listening, reading, and speaking skills and become comfortable in working in bilingual business environments.

10.15.22 VIRGINIA

10.15.22.1 Virginia Satellite Network (VSEN)

Created to alleviate educational disparities in Virginia, the Virginia Satellite Network (VSEN) delivers advanced-level courses in English, Japanese, Latin, U.S. history, statistics, and calculus to placebound students. These courses are also delivered to schools without qualified instructors and schools in which the numbers of qualifying students are too few to merit employment of full-time teachers.

10.15.23 WEST VIRGINIA

10.15.23.1 Satellite Network of West Virginia (SatNet)

The Satellite Network of West Virginia (SatNet) supports delivery of non-credit continuing education courses, professional development courses, and public affairs and public administration programs developed by an educational consortium consisting of state colleges and universities. In addition, SatNet transmits undergraduate

and graduate credit courses to supplement the programs offered by the State College
and University System.

10.15.24 WISCONSIN

10.15.24.1 Wisconsin Educational Communications Board (WECB)

The Wisconsin Educational Communications Board (WECB) operates an uplink
facility in Madison for broadcasting professional development programs and
advanced courses in science, economics, and mathematics to public-sector service
and educational agencies in Green Bay, Milwaukee, and Madison. The WECB also
broadcasts continuing education programs, re-certification programs, and undergraduate and graduate credit courses developed by the University of Wisconsin and the
Wisconsin Technical College System. These programs and courses are available on
Wisconsin Public Television and Milwaukee Public Television channels. Satellite
broadcasts enable students to participate in courses, regardless of work schedules,
travel restrictions, physical disabilities, and small on-site enrollments.

10.16 INTERNATIONAL SATELLITE TELECOMMUNICATIONS
INITIATIVES

10.16.1 AUSTRALIA

10.16.1.1 *FedSat*

Launched by the National Space Development Agency of Japan (NSDA), FedSat,
the Australian Centenary satellite, supports investigations in navigation and space
science and fosters UHF (Ultra-High Frequency) communications experiments. The
capabilities of experimental satellite components and computing equipment developed by the University of California at Los Angeles and the University of Newcastle
in promoting space science research are also assessed.

10.16.2 BULGARIA

10.16.2.1 Bulgaria Academic and Research Network (UNICOM-B)

UNICOM-B, the Bulgarian Academic and Research Network, employs satellite links
between Sofia and Vienna for provisioning access to information resources. Landline
links connecting Sofia to Amsterdam enable access to networks developed by the
European research community.

10.16.3 CANADA

10.16.3.1 Hastings and Prince Edward School District in Ontario and
the Lewisport and Gander School District in Newfoundland

Telesat Canada, International Datacasting Corporation, and Rebel.com plan to implement a satellite network for distribution of satellite programs and courses in the

Hastings and Prince Edward School District in Ontario and the Lewisport and Gander School District in Newfoundland. The satellite configuration complements and extends the reach of CA*net 2 and CA*net 3 to educational institutions in rural and remote communities.

10.16.3.2   Saskatchewan Communications Network (SCN)

The Saskatchewan Communications Network (SCN) uses satellite technology to broadcast continuing education and college credit courses produced in classroom studios on the SCN Training Network. Reception site personnel are responsible for purchasing, installing, operating, and maintaining satellite downlink facilities. Program participants include the Universities of Regina and Saskatchewan and the Saskatchewan Institute of Applied Science and Technology.

10.16.3.3   University of Ottawa Heart Institute

Satellite delivery of telehealthcare services in geriatric care, teleradiology, and tele-education to medical professionals, and satellite-supported videoconferences between primary care physicians and medical specialists, enrich healthcare options available to residents in the remote areas of Canada. The University of Ottawa Heart Institute plans to use a broadband hybrid ATM and satellite infrastructure to enable teleconsultations between patients and their local caregivers and cardiac specialists.

10.16.4   ESTONIA

10.16.4.1   Estonian Educational and Research Network (EENET)

EENET (Estonian Educational and Research Network) provisions services to scientific, cultural, and educational institutions via satellite links to NORDunet2 (Nordic Countries Network, Phase 2). EENET participants include K–12 schools, vocational schools, research centers, and universities; city and county government agencies; and the national museum, library, and theater. Estonia also participates in TERNA (the Trans-European Research and Education Network Association), FUNET (the Finnish University and Research Network), DANTÉ (Delivery of Advanced Network Technology to Europe), and BALTNET, the tele-education and teleresearch network in the Baltic States.

10.16.5   GEORGIA

10.16.5.1   Georgia Research and Educational Networking Association (GRENA)

The Georgia Research and Educational Networking Association (GRENA) supports implementation of a satellite network that enables transmissions at 128 Kbps (Kilobits per second) and provisions links to NRENs in the European Union. GRENA also supports utilization of wireless Ethernet networks to overcome operational constraints caused by the substandard condition of the country's in-place wireline infrastructure. Participants in GRENA include Tblisi State University, the Georgian University of Agriculture, the Georgian Technical University, and the Georgian Academy of Science.

10.16.6 HONDURAS

10.16.6.1 Catholic University of Honduras

The Catholic University of Honduras employs a satellite IP networking solution for provisioning connectivity to the Web and access to high-speed broadband applications. This infrastructure also supports delivery of teletraining programs and tele-education courses.

10.16.7 ISRAEL

10.16.7.1 Israel InterUniversity Computation Center (IUCC)

IUCC supports satellite connections between Tel Aviv and Chicago via INTELSAT for enabling interconnectivity to Internet2 (I2) at rates reaching 44.736 Mbps (T-3). Excess capacity supports transmission of conventional Web traffic.

10.16.8 LITHUANIA

10.16.8.1 Academic and Research Network of Lithuania (LITNET) and LITNET-2 (LITNET, Phase 2)

A satellite Earth station at the Vilnius University PoP (Point of Presence) enables international interconnections between LITNET (Academic and Research Network of Lithuania) and LITNET-2 (LITNET-Phase 2) and TEN-155 (Trans-European Network-155.52 Mbps) and supports transmission at 512 Kbps and higher rates. This interconnectivity provisions access to satellite-based scientific videoconferences, teletraining sessions, and tele-education programs, and fosters telecollaborative research. A terrestrial component consisting of a Frame Relay network operating at 2.048-Mbps (E-1) rates interconnects key Lithuanian cities. ATM trials conducted at Kaunas University of Technology and Vilnius Gediminas Technical University contribute to the development of a nationwide ATM wireline infrastructure for LITNET-2.

10.16.9 MEXICO

10.16.9.1 UNAM (Universidad Nacional Autonoma de Mexico)

UNAM employs satellite communications services for delivering tele-education courses to distant students. Slated to become the first Internet2 institution in Latin America, UNAM also employs ATM technology for enabling teleresearch, video-conferencing, and telemedicine applications.

10.16.10 POLAND

10.16.10.1 Polish National Research and Education Network (Polish NREN)

The Polish NREN uses a Warsaw-to-Stockholm satellite connection that provides access to NORDUNnet2 at rates reaching 2 Mbps. In addition, this NREN also employs a fiber optic terrestrial link for enabling ATM transmission at rates reaching 155.52 Mbps between Warsaw and Stockholm. The Warsaw Metropolitan Area Network (WARMAN) employs fiber optic links for enabling ATM connectivity throughout the city at 155.52 Mbps (OC-3).

10.16.11 UNITED KINGDOM

10.16.11.1 University of Wales

Satellite course delivery promotes new ways of sharing knowledge and subject matter expertise and contributes to innovative solutions for supporting tele-instruction. As an example, the University of Wales supports interactive satellite videoconferences that enable secondary school students and their teachers in Wales to learn foreign languages and customs from faculty in France, England, and Spain.

10.17 EUROPEAN COMMISSION TELEMATICS APPLICATIONS PROGRAM (EC-TAP)

10.17.1 EUROPEAN EDUCATIONAL TELEPORTS (EETP)

The EETP initiative supports development of a coordinated network of teleports in remote locations. Teleports are centers of learning that facilitate distribution of tele-education courses, teleseminars, teletraining sessions, and teleturoials to professionals. Remote locations are linked to teleports via satellite and EuroISDN technologies. Tele-education courses that combine satellite video communications and Web distance learning activities are also supported.

10.17.2 FLEXIBLE LEARNING ENVIRONMENT EXPERIMENT (FLEX)

Designed for preschool and primary school youngsters unable to participate in conventional classrooms, FLEX employs satellite technology for provisioning access to a personalized and interactive learning environment. With FLEX, teachers use individualized programs that combine Web resources and traditional educational materials and manage and monitor the progress of each student.

10.17.3 MEDICAL EMERGENCY AID THROUGH TELEMATICS (MERMAID)

MERMAID enables medical specialists to broadcast teletraining sessions on dealing with medical emergencies to paramedics working on oil platforms and ships at sea. In addition, MERMAID supports interactive teleconsultations between healthcare professionals and caregivers at remote sites. The mixed-mode infrastructure includes INMARSAT satellite links and terrestrial EuroISDN configurations.

10.17.4 SAFETY-NET

SAFETY-NET promotes delivery of satellite-based teletraining modules to employees working in hazardous conditions in the transportation, maritime, and nuclear industries. Satellite-based videoconferencing supports delivery of tele-education sessions on first aid, safety awareness, and health and safety regulations. Participants include employees, contractors, senior engineers, and general and line managers. Satellite and EuroISDN technologies contribute to the development of the SAFETY-NET infrastructure.

10.18 EUROPEAN COMMISSION ADVANCED COMMUNICATIONS TECHNOLOGIES AND SERVICES (EC-ACTS) INITIATIVES

Satellite communications services play a critical role in provisioning global coverage for second-generation and third-generation cellular systems. Satellite services are also expected to fill in the gaps where terrestrial system deployment is not feasible, thereby enabling global roaming and anytime and anywhere access to critical resources. Representative satellite initiatives supporting user mobility, integration with terrestrial ATM networks, cost-effective multiplexing, and innovative applications sponsored by the European Commission ACTS program are highlighted in the material that follows.

10.18.1 DIGITAL INTERACTIVE SATELLITE TERMINAL FOR MASTER ANTENNA SYSTEMS IN THE THIRD MILLENNIUM (S3M)

The S3M initiative involved development of low-cost, pre-operational equipment for distribution of satellite digital television signals through SMATV (Satellite Master Antenna Television). In SMATV environments, several users residing in a single building typically share the satellite interactive terminal. Designed primarily for SOHO settings, SMATV solutions support affordable access to diverse services, including broadband network applications and interactive television programming. S3M trials validated capabilities of the technical equipment and demonstrated the feasibility of using satellite networks to provision interactive television services and facilitate access to Web-based multimedia applications.

10.18.2 INTERACTIVE SATELLITE MULTIMEDIA INFORMATION SYSTEM (ISIS)

The ISIS project demonstrated the economical and technical feasibility of using satellite-supported interactive application delivery. Capabilities of a dual-band satellite link, employing the Ku-band in the spectrum between the 12 and 14 GHz frequencies on the downlink and the Ka-band in the spectrum between the 20 and

30 GHz spectral bands on the uplink, were verified. Approaches facilitating the integration of satellite networks with terrestrial networks for supporting tele-education, Web navigation, teleworking, and telemedicine were also identified. ISIS participants included the Universities of Florence and Salzburg.

10.18.3 MULTIMEDIA USER INTERFACE FOR INTERACTIVE SERVICES (MUSIST)

The MUSIST initiative promoted development and assessment of an intelligent multimodal user interface that conformed to DAVIC and DVB specifications and supported applications and services delivered by satellite networks and cable networks. The MUSIST interface provisioned reliable access to multimedia applications and enabled Web browsing and content retrieval via voice and data commands. The MUSIST interface was evaluated in testbed trials in France, Italy, and Germany.

10.18.4 SWITCHED ATM AND IP OVER SATELLITE (ASSET)

The ASSET project verified capabilities of the EuroSkyWay (ESW) broadband satellite network system in dependably transporting multimedia traffic on-demand to mobile and fixed terminals. Rates reaching 2 Mbps on the uplink and 33 Mbps on the downlink were supported. In addition, the ASSET project validated the performance of a mixed-mode satellite and ATM network implementation in enabling broadband applications such as interactive videoconferencing.

10.19 EUROPEAN COMMISSION INFORMATION SOCIETY TECHNOLOGIES (EC-IST) PROGRAM

10.19.1 BRAHMS

The BRAHMS project sponsors development and implementation of broadband satellite networks that optimize connectivity to integrated global broadband multimedia services. Voice, video, and data will be transmitted at rates of 150 Mbps. Approaches for merging satellite applications provisioned by LEO and GEO fleets, and methods for making these applications available to users via a unified interface will be developed for Direct-to-the-Home (DTTH) and Direct-to-the-Office (DTTO) networks. By integrating broadband fixed wireless access (FWA) services and UMTS functions, this initiative will also support development of integrated multimedia broadband network configurations.

10.19.2 IBIS

The IBIS initiative supports utilization of a Satellite Interactive Link System that enables affordable access to interactive television programming, Web services, and multimedia applications. IBIS will interwork with landline connections and support MPEG-2, and IP specifications. DVB-S (Direct Video Broadcast-Satellite) will facilitate transmissions on the downlink. DVB-RCS (Direct Video Broadcast-Return Channel for Satellite Service) will enable transmissions on the uplink.

10.19.3 MULTIKaRa

Designed to provision Direct-to-the-Home satellite services, the MULTIKaRa initiative sponsors development and evaluation of a multireceiving Ka-band antenna architecture provisioning very-high multibeam coverage. Operations in spectrum between the 18 GHz and the 31 GHz RF bands will be supported. Moreover, this multibeam antenna architecture will also facilitate reception of satellite transported interactive multimedia to enable distance education and telemedicine services in SOHO (Small Office/Home Office) environments.

10.19.4 SATIN (SATELLITE-UMTS IP NETWORK)

This SATIN (Satellite-UMTS IP Network) initiative supports development of IP architectures for optimizing T-UMTS (Terrestrial-UMTS) functions and operations to facilitate delivery of point-to-multipoint multicast applications. Procedures for employing the W-CDMA (Wideband- Code-Division Multiple Access) protocol to support Layer 1 or Physical Layer operations will be defined. Approaches for using advanced receivers operating at Layer 2 or the Data-Link Layer of the OSI (Open Systems Interconnection) Reference Model will also be explored.

10.19.5 SATIN (SATELLITE-SUPPORTED INTERNET TERMINALS)

A strategy for implementation of a satellite-based kiosk network that provisions Internet access is explored in the SATIN (Satellite-Supported Internet Terminals) initiative. This project will facilitate high-performance, high-speed voice, video, and data delivery in the downstream direction from a centralized hub to distributed kiosks that function as satellite terminals. Kiosks will be situated in heavily trafficked pedestrian areas such as bus depots, train stations, and city centers. For the upstream direction, kiosk users will transmit queries via low-cost VSAT connections to the centralized hub.

10.20 SATELLITE NETWORK IMPLEMENTATION CONSIDERATIONS IN EDUCATIONAL VENUES

Satellite network initiatives in the tele-education domain facilitate access to new student populations in distant locations, sustain transborder collaboration and research, and provide equal educational opportunities for rural students. These initiatives also bring advanced-level courses to students situated in areas that are sparsely populated, and promote resource sharing and curricular enhancement and enrichment. Satellite networks support interactive videoconferences, teletraining sessions, telecollaborative teleresearch, teleworkshops, and teleseminars. Moreover, satellite videoconferences provision access to subject specialists and authorities in real-time and facilitate development of interactive learning environments.

Planning for a satellite tele-instructional implementation in the school and university setting involves an analysis of mission-critical and anticipated curricular requirements as well as a careful examination of student and faculty requirements and institutional goals and objectives. A program for enabling faculty and staff to effectively implement satellite-supported modules in the classroom must also be put into place. The planning process also involves establishing budget allocations and assessing the feasibility of integrating satellite networks with the in-place terrestrial infrastructure.

Other considerations associated with satellite deployment include determining equipment requirements, administrative and operating costs, geographic service availability, and satellite network capabilities in supporting seamless real-time voice, video, image, and data delivery.

Satellite teleprograms feature remarkable benefits for participants. However, a broad array of educational, technical, legal, jurisdictional, fiscal, and standards issues remain unresolved. Moreover, satellite instruction poses administrative challenges. Satellite telecourse delivery by consortia and partnerships requires adjustment of credit hours awarded and academic calendars so that starting and ending dates and holiday and vacation periods are synchronous. Expectations concerning faculty curricular contributions, workload, technical responsibilities, compensation, and office hours must also be clarified. Satellite courses and programs selected for deployment should accommodate learner requirements and justify expenditures.

Because satellite signals travel out to space and back to Earth again, satellite transmissions are associated with propagation delays that result in disruptions in video, voice, and data transport.

Despite advanced planning, solar flares, atmospheric and interstellar noise, cosmic radiation, interference from electronic devices, and precipitation and rain absorption in high frequencies employed by satellites, can degrade network performance, impede data throughput, and affect quality of service (QoS) guarantees. Moreover, satellite operations are subject to extreme temperature fluctuations, depending on the satellite's orbit in relation to the sun. Designing a satellite solution for enabling tele-instruction therefore also involves implementing a plan for addressing technical issues associated with satellite transmission and reception, and creating backup course modules in case of catastrophic satellite failures. Security mechanisms such as encryption must also be put into place because satellite transmissions traveling through the air can readily be intercepted.

Transmission of satellite content can be unicast to a single user, multicast to a specified group of users, or broadcast to all users. Satellite operators require licenses from countries in which service is provided. The proliferation of Web-based multimedia applications and the accelerating demand for telecommunications capabilities and curricular innovations contribute to the profusion of current and projected satellite initiatives.

10.21   SUMMARY

Satellite communications technology is a viable mobile and personal communications solution for supporting global services and broadband applications via second- and third-generation terminals and portable communicators. Allocations for satellite services include the 1.980 to 2.010 GHz frequency bands and the 2.170 to 2.200

GHz frequency bands. Satellite configurations are readily upgradeable. Costs for satellite networking and data transmission are distance independent. Moreover, Odyssey, Inmarsat, Globalstar, and Teledesic are designed to facilitate provision of complementary telephony services and value-added roaming services for second- and third-generation cellular subscribers.

In this chapter, satellite tele-education and telemedicine initiatives, government-sponsored projects, and research and scientific programs are described. As indicated, satellite utilization in the telemedicine domain expands specialist availability to patients, shortens waiting times for treatment in emergencies, and provisions continuing medical education to practitioners at remote and isolated locations. NASA also examines satellite services and capabilities through NASA-NREN and NGI deployments. NASA-NREN and NGI are high-performance, high-speed, next-generation internetworking testbeds supported by hybrid network configurations consisting of satellite and terrestrial components.

Effective satellite implementations provide functionality and flexibility in accommodating user requirements and compare favorably with competing telecommunications systems. Multimode and multiband handheld wireless communicators that take advantage of satellite capabilities are in development. These systems provide telephony coverage where terrestrial system deployment is not physically feasible or economically viable, and serve as integral components of the third-generation infrastructure. In the future, applications combining the multimedia capabilities of S-UMTS and the location and positioning capabilities of Global Positioning Systems (GPS) are expected to provide telephony services via handheld communicators. Wireless communicators equipped with GPS capabilities designed for automobiles can provision drivers with personalized information, security services, and location-specific travel data. Through the incorporation and integration of satellite services into hybrid networks, next-generation mobile systems are expected to enable users on the move to access, manipulate, and distribute data, voice, and video at anytime. Teleshopping, video-on-demand, person-to-person telephony, multipoint videoconferencing, as well as in-car navigation, news retrieval, teletraining, and distance learning, are representative of the broad array of network applications and services supported by satellite network solutions.

10.22 SELECTED WEB SITES

European Commission. Information Society Technologies (IST) Program. Home Page.
Available: http://www.cordis.lu/ist/home.html
European Commission. The ACTS Information Window.
Available: www.de.infowin.org/
European Spaceway Agency. ARTES (Advanced Research in Telecommunications System) Program Home Page.
Available: http://telecom.estec.esa.nl/artes/home/
European Telecommunications Standards Institute (ETSI). Working Group on Aeronautical Satellite Earth Stations. Last modified on September 25, 2001.
Available: http://portal.etsi.org/ses/SESaes_Tor.asp

European Telecommunications Standards Institute (ETSI). Working Group on Broadband Satellite Multimedia. Last modified on September 25, 2001.
Available: http://portal.etsi.org/ses/SESBSM_tor.asp

European Telecommunications Standards Institute (ETSI). Working Group on Harmonized Standards for Little LEOs. Last modified on September 25, 2001.
Available: http://portal.etsi.org/ses/SEShsleo_Tor.asp

European Telecommunications Standards Institute (ETSI). Working Group on Satellite-UMTS. Last modified on October 18, 2001.
Available: http://portal.etsi.org/ses/SESs-umts_ToR.asp

Globalstar. Technology.
Available: http://www.globalstar.com/pages/tech.html

Hughes Network Systems. Welcome to DirecPC.
Available: http://www.direcpc.com/index2.html

National Aeronautics and Space Administration (NASA). About NREN. Last modified on March 3, 2001.
Available: http://www.nren.nasa.gov/about/index.html

National Aeronautics and Space Administration (NASA). About the ACTS Project: Overview.
Available: http://acts.grc.nasa.gov/about/

Norsat International. Global Leadership in Satellite Networks. Home Page.
Available: http://www.norsat.com/

Ohio Consortium for Advanced Communications Technology (OCAT). Consortium Agreement. March 28, 2001.
Available: http://www.csm.ohiou.edu/ocact/Agreement-3-28-01a.pdf

Satellite Educational Resources Consortium. What is SERC? Last modified on April 19, 2001.
Available: http://www.serc.org/whatis/index.htm

South Carolina ETV. Statewide Distance Learning Courses. Last modified on May 7, 2001.
Available: http://www.scetv.org/k12/dist.html

Stratys Learning Solutions. About Stratys Learning Solutions.
Available: http://www.ntu.edu/home/aboutus.asp

Teledesic Company. About Teledesic.
Available: http://www.teledesic.com/about/about.htm

United States Coast Guard Navigation Center. GPS.
Available: http://www.navcen.uscg.gov/gps/defaulttxt.htm

United States Department of Education. Star Schools. What is the Star Schools Program? Last modified on January 11, 1999.
Available: http://www.ed.gov/prog_info/StarSchools/whatis.html

University of Wisconsin Extension. Distance Education Clearinghouse. Last modified on September 24, 2001.
Available: http://www.uwex.edu/disted/

11 Next-Generation Networks

11.1 INTRODUCTION

The remarkably rapid evolution of the commodity or public Internet contributes to vastly increased expectations for bandwidth on-demand, QoS (Quality of Service) guarantees, gigabyte and terabyte transmission rates, and instantaneous connectivity. Saturated by bottlenecks, the commodity Internet is incapable of provisioning enough bandwidth to support current and leading-edge broadband applications. Demand for high-capacity networks that accommodates the proliferation of new generations of bandwidth-intensive applications and services contributes to the emergence of next-generation networks that feature sophisticated wireline and/or wireless technologies, protocols, and architectures.

11.2 PURPOSE

In this chapter, broadband network testbeds, trials, and projects that promote the development of next-generation telecommunications applications and services are explored. Distinguishing characteristics of telecommunications technologies that provision the infrastructure for next-generation networks are described.

This chapter begins with an examination of the NGI (Next-Generation Internet) and major NGI initiatives. The contributions of research centers, supercomputing facilities, educational institutions, government agencies, and industries that sponsor development of the NGI and carry out government-funded projects to facilitate NGI implementations are highlighted. Distinctive attributes of NRENs (National Research and Education Networks) that employ advanced network technologies, architectures, and protocols to overcome Internet constraints and accommodate accelerated network performance demands are described. Capabilities of next-generation configurations that enable secure transborder teleresearch, telemedicine services, distance education programs for student populations and lifelong learners, and sophisticated applications are reviewed. Technical features and functions of information grids and digital library configurations are presented.

11.3 NEXT-GENERATION INTERNET (NGI)

11.3.1 NEXT-GENERATION INTERNET (NGI) OVERVIEW

A multi-agency federal research and development program, the NGI (Next-Generation Internet) initiative promotes experimentation with next-generation networking technologies and development of next-generation network testbeds to connect universities and federal research centers throughout the United States at high-speed rates. NGI information grids interconnect high-performance supercomputers for enabling complex computations and simulations and advanced data analyses. In addition, the NGI supports testbed implementations to evaluate the performance of advanced telecommunications technologies and assess the capabilities of next-generation digital libraries, E-commerce and E-government services, and the viability of telemedicine and tele-education programs.

Sponsored by the University of Southern California Information Sciences Institute, the NGI Multimedia Applications and Architecture (NMAA) initiative examines procedures for integrating HDTV and Internet technologies and sponsors trials benchmarking Internet capabilities in carrying compressed and uncompressed HDTV via Internet2 and the SuperNet testbed. Strategies for creating a digital amphitheater that enables a thousand individuals to share a virtual learning experience are also explored.

As with Internet2, NGI trials facilitate evaluation of next-generation digital libraries, E-commerce services, and E-government applications, and support development of new generations of network security mechanisms and advanced tools for creating shared research and educational virtual environments. Internet2 and NGI also support deployment of a National Tele-Immersion Initiative (NTII) that facilitates interactions between participants and between participants and computer-generation models through the integration of networking and media technologies in distributed and fully immersive telecollaborative environments.

11.3.2 NGI PARTICIPANTS

The U.S. Departments of Energy, Commerce, and Defense and the National Science Foundation (NSF) participate in the NGI initiative. In addition, federal agencies such as DARPA (U.S. Department of Defense Advanced Research Projects Agency), NASA, NIH (National Institutes of Health), and NIST (National Institute of Standards and Technology) take part in the NGI Program.

11.3.3 NGI TESTBEDS AND RESEARCH INITIATIVES

11.3.3.1 NGI Testbeds

The NGI consists of multiple network testbeds for enabling innovative and practical computational and networking solutions and advanced applications in healthcare, the environment, education, manufacturing, crisis management, and basic science. Moreover, NGI research initiatives promote deployment of a next-generation mesh architecture that supports QoS (Quality of Service) guarantees, IP multicasts, wide area Web caching services, and high-speed parallel computations. Research centers and universities conduct NGI-funded projects and scientific investigations in topical areas such as digital libraries, telesurgery, telehealthcare, tele-education, distributed computing, bioinformatics, and space science.

11.3.4 NGI INFRASTRUCTURE

The NGI sponsors development of NGI Exchanges (NGIXs) to interconnect next-generation networks established by government agencies such as DARPA, NSF, and

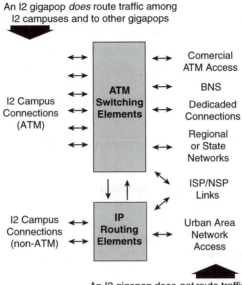

An I2 gigapop *does* route traffic among I2 campuses and to other gigapops

I2 Campus Connections (ATM)

ATM Switching Elements

Comercial ATM Access

BNS

Dedicaded Connections

Regional or State Networks

ISP/NSP Links

I2 Campus Connections (non-ATM)

IP Routing Elements

Urban Area Network Access

An I2 gigapop does *not* route traffic among non-I2 networks and providers

FIGURE 11.1 An I2 GigaPoP.

NASA. In an effort that parallels NGI employment of NGIXs, Internet2 (I2) implements GigaPoPs (Gigabit Points of Presence) to interlink distributed I2 sites (Figure 11.1).

The NGI is designed to support interconnections between high-performance networks such as Internet2 (I2), ATDNet (Advanced Technology Demonstration Network), the NASA Research and Education Network (NASA-NREN), DREN (Defense Research and Engineering Network), SuperNet, and ESnet (Energy Sciences Network) in a nationwide mesh internetwork configuration.

11.3.5 NGI NETWORK INITIATIVES

11.3.5.1 Defense Research and Engineering Network (DREN)

Developed by the U.S. Department of Defense (DoD), the Defense Research and Engineering Network (DREN) is a Virtual Private Network (VPN) that employs a mix of technologies including Ethernet, Fast Ethernet, Gigabit Ethernet, ATM-over-SONET, and IP-over-SONET for enabling high-performance multimedia applications. The DREN configuration interlinks DoD HPC (High-Performance Computing) Centers and fosters access to HPCMP (High-Performance Computing Modernization Program) resources. DREN employs Kerberos for transmitting encrypted messages and SecureIDs to authenticate the identities of network users, and provisions external links to peer-level networks via Metropolitan Area Exchanges (MAXs) and Federal Internet Exchanges (FIXs).

The interim Defense Research Engineering Network (iDREN) served as the foundation for the DREN testbed. Launched in 1993, the iDREN platform featured a T-3 backbone network that supported links to the Air Force Supercomputer Network (AFSCNET) and the Army SuperComputer Network (ASNET).

Operational through 2002, NGI supports network research and development of advanced network technologies and high-speed network testbeds that support gigabit transmission rates, provision high-performance connections, enable end-to-end QoS assurances; and interlink a minimum of 100 university sites. NGI testbeds benchmark the performance of network protocols, software and hardware, security policies and protocols, and management services. NGI research complements I2 investigations. NGI also depends on I2 for developing an advanced campus, local, and regional networking infrastructure to support NGI projects and programs.

11.3.5.2 National Aeronautics and Space Administration (NASA) Integrated Services Network (NISN)

The NASA Integrated Services Network (NISN) is a multiservice, multimedia, integrated configuration that replaced special-purpose networks such as the NASA Communications Network (NasCom), the Earth Observation Data Information System (EODIS), the Aeronautics Network (AEROnet), and the International Communications Network (ICN). A federal network initiative, the NISN streamlines NASA operations and eliminates redundant networking services.

11.3.5.3 National Laboratory for Applied Network Research (NLANR)

A NSF initiative, the National Laboratory for Applied Network Research (NLANR) provisions technical support for NGI implementations and management services for federal infrastructure installations and HPNSPs (High-Performance Network Service Providers) such as I2 and STAR TAP (Science, Technology and Research Transit Access Point). The Distributed Applications Support Team (DAST) enables implementation of advanced applications based on technologies such as CORBA (Common Object Request Broker Architecture) and employs the Globus toolkit for providing middleware services that support application security.

11.3.5.4 NASA National Research and Education Network (NASA-NREN)

An NGI participant, the NASA National Research and Education Network (NASA-NREN) supports network implementations in fields that include the Earth sciences, astrophysics, astrobiology, telemedicine, aeronautics, and space exploration. NASA-NREN initiatives are developed in telecollaborative sessions by researchers, educators, scientists, and industry partners at geographically distributed sites. The NASA-NREN infrastructure supports IPv6 services and ATM delivery of multimedia applications with QoS guarantees. This multiservice platform also facilitates SONET (Synchronous Optical Network), WDM (Wavelength Division Multiplexing), and DWDM (Dense WDM) operations, and enables terrestrial and satellite gigabyte network deployments.

11.3.5.5 QUALIT Initiative and the QBone Project

Sponsored by the U.S. Department of Energy (DoE) NGI Program, the QUALIT testbed enables evaluation of QoS capabilities provisioned by MPLS (MultiProtocol

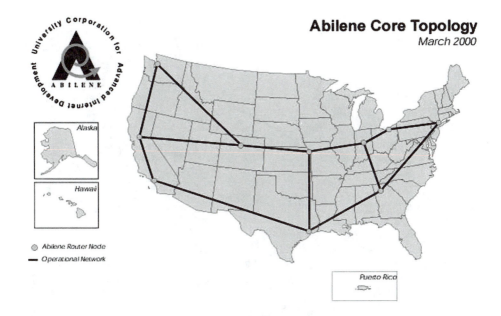

FIGURE 11.2 The architecture of the Abilene Network.

Label Switching), IP-over-ATM, and IPv6 in enhancing next-generation network performance. The QUALIT testbed also supports assessment of QoS assurances provisioned by the I2 QBone network in facilitating scientific telecollaborative research, interactive virtual classes, and dependable delivery of high-performance applications to multiple independent networks. (See Figure 11.2.)

11.3.5.6 SuperNet

A core NGI network, SuperNet employs IP-over-WDM (Internet Protocol-over-Wavelength Division Multiplexing) technology for supporting bandwidth-intensive services at 2.488 Gbps (OC-48) and higher rates that include delivery of uncompressed HDTV-over-IP across the United States. SuperNetresearch network testbeds include ONRAMP (Research and Demonstration of a Next-Generation Internet Optical Network for Regional Access using Multiwavelength Protocols), BoSSNET (Boston to Washington, D.C. Fiber Optic Network), ATDNet (Advanced Technology Demonstration Network), MONET (Multiwavelength Optical Network), and NTON II (National Transparent Optical Network, Phase 2). SuperNet testbed participants in the BART (Bay Area Rapid Transport) SuperNet testbed segment include the University of California at Berkeley (UC Berkeley), Lawrence Livermore National Laboratory, GST, and Sprint. ONRAMP interlinks network nodes such as MIT and AT&T. The ATDNet and MONET configurations support transmissions at 20 Gbps via a WDM dual ring topology. NTON II enables transmission rates at 40 Gbps via four wavelengths or colors of light. (See Figure 11.3.).

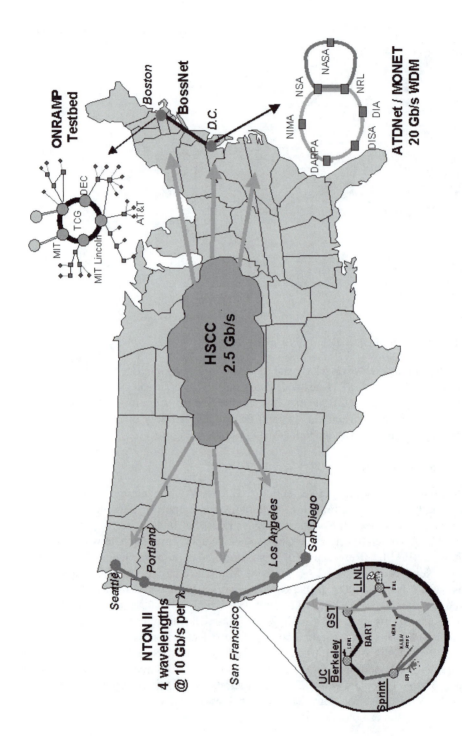

FIGURE 11.3 The architecture of the SuperNet Testbed. (Courtesy of the High-Speed Connectivity Consortium (HSCC)).

The SuperNet cross-country testbed verifies capabilities of automated network monitoring and management tools. In contrast to production networks such as ESNet and NASA-NREN that benchmark capabilities of next-generation applications, SuperNet supports research trials for assessment of networking technologies.

11.4 CONTRIBUTORS TO THE DEVELOPMENT OF A BROADBAND GLOBAL INTERNETWORK

This section examines distinctive characteristics of next-generation networking initiatives and programs that contribute to establishing the framework for a sophisticated broadband global network platform.

11.4.1 HIGH-PERFORMANCE INTERNATIONAL INTERNET SERVICES (HPIIS) PROGRAM

The NGI employs a broadband infrastructure for interlinking research centers, scientific laboratories, and academic institutions across the United States to enable high-performance computing services and telecollaborative scientific and practical investigations. Sponsored by the NGI, the High-Performance International Internet Services (HPIIS) Program promotes the interconnectivity of advanced networks in the United States with internationally equivalent networks. The HPIIS Program benchmarks the capabilities of network protocols such as IPv6 and RSVP (Resource Reservation Protocol) and funds high-performance international networking operations supported by Euro-Link, TransPAC (Trans-Pacific Network), APAN (Asia-Pacific Advanced Network), and MIRnet (Joint United States and Russian High-Performance International Internet Services). Moreover, the HPIIS Program facilitates global interoperability of broadband networks by providing a common link or interconnection point via the STAR TAP (Science, Technology and Research Transit Access Point) SBC/AADS (SBC/Ameritech Advanced Data Services) NAP (Network Access Point) in Chicago.

11.4.2 SCIENCE, TECHNOLOGY, AND RESEARCH TRANSIT ACCESS POINT (STAR TAP) AND STARLIGHT

11.4.2.1 STAR TAP Fundamentals

The Science, Technology and Research Transit Access Point (STAR TAP) features a high-performance network infrastructure to enable interconnectivity of advanced international networks and next-generation broadband configurations throughout the United States via the SBC/AADS (Ameritech Advanced Data Services) NAP in Chicago. The SBC/AADS NAP transports ATM cells between NSPs (Network Service Providers). Rates of transmission at 155.52 Mbps (OC-3) and 622.08 Mbps (OC-12) are supported. The SBC/AADS NAP in Chicago and STAR TAP are examples of Next-Generation Internet Exchanges (NGIXs).

STAR TAP provisions access to massive datasets, information grids, and distributed technical resources to support complex transborder research initiatives. As an example, GEMnet (Global Electum Cyber Society Network) links universities in Japan to MREN (Metropolitan Research and Education Network) and Abilene via STAR TAP and the SBC/AADS NAP for enabling high-performance network research projects. The Rechenzentrum University of Stuttgart also participates in telecollaborative research projects with the Pittsburgh Supercomputing Center and the Sandia National Laboratories in Albuquerque, New Mexico, via links to STAR TAP and the SBC/AADS NAP.

Sponsored by Florida International University (FIU) and Global Crossing, the AMPATH (AmericasPATH) initiative supports the interconnectivity of NRENs in Central America, South America, the Caribbean Region, and Mexico to peer-level networks in the United States via interconnection points such as STAR TAP and the SBC/AADS NAP.

STAR TAP network operations are provided by SBC/Ameritech Advanced Data Services (AADS). The Electronic Visualization Laboratory (EVL) at the University of Illinois at Chicago and the Argonne National Laboratory provision STAR TAP management services.

11.4.2.1.1 STAR TAP Peering Relationships

STAR TAP connects with the SBC/AADS Network Access Point (NAP) in Chicago to enable U.S. research centers and universities to participate in telecollaborative projects and scientific investigations with international peer-level entities. As noted in Chapter 2, peering between networks indicates an agreement to exchange network traffic via direct links. If peering relationships are not established, networks must exchange traffic at designated public network exchange points.

In the United States, participants in the Internet2 (I2) initiative and networks affiliated with the NGI, such as vBNS+, Abilene, ESnet, DREN, NISN, and the NASA-NREN, connect via the SBC/AADS Network Access Point to STAR TAP. Launched in 1997, STAR TAP serves as the primary international connection point for high-performance computing networks such as REUNA (Chile National University Network), BELnet4 (NREN of Belgium, Phase 4), RNP2 (NREN of Brazil, Phase 2), and SINET (Science Information Network of Japan). STAR TAP NREN partners also include Renater2 (NREN of France, Phase 2), SingAREN (Singapore Advanced Research and Education Network), and TANet2 (NREN of Taiwan, Phase 2). Moreover, NORDUnet2 (Nordic Countries Network, Phase 2), SURFnet5 (NREN of the Netherlands, Phase 5), and KOREN (Korean Advanced Research Network) are international NREN partners as well.

11.4.2.2 StarLight

A next-generation STAR TAP initiative, StarLight employs an advanced optical fiber infrastructure that supports high-speed, high-performance applications such as video streaming, Web caches, information grid services, advanced data mining, and sophisticated analyses and measurements of network performance. StarLight also supports the Globus toolkit, GMPLS (General MultiProtocol Label Switching), and the Optical Border Gateway Protocol (OBGMP) to facilitate optical network management functions.

StarLight initiated operations in 2001 by providing services for SURFnet5. StarLight is also scheduled to provision services for the National Center for Supercomputing Applications (NCSA) at the University of Illinois at Urbana-Champaign, the Universities of Chicago and Illinois at Chicago, and the Illinois Institute of Technology. The aforementioned institutions participate in the state-sponsored I-WIRE (Illinois Wired-Wireless Infrastructure for Research and Education) configuration. Designed as the successor to STAR TAP, StarLight interconnects with STAR TAP, Abilene, and ESnet (Energy Sciences Network), and supports rates at 2.488 Gbps (OC-48) with a planned expansion to 10 Gbps (OC-192). Moreover, StarLight establishes gateways for supporting applications enabled by technologies such as SONET (Synchronous Optical Network), POS (Packet-over-SONET/SDH), Gigabit Ethernet, and 10 Gigabit Ethernet. Communications carriers including Qwest, Global Crossing, Teleglobe, SBC/Ameritech, and AT&T participate in StarLight projects.

11.4.3 ASIA-PACIFIC ADVANCED NETWORK (APAN)

11.4.3.1 APAN Technical Features and Functions

Established in 1997, the Asia-Pacific Advanced Network (APAN) conducts research in advanced networking services and applications throughout the Asia-Pacific Region. A multinational configuration, APAN supports initiatives in fields that include bioinformatics, distributed computing, telemanufacturing, and remote robotic control. The APAN infrastructure features intercontinental hub systems that are interlinked to the APAN backbone network. Also called exchange points (XPs), APAN hubs are situated in Seoul, Tokyo, and Singapore. Rates at 44.736 Mbps (T-3) and 155.52 Mbps (OC-3) are supported.

APAN supports connections via STAR TAP to the SBC/AADS NAP in Chicago where APAN connects to vBNS+. APAN also supports connections to Internet2, NORDUnet2, CA*net II (Canadian Network for the Advancement of Research, Industry, and Education, Phase 2), CA*net3 (Canadian Network for the Advancement of Research, Industry, and Education, Phase 3), and GÉANT, the next-generation trans-European research network.

A research and experimental network provisioning a wide range of services, APAN fosters development of next-generation networking technologies and implementation of a high-performance global IP multicast network infrastructure. APAN also collaborates with CANARIE (Canadian Network for the Advancement of Research, Industry, and Education) and DANTÉ (Delivery of Advanced Network Technology to Europe, Ltd.) in planning and testing capabilities of new generations of broadband applications and services.

11.4.3.2 APAN Consortium (APANC) and APAN Working Groups

APAN Consortium (APANC) participants include communications carriers and government agencies in Australia, China, Japan, Korea, Malaysia, and Singapore. Sponsored by the APAN Consortium, APAN Working Groups conduct research in fields that include IPv6, bioinformatics, digital libraries, and medical informatics. These Working Groups assess the viability of network security mechanisms and network protocols such as RSVP and DiffServ, and support development of advanced multimedia applications featuring video-on-demand (VOD), virtual reality elements, and streaming audio and video to enable manufacturing, telesurgery, and tele-education services.

11.4.3.3 APAN Initiatives

The APAN Consortium works with the Asian Internet Interconnection Initiatives (AI3) in provisioning satellite links to the Internet at rates between 1.5 and 2 Mbps. The Satellite Internet Competency Unit in Singapore, the Nara Institute of Science and Technology, and the Institute of Technology at Bandung, Indonesia, take part in this initiative.

The National Partnership for Advanced Computing Infrastructure (NPACI), the National Center for Supercomputing Applications (NCSA), and Indiana University support international internetworking connections from vBNS+ to APAN via the SBC/AADS NAP in Chicago and STAR TAP to enable joint participation in tele-collaborative research initiatives. For example, APAN and I2 institutions take part in High-Performance International Internet Services (HPIIS) Program projects in research areas that include wide area parallel computing, biomedicine, bioinformatics, astronomy, and high-energy physics. In addition, collaborative approaches for measuring the performance of broadband services and the effectiveness of broadband operations are explored. Procedures leading to a virtual university, telemedicine simulations, and weather prediction models are under consideration.

11.4.4 EURO-LINK

11.4.4.1 Euro-Link Consortium

The Euro-Link Consortium supports implementation of next-generation networks, technologies, and applications to provision high bandwidth, support QoS guarantees, and enable instantaneous access to complex scientific applications. Sponsored by Denmark, Finland, Iceland, Norway, and Sweden, NORDUnet2 (Nordic Countries Network, Phase 2) is a Euro-Link participant. In addition, the Electronic Visualization Laboratory (EVL) at the University of Illinois at Chicago and the Israeli Inter-University Computation Center (IUCC) also take part in the Euro-Link Consortium.

11.4.4.2 Euro-Link Initiatives

The Euro-Link infrastructure facilitates development of high-performance computing initiatives. As an example, the Electrotechnical Laboratory in Japan, the Real World Computing Partnership, Argonne National Research Laboratory, and the University of Chicago employ the Euro-Link platform to facilitate development of a collaborative repository of observed astronomical data. The High-Performance Computing Center in Stuttgart and Sandia National Laboratories support distributed parallel supercomputing research, deployment of VR (virtual reality) environments, and distributed simulations via the Euro-Link infrastructure as well. The Euro-Link platform also enables the telecollaborative resolution of complex problems by scientists in Japan, Germany, and the United States.

11.4.5 MIRnet (Joint United States and Russian High-Performance International Internet Services) and the MIRnet Consortium

11.4.5.1 MIRnet Features and Functions

A collaborative initiative between the United States and Russia, MIRnet is a high-performance international broadband network that supports telecollaborative projects sponsored by scientists and educators in Russia and the United States. MIRnet participants access vBNS+ advanced high-performance network services via links from a Moscow PoP (Point of Presence) to intermediate cross-connect points in Copenhagen and New York and then to STAR TAP. From STAR TAP, links to the SBC/AADS NAP and then to the HPIIS policy router establish connections for enabling MIRnet projects. The MIRnet infrastructure supports telecollaborative initiatives in medical imaging, tele-education, data visualization, distributed computation, and environmental engineering, as well as telecollaborative development of approaches for controlling nuclear materials. Participants in the MIRnet Consortium include the Universities of Minnesota and Tennessee at Knoxville, the Russian Institute for Public Networks, Moscow State University, the Russian Academy of Science, the National Science Foundation, and the Ministry for Science and Technology of the Russian Federation.

11.4.6 NORDUnet2 (Nordic Countries Network, Phase 2)

11.4.6.1 NORDUnet2 Features and Functions

Sponsored by Nordic government agencies and the Nordic Council of Ministers, NORDUnet2 (Nordic Countries Network, Phase 2) is an academic backbone network implemented by the Nordic Countries. NORDUnet2 sponsors a maritime distance learning program. In the telemedicine domain, NORDUnet2 brings medical experts and patients together for teleconsultations, regardless of geographical location. NORDUnet2 enables IP multicasts, Quality of Service (QoS) delivery guarantees, advanced network management services, and technical support. Moreover, NORDUnet2 provisions Internet links for NRENs in Estonia, Lithuania, Poland, Russia, and the Ukraine, and security services for member networks via the NORDUNet2 Computer Emergency Response Team (CERT).

11.4.6.2 NORDUnet2 Operations

The NORDUnet2 infrastructure employs IP-over-SDH (Synchronous Digital Hierarchy) and uses ATM technology to provide MBS (Managed Bandwidth Service) for NORDUnet2 institutions. A participant in the GÉANT Network, NORDUnet2 connects to vBNS+ via STAR TAP and the SBC/Ameritech Advanced Data Services (AADS) NAP (Network Access Point) in Chicago, and to the Abilene Network via a Teleglobe and Qwest PoP (Point of Presence) in New York. NORDUnet2 maintains peer-level relationships with research networks that include CA*net3, SingAREN, ESnet, NACSIS (National Center for Science and Information Systems of the Japanese Ministry of Education, Science, Sports and Culture), and SINET (Science Information Network of Japan). NORDUnet2 participants include FUNET (Finnish University and Research Network), DARENET (NREN of Denmark), IsNet (NREN of Iceland), and UNINETT (NREN of Norway).

11.4.7 TRANSPAC (TRANS-PACIFIC) NETWORK

11.4.7.1 TransPAC Network Initiatives

The Trans-Pacific (TransPAC) Network enables implementation of a high-speed inter-national telemedicine configuration that interlinks the National Cancer Center, the University of North Carolina at Chapel Hill, Duke University, the University of Hawaii, and APAN institutions. The TransPAC Network also enables development of tele-immersive VR environments, a virtual university, and telecollaborative research initia-tives between APAN institutions and vBNS+ participants. The TransPAC GOIN (Global Observation Information Network) Project supports distributed problem-solving simu-lations. In an initiative sponsored by the Committee of Earth Observation Satellites (CEOS) and GOIN, the TransPAC Network generates computational analyses of com-plex geological data for reports that enable scientists to issue disaster warnings in the event of wildfires, tornadoes, hurricanes, heavy snowstorms, and torrential rainfall.

Keio University, the Nara Institute of Science and Technology, and the University of Wisconsin delivered a graduate course in computer networks via the TransPAC infrastructure in 1999. This course supported IP multicasts, real-time interactive lectures, and lectures on-demand. In addition, the TransPAC platform enables telecollaborative initiatives in high-energy physics, telemedicine, computational science, biology, mete-orology, and visualization, and the deployment of the APBionet (Asia-Pacific Bioinfor-matics Network). Moreover, the TransPAC Network supports development of Web cache meta-network services to provide visualizations of large-scale datasets for problem-solving activities and interactive simulations in neuroscience, biotechnol-ogy, and the life sciences.

11.4.7.2 TransPAC Consortium

TransPAC (Trans-Pacific Network) Consortium participants include Indiana Univer-sity and APAN institutions such as the Australian National University (ANU), the Japanese Advanced Institute of Science and Technology (JAIST), and the Japanese School of the Internet (SOI). Seoul National University, the National University of Singapore, the Korean Advanced Institute for Science and Technology, and the University of Tokyo take part in TransPAC projects as well. Entities enabling Trans-PAC initiatives include the WIDE (Widely Integrated Distributed Environment) Project, JGN (Japanese Gigabit Network), APAN, I2-DVN (Internet2-Digital Video Network), STAR TAP, NISN, and NASA-NREN.

11.5 NGI GRIDS

11.5.1 GRID FUNDAMENTALS

A Grid is a continuous and persistent national computing infrastructure that enables distributed research by teams of scientists. A core component in the NGI Program, 21st century grids integrate visualization environments, high-performance comput-ers, massive databases, and remote instrumentation into broadband mesh networks. Grid technologies support the development of large-scale distributed metacomputing initiatives that support seamless access to massive terabyte and petabyte datasets, scientific collaboratories, and distributed computational resources. Formed by consortia of educational institutions, research centers, and/or regional and national government agencies, grids process large numbers of independent transactions in distributed computing environments and address problems associated with collecting, managing, and processing vast amounts of voice, video, and data.

Typically, distributed over wide areas, grids enable implementation of complex scientific simulations and development of computational infrastructures and interactive VR tele-immersive environments. In addition, grids provision access to vast data archives maintained in massive digital storage repositories, sophisticated computational resources, and toolsets for resolving complex problems across a broad spectrum of disciplines, including high-energy physics, chemistry, astronomy, the life sciences, climatology, genetics, and space exploration.

11.5.2 GRID IMPLEMENTATIONS

11.5.2.1 Accelerated Strategic Computing Initiative (ASCI)

The U.S. Department of Energy (DoE) sponsors the Accelerated Strategic Computing Initiative (ASCI). This initiative fosters development of balanced terabyte computing platforms that will be available by 2004 by using the fastest supercomputer in the world — specifically, the ASCI White supercomputer at the Lawrence Livermore National Laboratory. Operational since 2000, the ASCI White supercomputer supports scientific investigations, development of problem-solving environments and next-generation network infrastructures, and creation of advanced grids for enabling sophisticated visualizations.

11.5.2.2 Center for Research on Parallel Computation (CRPC)

The Center for Research on Parallel Computation (CRPC) is an NSF Science and Technology Center that fosters development of next-generation parallel computing processes and techniques for resolving complex scientific and engineering problems. The CRPC works with the NASA Information Power Grid (IPG) Project, the National Computational Science Alliance, and the Universities of Chicago and Wisconsin in sponsoring the Globus testbed. This testbed enables design and implementation of the Globus Grid, a very high-bandwidth, high-capacity metacomputing infrastructure. CRPC participants include the California Institute of Technology (Cal Tech), the Los Alamos National Laboratory, the Northeast Parallel Architecture Center (NPAC) at Syracuse University, the North Carolina Supercomputing Center (NCSC), and the National Center for Supercomputing Applications (NCSA).

11.5.2.3 Grid Tele-education Initiatives

In the tele-education domain, CRPC supports the Living Schoolbook Project. Developed by NPAC and the School of Education at Syracuse University, the Living Schoolbook Project enables students and their teachers in New York school districts to utilize advanced educational resources for classroom projects that are only accessible via next-generation network testbeds, multimedia information servers, and supercomputers.

11.5.4 GRID ORGANIZATIONS AND ACTIVITIES

11.5.4.1 Grid Forum

The Grid Forum encourages the development and deployment of distributed computing and communications infrastructures, and conducts research leading to sophisticated grid applications and technologies, an integrated grid architecture, and collaboration among members of the grid community. Participants in the Grid Forum include the Argonne National Laboratory, NASA Information Power Grid (IPG), Sun Microsystems, and Microsoft Corporation. The San Diego Supercomputer Center at the University of California at San Diego, the University of Virginia, Northwestern University, and NCSA also participate in the Grid Forum.

11.5.4.2 Multi-Sector Crisis Management Consortium (MSCMC)

Sponsored by the Alliance Center for Collaboration, Education, Science and Software (ACCESS), the Multi-Sector Crisis Management Consortium (MSCMC) enables development of an integrated global disaster information grid, leading-edge supercomputing technologies, and advanced research applications. These advanced services are designed to mitigate actual terrorist attacks involving the use of chemical, biological, or nuclear weapons and the adverse impact of natural disasters. Participants include the National Computational Science Alliance (The Alliance) and the Universities of Maine and Illinois at Chicago.

11.5.4.3 National Computational Science Alliance (The Alliance)

The Alliance, also known as the National Computational Science Alliance, is a partnership that includes computer and computational scientists and professionals at universities across the nation. The Alliance participates with scientists and technology teams nationwide in designing and implementing an integrated grid infrastructure that supports large-scale, distributed, parallel computing investigations; collaborative virtual environments; remote control of instrumentation; and real-time data acquisition. The Alliance also contributes to the development of discipline-specific computational workbenches for platform-independent, high-performance scientific initiatives in biology and chemistry. The Alliance sponsors high-performance computational initiatives with I2, MREN, NLANR, STAR TAP, vBNS+, and the National Center for Supercomputing Applications (NCSA) at the University of Illinois at Urbana-Champaign.

11.5.4.4 National Center for Supercomputing Applications (NCSA)

The National Center for Supercomputing Applications (NCSA) at the University of Illinois at Urbana-Champaign fosters development of high-performance supercomputing technologies, protocols, architectures, and toolsets that contribute to grid development and deployment. In addition, the NCSA conducts outreach educational activities at public school districts and a science center in central Illinois that support the inclusion of computing and networking applications in the K–12 curriculum. The NCSA also develops interactive simulation programs for science education. In addition, the NCSA collaborates with the Mathematics Department of the University of Illinois at Urbana-Champaign in developing a distance learning program for delivering math and calculus courses and teleworkshops on Internet utilization to students and their teachers at rural K–12 schools in Illinois.

11.5.4.5 Ohio Supercomputer Center (OSC)

The Ohio Supercomputer Center (OSC) creates virtual simulations for training medical residents to administer epidural blocks. A distributed triage project sponsored by the OSC in conjunction with Georgetown University Medical Center and the University of Hawaii enables delivery of telehealthcare treatment and services to individuals at remote geographical locations. OSC simulations verify wheelchair compliance with regulations specified by the Americans with Disabilities Act. For this study, individuals in wheelchairs are monitored as they navigate OSC virtual environments.

11.5.4.6 Pittsburgh Supercomputing Center (PSC)

Sponsored by Carnegie Mellon University (CMU), the University of Pittsburgh, and Westinghouse Electronic Company, the Pittsburgh Supercomputing Center (PSC) supports research projects in weather forecasting, earthquake modeling, and cardiology. In addition, the PSC develops simulations of the heart, lung, and brain for enabling medical research.

In 2001, the PSC initiated full-scale operations of the Terascale Computing System (TCS). The TCS enables research in advanced molecular biology, storm and earthquake predictions, detailed simulations of heart functions, and fluid-dynamic and electromagnetic properties of the Earth's magnetic fields.

In the distance education arena, the PSC sponsors national pilot programs for introducing Web services and applications in K–12 public schools. The PSC also works with the University of Pittsburgh and the Pittsburgh public schools in using advanced technologies to teach courses in foreign languages, writing, visual arts, science, and math; establishing a digital library; and creating a virtual art gallery for enhancement of K–12 classroom activities.

11.5.4.7 San Diego Supercomputer Center (SDSC)

The San Diego Supercomputer Center (SDSC) at the University of California at San Diego works with the California State Department of Transportation in using analytical and visualization resources to test road-paving materials for safety and durability. SDSC-sponsored computing applications support development of advanced medical treatments for heart disease, asthma, Alzheimer's disease, and cancer.

11.5.5 U.S. Grid Implementations

11.5.5.1 Corridor One

The Corridor One grid initiative supports utilization of high-performance technologies for developing a sophisticated prototype of an integrated distance visualization environment. Corridor One also evaluates capabilities of next-generation technologies, transmission protocols, and adaptive networks in enabling visualization of large-scale datasets for telecollaborative problem-solving activities. Corridor One participants include the Lawrence Berkeley National Laboratory (LBNL), the Universities of Illinois and Utah, and Princeton University

11.5.5.2 ESnet/MREN Regional Grid Experimental NGI Testbed (EMERGE)

Sponsored by the U.S. Department of Energy (DoE), the EMERGE (ESnet/MREN Regional Grid Experimental NGI Testbed) initiative employs an IP-over-ATM regional network platform to support NGI applications. EMERGE fosters development of directory services via the Lightweight Directory Access Protocol (LDAP) and dependable delivery of multimedia applications by using RTP (Real-time Transport Protocol) for time stamping, content identification, and sequence numbering. EMERGE also utilizes RTCP (Real-Time Control Protocol) to support a suite of advanced network services that work in concert with grid technologies in fields such as high-energy physics.

Developed as part of the EMERGE initiative, an Earth system grid supports telecollaborative analyses of petabyte datasets. Participants in EMERGE include the Universities of Chicago, Wisconsin, and Illinois at Chicago; NCSA at the University of Illinois at Urbana-Champaign; the International Center for Advanced Internet Research (iCAIR) at Northwestern University; and the NASA Ames, Argonne, and Lawrence Berkeley National Laboratories.

11.5.5.3 Globus Grid

A distributed computational infrastructure, the Globus Grid provisions high-performance access to massive amounts of data. The Globus Grid supports shared use of integrated, high-performance supercomputers, datasets, networks, and scientific instruments that are managed and owned by diverse data centers, scientific organizations, research centers, and educational institutions.

11.5.5.3.1 Globus Ubiquitous Supercomputing Testbed (GUSTO)

The primary Globus testbed is GUSTO (Globus Ubiquitous Supercomputing Testbed). As with power grids that make electricity available, GUSTO makes supercomputing power available to users on-demand. With GUSTO, users of advanced networks such as I2 seamlessly access the power of linked supercomputers, VREs (Virtual Reality Environments), and gigantic terabyte and petabyte data archives. The GUSTO initiative enables astrophysicists to investigate the structure of the universe; scientists to remotely visualize Einstein's gravity wave equations; and engineers to develop transportation vehicles of the future. The GUSTO testbed also supports distributed supercomputing computations for problem resolution and utilization of extraordinarily fast desktop computers for enabling chemical modeling and symbolic algebra research.

11.5.5.3.2   Globus Grid Projects

The Globus Grid is an enabler of tele-immersive applications at the University of Illinois at Chicago Electronic Visualization Laboratory (EVL) and advanced computations at the University of Wisconsin. In addition, the Globus infrastructure fosters construction of the European Data Grid and the NEESgrid (NSF Network for Earthquake Engineering Simulation Grid). Built with the Globus Toolkit, the NEESgrid is a virtual laboratory for earthquake engineering and supports simulations and analyses of the effects of seismic activities on buildings and bridges. Globus Grid components also support trans-Atlantic and trans-Pacific supercomputing applications for enabling scientific data analyses and collaborative computing initiatives. The Globus Grid contributes to the development of automated processes for supporting resource allocations, resource management, and application performance guarantees.

11.5.5.3.3   Globus Toolkit and the Grid Physics Network (GriPhyN)

The Globus Toolkit consists of services and libraries for developing additional toolkits that employ geographically distributed resources. Developed by Argonne National Laboratory and the University of Southern California Information Sciences Institute (ISI), the Globus Toolkit enables Grid initiatives such as the Grid Physics Network (GriPhyN). Sponsored by The Alliance, NASA, and the National Partnership for Advanced Computational Infrastructure (NPACI), GriPhyN facilitates utilization of a Globus-based toolkit by virtual communities of scientists responsible for analyzing vast amounts of data generated by physics and astronomy experiments. GriPhyN participants include Cal Tech and the Universities of Florida and Georgia.

11.5.5.4   Information Power Grid (IPG)

Developed by NASA, the Information Power Grid (IPG) consists of interlinked storage devices, scientific instruments, clusters of computers, and supercomputers at NASA Ames, Glenn, and Langley National Laboratories that enable multisite, multi-user collaborative research. The IPG platform benchmarks the performance of middleware such as the Globus Toolkit and supercomputing applications and services.

11.5.5.5   International Grid (iGRID)

Sponsored by organizations that include the NSF and the Internet Society (ISOC), the International Grid (iGrid) features an advanced metacomputing infrastructure for enabling international telecollaborative research and advances in grid technologies and applications. iGrid initiatives employ the Globus Toolkit and CAVERN (Cave Automatic Virtual Environment Research Network) software to support tele-immersive research in CAVEs (CAVE Automatic Virtual Environments).

At a series of supercomputing conferences held in 1998 and 2000, iGrid simulations illustrated the capabilities of distributed VR technologies in enabling telecollaborative product design in shared virtual workspaces and the manufacture of medical prostheses. In addition, iGrid projects demonstrated the capabilities of wireless and wireline ATM networks in delivering teleseminars and distance education telecourses in ecology, general physics, and foreign languages. Also an iGrid initiative, 3DIVE (3-D or Three-Dimensional Interactive Volume Explorer) enabled multiple participants to manipulate, explore, and describe high-volume biological datasets generated by MRI (Magnetic Resonance Imaging) scans. Developed by institutions in the United States and Singapore, GiDVN (Globe Internet Digital Video Network), also an iGrid project, facilitated implementation of an advanced digital video platform for enabling real-time communications between scientists via high-speed, high-capacity networks such as TransPAC. The TransPAC platform promotes iGrid (International Grid) research by provisioning access to distributed bandwidth-intensive data, visualizations of complex datasets, and massive computing resources maintained by APAN, the WIDE (Widely Integrated Distributed Environment) Project, and the Japanese Gigabit Network (JGN). Participants in iGrid initiatives include the National Universities of Japan and Singapore; Waseda University; the San Diego Supercomputer Center; NPACI (National Partnership for Advanced Computational Infrastructure); the Universities of Hawaii and Tokyo; and Indiana, Purdue, Moscow State, and Osaka Universities.

11.5.5.6 NPACI (National Partnership for Advanced Computational Infrastructure) Grid

Sponsored by NSF, the National Partnership for Advanced Computational Infrastructure (NPACI) supports the design and implementation of the NPACI Grid. An advanced metacomputing infrastructure for enabling scientific research, the NPACI Grid consists of two subgrids, specifically, the Computational Grid and the Access Grid. The NPACI subgrids facilitate utilization of high-performance computers, storage devices, and supercomputing resources to support parallel and distributed computation projects and initiatives in fields such as climatology, molecular biology, environmental hydrology, and cosmology

11.5.5.6.1 The Alliance

The Access Grid is an interactive virtual workspace for grid communities that supports group-to-group communications and formal and informal interactions. This grid brings groups together in real-time for tele-education, teletraining, teletutorials, telemeetings, teleseminars, and telecollaborative research regardless of their actual physical locations. In addition, the Access Grid supports resource allocation, user authentication, and computational operations at specified performance levels across multiple sites and enables scientific and educational research projects.

11.5.5.6.2 Computational Grid

The Access Grid complements the Computational Grid. As with the Access Grid, the Computational Grid employs bandwidth-intensive storage devices, high performance computers, and visualization systems for complex problem resolution and compelling experiences. The Computational Grid supports development of a shared workspace for enabling telecollaborative research in real-time by investigators at diverse locations.

11.5.5.7   National Technology Grid

The National Technology Grid supports development and implementation of a next-generation computational infrastructure that integrates high-speed, high-performance networks, storage systems, and grid clusters in a mesh configuration to accommodate multifaceted interdisciplinary research requirements. The National Technology Grid enables utilization of advanced visualization environments for complex problem resolution, remote instrumentation control, data mining, and implementation of distributed Storage Area Networks (SANs).

In addition, the National Technology Grid supports deployment of VR projects in the arts and the humanities; telecollaborative research leading to cancer treatment and prevention; and development of advanced VR techniques for enabling toxic waste cleanup, airline scheduling, and oil reservoir management. Participants in the National Technology Grid include the Center for Research on Parallel Computation (CPRC); the National Partnership for Advanced Computing Infrastructure (NPACI), the National Center for Supercomputing Applications (NCSA); and the Universities of Tennessee, Maryland, Texas, Boston, and Indiana.

11.5.5.8   Distributed Terascale Facility (DTF)

Featuring an extremely high-speed, high-performance computing infrastructure, the NSF-sponsored Distributed Terascale Facility (DTF) is regarded as the largest computing platform in the world dedicated to open scientific research. TeraGrid or the DTF infrastructure supports advanced simulations of complex phenomena and interconnects massive data collections and data repositories, remote instruments, high-performance information grids, and supercomputers to enable next-generation visualizations of massive datasets and innovative scientific research and discoveries in fields such as biology, astronomy, and high-energy physics. DTF enables 13.6 teraflops or trillions of calculations every second, supports transmissions at 40 Gbps with upgrades to 80 Gbps, and facilitates the management and storage of more than 450 terabytes of data. In contrast to the TCS (Terascale Computing System) maintained at the Pittsburgh Supercomputing Center, the DTF is situated at distributed sites.

The DTF interlinks the National Center for Supercomputing Applications (NCSA) at the University of Illinois at Urbana-Champaign, Argonne National Laboratory, the San Diego Supercomputer Center at the University of California at San Diego, and the California Institute of Technology (Cal Tech). The DTF enables connections to Abilene, CalREN-2, and StarLight and to research and educational institutions worldwide via the StarLight interconnect in Chicago. The DTF implementation also uses Qwest facilities in Chicago, San Diego, and Los Angeles and the I-WIRE (Illinois Wired-Wireless Infrastructure for Research and Education) optical infrastructure. The TCS and DTF are the key components of an emerging cyber-infrastructure.

11.5.6   INTERNATIONAL GRID INITIATIVES

11.5.6.1   DataGrid and Global DataGrid

Sponsored by the European Commission, the Global DataGrid Project supports the establishment of a massive research network as a foundation for the Global DataGrid.

To facilitate Global DataGrid implementation, the European Commission Information Societies Technologies (EC-IST) Program sponsors the DataGrid initiative to identify core network elements and requirements and facilitate an assessment of the capabilities of an IP-over-WDM (Internet Protocol-over-Wavelength Division Multiplexing) optical metropolitan network infrastructure in supporting a multifunctional, multiservice grid environment. The DataGrid supports transmission rates at 40 Gbps and facilitates an evaluation of the capabilities of core infrastructure components for the Global DataGrid project. Participants in the DataGrid initiative include the National Institute of Telecommunications, the University of Essex, and the National Technical University of Athens.

The Global DataGrid is designed to support shared applications among multiple participants at distributed locations and advanced scientific projects in disciplines such as biology, physics, and the earth sciences. Additionally, the Global DataGrid will facilitate utilization of supercomputers that are interlinked in a mesh configuration to perform complex scientific computations. Participants in the Global DataGrid, also known as the WWG and the Worldwide Grid, include the United Kingdom Particle Physics and Astronomy Research Council, the European Space Agency, the Dutch National Institute for Nuclear Physics and High-Energy Physics, and the Swedish Natural Science Research Council. Sponsored by the European Commission and the NSF, the CERN (European Organization for Nuclear Research) Computational Grid Testbed also promotes implementation of Global DataGrid components.

Plans for creating the EuroGrid Forum and the EuroGrid Consortium to support implementation of the Global DataGrid are in development as well. To promote Global DataGrid participation, the European Commission Information Societies Technologies (EC-IST) Program intends to support transmission rates at 2.488 Gbps (OC-48) for every NREN in the European Union.

11.6   NGI TELEMEDICINE PROJECTS

11.6.1   NATIONAL LIBRARY OF MEDICINE (NIH)/NATIONAL INSTITUTES OF HEALTH (NIH)

The National Library of Medicine (NIH)/National Institutes of Health (NIH) sponsor innovative NGI testbed trials, pilot tests, and full-scale implementations in telemedicine and telehealthcare through the United States. These demonstration projects illustrate the role of the NGI in supporting medical education, healthcare, and medical research. As an example, NLM/NIH NGI projects facilitate transmission of bandwidth-intensive high-resolution images such as x-rays and MRI (Magnetic Resonance Imaging) scans between isolated healthcare facilities and major healthcare centers where on-staff radiologists can determine the course of radiotherapy treatment. These trials also enable real-time teleconsultations between on-call physicians and medical experts in treating life-threatening injuries to trauma patients and clarify approaches for enabling the secure transmission of confidential and sensitive electronic patient data.

Moreover, NLM/NIH NGI projects evaluate benefits associated with providing telehealthcare services to patients in route to trauma centers, delivering tele-educational programs to families of high-risk newborns with very low birth weights, and using telemedicine networks to reduce isolation of rural primary care physicians. Methods for provisioning remote healthcare services to elderly patients with Alzheimer's disease over the NGI are evaluated. Approaches for using new medical imaging devices for improved detection of cancerous lesions via the NGI are investigated. The merits of using medical image reference libraries and TIE (Telemedicine Information Exchange), an online telemedicine repository, for provisioning access to digital images, data, and video representing normal and pathological medical conditions via the NGI are evaluated. Capabilities of the NGI in enabling telemammography, telesurgery, and health science education and processing massive amounts of epidemiological data for tracking viruses are also assessed. Representative NGI implementations in the telemedicine area are highlighted in this section.

11.6.2   NLM/NIH TELEMEDICINE PROJECTS AND INITIATIVES

11.6.2.1   California

11.6.2.1.1   Stanford University School of Medicine
The Stanford University School of Medicine supports real-time teleteaching sessions in surgery and human anatomy that are delivered over the NGI.

11.6.2.2   Indiana

11.6.2.2.1   Indiana University (IU) School of Medicine
Developed by the Indiana University (IU) School of Medicine, the Indianapolis Network for Patient Care (INPC) enables access to digital patient records, medical library resources, and prescription data from community pharmacies in hospital emergency rooms and trauma care centers. Additional connections to clinical laboratories and the Indiana State Public Health Department enable clinicians to track immunization data and access reports of communicable diseases. IU plans to convert the INPC configuration into a NGI testbed for evaluating the capabilities of advanced networking technologies in supporting videoconferencing between patients in nursing homes and healthcare providers.

11.6.2.3   Iowa

11.6.2.3.1   University of Iowa Rural Telemedicine Initiative
The University of Iowa National Laboratory for the Study of Rural Telemedicine evaluates capabilities of videoconferencing teleconsultations over the NGI between healthcare specialists and patients with special needs, including children with heart conditions or disabilities and persons with mental illness. In addition, the merits of delivering healthcare information to television sets equipped with special devices in the homes of patients with diabetes for enabling these patients to effectively manage their disease are explored.

11.6.2.4   Louisiana

11.6.2.4.1   Louisiana State University (LSU) Medical Center

LSU (Louisiana State University) Medical Center supports implementation of a trauma care system via an NGI telemedicine network that provisions instantaneous access to the Trauma Team at the Medical Center of Louisiana at New Orleans for supporting online emergency medical assistance. In addition, tele-education and teletraining sessions designed for emergency medical personnel are also supported. Approaches for transporting digitized patient records over the NGI that conform to an electronic format developed by the U.S. Department of Defense, the Indian Health Service, and the LSU Medical Center are also under consideration.

11.6.2.5   Maryland

11.6.2.5.1   Johns Hopkins University

The Applied Physics Laboratory at Johns Hopkins University evaluates NGI services in supporting oncology treatment.

11.6.2.5.2   University of Maryland at Baltimore

The University of Maryland at Baltimore supports implementation of an advanced mobile telemedicine testbed over the NGI that is called the Mobile Telemedicine System (MTS). The MTS supports transmission of vital signs and video images of patients at accident scenes or in ambulances that are on the way to trauma centers to on-site medical specialists for improving the quality and timeliness of diagnostic services and healthcare treatment.

11.6.2.6   Massachusetts

11.6.2.6.1   Children's Hospital in Boston

Children's Hospital in Boston conducts an NGI initiative for enabling development of Personal Internetworked Notary and Guardian (PING) infant healthcare records that are available to postpartum mothers at work, home, or school from any Internet-linked device. The PING record is encrypted and can only be accessed by authorized individuals.

11.6.2.7   Missouri

11.6.2.7.1   University of Missouri-Columbia School of Medicine

Sponsored by the University of Missouri-Columbia School of Medicine, a telehealth-care network that operates in conjunction with the NGI provisions links between rural primary care physicians situated in three isolated communities and medical specialists at the University Health Science Center. The effectiveness of this network in enabling rural practitioners to remain in remote locations while also accessing online medical records and patient data and conferring with healthcare providers at a distance to determine patient treatment protocols is assessed.

11.6.2.8 Washington

11.6.2.8.1 Washington University in St. Louis

Washington University in St. Louis sponsors a high-performance broadband tele-medicine network via the NGI platform to interlink municipal hospitals in St. Louis for enabling healthcare specialists and hospital pharmacists to monitor and prevent adverse drug interactions among the elderly. In addition, this NGI demonstration project provisions access to patient treatment records, drug orders, and results of laboratory tests that are designed to detect adverse drug reactions in at-risk patients.

11.6.2.9 New York

11.6.2.9.1 Columbia University

At Columbia University, procedures for distributing data on disease prevention and home healthcare management of chronic illnesses over the NGI are in development. Upon completion of this project, patients will be able to communicate with healthcare providers, access and review their own medical records, and routinely enter their glucose levels, pulmonary test results, and blood pressure into digitized medical record files. Techniques for safeguarding the confidentiality of digitized patient healthcare records will also be evaluated.

11.6.2.10 North Carolina

11.6.2.10.1 East Carolina University (ECU) School of Medicine

The viability of using IP video-over-the-NGI to facilitate real-time broadband video telemedicine applications and patient cardiology tele-education is evaluated in a series of trials conducted at the East Carolina University (ECU) School of Medicine.

11.6.2.11 Pennsylvania

11.6.2.11.1 University of Pennsylvania

The University of Pennsylvania sponsors development of an NGI testbed to dem-onstrate the capabilities of a national breast imaging archive and broadband network infrastructure in supporting access to and retrieval of digital mammography records for medical and clinical research and improvements in breast cancer screening methods. A multilevel security system embedded in this system is designed to ensure patient confidentiality and privacy.

11.7 UNIVERSITY CORPORATION FOR ADVANCED INTERNET DEVELOPMENT (UCAID) I2 (INTERNET2) NGI PROGRAMS AND INITIATIVES

Sponsored by UCAID (University Corporation for Advanced Internet Development), I2 is a nationwide effort by universities and research centers to develop sophisticated services, applications, and technical solutions for everyday use on the next-genera-tion commodity or public Internet. Approaches for enabling dependable and reliable transmission of voice, video, and text; multimedia streams; and sophisticated medical images are in development.

Strategies for using advanced software tools, technologies, and applications in K–12 schools, libraries, museums, and community colleges are investigated in UCAID-sponsored K–20 initiatives. For instance, Virginia Tech plans to use I2 to examine tactics that facilitate VR innovations developed by EVL and PACI. This initiative enables students in two schools in Virginia to interact with an instructor in another location in a seamless virtual environment and learn about chemistry. Also sponsored by Virginia Tech, a follow-on project to the Jason initiative will enable students in grades K–12 at schools in Virginia to participate in multipoint interactive videoconferences to discuss the Jason Project. The I2 backbone will provision satellite services and interconnectivity. Participants include the Virginia Community College System and the Science Museum of Virginia.

MOREnet (Missouri Research and Education Network) supports utilization of advanced digital resources and development of I2-based virtual learning environments that enable K–12 students to strengthen critical thinking and communications skill by participating in historical programs centering on the Lewis and Clarke period of discovery. Virtual St. Louis employs VR resources to enable students to learn about the development of the City of St. Louis through ten-year periods that extend from 1890 to 1990.

UCAID also sponsors the Quilt initiative for providing access to advanced networks throughout the educational and research community and contributes to the development of the NGI through I2 projects such as the Internet2-Digital Video Network (I2 DVN) and the Internet2-Middleware Initiative (I2-MI). Through UCAID support, I2 entities collaboratively work with the ResearchChannel Consortium, CAVENER and ViDeNet in developing methods for distributing high-quality on-demand video via the Internet and I2 deployments. These initiatives are examined in this section.

Sponsored by UCAID (University Corporation for Advanced Internet Development), I2 is a nationwide effort by U.S. universities to upgrade network connections. I2 configurations employ STAR TAP for enabling links to international research and education networks. Academic and scientific communities employ the I2 platform for conducting telecollaborative research projects that contribute to advanced applications in tele-education and telemedicine. Distinctive attributes of the Internet2-Digital Video Network (I2-DVN) and the Internet2-Middleware Initiative (I2-MI) in supporting NGI development are highlighted in this section.

11.7.1   INTERNET2-DIGITAL VIDEO NETWORK (I2-DVN)

11.7.1.1   I2-DVN Applications

The Internet2-Digital Video Network (I2-DVN) promotes development of a nationwide network infrastructure for enabling implementation of advanced digital video services. A popular format in the academic domain, digital video is typically featured in documentaries, course content, interactive lectures, videoconferencing, and video-on-demand (VOD). I2-DVN applications can be transmitted via the NGI to distributed user communities. Techniques for transmitting animations, simulations, and VR movies with audio sound tracks are in development as well.

I2-DVN sponsors include the ResearchChannel Consortium, SURA (Southeastern Universities Research Association), MREN (Metropolitan Research and Education Network), and iCAIR (International Center for Advanced Internet Research) at Northwestern University. iCAIR also participates in the Global Internet Digital Video Network (GiDVN) and the I2-Distributed Storage Infrasturcture (I2-DSI) initiative.

11.7.2   INTERNET2-MIDDLEWARE INITIATIVE (I2-MI)

Middleware refers to data and tools that enable applications and services to access and use networked resources. The I2-MI (Middleware Initiative) facilitates grid applications, digital library projects, security and authentication services, and secure multicast operations. The I2-MI initiative also supports development of core middleware applications for enabling resource scheduling, bandwidth brokering, and transaction and message handling at I2 institutions.

11.7.3   INTERNET2-DISTRIBUTED STORAGE INFRASTRUCTURE (I2-DSI)

The I2-DSI (I2-Distributed Storage Infrastructure) initiative facilitates implementation of a distributed metacomputing infrastructure comprised of servers with massive storage capabilities that also support NGI applications and services. This infrastructure performs repetitive hosting functions for enabling authenticated I2 users to access Web-based multimedia content and applications via a series of I2-DSI channels.

Developed by I2 institutions and research centers, I2-DSI channels feature a comprehensive repository of multimedia content and advanced applications relating to such topical areas as the Mars Polar Lander Mission. In addition, the I2-DSI initiative features an extendible and scalable architecture for provisioning access to sophisticated digital libraries composed of massive collections of broadcast quality videos, mathematical software, and geospatial datasets.

I2-DS (I2-Distributed Storage) sites are situated at the North Carolina Supercomputing Center, the Earth Resources Observation Systems (EROS) Data Center, and the Universities of Hawaii at Manoa and Tennessee at Knoxville. In conjunction with the I2-DSI initiative, IBM supports development of Web Cache Managers that feature a combined capacity of nearly six terabytes.

11.7.4   RESEARCHCHANNEL

Sponsored by the Research Channel Consortium, the ResearchChannel supports initiatives demonstrating I2 capabilities in delivering broadcast-quality video via the NGI and explores approaches for enabling on-demand video delivery from an extensive video library via the commodity Internet. Educational and research institutions affiliated with the ResearchChannel Consortium include Carnegie Mellon, Tufts, Duke, Rice, Stanford, and Princeton Universities and the Universities of Alaska at Fairbanks, Chicago, Colorado, Pennsylvania, Hawaii, Washington, and Virginia.

11.7.5   CENTER FOR ADVANCED VIDEO NETWORK ENGINEERING AND RESEARCH (CAVNER)

The Center for Advanced Video Network Engineering and Research (CAVNER) sponsors advanced research initiatives in conjunction with I2 institutions for enabling implementation of next-generation video technologies and applications such as videoconferencing, real-time broadcasts, uncompressed video, and video-on-demand (VOD) that also enrich NGI programs and services. CAVNER implementations support real-time telecollaborative research between investigators at geographically separated sites; facilitates delivery of IP multicasts and High-Definition Television (HDTV) programming; and enables videostreaming, establishment of video archives. CAVNER features an open architecture that supports multivendor network operations. CAVNER also facilitates transmission of uncompressed video and data via DWDM (Dense Wavelength-Division Multiplexing) networks.

11.7.6 ViDeNet

Sponsored by CAVNER and SURA (Southeastern Universities Research Association), the ViDe (Video Development) initiative supports development of a global ITU-T (International Telecommunications Union-Telecommunications Standards Sector) H.323 videoconferencing network called ViDeNet that provisions seamless delivery of stored or real-time streaming video with QoS guarantees. Moreover, ViDeNet enables transmission of video-over-IP and voice-over-IP services via I2, NRENs (National Research and Education Networks), and the commodity Internet. In addition, ViDeNet facilitates video delivery to wireless IP devices and desktop videoconferencing systems and promotes development of next-generation video initiatives for real-time tele-education projects, teletraining sessions, telesurgery, and virtual hospital grand rounds that benefit NGI projects. The Universities of Tennessee and North Carolina at Chapel Hill, New York State Education and Research Network, Year 2000 (NYSERNet 2000), and the Georgia Institute of Technology manage ViDeNet operations.

11.8 CONNECTIONS TO THE INTERNET

11.8.1 Computer Science and Engineering Directorate (CISE)

Sponsored by the NSF, the Computer Science and Engineering Directorate (CISE) supports a multiservice network research program for enabling implementation of next-generation networks and complex applications. Within CISE, the Advanced Networking Infrastructure and Research (ANIR) Division and the Advanced Networking Infrastructure (ANI) Program promote multidisciplinary, multiservice networking initiatives. ANIR and ANI contribute to the development of a global high-performance, next-generation network infrastructure and sponsor network testbeds for evaluating the capabilities of multimedia applications and verifying performance of advanced networking technologies.

11.8.2 Advanced Networking Infrastructure and Research (ANIR) Division

The ANIR Division sponsors the Connections to the Internet Program and the High-Performance Network Server Provider (HPNSP) initiative. The HPNSP provisions high-speed networking services in the broader community, and links to vBNS+ and to Next-Generation Internet Exchanges (NGIXs) such as the SBC/Ameritech Advanced Data Services (AADS) NAP in Chicago. Financial awards enable recipients to establish connections to advanced networks and conduct projects that contribute to the development of the NGI and innovative multimedia applications in diverse fields such as forestry, mining, land use, cognitive psychology, tropical cyclone forecasting, image processing, and geometric modeling and visualization. Representative ANIR initiatives are highlighted in this section.

11.8.2.1   ANIR Projects

The University of Alaska and the Arctic Region Supercomputing Center establish OC-3 links to vBNS+ to support an Arctic regional digital library, weather forecasts, and ocean and sea ice modeling. California State University at Hayward also supports OC-3 links to vBNS+ for enabling remote operation of scientific instruments, QoS (Quality of Service) routing for multipoint Internet teleconferences, and real-time visual models of satellite guidance systems.

Florida Agricultural and Mechanical University (FAMU) establishes OC-3 links to vBNS+ via the Florida GigaPoP for conducting research in distributed database processing, interactive simulation environments, and distributed real-time computing.

Florida Atlantic University (FAU) proposes to employ DS-3 links to vBNS+ to implement projects in medical and scientific imaging, HDTV, digital mammography, and visualization of activity in the human brain. FAU also sponsors research initiatives leading to interdisciplinary computational simulations and implementation of an independent oceanographic sampling network. Mississippi State University fosters connections to vBNS+ at DS-3 to enable distance learning applications, experimentation in wide area parallel computing, and participation in collaborative visualization projects. North Dakota State and South Dakota State Universities, the South Dakota School of Mines and Technology, and the Universities of North Dakota and South Dakota employ DS-3 links to vBNS+ to support projects in the life sciences, cosmology, and Upper Missouri River basin hydrology.

In Ohio, Wright State University supports DS-3 connections to vBNS+ for deployment of projects in multimedia data warehousing, asynchronous and synchronous collaboration in distributed VEs (Virtual Environments), and molecular modeling and visualization. The University of Oklahoma and Oklahoma State University provision OC-3 links to Abilene to conduct telemedicine research and projects in numerical weather prediction and distance education. The University of Oregon establishes DS-3 links to vBNS+ to conduct research in geological sciences, high-energy physics, and astronomy.

11.9   DIGITAL LIBRARIES (DLS)

11.9.1   DIGITAL LIBRARIES FEATURES AND FUNCTIONS

Also called electronic libraries and virtual libraries, digital libraries (DLs) employ high-performance, high-speed broadband network configurations for provisioning universal access to distributed multimedia information resources. Digital library

(DL) implementation involves a clarification of procedures for developing and maintaining a secure, scalable, and extendible DL infrastructure that accommodates a diverse array of resources for enabling users to readily retrieve desired data. A DL requires a storage system that maintains vast amounts of information in diverse formats, advanced software for browsing and navigation, intelligent user interfaces, and multilingual services. DL implementations also establish Acceptable Use Policies (AUPs); guidelines for reference services, collection development, and intellectual property rights; and copyright procedures for material in digital formats. DLs support curricular enhancements in disciplines that include the earth sciences, mathematics, engineering, computer science, the arts, the humanities, microbiology, history, and high-energy physics.

11.10   FEDERAL DIGITAL LIBRARY (DL) INITIATIVES

11.10.1   D-LIB Forum (Digital Libraries Forum)

Sponsored by the U.S. Department of Defense Advanced Research Projects Agency (DARPA), the D-Lib Forum promotes the creation of an internetworked global digital library system. This system consists of diverse library collections in all types of formats and complies with the Open Archival Information Standard (OAIS) developed by the International Standards Organization (ISO). The D-Lib Forum also promotes utilization of quantitative research methods for benchmarking performance of DL services and evaluating capabilities of DL component technologies.

11.10.2   Digital Libraries Initiative-Phase 1 (DL-1) and DL-2 (DL-Phase 2)

Operational from 1994 to 1998, DL-1 (Digital Libraries Initiative-Phase 1) established approaches for digitizing special collections, such as historical photographs, archival records, sheet music, and museum images, and developing networked digital libraries in disciplines that included earth and space sciences, economics, biosciences, geography, the humanities, and the arts. DL-1 sponsors included NSF, DARPA, NLM (National Library of Medicine), the Library of Congress, NASA, the FBI (Federal Bureau of Investigation), and the Smithsonian Institution.

Initiated in 1999, DL-2 builds on earlier Phase 1 efforts in moving forward with methods for establishing a broadband network infrastructure and storage repositories to enable on-demand access to DL content such as photographic images; 3-D simulations; and historical, museum and scientific slide collections. In Phase 2, approaches for creating digital representations of multimedia; algorithms and intelligent systems for cataloging, classifying, and indexing DL content; linking information objects and documents; and preserving metadata are established. Guidelines for information retrieval, natural language analysis, and knowledge management of large-scale digital collections are developed. Academic institutions that participate in DL initiatives include Johns Hopkins, Kentucky, Pennsylvania, California at Davis, Washington, and Michigan State Universities and the Universities of Michigan and Texas at Austin.

11.11   U.S. DIGITAL LIBRARY INITIATIVES

11.11.1   CALIFORNIA

11.11.1.1   California Digital Library (CDL)

The California Digital Library (CDL) supports access to statewide multimedia resources and development of an experimental infrastructure for interlinking distributed archival collections. Approaches for metadata management are under consideration.

11.11.1.2   Stanford University Digital Library Technologies Project

The Stanford University Digital Library Technologies Project supports dependable access to electronic library resources via wireline and wireless networks. Protocols for implementation of a distributed replicated Web cache, strategies for using the 3Com Palm Pilot for library applications, approaches for intellectual property protection, query translators for enabling library users to compose search requests, and filtering tools for identifying relevant documents are in development. The San Diego Supercomputer Center (SDSC), the University of California at Santa Barbara (UC Santa Barbara), and the University of California at Berkeley (UCB) participate in this initiative.

11.11.1.3   University of California at Berkeley (UCB)

Sponsored by NSF, DARPA, and NASA, the University of California at Berkeley (UCB) digital library project supports development of technologies and tools for accessing massive distributed digital collections of video maps, aerial photographs, geographical data, botanical datasets, environmental reports, and satellite images. Tools for content analysis, annotation, multimedia indexing, image analysis, and seamless retrieval of DL resources are in development.

11.11.1.4   University of California at Santa Barbara (UC Santa Barbara)

The University of California at Santa Barbara (UC Santa Barbara) Alexandria Digital Library (ADL) project supports implementation of distributed digital libraries containing geospatial multimedia resources such as map records, images, gazetteers, aerial photographs, scientific datasets, and texts that provision information in earth and social sciences. A successor project to ADL, ADEPT facilitates development of comprehensive meta-information systems and geospatial collections and uses Planet Earth as a metaphor for organizing and presenting DL resources on topics that include the environment and water resource management.

11.11.2   ILLINOIS

11.11.2.1   University of Illinois at Urbana-Champaign

The Engineering Library at the University of Illinois at Urbana-Champaign sponsors the DeLiver (Desktop Link to Virtual Engineering Resources) initiative to enable access to the full text of articles published since 1995 in computer science, engineering, and physics journals.

11.11.3 MARYLAND

11.11.3.1 University of Maryland at College Park

The University of Maryland at College Park conducts the Digital Libraries for Children initiative. This initiative promotes the design and implementation of a digital library specifically for youngsters and features images, text, video, and audio resources pertaining to animals.

11.11.4 NEW YORK

11.11.4.1 Columbia University

Sponsored by Columbia University, the PERSIVAL (Personalized Retrieval and Summarization of Image, Video, and Language Resources) project enables access to online medical literature and consumer health information for patients and healthcare providers. The DL PERSIVAL initiative employs secure digitized patient records at Columbia Presbyterian Medical Center (CMPC) for construction of a model that predicts user information needs and interests. Pilot tests are carried out at the CMPC Heart Failure Center and the CMPC Diabetes Center. In addition to PERSIVAL, Columbia University sponsors a DL initiative in the earth sciences that supports access to multimedia resources, enables links to digital research publications, and promotes development of an online technology system for presenting information on Planet Earth.

11.11.5 PENNSYLVANIA

11.11.5.1 Carnegie Mellon University (CMU)

Carnegie Mellon University (CMU) conducts Phase 1 and Phase 2 of the Informedia Digital Video Library initiative. This initiative fosters development of sophisticated video collages for enabling users to effectively browse multimedia documents; new approaches for automated audio and video indexing; procedures for enabling comprehensive multimedia visualizations; and innovative search and retrieval operations.

11.11.6 TEXAS

11.11.6.1 University of Texas at Austin

The University of Texas at Austin implements a digital repository of skeletal elements from small- and large-sized species for scientific study and research. A discovery user interface enables novice and advanced users to readily access source materials.

11.11.7 VIRGINIA

11.11.7.1 Virginia Polytechnic Institute and State University (Virginia Tech)

Virginia Tech supports the digitization of comprehensive collections of dissertations and theses; collections of instructional materials maintained by the Computer Science Teaching Center; and a repository of material related to Web transactions for the Open Archives Initiative (OAI), formally known as the Universal Preprint Service Project.

11.12 INTERNATIONAL DL INITIATIVES

As with U.S. DL initiatives, international DL projects support access to rapidly growing digital collections of maps, archival records, video and audio recordings, and historical images and texts. Like their U.S DL counterparts, international DL initiatives employ state-of-the-art technology, knowledge management techniques for effective information retrieval, and high-speed broadband networks for provisioning dependable access to and delivery of DL resources.

11.12.1 JAPAN

11.12.1.1 National Diet Electronic Library

The Japanese National Diet Electronic Library supports development of DL projects focusing on children's books and Japanese legends. A full-text database provisions access to these multilingual resources.

11.12.2 NEW ZEALAND

11.12.2.1 New Zealand Digital Library

Developed by the University of Waikato, the New Zealand Digital Library maintains extensive digital collections of computer science technical reports and literary works.

11.12.3 UNITED KINGDOM

11.12.3.1 United Kingdom (U.K.) Electronic Libraries Program

The United Kingdom (U.K.) Electronic Libraries Program sponsors the MODELS (Moving to Distributed Environments for Library Services), EDDIS (Electronic Document Delivery), and ESPERE (Electronic Submission and Peer Quality Review Project) initiatives to facilitate DL implementation, and NETSKILLS to support user training. In addition, the U.K. Electronic Libraries Program sponsors the Distributed National Electronic Resource (DNER) project to provision access to Web resources such as textbooks, maps, music scores, geospatial images, videos, and sound recordings. The U.K. Electronic Libraries Program also implements the JTAP (Joint Information Systems Committee Technology Applications Program) initiative to foster development of the Data Archive, a national resource center for digital data in the sciences and humanities. Furthermore, the JTAP project supports development of comprehensive digital collections of U.K. census statistics, government surveys, digital maps, and scientific datasets; sponsors the Resource Discovery Network (RDN); and fosters distributed access to advanced Web resources via subject gateways. Participants in the U.K. Electronic Libraries Program include the British Library, the Cooperative Academic Information Retrieval Network for Scotland, and SEREN (Sharing Educational Resources in an Electronic Network).

11.12.4   DB2 Digital Library Initiative

Developed by IBM, the DB2 Digital Library Initiative supports conversion of multimedia resources into digital formats for distribution via public and/or private networks. Designed for a diversified clientele, the DB2 Digital Library Initiative facilitates access to digitized collections of CBS News programs and voice, video, and/or data resources at the Library of Congress, the Vatican Library, the National Palace Museum in Taiwan, and the State Hermitage Museum in Russia. Participants in this international initiative include the Indiana University School of Music, the National Library of the Netherlands, the Osaka National Museum of Ethnology.

11.13   EUROPEAN COMMISSION TELEMATICS FOR LIBRARIES PROGRAM

11.13.1   Development of a European Service for Information on Research and Education-I (DESIRE-Phase I) and DESIRE-II (DESIRE-Phase II)

Sponsored by the European Commission Telematics for Library Program, DESIRE-I (Development of a European Service for Information on Research and Education-Phase I) supported implementation of secure and confidential search services that used an automated Web metadata indexing system for locating specific information on the Internet. DESIRE-I established subject-based search services that reflected the classification and description of network resources and created regional search services that were based on metadata indexes generated by automated Web crawlers

A follow-on initiative to DESIRE I, DESIRE-II (DESIRE-Phase II) supports development of approaches to identify, select, and retrieve Web-based multimedia resources via a high-speed ATM infrastructure. DESIRE-II facilitates design of a common user interface and a mesh caching system that provisions access to frequently consulted Web resources, facilitates speedy retrieval, and conserves available bandwidth for enabling additional applications. The mesh caching system also provides a foundation for a pan-European caching network. In 2000, DESIRE-II produced a toolkit that supports metadata queries and features automatic classification and indexing tools. The Universities of Twente, Newcastle-upon-Tyne, Bath, Utrecht, and Bristol and Queens University of Belfast participate in DESIRE initiatives.

11.14 EUROPEAN RESEARCH CONSORTIUM FOR INFORMATICS AND MATHEMATICS (ERICM)

A sponsor of high-performance networking initiatives in the European research community, ERICM supports investigations in fields that include digital libraries, programming languages, environmental modeling, network technologies, user interfaces, and electronic commerce. ERICM also sponsors MBS (Managed Bandwidth Service) initiatives.

11.14.1 ERICM HIGH-PERFORMANCE NETWORKING PROJECTS

11.14.1.1 *CRUCID*

CRUCID promotes development of information technologies for water resources management and high-performance computing networks that generate models and simulations of semi-arid areas. CRUCID simulations enable flood prevention and environmental impact studies.

11.14.1.2 ESIMEAU

ESIMEAU supports implementation of an advanced water resource management information system that integrates spatial geographic databases and data management tools for enabling planners to make informed decisions on water-related issues.

11.14.1.3 *SIMES*

SIMES supports implementation of a sophisticated multimedia monitoring system for tracking the utilization of renewable resources and the effects of human activities on the environment in the sub-Saharan region of Africa.

11.14.1.4 *THETIS*

THETIS fosters implementation of multimedia repositories and an advanced visualization system for providing coastal zone management services and monitoring topographical changes in the Mediterranean region.

11.14.2 ERCIM DELOS WORKING GROUP AND THE DIGITAL LIBRARY (DL) INITIATIVE

The ERCIM DELOS Working Group supports development of integrated, interlinked, large-scale, multidisciplinary repositories to serve as the framework for the ERCIM DL (Digital Library) initiative. The DELOS Working Group collaborates with NSF Task Forces in designing approaches to enable interconnections between the ERCIM DL and DLs in the United States and defines methods that facilitate intellectual property protection, resource indexing, and resource discovery in a globally integrated distributed library for the research community. Participants in the DELOS Working Group include the Swiss Federal Institute of Technology, the

Russian Academy of Science, and the Norwegian University of Science and Technology.

11.14.3 ERCIM MECCANO (Multimedia Education and Conferncing Collaboration over ATM Networks) Initiative

The ERCIM MECCANO (Multimedia Education and Conferencing Collaboration over ATM Networks) initiative supports implementation of a broadband IP-over-ATM network infrastructure that enables IP multicasts and telecollaborative research in mathematics, medicine, and language. Additionally, the MECCANO broadband network platform interoperates with SMDS (Switched Multimegabit Data Service), DBS (Direct Broadcast Satellite), and ISDN (Integrated Services Digital Network) technologies. However, the quality of audio and sound, the speed of transmission, and the degree of interactivity with SMDS, DBS, and ISDN is not equivalent to transmission speeds and services provisioned by ATM technology.

RENATER2 (NREN of France, Phase 2) provides technical support and manages network operations for MECCANO links between sites in the United Kingdom and France. Trials benchmarking the capabilities of MECCANO toolsets for novice users are conducted on CA*net II and SuperJANET4 (Super Joint Academic Network, Phase 4) testbed segments. Participants in the MECANNO project include the University of London; the School of Slavonic and Eastern European Studies; and the Universities of Bremen, Stuttgart, Mannheim, and Oslo.

11.14.3.1 MECANNO MBS (Managed Bandwidth Service)

The MECANNO (Multimedia Education and Conferencing Collaboration over ATM Networks) initiative established procedures in alpha and beta tests for setting up MBS accounts, defined operations performed by the ATM NOC (Network Operations Center) in support of MBS functions, and identified and resolved problems associated with internetworking NREN implementations. Alpha and beta trials via the MECANNO platform also confirmed MBS reliability. MBS alpha trials at the University College London, Essen University, and the University of Stuttgart verified MBS effectiveness in provisioning committed bandwidth for VPN implementations and MBS support of interactive telecollaborative research. In 1999, TEN-155 (Trans European Network-155.52 Mbps) conducted beta tests that verified MBS (Managed Bandwidth Service) capabilities in supporting virtual classroom sessions between schools situated in Ottawa and Edmonton in Canada; Dublin, Ireland; Basel, Switzerland; and Berlin, Germany.

11.15 TEN-155 (TRANS-EUROPEAN NETWORK-155.52 MBPS)

11.15.1 TEN-155 Features and Functions

A successor network to TEN-34, TEN-155 provisioned an international infrastructure for connecting next-generation networking initiatives such as APAN, I2, and

National Research and Education Networks (NRENs) in countries that included the Czech Republic, Iceland, Hungary, Norway, Switzerland, and Slovenia. To facilitate development of a pan-European network infrastructure, TEN-155 supported IP multicasts, high-speed multimedia transmissions, advanced telecollaborative research, and implementation of new networking technologies, and VPN (Virtual Private Network) deployments that used MBS (Managed Bandwidth Service). TEN-155 also operated connection points in Amsterdam and Frankfurt for peer-level traffic exchange and employed a combination of IP, ATM, and SDH (Synchronous Digital Hierarchy) technologies to support interoperable networking applications and projects. DANTÉ (Delivery of Advanced Network Technology to Europe, Ltd.) managed TEN-155 operations.

The DANTÉ Network Engineering and Planning Group supported migration from TEN-34 (Trans-European Network-34.368 Mbps) to TEN-155 and initially conducted pilot tests to verify capabilities of high-speed services and applications slated for TEN-155 implementation. In addition to DANTÉ, the Trans-European Research and Education Networking Association (TERENA), and the joint TERENA and DANTÉ Task Force (TF-TANT) conducted tele-education and tele-research applications in the context of the Quantum Test Program (QTP). In 1999, DANTÉ established a 44.736 Mbps (T-3) connection between TEN-155 NRENs and the Abilene network via STAR TAP and the SBC/AADS NAP (Network Access Point) in Chicago. DANTÉ and DFN (NREN of Germany) also formed a Consortium called CAPE (Connecting Asia-Pacific and Europe) to establish interconnectivity between TEN-155 NRENs and major international networks such as APAN. At the outset, TEN-155 implemented an IP network overlay on top of the ATM platform and supported best-effort packet delivery service. However, multimedia services depend on QoS (Quality of Service) guarantees. Because these guarantees were not provided in a best-effort IP network, TEN-155 subsequently deployed MECCANO-supported MBS (Managed Bandwidth Service).

DANTÉ enhanced the capabilities of the TEN-155 infrastructure by upgrading the central SDH ring linking NRENs in France, Germany, the United Kingdom, and the Netherlands to support increased rates at 622.08 Mbps (OC-12). EuroCERT (European Union Computer Emergency Response Team) provisioned security services for TEN-155 members. NREN participants in EuroCERT include ACONET (NREN of Austria), ARNES (NREN of Slovenia), DFN (NREN of Germany), Renater2 (NREN of France, Phase 2), SURFnet5 (NREN of the Netherlands, Phase 5), SuperJANET4 (Super Joint Academic Network, Phase 4), UNINETT (NREN of Norway), and RedIRIS (NREN of Spain).

It is interesting to note that TEN-155 NRENs transported commercial traffic and traffic generated by research centers and educational institutions. In contrast to TEN-155, the Abilene and the vBNS+ Internet2 (I2) backbone networks only carry high-performance research and education transmissions generated by I2 entities. The commodity or public Internet supports delivery of commercial traffic and traffic generated by educational institutions and research centers that are not I2 affiliates. (See Figure 11.4.)

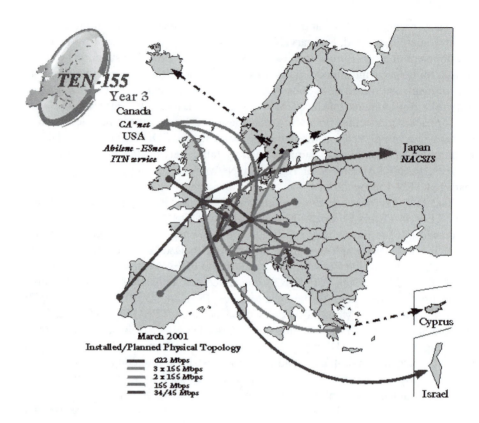

FIGURE 11.4 The TEN-155 configuration.

11.16 NATIONAL RESEARCH AND EDUCATION NETWORKS (NRENS) AND PEER-LEVEL REGIONAL NETWORKS

11.16.1 ARGENTINA

11.16.1.1 Internet2 Argentina

Sponsored by the Internet2 Argentina Foundation, Internet2 Argentina interlinks universities and research institutions via high-speed ATM connections. This network enables tele-education programs, supercomputing applications, and digital library projects.

11.16.2 AUSTRALIA

11.16.2.1 AARNET2 (Australian Academic and Research Network, Phase 2)

The Australian Academic and Research Network, Phase 2 (AARNet2) interconnects regional network hubs in the Australian Capital Territory, New South Wales, Queensland, the Northern Territory, Southern Australia, Victoria, Tasmania, and Western

Australia. AARNet2 hubs are interlinked to an ATM VPN. Also called the NREN (National Research and Education Network) of Australia, AARNet2 maintains links to international IP backbone networks; facilitates research associated with next-generation network technologies and applications; and enables initiatives such as voice-over-IP (VoIP) service for bypassing telephone toll rates.

In addition, AARNet2 provisions access to distributed Web spatial information systems in the fields of biological and environmental sciences and supports the implementation of standardized wireless systems at SOHO (Small Office/Home Office) venues. Network and computer security functions for AARNet2 are provided by the Australian Computer Emergency Response Team (AusCERT).

11.16.2.2   ABN (Australian Corporate Research Center for Advanced Computational Systems-Broadband Network)

An ATM regional network initiative, the Australian Corporate Research Center for Advanced Computational Systems-Broadband Network (ABN) supports transmission rates at 155.52 Mbps (OC-3) and provisions links to sites in Sydney, Adelaide, the Australian Technology Park, and the Bureau of Meteorology. ABN features a broadband infrastructure that supports DL initiatives, data mining operations, scientific research, and development of systems to support on-demand access to distributed data archives. In addition, the ABN Virtual Reality (VR) initiative facilitates utilization of 3-D immersive technologies for enabling industrial applications.

11.16.3   AUSTRIA

11.16.3.1   ACONET (Austrian Academic Computer Network)

Designed for scientific, academic, and research communities, the Austrian Academic Computer Network (ACONET), also known as the NREN of Austria, supports voice, video, and data services and applications via an ATM backbone network. This NREN fosters Fast Ethernet and Frame Relay connections and ATM-over-SDH metropolitan network implementations in Salzburg, Linz, Graz, and Vienna. SMDS provides emergency backup service in the event of ATM disruptions. Satellite and terrestrial connections foster information exchange between Vienna and Bucharest. The ACONET Internet Access Point (IAP) in Vienna is connected to the EBone (European Backbone) network.

11.16.4   BELGUIM

11.16.4.1   BELnet4
(National Research and Education Network of Belgium Phase 4)

BELnet4 (NREN of Belgium, Phase 4) is an ATM-over-SDH implementation that employs a star topology centered in Brussels. BELnet4 supports direct connections to academic and research networks at rates ranging from E-3 (34.368 Mbps) to OC-3 (155.52 Mbps). Schools, universities, research centers, libraries, and hospitals access BELnet4 via ADSL (Asynchronous Digital Subscriber Line), Fast Ethernet,

ISDN (Integrated Services Digital Network), or leased line connections. BELnet4 provisions MBone services, IP multicasts, interactive multimedia applications, MBS, and voice-over-IP (VoIP). In addition, BELnet4 supports trans-Atlantic connections via the New York PoP at 27 Mbps.

11.16.5 CANADA

11.16.5.1 Canadian Network for the Advancement of Research, Industry, and Education, Phase 2 (CA*net II), and CA*net3 (CA*net, Phase 3)

Sponsored by CANARIE, CA*net II (Canadian Network for the Advancement of Research, Industry, and Education, Phase 2) and CA*net3 (CA*net, Phase 3) are advanced networking configurations that support high-performance, high-capacity broadband services. CA*net II and CA*net3 capabilities are examined in Chapters 2 and 3.

11.16.6 CHILE

11.16.6.1 REUNA2 (Chile National University Network, Phase 2)

In 2000, IMPSAT Fiber Network, a STAR TAP provider, established initial connections between REUNA2 (Chile National University Network, Phase 2) and I2 networks via STAR TAP and the SBC/AADS NAP in Chicago. REUNA2 employs an ATM infrastructure for enabling videoconferencing, tele-education, interactive television broadcasts, IPv6 services, research in anthropology and ecology, and access to digital library collections.

11.16.7 CHINA

11.16.7.1 CERnet (NREN of China)

Funded by the Chinese Ministry of Education and the Chinese government, CERnet (NREN of China) enables broadband applications, IPv6 services, and educational and research initiatives in schools, universities, and scientific centers throughout the country. CERnet participants include Beijing and Shanghai Universities.

11.16.7.2 China High-Speed Network Testbed

The China High-Speed Network Testbed supports development of advanced technologies and next-generation networking applications. Beijing and Tsinghua Universities and the Chinese Academy of Sciences participate in this network initiative.

11.16.7.3 Joint Universities Computer Center (JUCC)

The Joint Universities Computer Center (JUCC) monitors network operations at institutions that include the Hong Kong University of Science and Technology, Hong

Kong Polytechnic University, and the Chinese and City Universities of Hong Kong. In addition, the JUCC provisions technical support for the Hong Kong Academic and Research Network (HARNET).

11.16.8 CROATIA

11.16.8.1 CARnet (Croatia Academic and Scientific Research Network or NREN of Croatia)

Featuring an ATM-over-SDH infrastructure, CARnet (Croatia Academic and Scientific Research Network, or the NREN of Croatia) interlinks universities and research centers in Croatia. Rates at 155.52 Mbps (OC-3) and 622.08 Mbps (OC-12) are supported. The CARnet infrastructure supports IP-over-ATM services and enables videoconferencing, VoIP, and VPN implementations. Moreover, CARnet maintains ATM links to the Zagreb Point of Presence (PoP). This PoP provides connections to the GÉANT Network via the EBone network node in Vienna.

11.16.9 CZECH REPUBLIC

11.16.9.1 TEN-155 CZ (NREN of the Czech Republic)

Developed by a consortium of academic institutions and the Academy of Science, TEN-155 CZ (NREN of the Czech Republic) employs an ATM infrastructure that supports tele-education projects, videoconferencing, and IP multicasts. In 2000, TEN-155 CZ initiated utilization of a fiber optic link enabling rates at 2.488 Gbps (OC-48) between network nodes in Prague and Brno. This fiber optic link employs Packet-over-SDH (POS) technology and works in concert with the in-place ATM infrastructure. TEN-155 CZ provided links to TEN-155 NRENs via a PoP in Prague at rates of 155.52 Mbps (OC-3) and a PoP in Germany at rates of 34.368 Mbps (E-3). TEN-155 CZ also supports connections to the GÉANT Network and vBNS+ via STAR TAP and the SBC/AADS NAP.

11.16.10 DENMARK

11.16.10.1 DARENET (NREN of Denmark)

DARENET (NREN of Denmark) provisions broadband services to universities and research institutions throughout Denmark and enables network applications for primary and secondary schools, universities, and vocational and technical centers.

11.16.10.2 DVUNI (Denmark Virtual University)

The Denmark Virtual University (DVUNI) is a tele-education initiative that provides access via a Web portal to post-graduate telecourses, academic teleprograms, and a digital research library.

11.16.11   ESTONIA

11.16.11.1   EENET (Estonia Educational and Research Network, or the NREN of Estonia)

EENET (Estonia Educational and Research Network, or the NREN of Estonia) supports connections to NORDUnet2 (Nordic Countries Network, Phase 2) at 4 Mbps. The NREN of Estonia provisions access to educational, scientific, cultural, and research applications, delivers teacher training courses, and provides links to network services for K–12 schools, research centers, and universities. EENET also participates in the GÉANT Network, CEEnet (Central and Eastern European Networking Association), and BALTnet (Baltic States Network).

11.16.12   FINLAND

11.16.12.1   FUNET (Finnish University and Research Network)

FUNET (Finnish University and Research Network) employs an IP-over-ATM-over-WDM (Wavelength Division Multiplexing) infrastructure for enabling transmission rates at 155.52 Mbps (OC-3). FUNET infrastructure upgrades for enabling rates at 2.488 Gbps (OC-48) are in development. FUNET supports links to I2 institutions and promotes access to supercomputing environments for enabling information grid activities. FUNET also participates in next-generation applications and programs such as the Scandinavian High-Performance Computing Center Network (HPC2N). The FUNET Network Drug Discovery Program promotes access to distributed molecular databases that form the foundation for drug production.

Accessible via FUNET, the DYNAMO (Dynamic Adaptive Modeling of the Human Body) initiative supports development of sophisticated mathematical models that contribute to an understanding of human physiological systems. Participants in DYNAMO include the Helsinki University of Technology and the Finnish Center for Scientific Computing.

FUNET utilizes an MBone multipoint videoconferencing system developed by the California Institute of Technology (Cal Tech) and CERN (European Organization for Nuclear Research) to deliver curricular services for the Finnish Virtual University. The Finnish Virtual University supports videoconferences, access to CERN lectures in high-energy physics that originate at the CERN webcast server, and real-time delivery of telecourses. FUNET TV broadcasts Finnish Virtual University programs and provides video-on-demand for curricular enrichment. The Slovak University of Technology and the Universities of Twente and Rome participate in the Finnish Virtual University initiative.

The Finnish Center for Scientific Computing also provides FUNET service to Finnish research centers, colleges, and universities. FUNET participants include the Helsinki Institute of Physics, the Oulu Institute of Technology, and Oulu and Tampere Universities.

11.16.12.2   Scandinavian HPC2N (High-Performance Computing Network)

Sponsored by Finland, Sweden, and Norway, the Scandinavian High-Performance Computing Network (HPC2N) supports links to shared supercomputing facilities that include the Finnish Center for Scientific Computing and the Swedish Center for Parallel Computers. A HPC2N participant, the Parallab (Parallel Computation Laboratory) is a state-of-the-art computing facility in Norway that fosters research in biotechnology and the natural sciences.

11.16.13   FRANCE

11.16.13.1   Renater2 (NREN of France, Phase 2)

Renater2 (NREN of France, Phase 2) interlinks LANs, MANs, and regional configurations into a high-performance national network that employs ATM-over-SDH technologies. Renater2 supports links to the GÉANT Network and previously participated in TEN-155. Sponsored by France Telecom and scientific, research, education, and government agencies, Renater2 facilitates development of next-generation network architectures and protocols and implementation of a high-performance infrastructure to verify IPv6 capabilities and multicast services.

Renater2 enables telecollaborative initiatives in astrophysics, astronomy, physical sciences, mathematics, and high-energy physics between French scientific institutions and universities and I2 participants, including the NASA Jet Propulsion Laboratory, Johns Hopkins and North Carolina State Universities, and the Universities of Kentucky and Chicago. Sponsored by Renater2, the Integrated Architecture for Networks and Services initiative promotes implementation of next-generation intranets and a second-generation Internet. A Renater2 project, EDISON (European Distributed Interactive Simulation Over Network) facilitates development of a distributed network environment for enabling interactive simulations that operate over a broadband infrastructure. The Renater2 Digital Video Group supports real-time multimedia videoconferences and video-on-demand (VOD) initiatives.

Renater1 provided the foundation for Renater2. The Public Internet Group, GIP, initiated Renater1 network management operations in 1993 and now monitors Renater2 network services. GIP participants include the French National Center for Scientific Research, the French National Space Agency, the French Atomic Energy Agency, and the French National Research Institute for Informatics.

11.16.13.2   French National Institute for Research in Computer Science and Control

The French National Institute for Research in Computer Science and Control is an enabler of advanced next-generation networks that support multimedia transmission via heterogeneous networking platforms, distribution of streaming video and audio, and IPv6 services. In addition, this institute develops procedures for introducing DBS (Direct Broadcast Satellite) technology in tele-education programs and sponsors research in computer science and applied mathematics.

11.16.14 GERMANY

11.16.14.1 DFN or Deutsche Forschungsnetz (NREN of Germany)

A nationwide broadband communications network, DFN or Deutsche Forschung-snetz (NREN of Germany) supports telemedicine, tele-education, and teleresearch initiatives. The DFN SDH-over-WDM platform enables real-time multimedia applications with QoS guarantees, IP multicasts, and SDH point-to-point connections for transmissions at 2.488 Gbps (OC-48). In addition, DFN WinShuttle supports dial-up and ISDN links for enabling schools, libraries, museums, and small businesses to access DFN services. DFN also provides links to I2 networks via the G-WiN (Gigabit-Wissenschaftsnetz) PoP in Washington, D.C.

An ATM VPN (Virtual Private Network), B-WiN (Breitband-Wissenschaftsnetz) supported core DFN operations until its replacement by Gigabit-Wissenschaftsnetz (G-Win). A high-speed broadband network, G-WiN employs an SDH-over-WDM (Wavelength Division Multiplexing) platform. Sponsored by DFN and GMD (National Research Center for Information Technology), DFN Testbed West inter-links institutions that include GMD, DLR (German Aerospace Center), and the Universities of Bonn and Cologne. DFN Testbed South connects sites in Berlin, Erlangen, and Munich. These gigabyte testbeds support pilot tests of new technologies.

11.16.15 GREECE

11.16.15.1 GRnet (NREN of Greece)

Previously associated with TEN-155, GRnet (NREN of Greece) supports links to GÉANT networking services. GRnet nodes are situated at sites that include Salonika, Patras, Larisa, Athens, and Heraklion. The National University of Athens, a GRnet participant, manages GRnet operations and provides technical support for the Athens Internet Exchange (AIX).

11.16.16 HUNGARY

11.16.16.1 HUNGARnet (NREN of Hungary)

Sponsored by the National Information Infrastructure Development Program, HUN-GARnet (NREN of Hungary) supports connections to the GÉANT Network. HBONE (HUNGARnet Internet Backbone Network) interlinks regional PoPs (Points of Presence) in a nationwide configuration; provides network services for Hungarian libraries, educational institutions, and research centers; and supports information transport via ATM, FDDI (Fiber Data Distributed Interface), microwave, leased-line, and optical fiber technologies. HUNGARnet participants include the Hungarian Academy of Sciences, the Hungarian National Committee of Technical Development, and the Hungarian Ministry of Culture.

11.16.17 Iceland

11.16.17.1 IsNet (Internet Iceland Inc., NREN of Iceland)

IsNet (Internet Iceland Inc., NREN of Iceland) supports academic and research applications in Iceland. This NREN employs a 44.736 Mbps (T-3) link to New York and a 2 Mbps link to NORDUnet2 for enabling access to international broadband network services.

11.16.18 Ireland

11.16.18.1 HEAnet (NREN of Ireland)

HEAnet (NREN of Ireland) is a broadband high-performance nationwide network that supports ATM, IP, Ethernet, and SDH technologies. HEAnet provides technical support for the Irish Neutral Exchange Point (INEX) and provisions links to the GÉANT Network and Internet2 (I2). Participants in HEAnet include the National Library of Ireland; the Limerick, Dundalk, and Waterford Institutes of Technology; and the Universities of Limerick and Ireland.

11.16.19 Israel

11.16.19.1 Israeli Internet-1 and Israeli Internet-2

Supported by the IUCC (Inter University Computation Center or MACHBA), the Israeli Internet-1 initiative provisioned connections to TEN-155 and Q-MED (Quality Network Technology for User-Oriented Multimedia in the Eastern Mediterranean Region) projects via an undersea optical fiber link that terminated at the London PoP (Point of Presence) and at Bar-Ilan University. Rates at 34.368 Mbps (E-3) were supported.

Currently in Phase 2, the Israeli Internet-2 configuration is an extension and enhancement of the IUCC (Inter University Computation Center or MACHBA) Israeli Internet-1 network. Israeli Internet-2 enables telecommuting, teleresearch, VR simulations, distance learning, and advanced scientific projects and features an IP-over-ATM infrastructure that provides IPv6 services. Israeli Internet-2 connects through STAR TAP to the SBC/AADS NAP in Chicago at 44.736 Mbps (T-3) via a satellite link and provides connections to the GÉANT Network, Euro-Link, Trans-PAC, and SingAREN (Singapore Advanced Research and Education Network).

11.16.19.2 Israeli Academic and Research Network

Supported by the IUCC, the Israeli Academic and Research Network interlinks Technion, Bar-Ilan, Haifa, Ben-Gurion, and Tel-Aviv Universities; the Weizmann Institute of Science; and the Hebrew University of Jerusalem in a configuration that extends to 240 kilometers. The Israeli Academic and Research Network also participates in Israeli Internet-2.

11.16.9.3   Israel One

A component in the Israeli Internet-2 Program, Israel One is an ATM distance education network initiative that interlinks 16 high schools representative of Israeli cultural diversity. The Israel One initiative enables teachers at a distance to help students use broadband technologies in high school classrooms and provisions interactive telecourses in English and math. Israel One projects are based on initiatives developed by the North Carolina School of Science and Mathematics for implementation on the NCIH (North Carolina Information Highway).

11.16.20   ITALY

11.16.20.1   GARR (NREN of Italy)

GARR (NREN of Italy) provisions telecommunications services that interlink universities and scientific research institutions in a configuration called MURST (Italian Ministry of Universities-Scientific and Technological Research) and supports connections to the GÉANT Network. Sponsored by major Italian academic institutions and research centers, INFN (Italian Institute for Particle and Nuclear Physics Research) provisions advanced networking services throughout Italy via the GARR-B broadband backbone network. INFN also enables next-generation metacomputing experiments for evaluating grid performance and transmission rates ranging from 622.08 Mbps (OC-12) to 2.488 Gbps (OC-48).

11.16.21   JAPAN

11.16.21.1   GEMnet (Global Electum Cyber Society Network)

A next-generation experimental testbed initiative sponsored by NTT (Nippon Telegraph and Telephone), GEMnet enables development of advanced networking technologies, applications, and services. GEMnet interoperates with MREN and CA*net2 via STAR TAP and the SBC/AADS NAP in Chicago, and maintains links to international NTT research centers. A scalable, extendible, and flexible networking solution, GEMnet employs an ATM-over-WDM infrastructure. Developed by NTT and the University of Malaysia, lightning surge protection devices enhance GEMnet system reliability.

GEMnet works in concert with wireline and wireless cable networks, Fiber-to-the-Home (FTTH), and satellite broadband residential access solutions; supports development of virtual ATM LANs; and promotes implementation of a seamless global mobile computing infrastructure based on the Personal Handyphone System (PHS).

In the distance education domain, GEMnet interconnects primary and secondary schools affiliated with Keio University and the Keio University Graduate School campus in New York. This project employs the Computer Aided Learning and Authoring Environment (CALAT) toolset for tele-instruction, supports interactive videoconferencing, and enables access to digital research libraries. In the telemedicine arena, GEMnet enables telediagnoses, teleradiology, and teleconsultations between physicians at the Keio University School of Medicine and the Cleveland Clinic. For telecommuters, GEMnet implements a VTOA (Voice-and-Telephony-over-ATM) gateway to enable access to high-quality VTOA services in virtual office environments.

In addition, NTT and the Kochi University of Technology support a joint initiative called Interspace to enable utilization of collaborative virtual workplace systems for telecommuters that correspond to actual physical communities. GEMnet and the Kochi Information Super Highway in southwestern Japan provision technical support for the Interspace initiative.

11.16.21.2 Japanese Gigabit Network (JGN)

Designed by the Telecommunications Advancement Organization (TAO) of Japan, the Japanese Gigabit Network (JGN) supports assessment of next-generation network technologies in pilot tests via an optical infrastructure that enables rates at 2.488 Gbps (OC-48). JGN participants include educational institutions, research centers, and private companies.

11.16.21.3 Networked Virtual Environments Collaborative Trans-Oceanic Research (N*VECTOR)

NTT and the Universities of Tokyo and Illinois at Chicago sponsor a collaborative research project called N*VECTOR. This trans-oceanic initiative employs a network architecture that enables sound, video, data, and imaging transmission in real-time with QoS assurances.

N*VECTOR also fosters implementation of an innovative 3-D VR system by interconnecting the Computer Augmented Booth for Image Automation developed by the University of Tokyo and the Cave Automatic Virtual Reality Environment (CAVE) developed by the Electronic Visualization Laboratory (EVL) at the University of Illinois at Chicago. This integrated system employs a high-performance network infrastructure to support shared virtual environments and telecollaborative activities. Individuals who are in physically separated locations can interact with one another in N*VECTOR-supported virtual environments as if they were physically together in the same room. EVL also implements the ImmersaDesk virtual reality (VR) system for enabling interactive graphics, advanced VR applications, and scientific visualizations in fields such as tele-education and telemedicine. Approaches for using video, audio, and database technologies to enable telecollaborative virtual environments for real-time data and art visualizations are in development. EVL also participates in STAR TAP, StarLight, and iGrid initiatives, and works with the National Center for Supercomputing Applications (NCSA) and iCAIR in sponsoring international collaborative research.

11.16.21.4 Real Internet Consortium (RIC)

The Real Internet Consortium (RIC) promotes deployment of advanced information networking technologies for enabling ultra-fast transmission rates in distributed networking environments. RIC projects foster development of next-generation QoS routing and multicast protocols that facilitate transmission of vast volumes of video, voice, and data to large numbers of users. Approaches for enabling network transmission at terabytes per second (Tbps) and petabytes per second (Pbps) rates are in development. RIC participants include Kyoto, Hiroshima, Kyushu, and Tokyo Universities; the University of Michigan; and the Tokyo Institute of Technology.

11.16.21.5 WIDE (Widely Integrated Distributed Environment) Network

The WIDE (Widely Integrated Distributed Environment) Network promotes development of an integrated large-scale distributed infrastructure to support global tele-collaboration and tele-exchange of technical information. The WIDE Network enables 6Bone trials for verifying IPv6 services, employs a multigigabit optical backbone network based on DWDM (Dense WDM) technology, and facilitates peer-level traffic exchanges with advanced broadband networks such as APAN and I2. Academic institutions affiliated with the WIDE Network include Keio, Osaka, and Stanford Universities; the Tokyo University of Technology; the Nara Institute of Science and Technology; and the Universities of Southern California and Wisconsin.

11.16.22 KOREA

11.16.22.1 APAN-KR (Asia-Pacific Advanced Network-Korea)

The Asia-Pacific Advanced Network-Korea (APAN-KR) sponsors high-speed, high-performance broadband networking projects in high-energy physics and enables VOD (Video-On-Demand), Web caching, and multicast services. Korea and Seoul National Universities participate in the APAN-KR Consortium.

11.16.23 MALAYSIA

11.16.23.1 Telekom Malaysia

Telekom Malaysia sponsors development of an optical fiber backbone network that supports voice, video, and data transport at rates between 2.488 Gbps (OC-48) and 10 Gbps (OC-192). This optical fiber network interoperates with ATM technology for enabling development of a Malaysian NREN that supports E-government, tele-medicine, and tele-education applications and functions.

11.16.24 MEXICO

11.16.24.1 CUDI (NREN or National Research and Education Network of Mexico)

The NREN of Mexico (CUDI) supports implementation of the Real Virtual Learning Space, a multifunctional, multiservice high-speed next-generation network platform for enabling telemedicine and tele-education applications. Plans for a high-speed I2 backbone link between CalREN-2 and CUDI are in development

11.16.25 Netherlands

11.16.25.1 SURFnet5 (NREN of the Netherlands, Phase 5)

SURFnet5 (NREN of the Netherlands, Phase 5) supports video-over-IP, IP multicasts, distance learning projects, development of a digital archives, and transmission rates at 2.488 Gbps (OC-48) via an ATM infrastructure. SURFnet5 also implements the FRIENDS (Framework for Integrated Engineering and Deployment of Services) initiative for provisioning access to broadband applications in an integrated networking environment. The feasibility of using cable network connections for connecting to SURFnet5 is examined in pilot tests at Nijmegen University. Capabilities of ADSL and GPRS (General Packet Radio Service) in supporting connectivity of SOHO (Small Office/Home Office) venues to SURFnet5 are explored as well. In 2001, Teleglobe began providing a 2.488 Gbps optical connection between the SURFnet5 backbone network at Hempoint in Amsterdam and STAR TAP via the SBC/AADS NAP in Chicago.

SURFnet5 also sponsors the GigaPort Network for evaluating the capabilities of an advanced optical communications infrastructure in provisioning multimedia services. The GigaPort Network is the core component of the GigaPort Project, an initiative that supports development of new networking services and applications.

11.16.26 Norway

11.16.26.1 UNINETT (NREN of Norway)

UNINETT (NREN of Norway) enables high-speed broadband tele-education and teleresearch applications, IP multicasts, and real-time videoconferencing services at rates of 2.488 Gbps (OC-48). Also sponsored by UNINETT, SUPERNETT-II participates in GÉANT, the next-generation pan-European network. SUPERNET-II currently supports ATM implementations in Tromso, Trondheim, Oslo, Kjellor, and Bergen and deployment of a national IPv6 testbed that enables network services at 1 Gbps. In addition, SUPERNETT-II maintains an IP DiffServ (Differentiated Services) network testbed that enables an evaluation of the IP DiffServ protocol in provisioning IPv6 services.

11.16.27 Poland

11.16.27.1 CYFRONET (Polish Academic and Research Network)

An ATM-based network, CYFRONET (Polish Academic and Research Network) supports voice, video, and data delivery at 155.52 Mbps (OC-3) and 622.08 Mbps (OC-12).

11.16.27.2 POL-34/155 (NREN of Poland-34.368 Mbps/155.52 Mbps)

POL-34/155 established connectivity to the TEN-155 network backbone via a 34.368 Mbps (E-3) link in 1999. Subsequently, the POL-34/155 infrastructure was upgraded to support rates at 155.52 Mbps (OC-3). POL-34/155 is slated to participate in the GÉANT network configuration.

11.16.28 PORTUGAL

11.16.28.1 RCCN (NREN of Portugal)

RCCN (NREN of Portugal) migrated from the TEN-34 network to TEN-1555 in 1999. RCCN currently supports migration to the GÉANT Network and enables transmissions at 622.08 Mbps (OC-12).

11.16.29 ROMANIA

11.16.29.1 RNC.ro (NREN of Romania)

Established by the Romanian National Agency of Science, Technology, and Innovation, the Romanian National Commission for Informatics, and the Romanian Ministry of National Education, RNC.ro (NREN of Romania) provisions networking services for academic and research communities. Slated to participate in the GÉANT Network, RNC.ro also belongs to CEEnet (Central and Eastern European Networking Association).

11.16.30 SINGAPORE

11.16.30.1 National University of Singapore

The National University of Singapore sponsors a diverse array of network applications and collaborative research trials and initiatives. For iGrid 2000, the National University of Singapore Center for Development of Teaching and Learning demonstrated the feasibility of transmitting a human anatomy lecture on-demand. The National University of Singapore IRDU (Internet Research and Development Unit) supports connectivity to APAN (Asia-Pacific Advanced Network), the Asia Internet Interconnection Initiative (AII), and the ASEAN Sciences and Technology Network. IRDU also takes part in the Asia-Pacific Networking Group (APNG). IRDU became the first Singapore node and the third Asian country following Japan and Kazakhstan to support connections to the 6Bone configuration. In addition, IRDU operates a 6Bone testbed and manages IPv6 operations for SREN (Singapore Research and Education Network) and SingAREN (Singapore Advanced Research and Education Network).

The National University of Singapore and Boston Children's Hospital support development of a complex decision model integrating information from multiple sources to improve the quality of clinical services and control costs in healthcare management. In another collaborative initiative, the National University of Singapore and the Electronic Visualization Laboratory (EVL) at the University of Illinois at Chicago implement 3-D simulations for enabling product development.

11.16.30.1.2 National University of Singapore and Massachusetts Institute of Technology

Sponsored by the National University of Singapore and the Massachusetts Institute of Technology (MIT), the Global Design Studio facilitates creation of a virtual workspace for enabling urban designers and architects to work collaboratively on historical reconstruction projects.

11.16.30.1.3 National University of Singapore and Virginia Tech

The National University of Singapore and the Virginia Polytechnic Institute and State University (Virginia Tech) support development of a dynamic digital library architecture for enabling information search, exchange, and retrieval, and fast access to globally distributed DL resources. This jointly sponsored DL initiative contributes to the establishment of a global digital library to support tele-education and lifelong learning initiatives.

11.16.30.1.4 National University of Singapore, Stanford University, and Brookhaven National Laboratory

The National University of Singapore, Stanford University, and the Brookhaven National Laboratory collect x-ray crystallographic data for determining the structure of bio-macromolecules. This initiative supports development of new drugs for combating diseases such as AIDS.

11.16.30.2 SingAREN
(Singapore Advanced Research and Education Network)

A multiservice broadband network, SingAREN (Singapore Advanced Research and Education Network) enables innovations in network technologies, architectures, and protocols that contribute to NGI development. These innovations also foster enhancements to SingaporeOne, the country's commercial broadband network. SingAREN participants such as the National University of Singapore foster development of distance education, telemedicine, and E-commerce initiatives, and support testbed trials and actual implementations for evaluating capabilities of classical IP (CIP)-over-ATM, DiffServ, IPv6, MPOA (MultiProtocol-over-ATM), MPLS (MultiProtocol Label Switching), and ADSL (Asynchronous Digital Subscriber Line) technologies.

Formalized in 1999, the CAN-Sing (Canada-Singapore) IT (Information Technology) MoU (Memorandum of Understanding) supports collaborative implementation of next-generation technologies over CA*net3 and SingAREN to facilitate development of advanced network services and applications in fields that include education, research, and industry. Moreover, an optical transmission system based on ATM-over-WDM technology to support SingAREN operations at 10 Gbps (OC-192) is in development.

SingAREN implements an experimental broadband network that operates at 155.52 Mbps (OC-3) and maintains 14 Mbps connections to STAR TAP and the SBC/AADS NAP and 155.52 Mbps links to CA*net2 and CA*net3. SingAREN uses the CANARIE trans-Atlantic connections for interworking with NRENs sponsored by member states in the European Union. The SingAREN Technology Center provides network management and technical support services.

11.16.31 SLOVAK REPUBLIC

11.16.31.1 SANET (NREN of Slovakia)

The SANET (NREN of the Slovak Republic) backbone network connects key cities in Slovakia including Kosice and Bratislava. SANET supports links to TEN-155 and facilitates connections to the GÉANT Network.

11.16.32 SLOVENIA

11.16.32.1 ARNES (NREN of Slovenia)

The ARNES (NREN of Slovenia) infrastructure employs leased lines and routers with nodes of concentration (NOCs) in small towns to support rates at 256 Kbps and in large towns to enable transmissions at 2 Mbps. In addition, ARNES maintained a 47.736 (T-3) Mbps link to TEN-155 and currently operates a 2 Mbps link to U.S. networks via STAR TAP and the SBC/AADS NAP. ARNES distributes information on network security via SI-CERT; encourages utilization of PGP (Pretty Good Privacy) public key encryption; and holds membership in FIRST (Forum of Incident Response and Security Teams).

11.16.33 SPAIN

11.16.33.1 Red IRIS (NREN of Spain)

In 1999, RedIRIS migrated to the TEN-155 network initiative via a 34.368 Mbps (E-3) ATM link that extended from Madrid to the TEN-155 PoP in Paris. Currently, RedIRIS provides multicast applications via MBone, employs ISDN technology for provisioning backup services, supports high-speed broadband teleservices, and enables connections to the GÉANT Network. In addition, RedIris operates an ATM link to Catalonia where dual fiber optic SDH rings interconnect Catalan research institutions. Participants in the Catalan initiative include the Catalan Research Foundation and Center for Supercomputing and the University of Barcelona.

11.16.34 SWEDEN

11.16.34.1 SUNET (NREN of Sweden)

SUNET provisions networking services for museums, research centers, and educational institutions that include the Karolinska Institute, Uppsala and Lund Universities, and the University of Goteborg. Rates of transmission at 155.52 Mbps are supported. SUNET also participates in NORDUnet2 and the Scandinavian High-Performance Computing Network (HPC2N). A HPC2N participant, the National Supercomputer Center (NSC) at Linkoping University in Sweden provisions access to high-performance computational resources for supporting virtual reality (VR) projects in physics, chemistry, fluid mechanics, computer science, and mathematics.

11.16.35   SWITZERLAND

11.16.35.1   SWITCH (NREN of Switzerland)

The SWITCH Network provisions teleservices via an IP overlay network that works in concert with an ATM-over-SDH infrastructure and supports a trans-Atlantic link to STAR TAP via the SBC/AADS NAP. Affiliated with the GÉANT Network, the SWITCH Network implements a Web cache that employs a hierarchy of cooperating proxies linked together for supporting networking operations. CERN (European Organization for Nuclear Research); the Hotel School of Lausanne; and the Universities of Basel, Bern, Lausanne, Geneva, Zurich, and St. Gallen take part in the SWITCH Network initiative. The successor network of SWITCH, SWITCHng (Next-Generation SWITCH) is currently in development.

11.16.36   TAIWAN

11.16.36.1   TAnet2
(National Research and Education Network of Taiwan, Phase 2)

An NREN that provisions networking services to K–12 schools and universities, TAnet2 also supports links to the National Museum of Natural Science, public libraries, and hospitals. These entities connect to TAnet2 via the TAnet2 GigaPoP at rates reaching 44.736 Mbps (T-3). TAnet2 employs an ATM-over-SDH infrastructure and supports streaming audio and video, video-on-demand (VOD), IPv6 services, and IP multicasts. In addition, TAnet2 maintains a peering relationship with vBNS+ via STAR TAP and the SBC/AADS NAP.

11.16.36.1.1   HiNet

Sponsored by Chunghwa Telecom, HiNet provisions basic and advanced telecommunications services throughout Taiwan and also manages TAnet2 operations and the Government Services Network (GSN). In addition, HiNet is an enabler of network connections to Australia, France, Korea, Japan, China, Singapore, and the United States.

11.16.37   THAILAND

11.16.37.1   Ginet (Thailand Government Network)

Ginet (Thailand Government Network) provisions network services to every province in Thailand via an ATM infrastructure that fosters rates at 155.52 Mbps (OC-3). NECTEC (National Electronics and Computer Technology Center) provides technical support for Ginet, fosters development of a national cache infrastructure, and facilitates implementation of SchoolNet, a virtual school networking initiative.

11.16.37.2   Thailand Information Superhighway Testbed

Built by a public and private partnership, the Thailand Information Superhighway Testbed employs an ATM backbone network that supports E-commerce services and applications.

11.16.37.3 ThaiSarn (Thailand National Academic and Research Network)

ThaiSarn (Thailand National Academic and Research Network) supports links to every state-owned university throughout the country and enables transmission rates at 622.08 Mbps (OC-12).

11.16.38 UNITED KINGDOM

11.16.38.1 SuperJANET 4 (Super Joint Academic Network, Phase 4)

A high-speed, high-performance NREN, SuperJANET4 capabilities are examined in Chapter 2.

11.17 GÉANT NETWORK

Sponsored by the European Commission and approximately 30 NRENs, the GÉANT Network is a pan-European next-generation research network that initiated operations on November 1, 2001. A successor to TEN-155, the GÉANT Network builds on TEN-155 foundations in providing advanced networking services to NRENs in the European Union. DANTÉ supports continued operations of TEN-155 testbeds, programs, and projects until migrations to the GÉANT Network are completed. To connect to the GÉANT Network, NRENs must support minimum rates of 34.368 Mbps (E-3). The GÉANT Network initially enables transmission rates at 2.488 Gbps (OC-48) with planned expansions to 10 Gbps (OC-192). With these expansions, the GÉANT Network will facilitate voice, video, and data transport at rates of 34.368 Mbps (E-3), 44.736 Mbps (T-3), 155.52 Mbps (OC-3 or STM-1), 622.08 Mbps (OC-12 or STM-4), 2.488 Mbps (OC-48 or STM-16), and 10 Gbps (OC-192 or STM-64).

GÉANT currently features ATM-over-SDH (Synchronous Digital Hierarchy) technology. However, plans for employing advanced optical infrastructure technologies such as WWDM and DWDM are in development. Capabilities of PONs (Passive Optical Networks) and 10 Gigabit Ethernet are also explored.

The GÉANT Network serves as a testbed for implementation of QoS assurances and IPv6 services that benefit the European education and research community. The GÉANT Network also benchmarks the performance of basic and premium IP services to accommodate specified QoS requirements. In addition, the GÉANT Network supports IP multicast distribution to facilitate effective broadcast delivery and videoconferencing applications.

Although the GÉANT Network employs ATM technology to provision transmissions with QoS guarantees, additional options for enabling QoS assurances are under consideration. Procedures to support migration from ATM-based MBS (Managed Bandwidth Service) to GCS (Guaranteed Capacity Service) based on MPLS (MultiProtocol Label Switching) and the BGMP (Border Gateway Management Protocol) to provide QoS assurances and enable secure VPN implementations are in development. In addition, GÉANT supports utilization of BGMP and MSDP (Multicast Source Discovery Protocol) for enabling seamless, dependable, and reliable inter-domain multicast distribution.

The GÉANT Network allocates Guaranteed Capacity Service (GCS) or subsets of network capacity to groups of users that conduct advanced research and value-added projects such as grid research. Approaches for using DWDM (Dense WDM) technology to provide Guaranteed Capacity Service (GCS) and developing a European Technology Access Point (EURO-TAP) interconnection point for enabling international networks to interface with peer-level European NREN counterparts are under consideration.

The GÉANT Network provisions value-added services such as Managed Bandwidth Service (MBS), IP multicasts, and IPv6 operations. A GÉANT Network initiative, the 6NET Project supports pilot trials over the GÉANT Network platform to test procedures that facilitate migration from IPv4 to IPv6 and clarifies approaches for provisioning IP Premium Service. The GÉANT Network testbed also supports development of tools to monitor the effectiveness of IP multicasts and measure traffic throughput. Moreover, he GÉANT Network facilitates implementation of procedures to reduce multiple redundant NREN international connections and trans-Atlantic links and ensure equitable traffic distribution across the GÉANT Network infrastructure. In addition, the GÉANT Network defines a core set of links that form the European Distributed Access (EDA) architecture. A connection to any GÉANT Network-based EDA functions as an effective link to every NREN on the GÉANT Network.

As with TEN-155, DANTÉ provisions technical support and network management services for the GÉANT Network. To streamline GÉANT Network operations, DANTÉ provides connections to research networks in other regions of the world and to the commercial Internet. DANTÉ will also provide basic GÉANT Network security services to protect network equipment against unauthorized access, forestall cybiner-intrusions, and prevent illegal use of network resources. Consisting of peer-level NRENs representing approximately 30 countries, the GÉANT Network Consortium determines GÉANT Network participants and projects. Currently, GÉANT NREN Consortium participants include ACONET (Austria), CA*net3 (Canada), CESnet (Czech Republic), DFN (Germany), GRnet (Greece), SINET (Japan), SingAREN (Singapore), SURFnet5 (Netherlands), POL 34/155 (Poland), ARNES (Slovakia), RedIRIS (Spain), and SWITCH (Switzerland). KOREN (Korea), SingAREN (Singapore), SINET (Japan), Abilene (U.S.), and CA*net3 (Canada) participate in GÉANT as well. The GÉANT Network employs a shared high-performance network core and a mesh topology to overcome potential disruptions and provisions networking services to countries that were not linked to TEN-155, including Estonia, Latvia, Lithuania, Romania, the Slovak Republic, and Bulgaria.

11.17.1 Task Force-Next Generation Network (TF-NGN)

Established by UKERNA (United Kingdom Education and Research Networking Association), the Task Force-Next Generation Networking (TF-NGN), a successor to TF-TANT, evaluates the capabilities of advanced networking technologies, services, and applications for implementation via the GÉANT Network. TF-NGN carries out pilot trials to evaluate the performance of technical solutions that provision Premium IP service, IPSec, IP QoS, advanced terabit IP routers transporting

MPLS-over-DWDM, GCS, IP multicasts, and IPv6. Pilot trials and gigabit experiments sponsored by TF-NGN are carried out in testbeds such as ATRIUM (Belgium and France), PlaGE (France), GRnet2 (Greece), GARR-G (Italy), Testnett (Norway), Pioneer (Poland), the United Kingdom (SuperJANET4), GigaPort (Netherlands) and StarLight (United States).

11.17.2 EUROPEAN COMMISSION INFORMATION SOCIETIES TECHNOLOGY PROGRAM (EC-IST)

11.17.2.1 SEQUIN (Service Quality Across Independently Managed Networks)

Sponsored by the European Commission Information Societies Technology (EC-IST) Program, the SEQUIN Project clarifies the meaning of the phrase "Quality of Service" (QoS) based on user requirements and capabilities of current and next-generation technologies. The SEQUIN Project also establishes a reliable method for determining predictable QoS guarantees or QoS assurances across multiple networks. Partners in the SEQUIN Forum include DFN (Germany), GRnet (Greece), RENATER2 (France), POL 34/155 (Poland), and DANTÉ.

11.18 SUMMARY

The public Internet is incapable of sustaining high transmission rates and provisioning on-demand bandwidth for multimedia applications. This inability drives implementation of next-generation network technologies developed by collaborative partnerships between government agencies, private- and public-sector organizations, and/or academic institutions and research centers.

As noted in this chapter, the NGI Program leverages the networking capabilities of I2 universities and research centers in developing a high-performance network fabric for enabling multimedia applications, middleware services, grid technologies, Web caching operations, IP multicasts, virtual testbeds, and advanced simulations. Participants in the NGI initiative include DARPA, NSF, NASA, NIST and the National Library of Medicine/National Institutes of Health. These entities also support advanced research and experimentation in next-generation NGI technologies and services to enable seamless wireline and wireless communications through participation in the National Science and Technology Council (NSTC) Interagency Working Groups (Wigs). These groups conduct network research and development to facilitate high-end computing, large-scale networks, human-computer interaction, information management, scalable and distributed applications, infrastructure support, and network security and investigate social, economic and workforce implications of information technology (IT).

As with NGI, I2 research initiatives and international NRENs that contribute to the development of next-generation networks to support QoS guarantees, sophisticated multicast multipoint services, global web caching, video-over-IP, voice-over-IP, multimedia applications, and advanced network protocols and architectures are explored. GÉANT Network capabilities in promoting joint next-generation imple-mentations are described. Regional, national, and international telecollaborative initiatives culminating in the establishment of new generations of tele-education programs, digital libraries, telemedicine applications, shared collaborative virtual reality environments, and massive information grids are highlighted.

11.19 SELECTED WEB SITES

Asia Pacific Advanced Network (APAN). APAN Home Page. Last modified on September 11, 2001.
Available: http://www.apan.net/

Coalition for Academic Scientific Computation (CASC). CASC Papers. Last modified on August 31, 2001.
Available: http://www.ncsc.org/casc/paper.html

Delivery of Advanced Network Technology to Europe, Ltd. (DANTÉ). GÉANT. The Next-Generation of European Networking. Last modified on August 31, 2001.
Available: http://www.dante.net/geant/

Delivery of Advanced Network Technology to Europe, Ltd. (DANTÉ). TEN-155 Introduction. Last modified on May 31, 2001.
Available: http://www.dante.net/ten-155/

Delivery of Advanced Network Technology to Europe, Ltd. (DANTÉ). The Quantum (Quality Network Technology for User Oriented Multimedia) Project. Last modified on May 7, 2001.
Available: http://www.dante.net/quantum/

Delivery of Advanced Network Technology to Europe, Ltd. (DANTÉ). The SEQUIN (Quality of Service for European Research) Project. Last modified on July 16, 2001.
Available: http://www.dante.net/sequin/index.html

Development of a European Service for Information on Research and Education (DESIRE) Consortium. Welcome to the DESIRE Project.
Available: http://www.desire.org/

DFN or Deutsche Forschungsnetz (NREN of Germany). Welcome. Last modified on September 20, 2001.
Available: http://www.dfn.de/willkommen/index.en.html

GARR (NREN of Italy). Research Activities within the GARR Network. Last modified on September 6, 2001.
Available: http://www.garr.it/r_d/garr-b-research-engl.shtml

HEAnet (NREN of Ireland). Welcome to HEAnet. Ireland's National Education and Research Network.
Available: http://www.heanet.ie/

International Federation of Library Associations and Institutions. Digital Libraries: Resources and Projects. Last modified on May 10. 2001.
Available: http://www.ifla.org/II/diglib.htm

MACHBA (Israel InterUniversity Computation Center). Internet-2 in Israel.
    Last modified on August 23, 2001.
    Available: http://www.internet-2.org.il/
Nordic Countries Network, Phase 2 (NORDUNet2). Distance Education and
    Lifelong Learning.
    Available: http://www.nordunet2.org/distance/index.htm
Partnerships for Advanced Computational Infrastructure (PACI). About PACI.
    Last modified on April 26, 2001.
    Available: http://www.ncsa.uiuc.edu/About/PACI/
QUALIT. University and DOE-Lab Interconnect Testbed.
    Available: http://qbone.internet2.edu/qualit.shtml
Singapore Advanced Research and Education Network (SingAREN). About
    SingAREN.
    Available: http://www.singaren.net.sg/html/about_singaren.html
Singapore One. Singapore One Network Infrastructure: Overview.
    Available: http://www.s-one.gov.sg/s1netinf/oview01.html
STAR TAP (Science, Technology and Research Transit Access Point). STAR
    TAP Applications.
    Available: http://www.startap.net/APPLICATIONS/
SWITCH (NREN of Switzerland). About SWITCH. Last modified on July 17,
    2001.
    Available: http://www.switch.ch/about/
Trans-Pacific Network (TransPAC). Last modified on August 16, 2001.
    Available: www.transpac.org
University Corporation for Advanced Internet Development (UCAID).
    Internet2: Advanced Networks.
    Available: http://www.internet2.edu/html/advancednets.html

12 Network Security

12.1 INTRODUCTION

The remarkable popularity of multimedia services and applications is accompanied by a significant expansion in the opportunities for network-related cyberintrusions that range from data interception and malicious code attacks to insider sabotage and theft and insertion of misinformation. Cyberbvandals determined to break into networked computers are always on the lookout for new ways to snoop around, eavesdrop, modify, destroy, or steal data.

12.2 PURPOSE

The pervasive use of communications technologies for enabling virtually ubiquitous and instantaneous connectivity to Web resources is accompanied by an expanding need for security solutions. This chapter examines various procedures, protocols, and techniques to counteract network incursions and safeguard data confidentiality and integrity. Authentication mechanisms are described. E-commerce security risks and solutions are indicated. Features and functions of school security policies are noted. Techniques for minimizing network vulnerabilities are explored. Approaches for dealing with system abuse, commodity fraud, and abrogation of personal privacy rights are indicated.

12.3 SECURITY INCIDENTS AND CYBERINVASIONS

Currently, network intrusions outpace security options available to create protected internetworking environments. System vulnerabilities have contributed to cyberintrusions at Web sites maintained by the San Diego Supercomputer Center (SDSC), the Pentagon, the FBI, the Defense Information Security Agency (DISA), and the U.S. Department of Justice.

Cyberhackers intent on disabling a network masquerade as legitimate users in order to read, monitor, and overwrite network files and assume control of network functions. Cyberintruders also exploit security holes and configuration errors for gaining access to sensitive data such as credit card numbers, hospital records, and personnel files. Even when countermeasures are in place, defensive plans are formulated in advance, and staff members are alert to possible offensives, cyberhackers still cause obvious damage through recklessly destroying or modifying data and making offensive or illegal information available online.

By exploiting capabilities of free network management products on the Web, such as SATAN (Security Administrator Tool for Analyzing Networks) and e-mail protocols such as SMTP (Simple Mail Transport Protocol) and POP (Post Office

Protocol), cyberhackers forge, delete, modify, and/or intercept e-mail and bring networking operations to a standstill. Distributed denial-of-service (DDOS) attacks result in economic disruption, monetary loss, and denial of public access to information at Web sites maintained by public and private entities. Moreover, cyberattacks caused by worms and viruses that are introduced into network systems by specific acts of vandalism, e-mail attachments to E-greeting cards, and the exchange of infected network files demonstrate the pervasiveness of Web cybercrime and underscore the need for strong security solutions.

12.4   SECURITY WARNINGS, ADVISORIES, AND FIXES

12.4.1   CENTER FOR EDUCATION AND RESEARCH IN INFORMATION ASSURANCE AND SECURITY (CERIAS) AT PURDUE UNIVERSITY

Sponsored by Purdue University, the Center for Education and Research in Information Assurance and Security (CERIAS) develops protocol capabilities to ensure secure database access; procedures for protecting educational data in the Web environment; approaches for copyright protection of digital video and images; and deployment of flexible and secure fault-tolerant network services. CERIAS is a successor to COAST (Computer Operations, Audit, and Security Technology). COAST developed security patches for critical systems and evaluated the viability of system monitoring tools such as COPS (Computer Oracle and Password System), a program that identifies security risks on Unix systems.

12.4.2   COMPUTER EMERGENCY RESPONSE TEAM/COORDINATION CENTER (CERT/CC)

Originally called CERT (Computer Emergency Response Team), CERT/CC (CERT/Coordination Center) is part of the Network System Survivability Program sponsored by the SEI (Software Engineering Institute) at Carnegie Mellon University (CMU). CERT/CC supports utilization of OCTAVE (Operationally Critical Threat, Asset, and Vulnerability Evaluation), a comprehensive risk assessment tool for identifying information resources at risk and fosters the development of comprehensive solutions for resolving network security problems and ensuring network perimeter protection. CERT/CC distributes security advisories, vendor bulletins, warnings, and reports describing security incursions, program vulnerabilities, virus and worm attacks, and scanning and probing activities via e-mail and postings at its Web site.

12.4.2.1   CERT Affiliates

AFCERT (Air Force CERT) conducts intrusion-detection operations, benchmarks the performance and vulnerabilities of Air Force computing systems, and develops procedures to enable secure network services. PCERT (Purdue University CERT) establishes campus security policies and responds to on-campus network security incidents.

AusCERT (Australian CERT) provides data on computer crime and its prevention to Internet users in Australia and New Zealand. As with CERT/CC, AusCERT serves as a trusted point of contact and provisions alerts advising

Internet users about security risks and attacks. AusCERT also collects statistics detailing security incidents.

12.4.3 FORUM OF INCIDENT RESPONSE AND SECURITY TEAMS (FIRST)

An international consortium of government, academic, and private-sector agencies, the Forum of Incident Response and Security Teams (FIRST) handles computer security incidents and promotes cybersecurity solutions. Participants in FIRST include CERT/CC, the National Technical Information Service (NTIS), and the U.S. Department of Energy.

12.4.4 INTERNATIONAL COMPUTER SECURITY ASSOCIATION (ICSA) AND THE INTERNET SERVICE PROVIDERS SECURITY CONSORTIUM (ISPSEC)

The International Computer Security Association (ICSA), the Internet Service Providers Security Consortium (ISPsec), distribute warnings, advisories, and alerts to the public to reduce the potentially destructive impact of such viruses as the Melissa Macro virus. After becoming embedded in a Microsoft Word document, the Melissa Macro virus initiates a massive deletion of program and data files.

12.4.5 INTERNET MAIL CONSORTIUM (IMC)

Mail spam or unsolicited bulk e-mail (UBE) such as transmission of multiple e-mail messages with large file attachments to unsuspecting users and e-mail bombs can overload network storage capacity, disable system services, and result in total shutdown of Web sites. To eliminate or reduce UBE, the Internet Mail Consortium (IMC), an international industry organization that includes Internet mail software vendors and hardware vendors among its membership, supports development of legislation and technical solutions.

12.4.6 NATIONAL SECURITY AGENCY (NSA)

The National Security Agency (NSA) conducts specialized activities in the field of foreign intelligence and develops training programs for network security practitioners. Established by the NSA, the Centers of Academic Excellence in Information Assurance Education develop education and training programs that enable security professionals to address security challenges threatening the integrity of the National Information Infrastructure (NII). Participants include the Universities of Idaho and California at Davis, and James Madison, George Mason, Iowa State, Idaho State, and Purdue Universities.

12.5 IDENTIFICATION MECHANISMS AND AUTHENTICATION MEASURES

12.5.1 PASSWORDS

Careless password selection and use are leading causes of network incursions. To identify passwords, cybercrackers run commercially available password cracking programs or try every entry in a dictionary. Stolen passwords empower cybervandals to alter information flow or content, send authenticated e-mail from invaded systems, subscribe to unwanted services, capture personal data about other users logging onto compromised systems, infiltrate distant systems to which local users connect, and destroy system files. The IETF (International Engineering Task Force) One Time Password Authentication Working Group supports development of a one-time password system that works with encryption algorithms for countering attacks that are committed by eavesdroppers who obtain logon identifications and passwords of authenticated users by collecting sensitive data that transits network connections.

12.5.2 BIOMETRIC SOLUTIONS

12.5.2.1 Biometric Technologies

Biometric identifiers based on technologies for DNA, voice printing, retinal and iris scans, fingerprint imaging, facial recognition, hand geometry, and typing rhythm are automated methods that verify a person's identity based on distinguishing physical traits. Biometric solutions compare measurable behavioral or physical characteristics with previously captured information. For example, an individual's thumbprint taken at the point of presentation can be compared to previously registered biometric data for enabling access to information appliances such as notebook computers, computer networks, an airport terminal, or a stadium football game. Facial imaging, also known as facial recognition, enables screening of previously identified terrorists at border crossings or ports of embarkation. Biometric identifiers based on unique physical traits also confirm identities of prisoners, legislators, university students, motorists, visitors at amusement parks, passport holders, students participating in school lunch programs, credit card holders, and parties in E-business transactions.

Implementation of biometric mechanisms raises concerns about due process, civil liberties, and privacy. It is also important to recognize that the accuracy of biometric implementations based on identifiers such as facial scans and fingerprints can be adversely affected by environmental factors such as background noise and lighting, erroneous database entries, and the condition of the networking equipment. Nonetheless, biometric characteristics cannot be lost, forgotten, or duplicated. These solutions also provide an additional layer of security in PKI (Public Key Infrastructure) transactions when used in conjunction with passwords or PINs (Personal Identification Numbers).

12.5.2.2 University of Southern California (USC)

The University of Southern California (USC) supports development of thumbprinting techniques for tracing intruders over a network. In addition, approaches for safeguarding network routing operations and tactics for recovering data from denial-of-service attacks are in development. Procedures for detecting misuse of computer systems based on security policies are also investigated.

12.5.2.3   Biometric Standards and Standards Organizations

12.5.2.3.1   *BioAPI (Biometrics Application Programming Interface)*
Consortium

The BioAPI Consortium supports development of biometric algorithms, methods for storing and managing biometric data, and a standard Application Program Interface (API). The standard API will enable software applications to communicate with a broad range of biometric technologies to facilitate the identification and authentication process in distributed computing environments. This standard is expected to generate cross-industry support and result in the deployment of biometric technologies throughout the commercial marketplace. The National Committee for Information Technology Standards (NCITS) promotes the ANSI (American National Standards Institute) endorsement of the BioAPI specification. Participants in the BioAPI Consortium include IBM, Microsoft, Miros, Novell, and Identicator Technology.

12.5.2.3.2   *Unites States Government Biometric Consortium*

Chartered by the National Security Policy Board through the Facilities Protection Committee, the U.S. Government's Biometric Consortium supports biometric research and works with the NIST (National Institute of Standards and Technologies) National Biometric Interoperability Performance and Assurance Working Group in promoting national and international implementation of biometric solutions. Consortium participants include biometric vendors, universities, and government agencies.

12.5.2.4   Biometric Marketplace

12.5.2.4.1   *BioID*

BioID products enable implementation of biometric authentication solutions based on face, voice, and lip movements to safeguard the integrity of corporate networks and Internet applications such as e-mail; protect PCs and laptops; and provide secure access to government buildings and courtrooms. BioID products also support a template of biometric reference patterns that range in size from 16 to 120 Kbps to validate an individual's identity that can be stored on a smart card.

12.5.3   Cookies

Cookies are unique identifiers or small bits of information that a Web site asks a browser to store on a PC hard drive. Cookies are regarded as valuable mechanisms for tracking a visitor's movements at a Web site, simplifying logon procedures, and customizing Web site content based on a visitor's past preferences. Because cookies are also used to monitor browsing behavior, keep records of graphics downloaded from the Web and newsgroups that are accessed, and extract e-mail addresses and user IDs, these mechanisms are considered an infringement on the freedom to explore cyberspace.

Rumors of Web sites sharing information collected by cookies contributed to the development and implementation by the IETF of guidelines that allow users to exercise greater control over the creation and collection of personal data on the Web. Packages such as NSClean for Netscape and IEClean for Microsoft Internet Explorer from Altus Software enable users to see, clear, and change information recorded by cookies on their PCs. These software products also support the use of aliases for enabling individuals browsing the Web to hide their true identities.

12.5.4 SMART CARDS

12.5.4.1 Smart Card Features and Functions

Smart cards authenticate user identities; support secure log-ons to computer network systems; and control access to Web sites, conference rooms, and computer labs. For example, the University of Calgary, Memorial University in Newfoundland, and Villanova and Xavier Universities issue smart cards to make sure that only verified members of the learning community use campus computing facilities, check out library books, enter campus residences, and participate in extracurricular campus activities. These cards feature magnetic strips or holograms on the back, a color photograph of the cardholder on the front, and the institutional logo. Smart cards used by students and staff at Aston, Exeter, and York Universities in the United Kingdom contain personal identifiers such as names, photographs, and university identification (ID) numbers. These cards enable utilization of university library services and function as electronic purses for payment transactions at campus shops as well as at local bars, retailers, and bingo establishments.

Smart cards can also contain distinctive biometric identifiers such as thumbprints or facial patterns for enabling strong authentication. Implemented in the European Union, SOSCARD (Secure Operating System Smart Card) is a next-generation smart card that supports the use of encryption technology. Smart cards require the use of a smart card reader and software for transforming information stored on a smart card into a digital format that can be readily compared with corresponding data stored in databases.

12.6 FIREWALLS

12.6.1 FIREWALL FEATURES, FUNCTIONS, AND OPERATIONS

Consisting of a combination of hardware and software, a firewall isolates and protects an internal network such as an intranet from an external or untrusted network such as the Internet where invasions can originate. Firewalls are security mechanisms that perform a variety of applications and services to enforce access control measures in accordance with in-place security policies. Firewalls direct traffic to the appropriate destination, hide systems that are vulnerable to Web attacks, and enable audit trails and traffic logs for monitoring network events and network transmissions. In addition, firewalls support virus and content scanning and packet filtering; enable automatic alarms that signal a cyberinvasion; and require user authentication mechanisms such as smart cards, a biometric identifier, one-time passwords, and digital certificates that employ public key encryption for accessing the protected network.

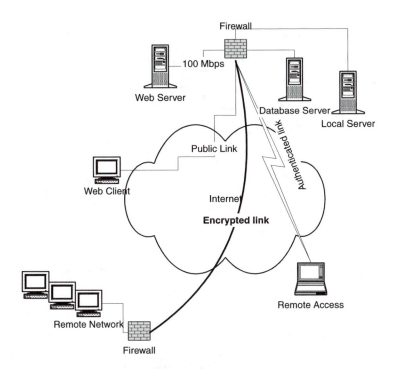

FIGURE 12.1 Firewall functions in maintaining the integrity of a remote network.

A dual-homed host firewall features an interface to facilitate a connection to an untrusted or external network and a second interface to enable a link to the trusted or internal network. A bastion-host firewall works in concert with filtering routers and packet filtering gateways so that outside or external connections are directed to the firewall and not to an internal network such as a VPN (Virtual Private Network) or an intranet. (See Figure 12.1.)

Personal firewalls such as BlackICE Defender and Norton Internet Security safeguard personal computing appliances, resources, servers, and workstations from vandalism and cyberinvasions. These tools also protect VPN operations and provision anti-virus support. In addition to firewalls, VPNs employ tunneling protocols such as IPSec (Internet Protocol Security) and L2TP (Layer 2 Tunneling Protocol) and security mechanisms such as encryption and authentication for enabling individuals to access VPN services and applications via the Web. (See Figure 12.2.)

Approaches for firewall protection are typically specified in a security policy. Generally, a firewall is administered from an attached terminal. In cases where strong security is required, firewall administration is performed directly from a firewall terminal, thereby eliminating problems associated with remote firewall access. As with other critical network resources, firewall files and databases must be backed up daily. In addition, physical access to the firewall must be controlled to prevent unauthorized changes to firewall operations.

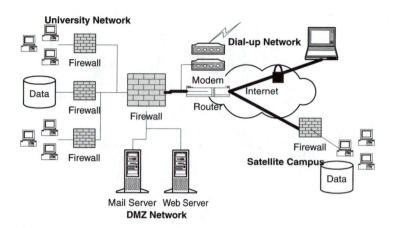

FIGURE 12.2 Firewall capabilities in safeguarding a VPN (Virtual Private Network).

12.6.2 CERTIFICATION OF FIREWALL PRODUCTS

Sponsored by TruSecure Corporation, the International Computer Security Association (ICSA) Labs certifies the availability, reliability, dependability, and performance of firewall products. Capabilities of firewall components such as firmware, hardware, application software, and utilities in limiting security risks and detecting intrusions are assessed. Firewall operations in safeguarding sensitive and confidential information maintained in an internal network from cyberattacks launched by an unprotected or external network are also evaluated. Procedures for supporting authentication and administrative functions, handling inbound and outbound traffic, and maintaining a log of network events are also reviewed. Firewall products certified by ICSA Labs are available from vendors such as IBM, Lucent Technologies, Nortel Networks, Alcatel, Sun Microsystems, CheckPoint, and Nokia.

12.6.3 FIREWALL MARKETPLACE

12.6.3.1 CheckPoint VPN-1/Firewall-1 Solution

The CheckPoint VPN-1/Firewall-1 solution enables dependable deployment of Web technologies and services without compromising security on the protected network and supports remote management of multiple firewalls at distributed locations.

12.6.3.2 Cisco Secure PIX Firewalls

Cisco Secure PIX firewalls that are designed for large-sized enterprises accommodate up to 500,000 concurrent connections and employ sophisticated authentication mechanisms and tools such as application gateways to block attacks and control network access for enabling strict security.

12.7 INTRUSION DETECTION SYSTEMS (IDSs)

12.7.1 IDS (INTRUSION DETECTION SYSTEM) FEATURES AND FUNCTIONS

IDS (Intrusion Detection System) installations enable institutions to conduct security assessments in order to identify and eliminate clusters of security holes and vulnerabilities that potentially allow unauthorized use of computer resources, changes to system files, and access to personal data. These systems implement reactive responses, communicate with firewalls and routers, disconnect user sessions, disable user accounts in the event of unauthorized network entry, and issue audio, e-mail, and pager security alerts to individuals responsible for network operations.

12.7.2 IDS STANDARDS ACTIVITIES

12.7.2.1 IETF (Internet Engineering Task Force) Intrusion Detection Working Group

Developed by the IETF (Internet Engineering Task Force) Intrusion Detection Working Group, the Intrusion Alert Protocol (IAP) supports the exchange of intrusion alert data across IP networks between sensors and analyzers. IAP is an Application Layer protocol that works in concert with HTTP (HyperText Transfer Protocol) and employs TCP (Transmission Control Protocol) as the Transport Layer. By enabling detection of a potential intrusion and notifying a network manager to take appropriate action, IAP enables secure transmission of sensitive data across best-effort IP networks, regardless of bandwidth constraints, latency, and congestion.

In addition, this IETF Intrusion Detection Working Group defines the Intrusion Detection Message Exchange Format (IDMEF) for investigating and reporting a suspicious event and generating security alerts. Approaches for exchanging data among separated IDSs and handling Distributed Denial of Service (DDOS) attacks in progress are under consideration. Procedures for configuring RMON (Remote Monitoring) devices to enable IDS functions are in development.

12.7.3 UNIVERSITY OF CALIFORNIA AT DAVIS (UC DAVIS)

The Security Lab at the University of California at Davis (UC Davis) sponsors development of an intrusion detection system (IDS) that works independently of the underlying operating system for enabling secure networking services in large-scale WANs (Wide Area Networks). The UC Davis Vulnerabilities Project has created a database called DOVES (Database of Vulnerabilities, Exploits, and Signatures) for classifying network attack mechanisms and security holes and categorizing intrusion detection functions.

12.7.4 IDS MARKETPLACE

IDS implementations such as Omniguard Intruder Alert from Axent Technology and Intrusion Detection Expert System from SRI International monitor network traffic patterns and packets with unusual or idiosyncratic formats, maintain audit trails, and track modifications to data files in real-time to intercept network breaches before systems are compromised. With a CheckPoint IDS, a cyberattack can be automatically terminated, recorded for forensic analysis, and/or logged to a database. NetRanger from Cisco Systems and Network Flight Recorder from NFR monitor multiple network systems in real-time, thereby enabling quick access to and analysis of critical security information.

Security Analyst from Intrusion.com generates security audits and works in concert with protocols and technologies such as MAPI (Microsoft Messaging Application Programming Interface), SNMP (Simple Network Management Protocol), and SMTP (Simple Mail Transport Protocol). Log file analysis tools such as Market Focus from Microsoft, Bazaar Analyzer Pro from Aquas Inc., and NetTracker from Sane Solutions monitor the actions of Web site visitors on protected networks.

12.8   CRYPTOSYSTEMS

12.8.1   ENCRYPTION FUNDAMENTALS

Encryption is an enabler of secure, private, and confidential communications and transactions between two parties at a distance. Cryptosystems such as RSA (Rivest, Shamir, and Addleman) and DES (Digital Encryption Standard) transform plaintext into ciphertext or a meaningless format by using sophisticated algorithms prior to transmission to safeguard data integrity. Decryption restores a message to plaintext or its original format. Moreover, cryptosystems employ public and private keys such as a random bit string or block of data for facilitating the encryption and decryption process.

In asymmetric cryptosystems, a public key supports the encryption process and a private key facilitates decryption. A public key for encrypting a message or file transmitted across a network is shared among two or more individuals. A secret or private key is known only to a message recipient. Because a private key is linked mathematically to a public key in asymmetric public key cryptosystems, RSA asymmetric public key cryptosystems use infinite numbers of keys to deter potential cybercrackers. In symmetric cryptosystems such as DES, the same secret or private key is used for encrypting and decrypting messages and files. Regardless of system format, if a key belonging to an authenticated sender and/or recipient is compromised, damaged, forgotten, or lost, then an assumption must be made that a cyber-invader can read encrypted information and forge a digital signature on a document.

12.8.2   ENCRYPTION SYSTEM SPECIFCATIONS, ALGORITHMS, AND SOLUTIONS

12.8.2.1   Advanced Encryption Standard (AES)

Developed by NIST (National Institute of Standards and Technology), the Advanced Encryption Standard (AES) defines a symmetric data encryption algorithm with key sizes of 128 bits, 192 bits, and 256 bits for every 128-bit block of data for ensuring protection of sensitive government information and provisioning strong cryptographic security. In the E-commerce domain, NIST supports utilization of cryptographic algorithms for enabling authentication services. NIST also fosters development of a cryptographic toolkit and standardized solutions for public key management, exchange and recovery, and develops DES (Digital Encryption Standard) system solutions.

12.8.2.2 Blowfish

Blowfish, an encryption algorithm that is freely available on the Web, enables strong encryption by employing a key of variable length that ranges in size from 21 to 448 bits for each 64-bit block. Reportedly faster than DES, the Blowfish encryption algorithm supports symmetric encryption implementations.

12.8.2.3 Digital Encryption Standard (DES)

Defined by the American National Standards Institute (ANSI) in a series of specifications that include X3.92 and X3.106, the Digital Encryption Standard (DES) is an encryption method that uses a private 56-bit key for every 64-bit block of data. With DES implementation, 72 quadrillion encryption keys are available. The key for each message is randomly generated. As with other private key cryptosystems, a sender and recipient must know and utilize the same secret key. Also known as strong encryption, triple DES implementations require utilization of three private keys in succession.

The National Institute of Standards and Technology (NIST) developed the DES solution in 1977 based on research performed at IBM. The DES specification maintains the integrity and confidentiality of unclassified and sensitive data associated with national security and critical military and intelligence missions. Federal organizations and agencies generate their own cryptographic DES keys in accordance with specified NIST procedures. With DES implementation, data are theoretically protected against all forms of attacks. Individuals and entities within the United States are also eligible to use DES cryptosystems.

The U.S. Government limits exports of DES software and hardware. The Office of Munitions Control, an agency in the U.S. Department of State, issues licenses for DES implementation to financial institutions and American subsidiaries with international operations. Despite these controls, versions of DES are reportedly available as shareware on the Web.

12.8.2.4 Internet Protocol Security (IPSec)

An IETF specification, IPSec (Internet Protocol Security) is a method for adding security to unprotected IP (Internet Protocol) transactions. IPSec supports the use of encryption and public key algorithms and provides a set of open standards for supporting data integrity, confidentiality, and authentication services. IPSec employs an Authentication Header for enabling verification of the identity of the message sender and an Encapsulating Security Payload for enabling data encryption. Moreover, IPSec facilitates dependable and reliable native IP applications such as IP telephony, IPv4 (Internet Protocol version 4) and IPv6 (Internet Protocol version 6), IP multicast services, and VPN implementations and supports E-commerce operations in multivendor environments. The IETF IP Security Protocol (IPSec) Working

Group develops security mechanisms and protocols such as the IKE (Internet Key Exchange) Security Protocol for enabling secure Web operations.

12.8.2.5   Internet Key Exchange (IKE) Security Protocol

A key management protocol that supports strong cryptography, the Internet Key Exchange (IKE) security protocol generates authentication and encryption keys and automates IPSec operations, thereby eliminating the need to manually specify IPSec parameters for secure communications. In addition, the IKE protocol employs Data Encryption Standard (DES) technology to encrypt data packets; foster dynamic authentication of IPSec-compatible devices such as firewalls, routers, and client information appliances; and provides Certification Authority (CA) support for extendible, reliable, and manageable IPSec implementations.

12.8.2.6   Kerberos

Developed at the Massachusetts Institute of Technology (MIT), Kerberos is an authentication protocol for safeguarding the transmission of passwords and other sensitive information in network traffic and, thereby preventing cybervandals from falsely claiming another individual's identity. Designed to provide strong authentication for applications in the client/server domain, Kerberos uses secret key cryptography and requires the secure communications of a shared secret key between a sender and recipient without discovery by a third party. In Kerberos implementations, an authenticated network user must also access a central server called a Key Distribution Center to acquire a validated Kerberos ticket or token in order to establish connections to a secure host.

Kerberos is available in commercial products from Cisco, Microsoft, and Sun Microsystems, and in the public domain. The IETF Kerberos Working Group clarifies Kerberos capabilities and develops additional mechanisms and tools for enhancing Kerberos security services.

12.8.2.7   MIME (Multipurpose Internet Mail Extensions) and
S/MIME (Secure/MIME)

MIME (Multipurpose Internet Mail Extensions) defines a standard format for sending e-mail messages that include audio, video, images, animation, and graphics in addition to text. Endorsed by the IETF, S/MIME (Secure/Multipurpose Internet Mail Extensions) is a protocol that adds encryption and digital signatures to Web-based MIME transmissions. S/MIME also requires utilization of digital certificates. S/MIMEv3 (S/Mime Version 3) supports applications sponsored by federal and local government agencies and private companies.

Approaches contributing to the utilization of S/MIMEv3 security labels for implementation of an organizational security policy are in development. In addition, mechanisms for defining additional key management techniques and procedures for distributing symmetric encryption keys to multiple message recipients are also investigated. Tactics for using CMS (Cryptographic Message Syntax) are explored as well. S/MIME employs the X.509v3 certificate format and triple DES (Data

Encryption Standard) for strong encryption. The IETF S/MIME Working Group clarifies S/MIME operations and defines methods leading to the enhancement of S/MIMEv3 PKI capabilities and operations.

12.8.2.8 Pretty Good Privacy (PGP)

PGP is a public key encryption system for enabling secure information transmission and reception across a public network such as the Internet. PGP also protects files in data warehouses and provides procedures for digitally signing and encrypting information. PGP supports symmetric encryption or utilization of an identical secret key for encrypting and decrypting a message. In addition, PGP employs asymmetric encryption or the utilization of a public key for message encryption and a private key for message decryption for workgroup activities.

With PGP, authors of postings in newsgroups can confirm the authenticity of their material. PGP implementation also enables recipients to verify that postings received were not altered, modified, or tampered with by a third party. PGP source code is freely available in the public domain. The IETF Open Specification for the PGP Working Group develops standards for PGP operations. FIRST participants use PGP for ensuring the confidentiality of e-mail messages and locally stored data files.

12.8.2.9 *OpenPGP*

Developed by the IETF OpenPGP Working Group, the OpenPGP specification is the latest version of PGP. As with PGP, OpenPGP supports message authentication and privacy. OpenPGP complements PGP operations by providing formats and algorithms for digitally signed and encrypted PGP information objects and provisions strong cryptographic and authentication services.

The OpenPGP Alliance supports utilization of the OpenPGP specification to promote development of compatible private e-communications systems. The Open-PGP Alliance also sponsors technical compatibility testing to ensure the interoperability of OpenPGP systems. In addition, this Alliance promotes utilization of the OpenPGP format by non-e-mail applications to enable privacy protection. Participants in the OpenPGP Alliance include SSH Communications Security, Zero Knowledge Systems, Qualcomm, and Hughes Communications.

12.8.2.10 Privacy Enhanced Mail (PEM)

Compatible with a wide range of protocols, user interfaces, and e-mail systems, Privacy Enhanced Mail (PEM) uses encryption and digital signatures for ensuring the transmission integrity of sensitive information and protecting the contents of e-mail messages from unauthorized disclosure. With PEM implementation, individuals cannot hide behind anonymous re-mailers.

12.8.2.11 RSA (Rivest-Shamir-Adleman) Solutions

Developed in the 1970s by Ron Rivest, Adi Shamir, and Leonard Addleman, RSA (Rivest-Shamir-Addleman) employs a public key encryption algorithm with a variable key length. This public key algorithm serves as the foundation for generating paired keys that are mathematically linked in asymmetric RSA cryptosystems. In a RSA implementation, a private key decrypts information encrypted by the public key. Created by a vendor consortium including the Massachusetts Institute of Technology (MIT), Apple, and Microsoft Corporation, the Public Key Cryptography Standard (PKCS) defines mechanisms for encrypting and signing data using RSA encryption.

12.8.2.11.1 RSA SecurID

Based on RSA algorithms and technologies, RSA security tools include RSA SecurID for two-factor authentication of teleworkers that access a VPN or intranet via direct dial-in connections or the Internet. BT (British Telecom) employs RSA SecurID authentication software in Ericsson smart phones for provisioning network authentication services and access to secure corporate network resources. RSA SecurID can also display a token code, thereby eliminating the need for individuals on the move to carry a separate token such as a smart card. In the healthcare domain, RSA SecurID ensures the secure exchange of medical records, lab results, and other confidential information between medical professionals at healthcare institutions.

12.8.2.11.2 RSA BSAFE

Developed by RSA Security, BSAFE enables the unified management of processing requests and the retrieval, revocation, storage, and validation of digital certificates and digital signatures to facilitate the design and utilization of PKI (Public Key Infrastructure) applications in multivendor PKI environments. BSAFE also operates in conjunction with RSA security certificate technologies for enabling secure transmission via the Internet.

RSA Security employs the Extensible Markup Language Key Management specification in its BSAFE development software and PKI products. This specification clarifies rules and transactions for effectively managing keys for digital signatures and provides a framework for encrypting data elements in XML (Extensible Markup Language) documents.

12.8.2.11.3 RSA Digital Certificates

RSA digital certificates conform to the ITU-T (International Telecommunications Union-Telecommunications Standards Sector) X.509v3 (X.509 version 3) Recommendation. Digital certificates authenticate a user's credentials when conducting Web transactions. Issued by a Certification Authority (CA), digital certificates are electronic credit cards that verify an individual's identity. Digital certificates contain an individual's public key and digital signature as well as the digital signature of the CA.

12.8.2.11.4 RSA Secure

RSA Secure is a commercially available software package for enabling secure PC and network transactions. RSA Secure also supports secure services for PDAs (Personal Digital Assistants), cellular phones, and pagers. Browsers from NSPs (Network Service Providers) such as Microsoft and Netscape and wireless technologies such as WAP (Wireless Application Protocol) and Bluetooth employ RSA Secure services as well.

12.8.2.12   Secure-HyperText Transfer Protocol (S-HTTP)

Developed by the IETF Task Force, S-HTTP (Secure-HyperText Transfer Protocol) is an extension to HTTP (HyperText Transfer Protocol). S-HTTP supports confidentiality, integrity, and authenticity of HTML (Hypertext Markup Language) documents by employing cryptographic algorithms. S-HTTP also enables verification of sender authenticity and message integrity. The SSL (Secure Sockets Layer) protocol establishes a secure connection between two peer entities. By contrast, S-HTTP enables secure data transmission by individual entities.

12.8.2.13   Secure Sockets Layer (SSL)

The SSL (Secure Sockets Layer) Protocol supports dependable and reliable connections between Web clients and Web servers, and employs RSA and DES cryptographic systems for enabling secure transmissions. In order for SSL to work properly, SSL-enabled browsers and SSL digital certificates for facilitating public key exchanges must be utilized. SSL digital certificates are available from Certificate Authorities (CAs) such as GTE and VeriSign. The SSL Handshake Protocol enables Web clients and Web servers to perform authentication functions and negotiate utilization of cryptographic keys and an encryption algorithm prior to information transport. Now in Version 3, SSL is endorsed by the IETF (Internet Engineering Task Force). A collaborative initiative supported globally by a core team of SSL developers, the OpenSSL initiative fosters development and implementation of an OpenSSL toolkit that supports standards-compliant SSL implementations.

12.8.2.14   Transport Layer Security (TLS) Protocol

Designed to enable secure client/server applications and prevent eavesdropping and tampering with message integrity, the TLS (Transport Layer Security Protocol) supports encryption of real-time communications over the Web. Regarded as the latest version of SSL, TLS supports encryption and certification functions. The TLS Working Group develops TSL specifications. TLSv1 (TLS version 1) is based on SSLv3 (SSL version 3).

In 1998, C2Net Software was the first company to implement TLS 1.0. Subsequently, Phaos Technology deployed SSLava2000, the first toolkit for building TLS solutions. The Phaos SSLava2000 Toolkit includes complete rapid development libraries for ensuring confidentiality, proof of identity, and data integrity.

12.9   DIGITAL SIGNATURES

12.9.1   Digital Signature Features and Functions

A digital signature refers to an electronic sound, process, or symbol attached to or associated with a record. Created electronically with public key encryption technology, a digital signature on a record signifies intent to do a legally binding act. The digital signature is linked to the record signed and to the signer, and therefore cannot be reproduced. The signing process is an act of affirmation. By employing a digital signature, the signer agrees to conduct transactions that involve legal consequences. The process of creating and verifying a digital signature provides a high level of assurance that the digital signature is genuine and belongs to the signer. Digital signatures support message integrity and minimize the occurrence of identity theft and forgery.

12.9.2 UNIFORM ELECTRONIC TRANSACTIONS ACT (UETA) AND THE E-SIGN BILL (ELECTRONIC SIGNATURES IN GLOBAL AND NATIONAL COMMERCE ACT)

Drafted by the National Conference of Commissioners on Uniform State Laws and approved in 1999, the Uniform Electronic Transactions Act (UETA) establishes a uniform framework for implementing legally binding electronic contracts, electronic records, and electronic signatures. According to the UETA, digital signatures that are appended to electronically transmitted and recorded transactions are as legally binding as conventional signatures on paper documents. In 2000, President Clinton signed the Electronic Signatures in Global and National Commerce Act. Also called the E-SIGN Bill, this legislation confirms the legally binding status of electronic signatures.

12.9.3 DIGITAL SIGNATURES AND ENCRYPTION

With a digital signature, two different but complementary encryption keys are employed. One key supports creation of a digital signature by transforming data into an unintelligible format or ciphertext. By contrast, another key returns the encrypted message to its original format or plaintext. With asymmetric encryption, the mathematically related key pairs are formally called an asymmetric cryptosystem. Public keys can be shared by a group of users for verifying a signer's signature. By contrast, the private key is computationally complicated and known only by the signer. This complex and sophisticated process counters forgery and theft of digital signatures.

12.9.4 DIGITAL SIGNATURE PROTOCOLS AND SPECIFICATIONS

12.9.4.1 Digital Signature Standard (DSS)

Established by NIST (National Institute of Standards and Technology), the Digital Signature Standard (DSS) specifies a format for digitally signing online contracts and agreements and verifying the authenticity of the digital signature. The signature cannot be forged unless the signer loses control of the private key and the key is therefore compromised. In the United States, federal government agencies are required to employ solutions based on the Digital Signature Standard and the Secure Hash Algorithm (SHA) to prevent data corruption and identity theft.

12.9.4.2 Simple Digital Security Scheme (SDSS)

Developed by the World Wide Web Consortium (W3C), the Simple Digital Security Scheme (SDSS) is a basic authentication protocol that functions as an extension to the HyperText Transfer Protocol (HTTP).

12.9.4.3   IETF/W3C/XML Digital Signature (DSig) Working Group

The joint IETF/W3C/XML Digital Signature (DSig) Working Group defines syntax and rules for XML-based digital signatures and approaches for verifying digital signature authenticity.

12.10   DIGITAL CERTIFICATES AND CERTIFICATION AUTHORITIES (CAS)

12.10.1   DIGITAL CERTIFICATES AND CERTIFICATION AUTHORITIES FEATURES AND FUNCTIONS

A digital certificate is an encrypted attachment to an electronic message that verifies an individual's identity so that this individual can participate in Web-based E-commerce transactions and sign legal documents and contracts. These certificates also support the creation of public-private key pairs and digital signatures for safeguarding data transported across the network.

Digital certificates are available from a CA (Certification Authority). A CA is a trusted third-party entity that issues and revokes digital certificates. Typically, the CA certifies the identity of an individual with information from financial institutions such as banks, credit card companies, and/or credit card registries. A Certificate Revocation List (CRL) refers to a digital list consisting of revoked certificates issued by a CA. Generally, digital certificates conform to the ITU-T X.509v3 Recommendation. Passwords and passphrases safeguard a PC-stored digital certificate.

To associate a key pair with a prospective customer, CAs also issue an electronic record or certificate that lists a public key and/or private key as the subject of the certificate. The certificate's primary function is to link a key pair with a particular subscriber. A recipient of an electronic transaction can use the public key listed in the certificate to verify the integrity of the transaction. The CA also provisions verification software and access to Certificate Revocation Lists (CRLs) maintained in archives and repositories.

12.10.2   DIGITAL CERTIFICATE SPECIFICATIONS AND SOLUTIONS

12.10.2.1   Access Certificates for Electronic Services Certification Authority (ACES CA)

ACES CA (Access Certificates for Electronic Services Certification Authority) supports implementation of digital signature and digital certificate technologies. DST (Digital Signature Trust) is an ACES-recognized CA that generates digital signatures and digital certificates for enabling individuals and entities to conduct commercial transactions electronically with the U.S. General Services Administration (GSA). The GPEA (Government Paper Elimination Act) and the Access America initiatives mandate the availability of comprehensive government services to all Americans via the Internet and contribute to the development and implementation of ACES CA.

12.10.2.2   National Center for Supercomputing Agency (NCSA)

The National Center for Supercomputing Agency (NCSA) employs a combination of solutions, including digital certificates and public key, private key, and personal pass phrases for safeguarding NCSA computing resources. NCSA participants are also required to access a (CA) Certificate Authority and obtain a certificate of authentication prior to connecting to secure NCSA networks. In addition, NCSA employs Secure Shell (SSH) authentication at major network sites. Endorsed by the IETF Secure Shell Working Group, SSH v2 (Secure Shell Version 2) allows the creation of RSA asymmetric key pairs for enabling strong encryption. Typically, SSH supports remote log-ons and encrypted Web sessions.

12.10.3   DIGITAL SIGNATURE MARKETPLACE

12.10.3.1   Communication Intelligence Corporation (CIC)

The Communication Intelligence Corporation (CIC) develops secure electronic signature solutions for E-commerce transactions. These solutions employ biometric measurements based on timing, speed, and style that characterize an individual signature. Moreover, the CIC supplies software technologies for enabling dynamic signature verification, multilingual handwriting recognition systems, and natural messaging solutions. The CIC also supports sign-on products for Pocket PCs to authenticate user identification prior to enabling access to the system.

12.11   PUBLIC KEY INFRASTRUCTURE (PKI)

12.11.1   PKI FEATURES AND FUNCTIONS

A *de facto* standard for implementing a secured infrastructure, a Public Key Infrastructure (PKI) implementation enables public and private entities to conduct E-commerce and E-business transactions via the Web in an environment of total trust. A PKI solution supports the utilization of a pair of public and private keys, a corresponding digital certificate, and authentication services. PKI deployments also enable key distribution, generation, and recovery operations and ensure nonrepudiation of agreements and data confidentiality. Moreover, PKI installations employ cryptosystems and services provisioned by CAs and Registration Authorities (RAs) to verify digital certificates and ensure secure management of public and private keys in business-to-business (B2B) transactions.

PKI digital certificates are generally stored in laptop computers or desktop PCs to prevent cybercrackers from employing these credentials to invade multiple distributed networks. A PKI configuration facilitates secure e-mail and intranet operations and employs a mix of security mechanisms such as smart cards, firewalls, and biometric identifiers for enabling dependable and reliable transactions.

12.11.2   PKI SPECIFICATIONS AND SOLUTIONS

12.11.2.1   ITU-T X.509v3 Recommendation

The major enabling standard for PKI is the ITU-T X.509 Recommendation. Approved in 2000 by Study Group 7 of the ITU-T, the X.509v3 Recommendation supports secure Web connections for enabling utilization of digital signatures in E-commerce transactions. This Recommendation works in conjunction with the PKI; employs attribute certificates that define user privileges in multiservice, multi-application, and multivendor environments; and supports enhancements to certificate processing and revocation services. Moreover, the ITU-T X.509v3 Recommendation specifies a framework for PMI (Privilege Management Infrastructure) to enable secure B2B (business-to-business) E-commerce applications.

12.11.2.2   IETF Public Key Infrastructure X.509 (PKIX) Working Group

The IETF Public Key Infrastructure X.509 (PKIX) Working Group supports the Certificate Management Protocol (CMP), the Online Certificate Status Protocol (OCSP), and the Certificate Management Request Format (CRMF) Protocol for managing PKI operations and services. Specifications for using digital certificates in legally binding nonrepudiation situations are also in development.

12.11.2.3   Minimum Interoperability Specification of PKI Components (MISPC)

NIST (National Institute of Standards and Technologies) specifies guidelines in the Minimum Interoperability Specification of PKI Components (MISPC) for supporting a federal PKI. NIST also operates a PKI Interoperability Testbed to evaluate the effectiveness of BCAs (Bridge Certification Authorities) in interlinking each federal PKI, regardless of policies, cryptographic algorithms, and architectures employed, into a federal PKI. NIST clarifies capabilities of XML digital signatures and supports utilization of an enhanced path-processing algorithm for ITU-T X.509 implementations to correct system defects.

12.11.2.4   Federal PKI Operations

The Federal PKI Steering Committee sponsors design and development of the Federal PKI to enable access to government services by authorized personnel and facilitates secure E-commerce transactions. Government agencies can access encrypted data supported by the Federal PKI in the event of emergencies. Federal PKI operations between public or private entities and U.S. federal government agencies require utilization of public key cryptographic solutions for ensuring transaction integrity, data confidentiality, participant authentication, and service nonrepudiation.

NIST defines security requirements for the Federal PKI architecture and establishes the use of S/MIMEv3 (S/MIME Version 3), the NIST version of S/MIME, for enabling secure e-mail exchange. NIST employs the PKI Interoperability Testbed for ensuring service and product conformance to the S/MIMEv3 specification.

12.11.2.5   Open Group PKI Deployments

An international vendor consortium, the Open Group promotes development and implementation of an integrated global PKI architecture that supports transnational

E-commerce applications. The Open Group also endorses the utilization of ITU-T X.509-compliant digital certificates, secure applications based on the CDSA (Common Data Security Architecture) developed by Intel for member states in the European Union, and interoperable PKI services in multivendor environments. Approaches for enabling seamless PKI key storage, recovery, distribution, suspension, revocation, reactivation, and management operations are in development.

12.11.3 PKI MARKETPLACE

12.11.3.1 Entrust Technologies PKI Solutions

PKI security solutions from Entrust Technologies support secure encrypted sessions and establishment of audit logs for guaranteeing the nonrepudiation of transactions. Entrust PKI solutions also enable the generation, signing, management, and revocation of X.509 digital certificates.

12.11.3.2 Baltimore Technologies

Developed by Baltimore Technologies, the Telepathy Product Suite enables mobile users to employ PKI architecture and PKI security solutions in the wireless domain. These solutions work in conjunction with WAP (Wireless Application Protocol) phones and PDAs (Personal Digital Assistants) for facilitating trusted wireless transactions and information exchange in a secure environment. In addition, the Telepathy Product Suite provisions WTLS (Wireless Transport Layer Security) for ensuring nonrepudiation of services; digital certificates for authenticating digital identities; and software tools for ensuring secure sessions between applications. By using the Telepathy PKI Registration System, a component in the Telepathy Product Suite, mobile device users retain their digital identities by employing digital certificates maintained in PKI systems. Also Telepathy Suite components, the Telepathy PKI Validation System and the Digital Signature Toolkit enable users to access multiple digital certificates and create wireless digital signatures for accommodating WAP requirements.

12.11.3.3 Identrus

Sponsored by a consortium of banks and financial institutions, including Chase Manhattan, Citigroup, Bank of America, and VeriSign, Identrus supports design and implementation of a global PKI framework based on open standards for enabling safe and secure E-commerce and E-banking services, including electronic funds transfers and electronic payments. Identrus solutions require the use of digital certificates issued by participating entities for conducting negotiations and arranging for payments in a trusted environment.

12.11.3.4 Xcert

Xcert develops PKI-compatible public key digital certificates. With Xcert solutions, an individual's identity can be authenticated prior to granting access to confidential information and sensitive data files. Government agencies and corporations utilize Xcert solutions to manage in-house virtual certification authorities that distribute public keys and assign digital certificates to trusted users. Public keys and certificates verify user identity and the authenticity of the digital signature and control access to centrally maintained electronic information files and archives. Private keys are stored on smart cards that can also be used to digitally sign documents and verify identity.

12.12 ELECTRONIC COMMERCE (E-COMMERCE) SECURITY CONSIDERATIONS

12.12.1 E-COMMERCE FUNDAMENTALS

The term "E-commerce" refers to commercial transactions over the Web. Initially, electronic interactions were limited to large-sized corporations such as airline carriers, banks, and major retail distributors with the resources, technology, and capital to invest in electronic infrastructures supporting virtual transactions. The popularity of the Web as a global marketplace contributed to the subsequent proliferation of E-commerce Web sites by public and private entities of all sizes. Although there is not a commonly accepted definition, the term "E-commerce" is used with increasing frequency. Typically, E-commerce refers to some form of Web electronic payment system between virtual buyers and virtual sellers in a virtual marketplace that enables the secure purchase and acquisition of virtual goods and services.

12.12.2 E-COMMERCE APPLICATIONS AND SERVICES

Currently, E-commerce Web sites enable commercial transactions in education, travel, fashion, product maintenance, textiles, entertainment, healthcare, tourism, transportation, insurance, real estate, law, business, banking, and music. Vortals, portals, electronic storefronts, and virtual shopping malls in the electronic commerce domain offer an unprecedented array of commodities, including artwork, antiques, books, computers, television sets, ceramics, jewelry, clothing, symphonic recordings, prepackaged software, cakes, candies, stamps, pets, toys, boats, cars, homes, and furniture.

Electronic commerce implementations include e-mail for business communications, electronic payment systems, electronic funds transfer, and Electronic Data Interchange (EDI) or the computer-to-computer transmission of digital data in standardized formats. E-commerce solutions support business-to-business (B2B) and business-to-consumer (B2C) transactions. These implementations require a network infrastructure that provides secure Web services for enabling consumers to purchase and Web site owners to sell tangible and intangible products. Communications solutions facilitating connectivity to Web-based E-commerce operations, applications, and services employ an array of narrowband and broadband wireline and wireless technologies such as POTS (Plain Old Telephone Service), ISDN (Integrated Service Digital Network), and ATM (Asynchronous Transfer Mode), cable networks, and DSL (Digital Subscriber Line) solutions.

12.12.3   E-COMMERCE OPERATIONS AND SECURITY RISKS

Web technical advancements enable E-commerce entrepreneurs to start innovative applications and activities in the worldwide electronic marketplace with minimal up-front investment and promote products and services directly to consumers at home and in the workplace in every facet of the economy, including the retail, communications, education, and information sectors. The E-commerce process involves intense competition in advertising, marketing, and supplying on-demand tangible and intangible commodities in an unprotected network environment.

E-commerce implementations are characterized by reliability problems; technical, legal, regulatory, and administrative challenges; and pervasive concerns about the security of electronic payments, information corruption, disclosure of private and sensitive data to untrusted third parties, and consumer exposure to fraud. Additional risks associated with the E-commerce process include misappropriation of funds, failure to credit payments, double spending or paying twice for the same commodity, DDOS attacks, and failure by vendors to supply advertised commodities subsequent to accepting payments.

Anonymity in the electronic marketplace enables cyberinvaders to mask their identities while stealing credit card numbers from a Web site or employing electronic payments for tax evasion and money laundering. An online campus billing office may in fact be a fake virtual storefront created to collect credit card numbers. As a consequence, security mechanisms and networking protocols that enable consumers to order and purchase virtual products in safe environments and authentication services for verifying the identities of each party to an E-commerce transaction are in development.

12.12.4   ELECTRONIC PAYMENT SYSTEMS

Electronic payment systems employed for Web commercial transactions feature cryptographic mechanisms, security protocols, authentication services, and tamper-resistant devices such as smart cards for enabling utilization of instruments such as virtual cash, tokens, electronic checks, debit cards, and credit cards to make micro-payments.

12.12.4.1   Authorize.Net

Authorize.Net enables consumers to use credit cards and electronic checks for purchasing items on the Web. Merchants use Authorize.Net to authenticate, process, manage, and settle E-commerce transactions.

12.12.4.2   CAFÉ (Conditional Access for Europe)

The CAFÉ (Conditional Access for Europe) initiative supports the use of secure electronic payment systems over the Web by consumers with CAFÉ-compliant electronic wallets. Electronic personal credentials that serve as passports, drivers' licenses, and house keys are in development. Academic participants in the CAFÉ project include Aarhus University, the University of Hildesheim, and the Catholic University of Leuven.

12.12.4.3 CyberCash

CyberCash payment solutions such as CyberCoin, PayNow, and the CyberCash wallet support secure encrypted credit card, debit card, electronic check, and micro-payment transactions on the Web and real-time authentication services. In 2000, the State of Oregon implemented a CyberCash solution for the Oregon Center for E-Commerce and Government that enables state residents to purchase permits, licenses, and state maps at state-sponsored Web sites.

12.12.4.4 DigiCash Ecash Solutions

Developed by DigiCash, Ecash solutions employ public key encryption technology for enabling micropayments for Web transactions.

12.12.4.5 Financial Services Technology Consortium (FSTC) Initiatives

Sponsored by the FSTC (Financial Services Technology Consortium), the Bank Internet Payment System (BIPS), Electronic Checks, and the Paperless Automated Check Exchange and Settlement (PACES) initiatives provision authentication and encryption services for enabling consumers to make secure Web payments. The FSTC also initiated development of the Secure Document Markup Language (SDML) specification for safeguarding the integrity of E-commerce exchanges. FSTC participants include Oak Ridge and Sandia National Laboratories, Columbia University, and the Polytechnic University of Brooklyn.

12.12.5 E-COMMERCE ORGANIZATIONS, SECURITY SPECIFICATIONS, AND SOLUTIONS

Entities transforming the Web into a global electronic marketplace free of fraud and deception include the Global Information Infrastructure Commission, CommerceNet, and Electronic Commerce Canada. The European Commission, the European Parliament, the European Telecommunications Standards Institute, the IETF (Internet Engineering Task Force), the International Electrotechnical Commission, and the ITU-T (International Telecommunications Union-Telecommunications Standards Sector) develop legal, technical, and commercial transborder E-commerce regulations for providing a trusted E-commerce environment as well.

Additional trade associations and private interest groups active in the E-commerce standards domain include the American Electronics Association, the Electronic Messaging Association, and the Software Publishers Association. The Global ECommerce Forum, originally known as First Global Commerce, is an international multivendor consortium that initiates E-commerce projects and promotes E-commerce infrastructure development.

12.12.5.1   ebXML (Electronic Business Extensible Markup Language)

The ebXML (Electronic Business Extensible Markup Language) initiative supports deployment of an XML global infrastructure that enables the secure use of E-business data by all parties involved in a transaction.

12.12.5.2   epf.net (Electronic Payments Forum)

An alliance of commercial entities, government agencies, universities, and standards organizations, the Electronic Payments Forum (epf.net) promotes the development and implementation of interoperable electronic payment systems for enabling global E-commerce services and applications.

12.12.5.3   Internet Law and Policy Forum

An international organization, the Internet Law and Policy Forum (ILPF) is an open forum that supports the discussion of legislative and policy issues that impact the growth and expansion of global E-commerce. The ILPF supports Working Groups on Content Regulation, Self-Regulation, Jurisdiction, and Electronic Authentication that contribute to the development of practical solutions for resolving transborder E-commerce legal issues and enabling consumer protection from fraud.

12.12.5.4   Mobile Electronic Transactions (MeT)

Sponsored by Motorola, Nokia, and Ericsson, the Mobile Electronic Transactions (MeT) initiative promotes the development of a uniform framework based on in-place standards to support secure, dependable, and reliable mobile E-commerce transactions.

12.12.5.5   Radicchio Initiative

A nonprofit organization consisting of certification service providers, and mobile carriers, Radicchio supports development of common standards to support secure M-commerce transactions. In addition, Radicchio supports development of a trusted PKI that works with wireless networks and personal handheld devices and enables secure electronic transactions at anytime and from anyplace.

12.12.5.6   Secure Electronic Transaction (SET)

Jointly designed by MasterCard International and Visa International, the Secure Electronic Transaction (SET) specification uses a blend of RSA and DES encryption for safeguarding online credit card transactions over the Internet. To protect consumers against online fraud, the SET specification establishes protocols for payment card operations over an open network and supports the use of digital certificates that are issued to cardholders and merchants in SET transactions to verify their identities. Software vendors, merchants, and financial institutions that provision SET-compliant products and services display the SET Mark.

12.12.5.7   United Nations Model Law on Electronic Commerce

Also called UNCITRAL, the Model Law on Electronic Commerce adopted by the United Nations Commission on International Trade Law features a set of internationally acceptable rules for addressing legal obstacles in E-commerce transactions and promoting development of a strong security environment.

12.12.5.8   World Wide Web Consortium (W3C)

The World Wide Web Consortium (W3C) sponsors an international effort for promoting global E-commerce that is hosted by the Massachusetts Institute of Technology (MIT) Laboratory for Computer Science in the United States, the National Research Institute of Information and Automation in the European Union, and Keio University in Japan. Formally called HTTP/1.1 (HyperText Transport Protocol/1.1), the W3C-sponsored Digest Authentication protocol supports deployment of identification mechanisms and authentication services for enabling secure E-commerce transactions between customers and merchants. W3C also supports the Joint Electronic Payment Initiative (JEPI) to enable secure electronic payments and the Digital Signature Initiative to standardize the format for signing digital documents.

12.12.6   EUROPEAN COMMISSION TELECOMMUNICATIONS APPLICATIONS PROGRAM (EC-TAP)

12.12.6.1   Interworking Public Key Certification Infrastructure for Commerce, Administration, and Research (ICE-CAR)

The ICE-CAR (Interworking Public Key Certification Infrastructure for Commerce, Administration, and Research) initiative fosters development of security technologies and solutions for safeguarding E-commerce network applications and services. ICE-CAR also promotes implementation of technically compatible and interoperable Public Key Infrastructures (PKIs). ICE-CAR research builds on the work of the ICE-TEL (Internetworking Public Key Certification Infrastructure for Europe) project. The ICE-TEL initiative established a foundation for enabling secure Web transactions, provided a framework for implementation of a Public Key Infrastructure (PKI) in Europe, and defined approaches for enabling interconnections between national CERTs (Computer Emergency Response Teams).

12.12.7   EUROPEAN COMMISSION ADVANCED COMMUNICATIONS TECHNOLOGIES AND SERVICES (EC-ACTS) PROGRAM

12.12.7.1   Secure Electronic Marketplace for Europe

The SEMPER (Secure Electronic Marketplace for Europe) initiative developed approaches for enabling secure Web E-commerce transactions and clarified commercial, legal, social, and technical requirements for implementing a dynamic virtual marketplace.

12.12.7.2 TELE-SHOPPE

The TELE-SHOPPE project enabled the integration of virtual reality technologies into virtual product displays to attract Web site visitors and promote the sales of virtual retail goods.

12.12.8 EUROPEAN COMMISSION INFORMATION SOCIETY TECHNOLOGIES (EC-IST) PROGRAM

12.12.8.1 DIGISEC

The DIGISEC project promotes development of a digital signature infrastructure that supports secure administrative operations and electronic commerce services. Smart cards and digital signatures for enabling E-business operations are tested with shoppers in actual business environments.

12.12.8.2 E-BROKER

The E-BROKER (Electronic Broker) project enables the design and development of a secure trading infrastructure to support safe and reliable data exchange on the extent of market demand and the availability of tangible and intangible goods and services in the marketplace.

12.12.8.3 FAIRWIS

The FAIRWIS project fosters design and deployment of a 3-D virtual European winery on the Web that includes a virtual exhibition of specially selected European wines. This VR prototype demonstrates the capabilities of the Web in supporting trans-European E-commerce services and virtual trade fairs for small- and medium-sized enterprises.

12.12.8.4 RESHEN

The RESHEN initiative supports secure data exchange and communications between individuals and their healthcare service providers in regional healthcare network configurations. This initiative also contributes to development of a trans-European PKI (Public Key Infrastructure) and clarifies procedures for establishing regional and transborder PKI implementations. Approaches for using Transport Transfer Protocols (TTP) for enabling PKI services are investigated as well.

12.13 PRIVACY ON THE INTERNET

Web sites collect information about consumers with and without their consent. In addition to gathering personal information about consumers online, electronic commerce companies, online vendors, and Network Service Providers (NSPs) also sell this information to advertising companies and telemarketing firms. These entities subsequently create electronic records profiling consumer browsing patterns and transaction-generated data for marketing specified products to targeted lists of online consumers.

As a consequence, TRUSTe and the Better Business Bureau award online privacy seals of approval to Web sites posting privacy policies. In addition, the FTC (Federal Trade Commission) Advisory Committee on Online Access and Security (ACOAS) publishes online guidelines for safeguarding personal data collected by commercial Web sites.

12.13.1   INTERNET PRIVACY COALITION (IPC)

The Internet Privacy Coalition (IPC) supports the right of individuals to communicate securely and privately on the Web without government restraints, interference, and/or restrictions. In addition, the IPC promotes the public availability of encryption tools and endorses legislative initiatives such as SAFE (Security and Freedom through Encryption) for making encryption tools available worldwide.

12.13.2   W3C PLATFORM FOR PRIVACY PREFERENCES (P3P) PROJECT

Developed by the World Wide Web Consortium (W3C), P3P (Platform for Privacy Preferences) enables Web site owners to post privacy policies at their Web sites in a standardized format so that these policies can be automatically retrieved by Web browsers for examination by Web site visitors. P3P also defines a format for enabling Web browsers to provision specified data to the P3P Web site based on visitor preferences.

Owners of P3P-compliant Web sites answer a standardized set of multiple-choice questions that provide an overview of the ways in which their Web sites handle personal information. P3P-compliant browsers retrieve P3P data and identify discrepancies between Web site visitors' privacy preferences and Web site owners' data collection procedures. P3P utilizes XML (Extensible Markup Language) for coding privacy policies and RDF (Resource Description Framework) for encoding metadata. P3P also uses APPEL (A P3P Preferences Exchange Language) for delineating sets of preferences in P3P policies. A Web site owner can employ a comprehensive P3P policy or multiple P3P policies that apply to various components of the Web site. The extent of Web site privacy depends on stated preferences of the Web site owner.

The P3P specification is complex. Nonetheless, P3P advocates expect that P3P tools for enabling Web site visitors and owners to build their own preferences will become widely available. P3P is an international solution that addresses consumer privacy issues and enables individuals to understand privacy policies of every P3P Web site visited.

The P3P specification is flexible, expandable, and works in concert with legislative and self-regulatory programs in the privacy domain. However, P3P does not limit the nature or volume of personal information collected. Although P3P enables consumer awareness of privacy policies, P3P does not establish minimum standards for privacy or facilitate increased privacy, nor is it equipped to determine whether Web site owners act in compliance with their own privacy policy procedures. Despite the aforementioned constraints, P3P enables the development of privacy solutions for creating an environment of trust on the Web. Participants in the P3P effort include the Independent Center for Privacy Protection, America Online, AT&T, the Internet Alliance, Microsoft, NEC, Netscape, and Nokia.

12.14 SCHOOL SECURITY POLICIES

12.14.1 FEATURES AND FUNCTIONS

The widespread implementation of Web services in schools and universities accelerates the corresponding need for security solutions to protect data integrity and confidentiality, and minimize exposure of network resources to accidental and/or unauthorized modification, corruption, and disclosure. Security policies for schools and universities generally safeguard the integrity of online records, institutional data, and networking resources, and clarify guidelines for the appropriate and responsible use of computing resources and equipment. In developing a security policy, a risk assessment is initially conducted to aid in the identification of security holes, network vulnerabilities, and critical resources to be protected, and the clarification of security requirements. A risk assessment also involves determining the consequences of unauthorized access to transcripts, registration records, course grades, and faculty research and procedures for dealing with cyberinvasions and insider sabotage.

Security policy development also involves identifying individuals who are allowed to use network resources, the extent of their privileges on the system, and their rights and responsibilities. Criteria for determining whether or not access to specific network sites are blocked or disabled are indicated as well. In addition, a security policy specifies guidelines for provisioning passwords, utilizing biometric identifiers, enabling authentication and authorization services, and employing Web filtering tools. Procedures for supporting information confidentiality, privacy, and reliable message delivery; reporting security breaches; and reacting quickly when under attack are also clarified.

Strategies for backup and disaster recovery from fire, floods, earthquakes, tornadoes, and/or hurricanes are established. Techniques for safeguarding on-site computing equipment from sabotage, vandalism, and theft are delineated. Personnel responsible for handling security breaches and contacts in case of emergency are designated. Sanctions for security violations are defined. Budgetary allocations for security services that function effectively on a daily basis are clarified.

The security policy should be reviewed and updated periodically to reflect network changes in order to prohibit information network tampering and cyberintrusions. This process raises security awareness and exposes neglected or overlooked security holes that potentially put networking services at risk. Procedures for fixing security vulnerabilities, managing e-mail for legal actions, and conducting audits and methods for preventing premature deletion of records and archives should also be indicated.

A security policy describes the features and functions of multiple security mechanisms, tools, and technologies in detecting security breaches and techniques for safeguarding computer software, hardware, and data from harm by external hackers and malcontent insiders.

Creating a security policy requires an understanding of the fundamentals of security technology. There is no single solution for countering intrusions. Individuals intent on discovering points of entry and system flaws can circumvent these mechanisms. Protection of network access points from subversion by intruders and insiders, particularly in a distributed computing environment, is difficult. Because any security scheme can be broken, the notion of security mechanisms safeguarding all nodes on a computer network from unauthorized access is illusory. The extent of protection needed is based on the perceived risk and degree of data sensitivity. Generally, security in the telelearning environment is a trade-off with expedience. Most users seem willing to accept a higher level of risk rather than forego the use of networking applications and services. No amount of planning for security-related concerns, however, will be effective if individuals are careless in taking the necessary precautions to use the technology properly.

12.14.2   ACCEPTABLE USE POLICIES (AUPs)

Continuous publicity surrounding the use of the Web as an alleged conduit for pornography, cybergambling, and criminal activities contributes to the development and implementation of Acceptable Use Policies (AUPs). AUPs regulate faculty, student, administrative, and staff use of computing resources in accordance with institutional missions and philosophies. Generally, academic AUPs advise Web site users that destruction or malicious modification of data, transmission of obscene material, and copyright violations can culminate in sanctions ranging from the cancellation of Internet privileges to expulsion. Typically, students, faculty, and administrators in a school or university must sign off on an institutional AUP, thereby indicating agreement to follow its guidelines prior to the initiation of their network accounts.

12.15   WEB FILTERS AND WEB RATING SOLUTIONS

12.15.1   WEB FILTERING TOOLS

The Web features a wide range of information that contributes to instructional enhancement and curricular enrichment, yet, it also carries information that is potentially harmful or illegal. The capability to share ideas on the Web can also result in potential exposure to cyberporn.

Parental and faculty concerns about inappropriate and objectionable online material contribute to the use of Internet filtering tools such as SmartFilter, Net Nanny, and Cybersitter to facilitate safe Web exploration, particularly in K–12 (Kindergarten through Grade 12) schools. These filtering tools block or limit access to Web sites deemed inappropriate in accordance with predefined lists of keywords and Web sites compiled by Web site publishers, parents, and educators. In addition, these tools limit the total time spent on the Web and also prohibit access to predefined newsgroups and forums. Internet filtering tools vary in the extent to which they support the First Amendment and the primacy of free expression. However, their intent is to establish standards and guidelines that enable users to filter out certain categories of Web content deemed objectionable.

12.15.2   Web Rating Systems

12.15.2.1   Platform for Internet Content Selection (PICS)

In responding to widespread concerns about the proliferation of indecent, violent, or inappropriate material on the Internet, the World Wide Web Consortium (W3C), a coalition of international research and educational institutions, developed the Platform for Internet Content Selection (PICS) specification. Designed to forestall government restriction on the free exchange of ideas on the Web, the PICS specification is composed of two components: the rating system and the rating label. The PICS rating system defines criteria for evaluating Web content. The PICS rating label describes the site's rating. The rating label appears on a Web page as part of the HTML (HyperText Markup Language) content or as part of the HTTP (HyperText Transfer Protocol) header.

By using PICS-compatible client software that is either part of the Web browser or a separate application, Web site visitors can determine what levels of content can be viewed at a particular Web site without personally reading Web site content. PICSRules is a metalanguage based on PICS labels that provides the technical framework for allowing or blocking access to Web sites.

The PICS specification supports Internet access without censorship controls. Web site authors, operators, and/or owners must take the first step, by submitting their content to a PICS rating system for evaluation. The PICS system also provides Web site visitors with the necessary information for setting an information appliance to browse only material considered appropriate.

PICS is an industry standard for Web rating systems. Web site owners, publishers, third parties, public and private entities, Network Service Providers (NSPs), and communications carriers including AOL Time Warner and Netscape support this specification. Microsoft Corporation bundles PICS services with the Content Advisor feature of Internet Explorer. Next-generation PICS solutions will support the authenticity of cryptographic signatures, enable Web sites to inform Web site viewers about their privacy policies, and feature indexing and searching capabilities.

12.15.2.2   Internet Content Rating Association (ICRA)

In response to mounting concern about the proliferation of offensive Web content, AOL Time Warner and Yahoo, Inc., agreed in 2001 to post voluntary content rating systems based on the ICRA (Internet Content Rating Association) solution. The ICRA system generates a descriptive tag that is embedded in the HTML code of the Web site. Users can then set their browser or operating system to select a variety of settings and sensitivity levels for blocking violent, nudity, and sexual content or other material deemed inappropriate.

The ICRA system supports the PICS standard. As with PICS, the ICRA solution enables labels or metadata tags to be associated with Web content. Like PICS, the possibilities of government censorship motivated the ICRA to develop the system. This content labeling advisory system enables parents and educators to identify appropriate Web sites for K–12 students, based on detailed Web site content described in responses to a questionnaire posted on the ICRA Web site that is completed by the Web site owner or Internet content provider. This questionnaire features specific questions relating to the nature, level, and intensity of the violence, offensive language, sex, and/or nudity contained within the site. After each Web page is assigned a rating label, Web browser software can be set to block access to Web sites with material deemed hateful, violent, pornographic, or otherwise inappropriate. The ICRA system is also based on the RSAC (Recreational Software Advisory Council) solution that was popularly known as RSACi (RSAC on the Internet).

12.16   NATIONAL LEGISLATIVE INITIATIVES

12.16.1   CHILDREN'S ONLINE PRIVACY PROTECTION ACT (COPPA)

According to the Children's Online Privacy Protection Act (COPPA), Web sites directed at children and general audiences that collect personal information from children under the age of 13 must obtain parental consent and post privacy policies. COPPA became effective on April 21, 2000. The FTC (Federal Trade Commission) monitors the Internet to ensure that Web sites for children comply with COPPA guidelines and determines whether legal action is warranted for Web sites with compliance problems. COPPA violators are subject to FTC law enforcement actions. Civil penalties are set at $11,000 per violation. In addition, the FTC works with the U.S. Department of Education in distributing information on COPPA guidelines to schools, maintains a Web site describing COPPA operations, and develops educational materials on online privacy for parents, students, and educators.

12.16.2   CHILDREN'S INTERNET PROTECTION ACT

Passed by the U.S. Congress in 2000 in response to growing public concern about online material deemed harmful to children, the Children's Internet Protection Act requires U.S. elementary and secondary schools and libraries to implement Internet filtering tools and technologies as a prerequisite to receiving universal service assistance. Those schools and libraries that fail to comply with this Act are required to repay federal funds already received. The University of Oregon Responsible Netizen Center for Advanced Technology in Education, the ACLU (American Civil Liberties Union), and the CDT (Center for Democracy and Technology) view this Act as a form of unconstitutional censorship. These groups also maintain that companies selling filtering products will be able to monitor student activities on the Web and sell this data to online marketers and advertisers.

12.16.3   NATIONAL PLAN FOR INFORMATION SYSTEMS PROTECTION

Issued in 2000 by President Clinton's administration, the National Plan for Information Systems Protection establishes a framework for safeguarding critical infrastructure resources and detecting cyberattacks and network intrusion by implementing the Federal Intrusion Detection Network (FIDNet) and the Cyberspace Electronic Security Act (CESA).

12.16.3.1   FIDNet (Federal Intrusion Detection Network)

A centralized intrusion detection monitoring system, FIDNet is designed to protect critical infrastructure resources on nonmilitary government federal computers from cyberattacks. However, in addition to recording activity deemed suspicious on government computers, FIDNet can also scan legitimate online exchanges.

12.16.3.2   CESA (Cyberspace Electronic Security Act)

With the Cyberspace Electronic Security Act (CESA), the federal government could obtain encryption keys for decrypting sensitive data entrusted to a third party for storage by court order. As an example, personal and sensitive digital documents as well as data stored by an individual in IBM World Registry E-business archives and personal cybervaults would no longer be protected.

12.16.4   PATRIOT (Provide Appropriate Tools Required to Intercept and Obstruct Terrorism) Act

Signed into law on October 26, 2001, the PATRIOT Act was developed in response to the September 11, 2001, terrorist attacks on the United States. This Law is designed to forestall cyberattacks on critical computer systems that can interfere with the integrity of the communications infrastructure and to eliminate serious delays that could be devastating to law enforcement investigations. The PATRIOT Act includes measures that broaden government powers to monitor electronic communications and conduct search and seizure operations to obtain electronic evidence. The PATRIOT Act also supports the use of the Carnivore tool for conducting electronic surveillance of individuals suspected of using the Internet for planning and executing cyberterrorism activities. Developed by the FBI, Carnivore scans e-mail and Internet traffic and searches for messages to and from targets of surveillance.

12.17   INTERNATIONAL LEGISLATIVE INITIATIVE

12.17.1   Council of Europe (CoE)

Developed by the Council of Europe (CoE), the Convention on Cybercrime Agreement addresses issues associated with transborder cybercrime. Although the draft Convention on Cybercrime Agreement is generally consistent with U.S. federal law, there are some notable differences. As an example, U.S. federal law criminalizes deliberate destruction of specified types of data and establishes a damage threshold of $5,000 for every malicious activity. By contrast, the CoE is not in favor of establishing a damage threshold for data destruction.

12.18   CYBER RIGHTS

12.18.1   Copyright and Intellectual Property Protection

The explosive growth of the Internet and popularity of online communications services contribute to the development of cyberspace policies and regulations relating to fair use and copyright infringement, freedom of expression, First Amendment and Fourth Amendment rights, cryptography, intellectual property protection, privacy protection, and E-commerce. Without enforceable legislation, a copyright owner's right to financially benefit from intellectual property available in digital format on the Internet is at risk. Approaches for extending the same legal protection that applies to educational and information products and their use in the physical environment to those works disseminated via the Web are in development.

The World Intellectual Property Organization (WIPO) works in cooperation with developing countries in furnishing model copyright, intellectual property protection, and privacy protection laws. Sponsored by a consortium of academic institutions, EDUCAUSE maintains a digital Information Resources Library on censorship, free speech, acceptable use, copyright, and intellectual property protection for academic institutions.

12.18.2   Center for Democracy and Technology (CDT)

The Center for Democracy and Technology (CDT) supports the preservation of public education, freedom of expression, individual privacy, freedom of association, constitutional civil liberties, democratic values, and the free-flow of information on the Web. The CDT opposes censorship, government surveillance on the Web, and utilization of Internet filtering tools, and monitors pending legislation on privacy, cybersecurity, federal regulations on cryptosystems, digital signatures, and authentication services to ensure that privacy protections are not eroded.

12.18.3   Electronic Frontier Foundation (EFF)

The Electronic Frontier Foundation (EFF) works in the public interest to safeguard fundamental civil liberties such as freedom of expression and protects individual privacy on the Internet. The EEF endorses legislation to protect, preserve, and extend First Amendment rights on the Web; advocates measures to ensure the right to use encryption technologies; conducts legal actions against anti-privacy initiatives such as digital wiretapping; and monitors the impact of the PATRIOT Act on civil liberties.

12.18.4   National Coalition against Censorship (NCAC)

An alliance of literary, religious, artistic, educational, labor, and civil liberties organizations, the National Coalition Against Censorship (NCAC) defends First Amendment values of freedom of inquiry, thought, and expression and opposes restraints on information access and censorship efforts in schools and libraries.

12.19   NETWORK MANAGEMENT PROTOCOLS

The capabilities of network management protocols in safeguarding networking applications, services, and operations are reviewed in this section. An innovative security solution developed at the University of Illinois at Urbana-Champaign that accommodates active network requirements is also highlighted.

12.19.1   SNMP (Simple Network Management Protocol)

An object-oriented remote networking management protocol, SNMP (Simple Network Management Protocol) is an accepted industry standard for network management. Developed by the IETF, SNMP works in concert with TCP/IP (Transmission Control Protocol/Internet Protocol) and defines guidelines for employing client/service architecture. A flexible, extendible, and scalable management solution, SNMP also establishes procedures for controlling network devices and managing their applications and services in multivendor network environments. SNMPv3 (Simple Network Management Protocol, version 3) is an extension to SNMPv2u and SNMPv2*. In contrast to earlier SNMP implementations, SNMPv3 supports data integrity, privacy, user authentication, and encryption services.

12.19.2   RMON (Remote Networking Monitoring) Devices

Also called instruments, probes, and monitors, remote network monitoring (RMON) devices are specifically dedicated to network management functions and collect operational statistics from Network Management Stations (NMS), support disaster recovery functions, and notify network administrators when a network problem, error, or other unique condition is detected. In addition, RMON devices generate reports of error conditions and provide value-added data for solving recurrent traffic problems. RMON fosters collection of statistics for benchmarking performance of the IP Differentiated Service (DiffServ) protocol. Core network elements such as routers, gateways, and bridges are controlled and supervised by RMON devices. These devices also access and retrieve relevant MIB (Management Information Base) data. An MIB is a virtual information store that contains a set of Management Objects (MOs) or network elements defined by RMON devices for enabling configuration, fault, and performance management operations.

12.19.3   Active Networks

The increased complexity of present-day networks and the pervasiveness of security threats drive the demand for adaptive or automatic functions that are integrated into network management systems and network analysis and monitoring tools to track network operations and performance. Active networks transform network packets into active elements, thereby enabling management services to evolve as packets transit the network. Active network configurations employ remotely programmable routers, disks, and sensors to improve response time, available bandwidth, and QoS (Quality of Service) performance and support interworking operations with other next-generation networks. A major barrier to active network deployment is the lack of an overall security system solution for enabling dependable active network operations. This problem is addressed in the Seraphim Project conducted by the University of Illinois Computer Science Department. The Seraphim Project supports development of a flexible security architecture that supports interoperable and dynamic security policies to ensure access control and reliable implementation of security measures. Designed to reduce security risks, the Seraphim architecture also enables interoperable security functions among diverse security domains and enables dependable security services in dynamic active network environments.

12.20   SUMMARY

The Web provisions access to media-rich resources in disciplines ranging from medicine, public policy, and music, to agriculture, high-energy physics, and education. However, the Web also carries potentially harmful content that can lead to e-mail harassment, unauthorized distribution of works with copyrights, illegal bomb and drug production, and incitement to hatred, discrimination, and violence.

Cybercrime is one of the fastest growing areas of criminal action. Communications networks support the exchange and delivery of personal information such as home addresses, social security numbers, and medical records that can be obtained illegally by cyberhackers through eavesdropping, packet sniffing, and interception. Every network with an external network connection is at risk. Cybercrimes include stolen identity, Web site defacement, and theft of national security data from government agencies. Gathering information on cyberintrusions and Web site hijackings is extraordinarily difficult, but even a conservative extrapolation from those reported indicates the problem is significant. Network users are not always aware of the risks of cyberintrusions or the extent of security incidents to which they are exposed.

In this chapter, capabilities of network security mechanisms and protocols are examined. Distinctive features of rating schemes and filtering tools for enabling an assessment of information content are described. Approaches for safeguarding E-commerce transactions from insider threats and cyberintrusions are explored. Strategies for creating a climate of trust that enables secure internetwork communications and E-commerce transactions are reviewed. Features and functions of PKI architecture and cryptographic solutions for protecting data from theft and/or unauthorized disclosure are examined. Procedures for recognizing and responding to security incursions and multiple methods for safeguarding information integrity from cyberbvandalism are explored. Recent initiatives in the legislative domain such as the U.S. Patriot Act are highlighted. Tactics for developing security policies in schools and universities are introduced and recent advances in active networks are noted.

12.21   SELECTED WEB SITES

Authorize.net. Home Page.
  Available: http://authorize.net/
BioAPI Consortium. Home page.
  Available: www.bioapi.com/
Business.com B2B Online E-commerce Resource Center.
  Available: http://www.b2b-ecommerce.ws/
Carnegie Mellon University. Software Engineering Institute. CERT/CC
  (CERT/Coordination Center). Home Page. Last modified on November 12,
  2001.
  Available: http://www.cert.org/

Center for Democracy and Technology. Home Page.
    Available: http://www.cdt.org/
Council of Europe
    Available: www.usdoj.gov/criminal/cybercrime/COEFAQs.htm
Electronic Freedom Foundation. About the EFF.
    Available: http://www.eff.org/abouteff.html
Electronic Privacy Information Center (EPIC). Home Page.
    Available: www.epic.org/
Financial Services Technology Consortium. About FSTC.
    Available: http://www.fstc.org/about/
Internet Content Rating Association. Home Page. Last modified on November
    18, 2001.
    Available: http://www.rsac.org
Internet Engineering Task Force (IETF). SNMPv3. Last modified on
    November 9, 2001.
    Available: http://www.ietf.org/html.charters/snmpv3-charter.html
Internet Engineering Task Force (IETF). Remote Network Monitoring. Last
    modified on November 15, 2001.
    Available: http://www.ietf.org/html.charters/rmonmib-charter.html
National Center for Missing and Exploited Children. Home Page.
    Available: http://www.missingkids.com/
National Coalition Against Censorship (NCAC). About NCAC. Last modified
    on November 17, 2001.
    Available: http://www.ncac.org/
National Fraud Information Center. Internet Fraud Watch.
    Available: http://www.fraud.org
National Institute of Standards and Technologies (NIST). NIST PKI. Program.
    Last modified on January 3, 2001.
    Available: http://csrc.ncsl.nist.gov/pki/
Purdue University. The Center for Education and Research in Information
    Assurance and Security (CERIAS) Home Page.
    Available: http://www.cerias.purdue.edu/
RSA Security. Home Page.
    Available: http://www.rsa.com/
U.S. Department of Defense Advanced Research Projects Agency (DARPA).
    Active Networks. Vision. Dr. Douglas Maughan.
    Available: http://www.darpa.mil/ito/research/anets/vision.html
University of Illinois at Urbana-Champaign. Department of Computer Science.
    Seraphim: Building Dynamic Interoperable Security Architecture for Active
    Networks.
    Available: http://devius.cs.uiuc.edu/Security/seraphim/
V.myths.com. Truth About Computer Virus Myths and Hoaxes.
    Available: http://www.vmyths.com
VeriSign. Home Page.
    Available: http://www.verisign.com/index.html

World Wide Web Consortium (W3C). Platform for Internet Content Selection (PICS). Home Page. Last modified on June 6, 2001.
Available: http://www.w3.org/PICS/

World Wide Web Consortium (W3C). Platform for Privacy Preferences (P3P) Project. Last modified on November 9, 2001.
Available: http://www.w3.org/P3P/

Index